Martzloff · A History of *Chinese ...*

LI SHANLAN
1811–1882

From the July 1877 issue of the *Gezhi huibian*
(The Chinese scientific and industrial magazine)

Jean-Claude Martzloff

A HISTORY OF
Chinese Mathematics

With Forewords by
Jaques Gernet and Jean Dhombres

With 185 Figures

 Springer

Jean-Claude Martzloff
Directeur de Recherche
Centre National
de la Recherche Scientifique
Institut des Hautes Études Chinoises
52, rue du Cardinal Lemoine
75321 Paris Cedex 05
France
e-mail: martz@ext.jussieu.fr

Translator:

Stephen S. Wilson
First Floor
19 St. George's Road
Cheltenham
Gloucestershire, GL50 3DT
Great Britain

Title of the French original edition:
Histoire des mathématiques chinoises. © Masson, Paris 1987

Cover Figure: After an engraving taken from the *Zhiming suanfa* (Clearly explained computational [arithmetical] methods). This popular book, edited by a certain Wang Ren'an at the end of the Qing dynasty, is widely influenced by Cheng Dawei's famous *Suanfa tongzong* (General source of computational methods)(1952). Cf. Kodama Akihito (2'), *1970*, pp. 46–52.
The reproductions of the Stein 930 manuscript and a page of a Manchu manuscript preserved at the Bibliothèque Nationale (Fonds Mandchou no. 191) were made possible by the kind permission of the British Library (India Office and records) and the Bibliothèque Nationale, respectively. For this we express our sincere thanks.

Corrected second printing of the first English edition of 1997, originally published by Springer-Verlag under the ISBN 3-540-54749-5

Library of Congress Control Number: 2006927803

ISBN-10 3-540-33782-2 Springer Berlin Heidelberg New York
ISBN-13 978-3-540-33782-9 Springer Berlin Heidelberg New York

Springer is a part of Springer Science+Business Media

springer.com

© Springer-Verlag Berlin Heidelberg 1997, 2006
Printed in Germany

Typesetting: Editing and reformatting of the translator's input files using a Springer TEX macro package
Production: LE-TEX Jelonek, Schmidt & Vöckler GbR, Leipzig
Cover design: Erich Kirchner, Heidelberg

Printed on acid-free paper   41/3100/YL   5 4 3 2 1 0

To France Alice

Foreword by J. Gernet

The uses of numbers, their links with the socio-political system, their symbolic values and their relationship to representations of the universe say a great deal about the main characteristics of a civilisation. Although our mathematics has now become the common heritage of humanity, our understanding of mathematics is essentially based on a tradition peculiar to ourselves which dates back to ancient Greece; in other words, it is not universal. Thus, before we can begin to understand Chinese mathematics, we must not only set aside our usual ways of thinking, but also look beyond mathematics itself. At first sight, Chinese mathematics might be thought of as empirical and utilitarian since it contains nothing with which we are familiar; more often than not it contains no definitions, axioms, theorems or proofs. This explains, on the one hand, earlier unfavourable judgements of Chinese mathematics and, on the other hand, the amazement generated by the most remarkable of its results. The Chinese have always preferred to make themselves understood without having to spell things out. "I will not teach anyone who is not enthusiastic about studying," said Confucius. "I will not help anyone who does not make an effort to express himself. If, when I show someone a corner, that person does not reply with the three others, then I will not teach him." However, the Chinese have indulged their taste for conciseness and allusion, which is so in keeping with the spirit of their language, to the extent that they detest the heaviness of formal reasoning. This is not a case of innate incapacity, since their reasoning is as good as ours, but a fundamental characteristic of a civilisation. Moreover, this loathing of discourse is accompanied by a predilection for the concrete. This is clearly shown by their methods of teaching mathematics, in which the general case is illustrated by operational models the possibilities of extension of which they record directly, via comparisons, parallels, manipulation of numbers, cut-out images, and reconstruction and rotation of figures. As J.-C. Martzloff notes, for the Chinese, numbers and figures relate to objects rather than to abstract essences. This is the complete antithesis to the Greeks, who rejected everything that might evoke sensory experience, and runs counter to the Platonic concept of mathematics as the theoretical science of numbers, an objective science concerned with the abstract notions of units and magnitudes "which enable the soul to pass from the ever-changing world to that of truth and essence." For the Chinese, on the other hand, numbers formed a component part of the changing world to which they adapt; for instance, there was no distinction between counting-rods and the divinatory rods which were used to create hexagrams from combinations of the *yin* and *yang* signs. Chinese diviners are credited with astonishing abilities as calculators. In China, there was a particularly close link between mathematics and the portrayal of cosmology and, as Marcel Granet wrote, numbers were used "to define and illustrate the organisation of the universe." This may explain the importance of directed diagrams and the fundamental role of position in Chinese algebra (which determines the powers of ten and the powers of the unknown on counting surfaces). The number 3 is sometimes used as an approximate equivalent of the number π because it is the number of the Heavens and the circle, in the same way that 2 is the number of the square and of the Earth. The set-square and compass are the attributes of Fuxi and Nüwa, the mythical founders of the Chinese civilisation and the persistence in Chinese mathematics of a figure such as the circle inscribed in a right-angled triangle (right-angled triangles form the basis for a large number of algebraic problems) cannot be simply put down to chance. Chinese mathematics was oriented towards cosmological speculations and the practical study of the hidden principles of the universe as much as towards questions with a practical utility. It can scarcely be distinguished from an original philosophy which placed the accent on the unity of opposites, relativity and change.

Martzloff provides not only an excellent analysis of the remaining testimonies to the long history of Chinese mathematics (many works have disappeared and many procedures which were only passed on by example and practice have vanished without trace) but a study of all aspects of its history, which covers contacts and borrowings, the social situation of mathematicians, the place of mathematics in the civilisation and Western works translated into Chinese from the beginning of the 17th century, including the problems involved in the translation of these works. There emerge an evolution with its apogee in the 12th and 13th centuries and a renaissance stimulated by the contribution of Western mathematics in the 17th and 18th centuries. This admirably documented book, in which the author has made every attempt not to "dress Chinese mathematics in clothes which it never wore," will be an indispensable work of reference for a long time to come.

Jacques GERNET

Honorary Professor at the Collège de France

Foreword by J. Dhombres

Since the encyclopedist movement of the 18th century which was in harmony with the ideas of the Enlightenment, we have got so used to viewing science as a common human heritage, unlike an individual sense of citizenship or a specific religion, that we would like to believe that its outward forms are universal and part of a whole, which, if it is not homogeneous is at least compulsory and unbounded. Thus, in a paradoxical return to ethnocentrism, it seems quite natural to us that, even though it means taking liberties with the writing of history, this science was that described by Aristotle's logical canons, Galileo's mathematical techniques and Claude Bernard's rational experimentalism.

Moreover, we have also assumed that, as far as mathematics is concerned, there is only one model, the evolution of which was essentially fixed from the origins of a written civilisation by the immutable order of the rules of the game, namely axioms, theorems and proofs displayed in a majestic architectural sequence in which each period added its name to the general scheme by contributing a column, an architrave, a marble statue or a more modest cement. One name, that of Euclid, whose *Elements* were used as a touchstone to test whether a work was worthy of being called "mathematical," has resounded from generation to generation since the third century BC. The model transcended mathematical specialisation (still suspected of favouring useless mysteries) since so many thinkers laboured to present their ideas *more geometrico*. They would have been insulted by the suggestion that they should replace this expression by another, such as "as prescribed by the School of Alexandria," which emphasized the geographical attachment. These thinkers believed that they proceeded in accordance with the universal rules of the human mind.

The civilisations of the Mediterranean Basin and, later, those of the Atlantic were not wrong to venerate the axiomatic method. They also knew how to yield graciously to mathematical approaches, such as the discovery of differential and integral calculus at the end of the 17th century, which were initially rightly judged to be less rigorous. Thus, apart taking an interest in another culture, and another way of thinking, not the least merit of a history of mathematics outside the influence of Euclid and his accomplices would be to improve our grasp of the strength and penetration of the Euclidean approach. To put it more prosaically, without risk of contradiction by French and Chinese gourmets, *doufu* and *haishen* taste better once one has tried *foie gras* and oysters!

Fortunately, there exist different types of mathematics, such as those which have developed continuously and fruitfully over approximately sixteen centuries in the basins of the Yellow and Blue Rivers. Should we still refuse these the right to the 'mathematical' label because there are as yet very few well-documented books about them? Certainly not, since we now have the present *Histoire des Mathématiques Chinoises* from the expert pen of Jean-Claude Martzloff. This enthusiastically describes the one thousand and one linguistic and intellectual pitfalls of the meeting between the end of the Ming culture and that of the Qing successors. This meeting involved Euclid or, rather, a certain Euclid resulting from the Latin version of the *Elements* generated in 1574 by Clavius. In fact, Clavius was the master of the Jesuit Ricci (otherwise known in Peking under the name of Li Madou) who translated the first six books of the mathematician from Alexandria into Chinese at the beginning of the 17th century.

Unfortunately, although the Jesuits placed the translation of mathematics before that of the Holy Scriptures, they did not have access to original Chinese mathematics such as the algebraic and computational works of the brilliant Chinese foursome of the 13th century Yang Hui, Li Zhi, Qin Jiushao and Zhu Shijie. What would they have made of this, when their own mathematical culture was so rich?

For it is a most surprising historical paradox that this meeting between the West and China took place at a time when a complete scientific reversal was under way in the West (the change occurred over a few short years). Sacrobosco's astronomy of the planets, a direct descendant of that of Ptolemy, which the 'good fathers' took with them on their long sea journey to distant Cathay, even when adapted in response to scholarly lessons received at the College of Rome where the Gregorian calendar was reformed in 1572, was very different from that given by Kepler in his *Astronomia Nova* in 1609. The theoretical and intangible reflections of the 14th-century mechanistic schools of Paris and Oxford were suddenly realised in the true sense, when they were applied in the real world by Galileo when he established the law of falling bodies. The West was seen to be on the outside in well-worn clothes, although the Far-East had forgotten its mathematical past. However, it is true that the *Suanfa tongzong* (General source of computational methods) which was issued in 1592, would not have disgraced a 16th-century collection of Western arithmetics! However, on both the Chinese and the Western sides, originality was to be found elsewhere.

It is because we are well aware of the originality of Galileo and Descartes that our interest turns to the above four Chinese 13th-century mathematicians. Their originality is so compelling that we are overcome with a desire to know more about how they thought and lived, the sum total of their results and the links between their works and their culture. In short, our curiosity is excited, and the merit of this book is that it leads through both the main characters and the main works.

But what is the intended audience of this book on the history of mathematics, given that its unique nature will guarantee its future success and longevity through the accumulation of specialised scholarly notices and, above all, more broadly, through reflexions by specialists in all areas? Is this book solely for austere scholars who use numerical writings to measure exchanges between the Indus and the Wei and between the Arabic-speaking civilisation and the Tang? Is it solely intended for those interested in the origin of the zero or the history of decimal positional numeration?

How narrow the specialisations of our age are, that it is necessary to tell ill-informed readers as much about the affairs and people of China as about modern mathematics, to enable them to find spiritual sustenance in the pages of this book. May they not be frightened by the figures or by a few columns of symbols. May they be attracted by the Chinese characters, as well as by the arrangements of counting-rods, since these determine a different policy in graphical space. Where can a mathematician or historian of China find so much information or a similar well-organised survey of sources? Where can one find such a variety of themes, ranging from the interpretation of the mathematical texts themselves to a description of the role of mathematics in this civilisation, which was strained by literary examinations from the Tang, preoccupied with the harmony between natural kingdoms, and partial to numerical emblems (as Marcel Granet breathtakingly shows in his *La Pensée Chinoise*)?

I shall only comment on a number of questions about this Chinese mathematics and a number of very general enigmas which have nothing to do with this exotic and quaint *enigma cinese*.

Firstly, there is the question of a difference in status between the mathematics of practitioners and that of textbooks intended for teaching purposes. Broadly speaking, as far as China is concerned, it is mainly textbooks which have come down to us, worse still, these are textbooks which belong to an educational framework which placed great value on the oral tradition and on the memorising of parallel, rhyming formulae. How could we describe 18th-century French mathematics if we only had access to the manuals due to Bézout, Clairaut or Bougainville? Moreover, should not textbooks be written in such a way that they adhere to the practice of mathematical research of a period, as Monge, Lagrange and Laplace deigned to believe during the French Revolution? Should greater importance be placed on metonymy, the passage from the particular to the general, based on the a priori idea that local success should reveal a hidden structure, even during the training procedure? Is a vague sense of analogy a sufficient basis on which to found an education at several successive theoretical levels? Thus, the history of the mathematics developed in Hangzhou, or any other capital, gives the teacher something to think about.

I have already mentioned the importance of the encounter between the West and the East in the 17th century, with which the French reader is familiar through such important works as J. Gernet's *Chine et Christianisme*, and J. D. Spence's *The Memory Palace of Matteo Ricci*. Unfortunately, these texts pass hurriedly over important scientific aspects. Thus, J.-C. Martzloff has provided an original contribution to an ongoing interrogation.

Finally, there is the question of whether or not the commentary plays a major role in the Chinese mathematical tradition. There is always a tendency to consider mathematicians as a taciturn breed; the very existence of a commentary on a mathematical text may come as a surprise. Arabic-speaking mathematicians distinguished between commentaries (tafsīr), "redactions" (or taḥrīr) and revisions (or iṣlaḥ). They may have done this because they were confronted with the Euclidean tradition which they transmitted and supplemented. The fact is that, within the framework of a theory, the axiomatic approach only ceases once the individual role of each axiom, the need for that axiom and its relative importance amongst the legion of other axioms have been determined. However, the Chinese, impervious to axiomatic concerns, added their own commentaries. At the beginning of the third century AD, Liu Hui, in his commentary on the Computational Prescriptions in Nine Chapters, gave one of the rare proofs of the Chinese mathematical corpus. Can one thus continue to believe that mathematical texts were treated like the Classics, with all the doxology accumulated over the generations, like a true Talmud in perpetual motion? Did mathematics feed so heartily on the sap secreted by a period, a culture or an anthropology that a commentary was necessary? The numerical examples chosen by mathematicians to construct the gates at the four cardinal points of a Chinese town, and the calculation of the tax base constitute a precise revelation of a lost world and are useful in archaeology. But beyond this, does not the mathematics developed by a generation reveal its innermost skeletal structure, much like an X-ray?

What a lot of questions now arise about this area of the history of Chinese mathematics, which at first seemed so compartmentalised, so technical, and scarcely worthy of the general interest of historians or, even less, the interest of those who study the evolution of mental attitudes. After studying general aspects of Chinese mathematics in the first part of his book, J.-C. Martzloff strikes an admirable balance by encouraging us to delve into the second part which concerns the authors and their works. In short, it is difficult not to be fond of his survey, which is solidly supported by bibliographic notes.

It is to be hoped that this first French edition will give rise to publications of the original Chinese texts (with translations) so that we would have a corpus of Chinese mathematics, in the same way that we are able to consult the Egyptian corpus, the Greek corpus and, to a lesser extent, the Babylonian corpus.

Jean DHOMBRES

Directeur d'Etudes à l'Ecole des Hautes Etudes
en Sciences Sociales
Directeur du Laboratoire d'Histoire des Sciences
et des Techniques (U.P.R. 21) du C.N.R.S., Paris

人皆知有用之用而莫知无用之用也

Everyone knows the usefulness of the useful,
but no one knows the usefulness of the useless.

Zhuangzi (a work attributed to
ZHUANG ZHOU (commonly known as ZHUANGZI)),
ch. 4, "The world of men"

Preface

Since the end of the 19th century, a number of specialised journals, albeit with a large audience, have regularly included articles on the history of Chinese mathematics, while a number of books on the history of mathematics include a chapter on the subject. Thus, the progressive increase in our knowledge of the content of Chinese mathematics has been accompanied by the realisation that, as far as results are concerned, there are numerous similarities between Chinese mathematics and other ancient and medieval mathematics. For example, Pythagoras' theorem, the double-false-position rules, Hero's formulae, and Ruffini-Horner's method are found almost everywhere.

As far as the reasoning used to obtain these results is concerned, the fact that it is difficult to find rational justifications in the original texts has led to the *reconstitution* of proofs using appropriate tools of present-day elementary algebra. Consequently, the conclusion that Chinese mathematics is of a fundamentally algebraic nature has been ventured.

However, in recent decades, new studies, particularly in China and Japan, have adopted a different approach to the original texts, in that they have considered the Chinese modes of reasoning, as these can be deduced from the rare texts which contain justifications. By studying the results and the methods explicitly mentioned in these texts hand in hand, this Chinese and Japanese research has attempted to reconstruct the conceptions of ancient authors within a given culture and period, without necessarily involving the convenient, but often distorting, social and conceptual framework of present-day mathematics.

This has led to a reappraisal of the relative importance of different Chinese sources; texts which until recently had been viewed as secondary have now become fundamental, by virtue of the wealth of their proofs. However, most of all, this approach has brought to the fore the key role of certain operational procedures which form the backbone of Chinese mathematics, including heuristic computational and graphical manipulations, frequent recourse to geometrical dissections and instrumental tabular techniques in which the position of physical objects representing numbers is essential. Thus, it has become increasingly clear that within Chinese mathematics, the contrasts between algebra and geometry and between arithmetic and algebra do not play the same role as those in mathematics influenced by the axiomatico-deductive component of the Greek tradition. It is now easier to pick out the close bonds between apparently unrelated Chinese computational techniques (structural analogy between the operation of arithmetical division and the search for the roots of polynomial equations, between calculations on ordinary fractions and rational fractions, between the calculation of certain volumes and the summation of certain series, etc.). It is in this area that the full richness of studies which focus on the historical context without attempting to clothe Chinese mathematics in garments which it never wore becomes apparent.

Beyond the purely technical aspect of the history of mathematics, the attention given to the context, suggests, more broadly, that the question of other aspects of this history which may provide for a better understanding of it is being addressed. In particular, we point to:

- The question of defining the notion of mathematics from a Chinese point of view: was it an art of logical reasoning or a computational art? Was it arithmetical and logistical or was it concerned with the theory of numbers? Was it concerned with surveying or geometry? Was it about mathematics or the *history* of mathematics?

- The important problem of the destination of the texts. Certain texts may be viewed as accounts of research work, others as textbooks, and others still as memoranda or formularies. If care was not taken to distinguish between these categories of texts, there would be a danger of describing Chinese mathematical thought solely in terms of 'Chinese didactic thought' or 'Chinese mnemonic thought.' The fact that a textbook does not contain any proofs does not imply that its author did not know how to reason; similarly, the fact that certain texts contain summary proofs does not imply that the idea of a well-constructed proof did not exist in China: one must bear in mind, in particular, the comparative importance of the oral and written traditions in China.

- The question of the integration into the Chinese mathematical culture of elements external to it. The history of Chinese reactions to the introduction of Euclid's *Elements* into China in the early 17th century highlights, in particular, the differences between systems of thought.

It is with these questions in mind that we have divided this book into two parts, the first of which is devoted to the context of Chinese mathematics and the second to its content, the former being intended to clarify the latter. We have not attempted to produce an encyclopedic history of Chinese mathematics, but rather to analyse the general historical context, to test results taken for granted against the facts and the original texts and to note any uncertainties due to the poorness of the sources or to the limitations of current knowledge about the ancient and medieval world.

Acknowledgements

I should firstly like to express my gratitude to Jacques Gernet, Honorary Professor at the Collège de France (Chair of Social and Intellectual History of China), Jean Dhombres, Directeur d'Etudes at the Ecole des Hautes Etudes en Sciences Sociales and Director of the U.P.R. 21 (C.N.R.S., Paris) for their constant support throughout the preparation of this book.

I am also very grateful to all those in Europe, China and Japan who made me welcome, granted me interviews and permitted me to access the documentation, including the Professors Du Shiran, Guo Shuchun, He Shaogeng, Liu Dun, Wang Yusheng, Li Wenlin, Yuan Xiangdong, Pan Jixing and Wu Wenjun (Academia Sinica, Peking), Bai Shangshu (Beijing, Shifan Daxue), Li Di and Luo Jianjin (Univ. Huhehot), Liang Zongju (Univ. Shenyang), Shen Kangshen (Univ. Hangzhou), Wann-Sheng Horng (Taipei, Shifan Daxue) Stanislas Lokuang (Fu-Jen Catholic University), Itō Shuntarō (Tokyo Univ.), Sasaki Chikara, Shimodaira Kazuo (Former President of the Japanese Society for the History of Japanese Mathematics, Tokyo), Murata Tamotsu (Rikkyo Univ., Tokyo), Yoshida Tadashi (Tohoku Univ., Sendai), Hashimoto Keizo, Yabuuchi Kiyoshi (Univ. Kyoto), Joseph Needham and Lu Guizhen (Cambridge), Ullricht Libbrecht (Catholic University, Louvain), Shōkichi Iyanaga, Augustin Berque and Léon Vandermeersch (Maison Franco-Japonaise, Tokyo), René Taton (Centre A. Koyré, Paris), Michel Soymié and Paul Magnin (Institut des Hautes Etudes Chinoises, Dunhuang manuscripts).

I should like to express my thanks to Professors Hirayama Akira (Tokyo), Itagaki Ryōichi (Tokyo), Jiang Zehan (Peking), Christian Houzel (Paris), Adolf Pavlovich Yushkevish (Moscow), Kobayashi Tatsuhiko (Kiryu), Kawahara Hideki (Kyoto), Lam Lay-Yong (Singapore), Edmund Leites (New York), Li Jimin (Xi'an), Guy Mazars (Strasbourg), Yoshimasa Michiwaki (Gunma Univ.), David Mungello (Coe College), Noguchi Taisuke, Ōya Shinichi (Tokyo), Nathan Sivin (Philadelphia), Suzuki Hisao, Tran Van Doan (Fu-Jen Univ., Taipei), Wang Jixun (Suzhou), Yamada Ryōzō (Kyoto) and Joël Brenier (who helped me to enter into contact with Wann-Sheng Horng) (Paris), Khalil Jaouiche (Paris), the late Dr. Shen Shengkun, Mogi Naoko, and Wang Qingxiang.

Finally, I should like to thank the Academia Sinica (Peking), the University of Fu-Jen (Taipei) and the Japanese Society for the Promotion of Science (JSPS, Nihon Gakujutsu Shinkōkai).

March 31st 1987   CNRS, Institut des Hautes
                                                   Etudes Chinoises, Paris

Contents

Part I. The Context of Chinese Mathematics

Part II. The Content of Chinese Mathematics

Abbreviations

CRZ	*Chouren zhuan*, Taipei (Shijie Shuju, reprinted 1982)
CRZ3B	*Chouren zhuan san bian.* Ibid.
CRZ4B	*Chouren zhuan si bian.* Ibid.
CYHJ	*Ceyuan haijing*
DKW	*Dai kanwa jiten* by T. Morohashi (Tokyo, 1960)
DicMingBio	*Dictionary of Ming Biography (1368–1644).* Goodrich and Fang (1), (eds.), *1976*
DSB	*Dictionary of Scientific Biography.* Gillipsie (1), *1970–1980.*
HDSJ	*Haidao suanjing*
Hummel	*Eminent Chinese of the Ch'ing Period* by A.W. Hummel (reprinted, Taipei (Ch'eng Wen), 1970)
j.	*juan*
JGSJ	*Jigu suanjing* (Wang Xiaotong)
JZSS	*Jiuzhang suanshu*
Li Di, *Hist.*	*Zhongguo shuxue shi jianbian.* Li Di (3′), *1984*
Li Yan, *Dagang*	*Zhongguo shuxue dagang.* Li Yan (56′), *1958*
Li Yan, *Gudai*	*Zhongguo gudai shuxue shiliao.* Li Yan (61′), *1954/1963*
Meijizen	*Meijizen Nihon sūgaku shi* Nihon Gakushin (1′), *1954–60*
MSCSJY	*Meishi congshu jiyao.* Mei Zuangao ed., 1874
Nine Chapters	*Jiuzhang suanshu*
QB	*Suanjing shishu.* Qian Baocong (25′), *1963*
QB, *Hist.*	*Zhongguo shuxue shi.* Qian Baocong, (26′), *1964*
RBS	*Revue Bibliographique de Sinologie* (Paris)
SCC	*Science and Civilisation in China.* Needham (2), *1959*
SFTZ	*Suanfa tongzong* (Cheng Dawei, 1592)
SJSS	*Suanjing shi shu* (Ten Computational Classics)
SLJY	*Shuli jingyun* (1723)
SSJZ	*Shushu jiuzhang* (Qin Jiushao, 1247)
SXQM	*Suanxue qimeng* (Zhu Shijie, 1299)
SY	*Song Yuan shuxue shi lunwen ji*, Qian Baocong *et al.* (1′), *1966*
SYYJ	*Siyuan yujian* (Zhu Shijie, 1303)
SZSJ	*Sunzi suanjing*
Wang Ling, *Thesis*	Wang Ling (1), *1956*
WCSJ	*Wucao suanjing*
XHYSJ	*Xiahou Yang suanjing*
YLDD	*Yongle dadian*
ZBSJ	*Zhoubi suanjing*

ZQJSJ   *Zhang Qiujian suanjing*
ZSSLC-P   *Zhong suan shi luncong.* Li Yan (51′), *1954–1955*
ZSSLC-T   *Zhong suan shi luncong.* Li Yan (41′), *1937/1977*

Remarks

- An abbreviation such as *JZSS* 7-2 refers to problem number 2 of chapter 7 of the *Jiuzhang suanshu* (or to the commentary to that problem).

- DKW 10-35240: 52, p. 10838 refers the entry number 52 corresponding to the Chinese written character number 35240 in volume 10 of the *Dai kanwa jiten* (Great Chinese–Japanese Dictionary) by MOROHASHI Tetsuji (Tokyo, 1960), page 10838.

- Pages numbers relating to the twenty-four Standard Histories always refer to the edition of the text published by Zhonghua Shuju (Peking) from 1965.

- Certain references to works cited in the bibliographies concern reprinted works. In such a case, as far as possible, the bibliography mentions two years of publication, that of the first edition and that of the reprint. Unless otherwise stated, all mentions of pages concerning such works always refer to the reprint. For example, "GRANET Marcel (1), 1934/1968. *La Pensée Chinoise.* Paris: Albin Michel" is cited as "Granet (1), *1934*" but the pages mentioned in the footnotes concern the 1968 reprint of this work.

Author's Note

The present English translation is a revised and augmented version of my *Histoire des mathématiques chinoises*, Paris, Masson, 1987. New chapters have been added and the bibliography has been brought up to date. I express my thanks to the translator, Dr. Stephen S. Wilson, and to the staff of Springer, particularly Dr. Catriona C. Byrne, Ingrid Beyer and Kerstin Graf. I am also much indebted to Mr. Karl-Friedrich Koch for his carefull collaboration and professionalism. Mr. Olivier Gérard has been helpful at the early stage of the composition of the book. Last but not least, many thanks to Ginette Kotowicz, Nicole Resche and all the librarians of the Institut des Hautes Etudes Chinoises, Paris.

Part I

The Context
of
Chinese Mathematics

1. The Historiographical Context

Works on the History of Chinese Mathematics in Western Languages

Prior to the second half of the 19th century, in Europe, almost nothing was known about Chinese mathematics. This was not because no one had inquired about it, quite the contrary. Jesuit missionaries who reached China from the end of the 16th century onwards reported observations on the subject, at the request of their contemporaries, but their conclusions were invariably extremely harsh. The comments of Jean-Baptiste Du Halde summarise them all:

As for their geometry, it is quite superficial. They have very little knowledge, either of theoretical geometry, which proves the truth of propositions called theorems, or of practical geometry, which teaches ways of applying these theorems for a specific purpose by means of problem solving. While they do manage to resolve certain problems, this is by induction rather than by any guiding principle. However they do not lack skill and precision in measuring their land and marking the limits of its extent. The method they use for surveying is very simple and very reliable.[1]

In other words, in their eyes, Chinese mathematics did not really exist. But certainly, one might assume a priori that Leibniz had some idea about Chinese mathematics. However, according to Eric J. Aiton (specialist on Leibniz, Great Britain), the enormous mass of manuscripts of the sage of Leipzig contains nothing on this subject.[2] All that can be said is that Leibniz succeeded in reconciliating the numerological system of the *Yijing* with his own binary numeration system. But, on the one hand, in China itself, as far as we know, neither the numerologists nor the mathematicians had ever dreamed of such a system and, on the other hand, as Hans J. Zacher showed, Leibniz was well aware of the 'local arithmetic' of John Napier (1617), which already contained the idea of the binary system.[3]

The European ignorance of Chinese traditional mathematics was still to last for a long time. Significantly, in his *Histoire des Mathématiques* (first ed. Paris, 1758), J. F. Montucla did not forget to present Chinese mathematics;

---

[1] Du Halde (1), *1735*, II, p. 330. See also Semedo (1), *1645* and Lecomte (1), *1701* (cited and analysed in Jaki (1), *1978* (notes 58 ff., p. 119)) as well as the letter from Parrenin to Mairan (cited in Vissière (1), *1979*, p. 359).

[2] Personal communication.

[3] Cf. Zacher (1), *1973*.

however, in spite of the wealth of his information, he finally could not manage to quote anything else but Chinese adaptations of European mathematical works due to Jesuit missionaries without even mentioning any autochthonous mathematical work whatsoever. While he merely repeats Du Halde's views on Chinese astronomy, the famous historian of mathematics develops at length his critical views on Chinese astronomy, chronology and calendrics. His list of Chinese adaptations of European works occupies two pages and contains 19 titles.[4]

In fact that is not surprising, since at the end of the 16th century, Chinese autochthonous mathematics known by the Chinese themselves amounted to almost nothing, little more than calculation on the abacus, whilst in the 17th and 18th centuries nothing could be paralleled with the revolutionary progress in the theatre of European science. Moreover, at this same period, no one could report what had taken place in the more distant past, since the Chinese themselves only had a fragmentary knowledge of that. One should not forget that, in China itself, autochthonous mathematics was not rediscovered on a large scale prior to the last quarter of the 18th century.

The echo of this belated resurrection of the mathematical glories of the Chinese past did not take long to reach Europe. In 1838, the mathematician Guillaume Libri (1803–1869), who had heard of it from the greatest sinologist of his time – Stanislas Julien (1797–1873) – briefly introduced the contents of the *Suanfa tongzong* (1592) which was then, as he wrote, "the only work of Chinese mathematics known in Europe to which the missionaries have not contributed."[5] From 1839, Edouard Biot issued a series of well-documented studies, notably on Chinese numeration and on the Chinese version of Pascal's triangle.[6] Finally, from 1852, learned society would have had access to an article giving a synthesis on the subject, the *Jottings on the Science of the Chinese: Arithmetic*[7] by the Protestant missionary Alexander Wylie (1815–1887), who was in a position to know the question well, since he lived in China and was in permanent contact with the greatest Chinese mathematician of the period Li Shanlan (1811–1882). For the first time, this contained details of: (i) 'The Ten Computational Canons' (*SJSS*) of the Tang dynasty, (ii) the problems of simultaneous congruences (the 'Chinese remainder theorem'), (iii) the Chinese version of Horner's method, and (iv) Chinese algebra of the 13th century.

This article was translated into several languages (into German by K. L. Biernatzki in 1856,[8] and into French by O. Terquem[9] and by J. Bertrand.[10] Being more accessible than the original which had appeared in an obscure

---

[4]Montucla (1), *1798*, I, pp. 448–480.

[5]Libri (1), *1838*, I, p. 387.

[6]Articles by E. Biot on Pascal's triangle in the *Journal des savants* (1835), on the *Suanfa tongzong* and on Chinese numeration in the *Journal Asiatique* (1835 and 1839, resp.) (full references in the bibliography of *SCC*, III, p. 747).

[7]Wylie (1), *1966* (article first printed in the *North China Herald*, Shanghai, 1852).

[8]Biernatzki (1), *1856*.

[9]Terquem (1), *1862*.

[10]Bertrand (1), *1869*.

Shanghai journal,[11] these translations had a great influence on the historians of the end of the 19th and the beginning of the 20th centuries, Hankel, Zeuthen, Vacca and Cantor[12]. But since they contained errors, and since the latter did not have access to the original Chinese texts, grave distortions arose: these inconsistencies were systematically attributed to the Chinese authors rather than to the translators![13]

However, it was not long before the works of Wylie were overtaken, since in 1913 there appeared a specialised work devoting 155 pages to the history of Chinese mathematics alone, *The Development of Mathematics in China and Japan.*[14] Its author, the Japanese historian Mikami Yoshio (1875–1950) had taken the effort to write in English, thus he had a large audience.[15] Naturally, he was able to read the original sources, but in those heroic days, he had immense difficulties in gaining access to them due to the inadequacies of Japanese libraries at that time;[16] it seems that he faced a similar handicap as far as the European sources were concerned and essentially only cites European authors through the intermediary of Cantor's work. This doubtless explains why his work is essentially based on the important *Chouren zhuan* (Bio-bibliographical Notices of Specialists of Calendrical and Mathematical Computations) by Ruan Yuan (1799) and to a lesser extent on the Chinese dynastic annals. This is the reason for the factual richness of his book (see, for example, the chapter on the history of π),[17] but also for its evident limits due to the over-exclusive use of this type of source. Moreover, Mikami does not always distinguish myths from real historical events.

Subsequently, throughout the first half of the 20th century, Western research was to mark time: the most characteristic writings of this period (with the exception of those of the American mathematical historian D. E. Smith[18] who worked with Mikami) are those of the Belgian Jesuit L. van Hée (1873–1951).[19] He, like L. Sédillot,[20] defended without proof the thesis that, as far as mathematics is concerned, the Chinese had borrowed everything from abroad: "But four times an influence comes from the outside. As if by magic, everything is set on its feet again, a vigorous revival is felt [...]."[21] Thus, his work, like that of those he inspired, should be used with caution.

In 1956, a researcher of the Academia Sinica, called Wang Ling submitted a thesis at Cambridge entitled *The* Chiu Chang Suan Shu *and the history of*

---

[11]See note 6, above.

[12]Cantor (2), *1880*, I.

[13]Libbrecht (2), *1973*, p. 214 ff.

[14]Published in Leipzig in 1913.

[15]Reprinted in 1974 in New York (Chelsea Pub. Co.). Cf. Mikami (4), *1913*.

[16]According to Ōya (2'), *1979*.

[17]Mikami, op. cit., p. 135 ff.

[18]This author has written a number of articles on the history of Chinese mathematics (references in J. Needham, *SCC*, III, p. 792) and Smith and Mikami (1), *1914*.

[19]On van Hée, cf. *SCC*, III, p. 3 ff. and Libbrecht, op. cit., pp. 318–324.

[20]Biography of Sédillot in Vapereau (1), *1880*, p. 1651.

[21]Cf. van Hée (2), *1932*, p. 260.

Chinese Mathematics during the Han dynasty.[22] This non-exhaustive study by
Wang Ling relates to the following subjects: dating of the *JZSS*, the Chinese
numeration system, the handling of fractions, the calculation of proportions,
Horner's method, the general characteristics of Chinese mathematics. It was
particularly well documented and reliable and surpassed by far everything then
existing in Western languages on the subject. Unfortunately, this fundamental
text was never published. However, in 1959, after having worked in collaboration with Wang Ling, the British biochemist Joseph Needham[23] (1900–1995)
published at Cambridge a history of Chinese mathematics as part of his *Science
and Civilisation in China*, a monumental project to compile a total history of
Chinese scientific and technical thought.[24] From the point of view of volume, this
new contribution was comparable with that of Mikami, but in all other respects
it was markedly different. Firstly, whilst the former had only had access to
relatively limited documentation, the latter had assembled a great deal more
material. Secondly, the two authors did not share the same historiographical
notions. Mikami proceeded empirically arranging his material around books and
authors, J. Needham on the other hand, arranged his material as a function of
a philosophy of history based on the following main theses:

(i) From time immemorial only a single universal science, teleologically
    structured from the beginning according to categories of thought com-
    parable with those of modern science, has existed: "Throughout this series
    of volumes it has been assumed all along that there is only one unitary
    science of nature. [...] man has always lived in an environment essentially
    constant in its properties and his knowledge of it, if true, must therefore
    tend toward a constant structure."[25]

(ii) All peoples have made their own particular contribution to the fabric of
    modern science: "What metaphor can we use to describe the way in which
    the mediaeval sciences of both West and East were subsumed in modern
    science? The sort of image which occurs most naturally [...] is that of
    rivers and the sea [...] and indeed one can well consider the older stream
    of science in the different civilisations like rivers flowing in the ocean of
    modern science. *Modern science is indeed composed of contributions from
    all the peoples of the Old World* [our italics], and each contribution has
    flowed continuously into it, whether from Greek and Roman antiquity,
    or from the Arabic world, or from the cultures of China and India."[26]
    (J. Needham was not the first to use such a hydrological metaphor. It had

---

[22]Wang Ling *Thesis*.

[23]Biography of J. Needham in Li Guohao et al. (1), *1982*, pp. 1–75.

[24]On Needham's work, cf. the Review Symposia published in *Isis*, *1984*, vol. 75, no. 276,
p. 171 ff. Review of Needham's history of Chinese mathematics in Libbrecht (4), *1980*. A
more general and well-balanced analysis of Needham's work has been recently published by
H. F. Cohen (Cohen (1), *1994*).

[25]*SCC*, IV, *1980*, p. xxxv.

[26]Needham (3), *1967*, p. 4.

already been used in the 19th century by William Whewell (1794–1866), one of the pioneers of the history of science).

Based on these presuppositions, he explained that the Chinese originality in mathematics should be researched from the direction of algebra, in contrast to Western mathematics which is characterised by its geometric genius.[27] Thus, with these premises he established a list of everything which, in his opinion, the West owed to China as far as mathematics is concerned: decimal notation, algebra, Horner's method, indeterminate analysis,[28] etc. Furthermore, believing that mathematics occupies a key position amongst the sciences, in the same chapter, he posed an ambitious question, far beyond the limited framework of the history of mathematics, namely: why has modern science not developed in China, when, paradoxically, until the end of the Middle Ages, that country was ahead of other civilisations?[29]

The works of Needham commanded broad authority. In his history of medieval mathematics, which first appeared in Russian in 1961, then in German and Japanese translations (1964 and 1970, respectively), the Soviet historian A. P. Yushkevich dedicated a hundred pages to the history of Chinese mathematics, basing himself broadly on recent original Chinese works (notably those of Li Yan) and on the works of the British biochemist. In addition, in his book, Yushkevich makes several conjectures about Chinese mathematical reasoning, although most of his predecessors merely affirmed the purely empirical nature of Chinese mathematics. Whence the great interest in the analyses of this historian.[30] Over the following years, certain Chinese historians echoed the theses of Needham.[31] More recently, other historians have pondered in a general way over the difficulties raised by the historiography of Needham. In particular, Nathan Sivin insisted several times on the need not to account for isolated discoveries but rather to understand Chinese science in its global historical context.[32] More specifically, regarding delicate questions on the technical aspect of the history of mathematics (origin of Pascal's triangle, decimal fractions) researchers such as A. S. Saidan believe that some of Needham's analyses and conclusions (affirmations of Chinese priorities) should be reexamined.[33]

Until now, no one has undertaken a systematic approach to this;[34] however, from 1960 to 1980 a series of specialised works involving research in areas left fallow until then provided a clearer view. To mention only the most

[27]*SCC*, III, p. 150 ff.

[28]Ibid., p. 146 ff.

[29]Ibid., p. 150. Cf. the critical remarks in Sivin (3), *1982*.

[30]Cf. Yushkevich (1), *1964*.

[31]Cf. The articles mentioned in Li Yan and Li Di (1'), *1980*.

[32]Cf. Sivin (2), *1977*, p. xi.

[33]Saidan (1), *1978*, p. 485: "The writer has the feeling that some of Needham's assertions are in need of further objective verification. But concerning priority to the decimal idea, the Chinese claim is strong."

[34]See however Libbrecht (4), *1980*.

important stages of this advance, we note that in his thesis in 1963,[35] the French engineer Robert Schrimpf brought to light the very important fact that the commentaries on the *Jiuzhang suanshu* (Computational Prescriptions in Nine Chapters) contained proofs of mathematical results, which meant that it was in part no longer necessary to resort to hypothesis in order to understand the achievements of the Chinese of the Han. Later (1975–1979), in the same vein, the Dane D. B. Wagner analysed the proofs of Liu Hui (end of the third century AD) concerning the calculation of the volume of certain solids (notably the volume of the sphere and of the pyramid).[36] In a different connection, other historians began a systematic translation of the major works of Chinese mathematics. Initially, the texts most often considered were those of the collection of 'The Ten Computational Canons' (*Suanjing shi shu*) of the Tang dynasty.[37] The *Jiuzhang suanshu*, which is the second of the 'Ten Canons' was translated three times: once into Russian by Madame E. I. Beryozkina (1957),[38] secondly into French by R. Schrimpf[39] (thesis remained unpublished, 1963) and thirdly into German by K. Vogel.[40] In the 1970s historians poured over the works of the 13th century mathematicians. Madame Lam Lay-Yong of Singapore translated the *Yang Hui suanfa* (Yang Hui's Methods of Computation) (ca. 1275) in full,[41] and Jock Hoe, resolving the problem of translating Chinese mathematical texts in a very original way,[42] translated the whole of the *Siyuan yujian* (Jade Mirror of the Four Origins) (1303). In order to remain faithful to the spirit of ancient Chinese texts, often obscured in translation by heavy paraphrases which "completely obliterate the preciseness and conciseness of the original," J. Hoe recommends resorting to translations into a 'semi-symbolic' language which has the advantage of rendering the terse Chinese formulations by similarly terse and 'telegraphic' English formulations rather than by verbose equivalents which give a false impression of what the Chinese original texts really are.[43] In addition, in a first class work, *Chinese Mathematics in the Thirteenth Century, the* Shu-shu chiu-chang *of Ch'in Chiu-shao*, U. Libbrecht ((2), *1973*) made a brilliant attack on the comparative history of the 'Chinese remainder theorem,' at the same time resiting Chinese mathematics of the 13th century in its proper social and cultural framework. In a completely different direction of research, but of equally essential interest, we also note the useful work of F. Swetz ((3), *1974*), *Mathematics Education in China, its Growth and Development*. It goes almost without saying that it would be desirable if these pioneering works were to be followed as of now by numerous others.

---

[35]Schrimpf (1), *1963*.

[36]Wagner (1), *1975*; (2), *1978*; (3), *1978*; (4) *1979*.

[37]Works of Schrimpf (op. cit.); Beryozkina (1), *1981*; Ang Tian-Se (1), *1969*; Brendan (1), *1977*; Ho Peng-Yoke (2), *1965*; Swetz (1), *1972*; Swetz and Kao (1), *1977*.

[38]*Istoriko-matematicheskie Issledovaniya*, Moscow, vol. 10, *1957*, pp. 427–584.

[39]Schrimpf (1), *1963*.

[40]Vogel (1), *1968*.

[41]Lam Lay-Yong (6), *1977*.

[42]Hoe (1) and (2), *1976* and *1977*.

[43]Hoe, (2), p. 23.

Works on the History of Chinese Mathematics in Japanese

Japanese studies of the history of Chinese mathematics began somewhat belatedly at the beginning of the 20th century[44] and were often conceived not in their own right, but as an introduction to the history of Japanese autochthonous mathematics.

As a result, sometimes articles which appear to treat only Japanese questions also in fact touch on Chinese questions. For example, in a book apparently devoted solely to the history of the *wasan* (traditional Japanese mathematics), Katō Heizaemon considers in detail the question of series developments in the work of Minggatu (?–1764).[45] In a series of articles published between 1932 and 1934 and devoted to the relationship between Seki Takakazu and Japanese mathematicians of the Osaka and Kyoto regions, Mikami analyses (apparently for the first time) the problem of the volume of the sphere in China, in order to determine the possible influence of Chinese techniques on Japanese mathematics.[46] This aspect of Chinese mathematics has recently been rediscovered (1978) independently of Mikami.[47]

The most important Japanese works as far as the history of Chinese mathematics is concerned include those due to Mikami Yoshio (1875–1950), Ogura Kinnosuke (1885–1962), Takeda Kusuo, Yabuuchi Kiyoshi, Kodama Akihito and Yamazaki Yoemon.

Mikami Yoshio, whom we have already mentioned, wrote three books, together with some 20 articles devoted solely to the history of Chinese mathematics.[48] His works, which sometimes take in points of highly specialised erudition, reveal a mind preoccupied with synthesis and inclined to pose questions of very general import (relationship between mathematics and art, place of the mathematician in society, influence of war on the development of mathematics,[49] circulation of ideas between civilisations).[50]

In an analogous direction, Ogura Kinnosuke[51] placed the accent on socio-economic problems. Two of his articles, devoted respectively to the social aspect of Chinese mathematics[52] and the universalisation of mathematics in the Far East[53] are still famous in Japan. In the first of these, he used the text of the *Jiuzhang suanshu* as a source for the economic history of the Han, and in the second he compared the history of mathematics in the Japan of the Meiji era

---

[44]Ōya (2'), *1979*.

[45]Katō (1'), *1969*.

[46]Mikami (7'), *1932–1934*.

[47]Wagner (3), *1978*.

[48]Bibliography of Mikami in Yajima (1), *1953*.

[49]See no. 189 of Yajima (1), *1953* (bibliography of Mikami).

[50]On this, see his criticisms of the ideas of the European mathematical historians Cantor, Loria, van Hée and Kaye (nos. 13, 24, 79, 86, 64 of Yajima's bibliography, respectively).

[51]Biography in Itō *et al.* (1'), *1983*, p. 138.

[52]Ogura (3'), *1978*, p. 185 ff.

[53]Ibid., p. 206.

and in the China of the end of the Qing. He explained that, since the beginning
of the Meiji era, Japan, unlike China had adopted the notation and the style
of Western mathematics directly, without retaining anything whatsoever of
traditional Japanese mathematics.

In a completely different area, Takeda Kusuo and Kodama Akihito were
interested in the history of the Chinese mathematical texts themselves. The
former established the genealogy of certain Chinese arithmetics of the Ming
period[54] and the latter reprinted ancient Chinese texts going back to the 14–
15th and 16th centuries with copious bibliographical notes.[55]

Synthesising the most recent Chinese and Japanese works, Yabuuchi Kiyoshi
compiled the history of Chinese mathematics dynasty by dynasty.[56]

We would also note the works of Yamazaki Yoemon and Toya Seiichi[57] on
the history of the abacus, not forgetting the articles by Japanese historians
which appear regularly in the journal *Sūgaku shi kenkyū*.

Finally, recently, Japanese research has caught up with the latest Chinese
preoccupations by tackling the problem of translating the *Jiuzhang suanshu* and
its commentaries. Ōya Shinichi, then the mathematician Shimizu Tatsuo first
published a translation of the problems and the rules for solving them.[58] Later,
in 1980 Kawahara Hideki translated Liu Hui's commentary for the first time *in
extenso*.[59]

Works on the History of Chinese Mathematics in Chinese

The Chinese have long been interested in the history of their mathematics.
The *Suishu* (History of the Sui Dynasty), which was presented to the throne
in 656, already includes a paragraph on the history of π.[60] More recently, in
the 18th and 19th centuries, historical questions have preoccupied Chinese
mathematicians to at least as great an extent as mathematics itself (see the
mainly philological work of Dai Zhen (1724–1777) on the *SJSS*, of Li Huang
(?–1812) on the *JZSS*, of Luo Shilin (1774–1853) and Shen Qinpei (fl. 1807–
1823) on the *Siyuan yujian* (1303), and of Ruan Yuan (1764–1849) on the
calendarist-mathematicians). Access to these works is still indispensable today,
for a number of reasons: (i) they reveal the conceptions of mathematics in China
which prevailed during different periods, (ii) they enable us to understand that,
even today, certain preoccupations of Chinese historians (analytical taste for

---

[54]Takeda (1'), (2'), (3'), *1953* and *1954*.

[55]Kodama (1'), (2'), *1966* and *1970*.

[56]Yabuuchi (2') to (5'), *1963* to *1970*. This author also wrote a short book on the history
of mathematics (Yabuuchi (6'), *1974*).

[57]Yamazaki (1), (2), *1959* and *1962*.

[58]Ōya (3'), *1980*, pp. 97–164 and the journal *Sūgaku seminā* (issues from February 1975 to
April 1976).

[59]Kawahara (1'), 1980.

[60]*Suishu*, j. 16, p. 388.

specific detail, tendency to present the information in the form of citations with a view to exhaustiveness considered as a good thing in its own right, predilection for moralising historiography polarised in terms of praise or blame) are part of a long tradition.[61] However, the fact remains that, in the 20th century, historians differ from their peers in the past in one essential way in that, unlike the latter, they have almost all received a high-level scientific training. The most important contemporary historians include Li Yan (1892–1963) who was a railway engineer,[62] Qian Baocong (1892–1974) who was a professor of mathematics with an honorary DSc from the university of Birmingham,[63] and Zhang Yong (1911–1939) who studied mathematics and physics at Göttingen.[64] If one had to compare these historians with European historians, one might say that the former bear greater resemblance to P. Ver Eecke or J. Itard than to G. Sarton, since they approach their subject as enlightened amateurs rather than professional historians.

Moreover, almost all of the history of Chinese mathematics is due to the above, who account for 90 percent of the 300 articles which appeared on this subject between 1900 and 1940.[65] What is more, these articles cover all areas, including biography, bibliography, critical studies of ancient texts, history of specific problems, syntheses, etc.

In 1955, China founded a team for research into the history of mathematics. At first, this was limited to three members, namely: Li Yan, Qian Baocong (both mentioned above) and Yan Dunjie (1917–1988), a self-taught historian who first published in 1936.[66] Although he worked for a long time under particularly difficult material conditions, exercising in turn the profession of office clerk, accountant and proof-reader,[67] Yan Dunjie stands out as one of the most solid researchers in the discipline, since he perceived the history of Chinese mathematics from a truly historical angle from a very early stage. His output is even more valuable, since even today a considerable number of works are based on a fundamentally one-dimensional, anachronistic approach which, without further ado, confuses mathematics and the history of mathematics. In particular, Yan Dunjie was responsible for observations on the influence of the development of paper money on the mathematics of the Song and Yuan periods, on the influence of poetical terminology on the technical vocabulary of mathematics, on the mathematical notions contained in certain Chinese novels, on the relationship between mathematics and medicine (cf. his analysis of the problem 3-36 of the *Sunzi suanjing*, below, p. 138) and on the mathematical foundations of calendrical astronomy. From 1957, several other researchers were

---

[61]We are not referring here to the overall Chinese historiographical tradition, but to one of its important streams, which is reasonably well represented by works such as the *Chouren zhuan*.

[62]Biography of Li Yan in Wong Ming (1), *1964*.

[63]Biographical elements in Li Di, *Hist.*, p. 411.

[64]Ibid., p. 412. See also Li Yan (50′), *1954*, p. 135 ff.

[65]According to Li Di, *Hist.*, p. 412.

[66]Idem.

[67]Wang Yusheng (3′), *1989*.

appointed to the same team.[68] Outside this restricted cadre, numerous other historians (more than 20, mainly university professors) have also taken an interest in the subject.[69]

Areas currently under exploration and programmes now under way include:

- Study of the *Jiuzhang suanshu* (works of Wu Wenjun[70] on the modes of reasoning used in geometry, of Li Jimin on fractions, of Guo Shuchun on proportionality and of Li Di on the dating of the text, etc.).

- Translation into modern Chinese or Western languages of the main ancient Chinese mathematical works (Bai Shangshu,[71] Shen Kangshen,[72] Guo Shuchun).[73]

- Study of the work of Chinese mathematicians of the Qing period (Liu Dun,[74] Luo Jianjin,[75] Shen Kangshen[76]).

- Study of the sources of Chinese adaptations of European works (Bai Shangshu[77]).

Finally, in Taiwan, we note the many articles and books by Horng Wann-Sheng,[78] which consider very diverse problems from different angles (history and teaching of mathematics, history of Chinese mathematical reasoning, Chinese mathematics and Confucianism, etc.). Horng Wann-Sheng is also the author of an excellent thesis on Li Shanlan, presented in New York under Prof. Joseph Dauben (March 1991).[79]

---

[68]Including, in particular: Mei Rongzhao, Du Shiran, He Shaogeng, Liu Dun. See Wu Wenjun (2'), *1985*, I, p. 1.

[69]Including, in particular: Shen Kangshen (Hangzhou), Li Jimin (Xi'an), Li Di (Huhehot), Bai Shangshu (Peking), Liang Zongju (Shenyang).

[70]Cf. Wu Wenjun, (1'), *1982*.

[71]Bai Shangshu (6'), *1985*.

[72]Project to translate the *Jiuzhang suanshu* into English.

[73]Project to translate the *Jiuzhang suanshu* and its commentaries into French, in collaboration with K. Chemla, researcher at C.N.R.S.

[74]Liu Dun, (1') to (4'), *1986* to *1989*.

[75]Luo Jianjin, (1') to (6'), *1982* to *1988*.

[76]Shen Kangshen, (3'), *1982* (a study on ancient Chinese mathematical terminology).

[77]Bai Shangshu (1') and (4'), *1963* and *1984*.

[78]Horng Wann-Sheng, (1') to (4'), *1981* to *1993*.

[79]Horng Wann-Sheng (1), *1991*.

2. The Historical Context

As far as one can judge, Chinese mathematics, as an autonomous, highly distinctive area of knowledge, handed down in writing in specific documents, is not highly ancient in origin. The corpora of Egyptian and Babylonian mathematics predate it by more than a millennium. Was it born at the time Greek science was at its peak? There is currently no evidence to confirm this; documents about ancient China never refer, either directly or indirectly, to a mathematical knowledge beyond the bounds of elementary arithmetic (numeration, operations, weights and measures, fractions). However, the Chinese of the period of the Warring States do appear to have been aware of Pythagoras' theorem and the similarity of right-angled triangles.[1]

In fact, the Chinese mathematical landscape begins to be more detailed from the time of the Former Han (208 BC – 8 AD). Various sources[2] tell us of the existence of the *Suanshu* (Computational Prescriptions) in 16 "volumina" (*juan*) by a certain Du Zhong,[3] together with the *Xu Shang suanshu*[4] (Computational Prescriptions of Xu Shang) in 26 *juan*, several decades before the start of the present era. But these works have not survived. The two oldest known Chinese mathematical texts, namely the *Zhoubi suanjing*[5] (Zhou Dynasty Canon of Gnomonic Computations) and the *Jiuzhang suanshu* (Computational Prescriptions in Nine Chapters), also date very approximately from the same period. These are both anonymous. The first of these is actually a complete book on quantitative cosmology, in which the inferences depend on calculations rather

---

[1]Cf. Li Yan, *Gudai*, p. 47, Chen Liangzuo (2'), *1978*, p. 288 ff. In their works on the history of Chinese mathematics, some authors such as Li Di ((3'), *1984*) include long descriptions of the geometrical drawings which appear on certain pieces of pottery (2nd–3rd millennia BC). However, it is not known whether these drawings have anything at all to do with mathematical preoccupations. Other authors speculate on the geometrical definitions of the Mohists (on this, see later, pp. 273 ff.).

[2]Li Yan, *Gudai*, p. 44 ff.

[3]Ibid., p. 47.

[4]Ibid., p. 45.

[5]Needham (*SCC*, III, p. 19) proposes a different translation of this title, namely: *The Arithmetical Classic of the Gnomon and the Circular Paths of Heaven*. To justify this translation he uses a gloss by Li Ji of the Song. However, in the source cited by J. Needham, the same Li Ji writes that *Zhou* refers to the dynasty of the Zhou. See: *Zhoubi suanjing yinyi* (Meanings and Pronunciations of [words] occurring in the *Zhoubi suanjing*), p. 55 (cited from *Suanjing shi shu*, I, Taipei: Shangwu Yinshuguan, *1978*.).

than on fantastic myths.[6] The authors were already using a type of decimal numeration and knew how to add, subtract, multiply and divide fractions and how to extract the square root of an arbitrary number. They also knew Pythagoras' theorem for (3,4,5) and (6,8,10) triangles. They used the value 3 for the ratio between the circumference and the diameter of the circle and knew how to manipulate similarity in the case of right-angled triangles. The *Jiuzhang suanshu* (which we shall refer to in what follows as the "Nine Chapters" or the *JZSS*) became, in the Chinese "tradition", the mandatory reference, the classic of classics. It is concerned in part with arithmetic (rule of three, proportional division, double-false-position), in part with geometry (calculation of surfaces and volumes (triangle, trapezium, circle, circular segment, ring, prisms, cylinders, pyramids on a square base, cone, frustra of the cone and the pyramid, tetrahedra)), and in part with algebra (linear systems) including the use of negative numbers. On reading it, one has the impression that the readers for whom it was intended had very specific interests, since the themes of surveying fields, trading in grain, taxes and excavation works crop up time and again. The historian Sun Wenqing, has conjectured that it was intended for the accountants of the Han administration.[7] This remark leads one to wonder about the social position of the mathematician in ancient China.

On this question, Mikami notes that there must have existed a connection between mathematicians and astronomers,[8] based on the fact that a global term *chouren*, was used to refer globally to both. He also remarks that one possible etymology of the character *chou* (to cultivate land, to measure the extent of lands) is very similar to that of the term "geometry" (to measure land).[9] Thus, Chinese mathematicians would have been land surveyors, like the Egyptian rope-stretchers, then also, by extension, surveyors of the heavens. However, other signs indicate that mathematics (or the practice of computations) would also have been associated with artisan circles.[10]

From the third to the sixth century, Chinese mathematics entered its theoretical phase. For the first time, it seems, importance was attached to proofs in their own right, to the extent that trouble was taken to record these in writing. Approximate values for the number π were then derived by computation and reasoning, rather than simply via empirical processes. At the end of the third century, Liu Hui obtained the value $157/50$ ($= 3.14$), while at the end of the fifth century, Zu Chongzhi found $355/113$. The volume of the sphere was calculated

---

[6]Details of this cosmology are given in Nakayama (1), *1969* and in Ho Peng-Yoke (2), *1966*, p. 43 ff.

[7]Sun Wenqing (1'), *1931*.

[8]Mikami (8'), *1934*, p.33.

[9]Cf. the well known allusion of Herodotus: "[Sesostris] divided the ground between all the Egyptians [...] If a river happened to take away part of someone's plot [...] [The Pharaoh] sent people to examine by how much the ground had decreased [...] It is my opinion that this gave rise to the invention of geometry." Quoted by Michel (1), *1950*, pp. 9–10.

[10]Since, in particular, Liu Hui's commentary on the *JZSS* alludes to the *Kaogongji* (The Artificers' Record), a section of the *Zhouli* (Records of the Rites of the Zhou Dynasty). See *JZSS* 4–24, in QB, I, p. 156.

using "Cavalieri's principle" and the volume of the pyramid by considering infinitely small quantities. All this was described in multiple commentaries of the *Jiuzhang suanshu* and seems to have acquired its momentum from the neo-Taoist current. At the same period, science and technology advanced remarkably: for example, geographical maps based on rectangular lattices of equidistant straight lines were constructed (by Pei Xiu, minister of public works under the first emperor of the Jin, at the end of the third century),[11] the precession of the equinoxes was discovered by the astronomer Yu Xi (fl. 307–338), Tao Hongjing (456–536) wrote the commentaries of the oldest pharmacopoeia (the *Shennong bencao jing* of the Han) and machines were developed by Zu Chongzhi (fifth century).[12]

Under the Sui dynasty (518–617), and above all under the Tang dynasty (618–907), mathematics was officially taught at the *guozixue* (School for the Sons of the State), based on a set of contemporary or ancient textbooks as written support. These textbooks included the *Zhoubi suanjing* and the *Jiuzhang suanshu* mentioned above, together with ten other texts dating from later than the Han, namely:

- the *Haidao suanjing* (Sea Island Computational Canon) by Liu Hui (mathematician, commentator of the *JZSS*, end of the third century),

- the *Sunzi suanjing* (Sunzi's Computational Canon) (fifth century?),

- the *Wucao suanjing* (Computational Canon of the Five Administrative Sections),

- the *Xiahou Yang suanjing* (Xiahou Yang's Computational Canon),

- the *Zhang Qiujian suanjing* (Zhang Qiujian's Computational Canon) (end of the fifth century),

- the *Zhuishu* (meaning unknown) by Zu Chongzhi,

- the *Wujing suanshu* (Computational Rules of the Five Classics),

- the *Shushu jiyi* (Notes on the Traditions of Arithmo-Numerological Processes),

- the *Sandeng shu* (the art of the three degrees – notation for large numbers based on three different scales),

- the *Jigu suanjing* (Computational Canon of the Continuation of Ancient [Techniques]) (seventh century) by the calendarist mathematician of the Tang, Wang Xiaotong.

Nowadays, these 12 books are often improperly referred to under the collective title of "The Ten Computational Canons."

---

[11] Cf. *SCC*, III, p. 200.
[12] Cf. Li Yan, *Dagang*, I, p. 60.

Overall, taking the *Jiuzhang suanshu* of the Han as a reference, the level of these textbooks (which are considered to be representative of their period) is low. However, they do include problems the solutions of which relate to the branch of mathematics known (to us) as number theory, namely the "Chinese remainder problem," the so-called problem of the hundred fowls, not forgetting equations of degree three or, more exactly, computational techniques for the extraction of generalised cube roots, which Wang Xiaotong knew how to construct.

None of the originals of these Ten Computational Canons has survived but, by chance, we do have a few fragments of arithmetics of the same period (first millennium AD). These are elementary texts which form part of the manuscripts discovered in Dunhuang at the beginning of this century.[13] One of these dates from 952 AD and contains a multiplication table for the numbers 1 to 9, omitting the product $a \times b$ when $b \times a$ has already been given,[14] together with a two-way table used to convert square *bu* (a step, a pace, unit of length), into *mu* (units of area).[15]

The period from the 10th to the 12th century is a void which is difficult for us to fill owing to a lack of original documents. However, according to various, partially overlapping bibliographies around 50 new works may be placed in this period.[16] From their titles, these appear to be probably of little importance, being presumably works on elementary arithmetic. However, a small amount of information may be gleaned elsewhere.

- The *Mengqi bitan*, the famous collection of notes of Shen Gua (1031–1095) includes a number of laconic paragraphs about mathematics. None of this attests to a particularly innovative activity, although a number of the formulae cited (approximation of the length of a circular arc, sum of a finite series)[17] are not given in the Ten Computational Canons.

---

[13] Analysis in Li Yan, *Gudai*, pp. 16 and 23 ff.; Libbrecht (5), *1982*; Martzloff (5), *1983*.

[14] Li Yan, ibid., p. 16; Libbrecht, ibid., p. 218.

[15] Li Yan, ibid., p. 27; Libbrecht, ibid., p. 211.

[16] Li Di, *Hist.*, pp. 148–151 (list obtained by perusing: *Songshi*, "Yiwenzhi;" *Chongwen zongmu*; *Suanfa tongzong*; *Suichu tang shumu*; *Yang Hui suanfa*; *Xin yixiang fayao*; *Zhizhai shulu jieti*; preface of the *Siyuan yujian* compiled by Zu Yi).

[17] Cf. Li Qun (1'), *1975*, p. 101 ff. In modern notation, Shen Gua's summation formula corresponds to

$$S = \sum_{k=0}^{n-1}(a+k)(b+k) = \frac{n}{6}\left[(2a+A)b + (2A+a)B\right] + \frac{n}{6}(B-b)$$

where a and b are given integers and $A = a + (n-1)$, $B = b + (n-1)$, respectively. More precisely, Shen Gua was interested in calculating the number of objects composing a solid made of n superposed rectangular layers having $(a+k)(b+k)$ objects each $(0 \le k \le n-1)$ respectively, each dimension being diminished by one unit from each layer to the next. Note that this formula includes as a special case, the formula for the sum of the squares of the first n numbers.

- Authors of the 11th and 12th centuries such as Jia Xian and Liu Yi appear to have been aware of Pascal's triangle and the Ruffini–Horner method, together with a number of techniques of computational algebra. Our knowledge of this subject is very imprecise, since it is based on somewhat allusive information, including extracts from 13th century works such as those by Yang Hui or Zhu Shijie, accessible to us through the medium of 19th century editions. For example, in his *Tianmu bilei chengchu jiefa* (1275), Yang Hui refers to the contents of an earlier book entitled *Yigu genyuan*, said to have been written by a certain Liu Yi who lived around 1113 AD.[18] Thus, we have the text of a score of algebraic problems of degree two due to this Liu Yi.[19] One of the prefaces of Zhu Shijie's *Sijuan yujian* contains a list of titles of "ancient" works, all of which are now lost, dealing with various algebraic techniques involving between one and four unknowns.[20] Pascal's triangle, which appears at the beginning of the same book, is referred to as the "ancient method" (*gu fa*).

On the other hand, the 50-year period between 1247 and 1303 is very much better documented. It saw the appearance of a succession of mathematical works at a level far higher than that of the Ten Computational Canons (*SJSS*), both from the point of view of the complexity of their algorithms and from that of the originality of their results. These works include:

- The *Shushu jiuzhang* (Computational Techniques in Nine Chapters) by Qin Jiushao, which is notable for its perfectly general algorithm for solving simultaneous congruences (generalised Chinese remainder theorem).

- The *Ceyuan haijing* (Mirror like the ocean, reflecting [the heaven] of circles [inscribed and circumscribed]) (1248) by Li Zhi, in which an algebra of rational fractions and generalised polynomials (including negative powers of the unknown) is developed.

- The *Yigu yanduan*[21] (1259) by the same author, which is devoted to the algebraic and geometric construction of 64 polynomial equations of degree less than or equal to 2, and which appears to have been designed as an introduction to the *Ceyuan haijing*.

- The *Suanxue qimeng* (Introduction to the Computational Science) (1299) by Zhu Shijie, which describes, in particular, the rudiments of a polynomial algebra similar to that of Li Zhi.[22]

- The *Siyuan yujian* (Jade Mirror of the Four Origins [= unknowns]) (1303) by the same author, which is concerned with the elimination of unknowns from systems of numerical polynomial equations in up to four unknowns.

---

[18] Cf. Crossley and Lun (1), *1987*, p. 128.

[19] Cf. Lam Lay-Yong (6), *1977*, pp. 83, 87, 112.

[20] Cf. Hoe (2), *1977*, pp. 117–118.

[21] The term *yanduan* is difficult to translate. It means an algebraic technique which depends both on computation and on geometric figures.

[22] Analysis in Lam Lay-Yong (7), *1979*.

In the same period, we must also mention the "spherical trigonometry" of the calendarist astronomer Guo Shoujing (1231–1316), not forgetting, in a very different vein, the much more elementary *Yang Hui suanfa* (1275) which includes only a few algebraic results and is almost solely concerned with elementary arithmetical formulae explained in a wealth of detail not found in the *SJSS*. This book also contains the oldest known Chinese magic squares of order greater than three.

This rapid overview attests to an extraordinary blossoming of algebra and, more generally, of all mathematics. It turns out that this flowering occurred at precisely the moment when China was entering into contact with the Islamic countries, at the time of the Mongol expansion. Under these conditions, should not the development of mathematics in China at this period be viewed as an epiphenomenon, a consequence of the diffusion of Islamic algebra and astronomy? We shall examine this question later and we shall see that Chinese mathematics and that of the Islamic countries were actually similar in many respects ("Ruffini–Horner method," decimal fractions, magic squares). However, these comparisons are meaningless as far as Chinese spherical trigonometry is concerned, since the latter is based on procedures which are totally alien to the Greco-Islamic tradition.

But the works of the Chinese mathematicians of the 13th century are very similar. One suspects that the latter must all have been aware of the mathematical works of their contemporaries; however, it is impossible to be completely certain about this, since in the texts which have come down to us these authors do not mention one another. Moreover often, the terms used in one or other of these books sound strange for mathematics; who could imagine that such curious expressions as *tianyuan* (celestial origin) and *dayan qiu yi shu*[23] (*dayan* computational process of finding unity – an allusion to a passage of the *Yijing*) might conceal algebraic and arithmetical ideas? Should one not interpret all this as the manifestation of a confusion between false and exact sciences? The answer is no, since the solutions obtained by these methods correspond well to the dicta of mathematics alone. Moreover, the problems which these methods may be used to solve are often so complicated (polynomial equations up to degree 14 with large coefficients (up to 15 digits)) that a derivation based on numerological manipulations would have very little chance of succeeding. Finally, our authors all explain, albeit laconically, how to proceed and their explanations are in no way those of a false science. But, one objects, why retain such a vocabulary? There is no easy answer; however, we note that in the China of the Song, those who studied astronomy and mathematics could not concern themselves with esoteric techniques at the same time, according, at any rate, to the "regulations relating to computational sciences taught at the *guozixue* in the Chongning era" (1102–1106):

All students must study the problems of the 'Nine Chapters' (*JZSS*), the 'Gnomon of the Zhou' (*ZBSJ*) together with the computational methods of the Sea Island Computational Canon (*HDSJ*), Sunzi's Computational Canon (*SZSJ*), the

---

[23] "Xici," part I, section 9.

Fig. 2.1. Portrait of Wu Jing inserted at the beginning of his *Jiuzhang suanfa bilei daquan* (1450).

Computational Canon of the Five Administrative Sections (*WCSJ*), Xiahou Yang's Computational Canon (*XHYSJ*) and the works of calendrical computation, divination (*sanshi*[24]) and astronomical astrology (*tianwen*) [...].[25]

In this context, it seems probable that mathematicians were also diviners; thus it is understandable that the vocabulary was carried across. Since, for those who practised numerological and arithmetical computation using similar operating procedures (manipulation of rods), it was much more natural to use the same words in all cases, rather then invent new terms ex nihilo, even though the computations did not have the same objective.

After the 13th century, Chinese mathematics entered a period of decline. Here, it is important to underline that the perception of the history of mathematics in terms of progress and decline is not only a modern idea, but also *Chinese*. At the beginning of the 16th century, Xu Guangqi (1562–1633) formulated it most explicitly and gave two explanations: (i) predominance of the *lixue* which led literati to neglect practical learning (*shixue*), (ii) confusion, under

---

[24]Cf. Kalinowski (1), *1983*; Ngo Van Xuyet (1), *1976*, p. 161 ff.; *SCC*, III, p. 141.
[25]Cited from Yan Dunjie (26'), *1982*, p. 62.

the Ming, between mathematics and numerology.[26] To be sure, soon, there was no one who could understand the nature of the *tianyuan* algebra. The possible reasons for this phenomenon include: (i) being insufficiently explicit, without oral teaching, the texts of the algebraists could no longer be penetrated without great effort by those who were no longer able to find out the information from the inventors themselves; (ii) many problems of the *tianyuan* knowledge could often be solved by simple and direct methods without the need for recourse to algebra (in other words, the content of some problems masked the originality of the new algebraic techniques); (iii) the particularly artificial and gratuitous nature of the problems may have been a hugely effective foil for those who judged everything by its economic usefulness.

But under the Ming dynasty (1368–1644), the tradition of the Nine Chapters (*JZSS*) was continued. In 1450, Wu Jing, Provincial Administration Commissioner *buzhengsi*[27] of the province of Zhejiang,[28] (Fig. 2.1) published an encyclopedic arithmetic containing, amongst other things, all the problems of the *Jiuzhang suanshu*: the *Jiuzhang suanfa bilei daquan* (Fully Comprehensive [Collection of] Computational Methods in Nine Chapters with [New Problems and Rules] Devised by Analogy with [Ancient Problems and Rules]). The abacus, a calculating instrument used by merchants, slowly supplanted the counting-rods used in the erudite practice of computation, and mathematics was restricted to commercial arithmetic. At the end of the 16th century, an obscure employee of the administration, Cheng Dawei, who had spent his life collecting arithmetical works published his popular *Suanfa tongzong* (General Source of Computational Methods) (1592), which has subsequently been reprinted numerous times, even in the 20th century, and has even reached Japan[29] and Vietnam.

At the end of the Ming (end of the 16th century, beginning of the 17th century) China entered a period of intellectual effervescence, characterised by great curiosity and independence of mind.[30] This trend rejected metaphysics in favour of placing a higher value on practical learning (*shixue*).

When the first Jesuit missionaries reached China at the end of the 16th century, they developed an original policy of evangelisation which involved seeking dialogue at the highest level rather than carrying out mass conversions. They became familiar with the language and with the classical authors, and presented themselves as "Western literati" (*xiru*).

Matteo Ricci (1552–1610), who was the first to apply this programme, rapidly realised at the same time the extreme difficulty of the christianisation of China and the importance of gaining a stable position in that country. He soon saw how to achieve this. By 1605, he had in fact understood the full advantage which the missionaries could draw from their scientific and technical know-how,

---

[26]See the preface of the *Tongwen suanzhi* (1614), compiled by Xu Guangqi.
[27]Cf. Hucker (1), *1985*, notice no. 4770.
[28]Cf. Li Yan, *Dagang*, II, p. 303; Li Yan, *ZSSLC-T*, I, p. 131; QB, *Hist.*, p. 134.
[29]Kodama (2′), *1970*.
[30]Cf. Gernet (1), 1972, p. 81.

particularly in astronomy, given the incompetence of those who were responsible for these matters in China, when he wrote:

> We should change the Chinese calendar, this would enhance our reputation, the doors of China would be even more open to us, our position there would be more stable and we would be freer.[31]

Ricci's view was particularly perceptive, since the question of reforming the calendar had already preoccupied the Chinese of the Ming for some time. By the middle of the 15th century, the latter had uncovered errors in the computation of the solstices and in the predictions of eclipses.[32] Various reform projects were proposed prior to the end of the 16th century, notably by the Prince Zhu Zaiyu (1536–1611), but these projects were not accepted. In 1596, the Bureau of Astronomy signalled its hostility to any change, since it feared that "rumours of astronomical miscalculations might lead to unrest or rebellion."[33] Ricci was neither an astronomer nor a mathematician, but he had studied at the Collegio Romano, where importance was attached to the teaching of science.[34] It is against this background that he undertook to translate a series of works by his former teacher C. Clavius (1538–1612) into Chinese.[35] By 1597–1598, Ricci had translated the first book of Euclid's *Elements* with one of his first disciples, Qu Rukui (this is the same person as the Chiutaiso or Kiutaiso [Qu Taisu] referred to by Ricci).[36] In 1599, when he was at Nanking, Ricci delivered this first translation to those who had commissioned it.[37] The first essay has not survived, but some years later, Ricci undertook to adapt Clavius's Latin commentary on Euclid's *Elements* (*Euclidis Elementorum, libri XV. Accessit liber XVI. De Solidorum Regularium cuiuslibet intra quodlibet comparatione [...].*) into Chinese.[38] With the assistance of one of his Chinese students, the Christian convert and high-level public official Xu Guangqi (1562–1633), he reached the end of Book 6 in 1607. The resulting work was entitled *Jihe yuanben.*[39] The translation technique involved Ricci explaining the contents of the original text orally to his collaborator who would then write down what he had understood (*koushou*). This is the same method which was used from the fifth to the eighth century to translate Buddhist texts.[40] Ricci and Xu Guangqi's translation respects the order of Clavius's work completely; however, it is much less verbose and does not cite the names of the authors of new corollaries such as Proclus, Peletier, or Commandino. Although the terminology of the *Elements* was almost entirely new, it has withstood the test of time; it is still extensively

---

[31] Tacchi Venturi (1), *1911–1913*, II, p. 285.

[32] Cf. Peterson (3), *1986*, p. 47 ff.

[33] *DictMingBio*, p. 369.

[34] Cf. Cosentino (1), *1970*; Krayer (1), *1991*.

[35] On Clavius, see the notice of the *DSB*; see also Knobloch (2), *1988*.

[36] Cf. Hummel, p. 199.

[37] D'Elia (2), *1956*, p. 166.

[38] First edition in 1574, reprinted in 1589, 1591, 1603, 1607, 1612.

[39] For the exact meaning of this title, cf. see below, p. 115.

[40] Cf. Zurcher (1), *1959*.

used nowadays in China, Korea and Japan.[41] The geometric figures were also innovative in terms of the Chinese tradition, since they were accompanied by written characters extracted from certain series, such as the denary series *jia, yi, bing, ding, . . .*, which played a role similar to that of Greek or Roman letters in the same situation. This new practice caught on very quickly.

Ricci and his collaborators also translated other elementary works on geometry, including, in particular, a book on the solution of right-angled triangles, the *Gougu yi* (Explanation of the *gougu*, that is of the right-angled triangle denoted not by any notion of angle but by two terms denoting the shortest *gou* and longest *gu* sides, respectively), another on the measurement of distances on the ground using the geometric square, the *Celiang fa yi* (The meaning of (various) methods of surveying), and a treatise on isoperimetric figures taken from Clavius's commentary on Sacrobosco's sphere, in which the name of Archimedes is mentioned for the first time in a Chinese text (the *Yuan rong jiao yi*). These subjects were not all new, but this was the first time in China that they were treated *more geometrico*, with proofs in the Euclidean style. In a different vein, Ricci also adapted Clavius' *Epitome Arithmeticae Practicae* into Chinese under the title *Tongwen suanzhi* (Combined Learning Mathematical Indicator) (1614). In this book, he systematically introduced written computation *bisuan* (lit. "computation with a brush"). This time, unlike Euclid's *Elements*, the text was the result of a compilation of Clavius' text and numerous earlier Chinese arithmetics.[42]

Ricci's policy, which involved using mathematics and, more generally, science and technology, to facilitate the evangelisation of China, caused many a stir, both within the Society of Jesus and amongst Chinese astronomers; however, as a result of its indisputable successes (predictions of eclipses crowned with success on 15 December 1610 and 21 June 1629),[43] it finally prevailed.

From 1630, twenty years after Ricci's death, the Italian and German missionaries Giacomo Rho and Adam Schall von Bell undertook a vast translation programme in order to prepare for the reform of the calendar:

We decided to translate the explanations of all the celestial movements and eclipses, in other words, all astronomy,[44] together with supplementary information about arithmetic, geometry and other parts of mathematics, supporting or relating to astronomy in some way, into Chinese. We divided the work into three parts. The first part included all the preliminary concepts and anything which could help astronomy in some way [. . .], the third part[45] was to facilitate computations with references to ready-made tables which would make the solution of triangles and any work which might distract mathematicians from studying the new order unnecessary [. . .]. All this

---

[41]Cf. Mei Rongzhao (2′), *1963*.

[42]Cf. Takeda (2′) and (3′), *1954*.

[43]Cf. Bernard-Maître (2), *1935*, p. 69.

[44]"all astronomy," in other words Tycho Brahe's system. See Sivin (1), *1973*; Hashimoto (2), *1988*.

[45]The second part concerns astronomy itself.

was contained in 150 small volumes, which took approximately five years of intensive work to produce.[46]

The resulting work was called the *Chongzhen lishu* (Chongzhen reign-period Astronomico-calendrical Treatise) (1630–1635).[47] As the above quotation explains, this text may be viewed as a work tool for practitioners of astronomy, who could find everything they might require in it. This explains its quite special orientation as far as mathematics is concerned, including the description of calculating instruments, trigonometric formulae and tables, and various mathematical facts about polyhedra, conics and the number π, as the following brief description of the main mathematical works included in the *Chongzhen lishu* shows:

The *Chousuan* describes the counting-rods invented by Napier (commonly known as Napier's rods) to facilitate multiplication. Source: Napier, *Rabdologiae, seu numerationis per virgulas*, Edinburgh, 1617.[48]

The *Biligui jie* (Explanation of the proportional compass), describes the tool of the engineer *par excellence*, the ancestor of the slide rule. Sources: works by Galileo.[49]

The *Dace* (lit. "the great measuring") (1631) is the first Chinese trigonometry (plane and spherical) (the term "great" refers here to the fact that the purpose of this trigonometry is the measuring of the great heavens rather than small earthly distances as in common surveying). It contains a definition of the "eight cyclotomical lines" *geyuan baxian*, in other words, the sine *zhengxian*, the cosine *yuxian*, the tangent *qiexian*, the cotangent *yuqie*, the secant *zhengge*, the cosecant *yuge*, the versed sine *zhengshi* and the versed cosine *yushi*. The book also explains the construction of trigonometric tables based on a method similar to that used by Ptolemy in his *Almagest* to construct his table of chords. Sources: Clavius, Pitiscus, Regiomontanus, Magini.[50]

The *Celiang quanyi* (Complete explanation of methods of planimetry and stereometry) (1635) contains a complete translation of Archimedes' short treatise on measurement of the circle (Fig. 2.2), Hero's formula for the area of a triangle and summary data about polyhedra (Fig. 2.3), conic sections (Fig. 2.4), and the stereometry of regular polyhedra. It also contains a complete treatise of spherical trigonometry. Moreover, it gives, without proof, the following bounds on π (Fig. 2.5), the accuracy of which leaves nothing to be desired (Fig. 2.5):[51]

$$3.14159265358979323846 < \pi < 3.14159265358979323847.$$

---

[46] Bernard-Maître (5), *1942*, p. 16.

[47] This work was subsequently renamed the *Xinfa suanshu* or *Xinfa lishu* (Treatise of Mathematics (resp. Mathematical Astronomy) According to the New Methods (i.e. according to the European methods)). Here "mathematics" *suan* and "mathematical astronomy" *li* are synonymous.

[48] Cf. William Frank Richardson's recent translation of J. Napier's Rabdology (Charles Babbage Institute Reprint Series for the History of Computing), 15, Cambridge Mass. and London: MIT Press, *1990*.

[49] See Appendix I, p. 384.

[50] Cf. Bai Shangshu (1'), *1963*.

[51] *Celiang quanyi*, j. 4. Source: Clavius (4), *1604*, p. 219.

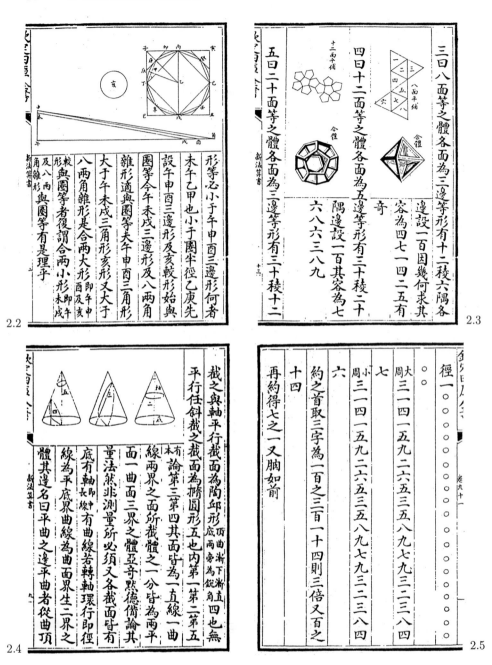

2.2

2.3

2.4

2.5

Fig. 2.2. The Chinese translation of Archimedes' treatise on the measurement of a circle in the *Celiang quanyi.*

Fig. 2.3. Polyhedra in the *Celiang quanyi.*

Fig. 2.4. Conic sections in the *Celiang quanyi.*

Fig. 2.5. The computation of π in the *Celiang quanyi.*

In 1653, the Polish Jesuit missionary Nikolaus Smogulecki (1610–1656)[52] introduced logarithms.[53]

From the second half of the 17th century, this new knowledge was intensively and critically studied by a small group of Chinese who sought to master Western astronomical techniques. The most representative person of this period is incontestably Mei Wending (1633–1721). Mei Wending, who came from a learned family with a fame going back to the Song[54] devoted his life to mathematics and astronomy (the collection of his works is called the *Lisuan quanshu* (Complete[55] works on mathematics and calendrical astronomy)). Although he had never worked in the Manchu administration, he succeeded in focusing great interest on the subject of his study, alternately teaching mathematics and entering into contact directly (often at the expense of long journeys across China) or by letter with very many personalities. He was involved in the official publication of the *Mingshi* (Ming History) monograph on calendrical astronomy. He also interested his large family in his research. During his tour of inspection in the South of China (1705), the Emperor Kangxi noticed him and named his grandson, Mei Juecheng (1681–1763) compiler in chief of the monumental mathematical encyclopedia of the end of his reign, the *Shuli jingyun* (1723).[56]

Compared with the European developments of the same period, the mathematical work of Mei Wending appears elementary. Calculating instruments, plane and spherical trigonometry, planimetry and stereometry are studied from all angles, but without symbolic algebra, with proofs using the arithmetic of proportions, similarity, Pythagoras' theorem and relationships which may be read off directly from figures. Euclid's geometry is completely transfigured in Mei Wending's three-dimensional figures, which take no account of perspective, and in his immersion in numerical computation.

In fact, Mei Wending studied regular and semi-regular polyhedra together with their various inscribed and circumscribed figures with computation and measurement in mind.[57] At the same time, Mei Wending rehabilitated ancient Chinese techniques such as the *fangcheng* method for solving linear systems. A major part of his work was incorporated in the *Shuli jingyun*.[58]

At the same period, but after much vicissitude (condemnation to death then rehabilitation of the German Jesuit A. Schall in 1664, disputes between the Belgian F. Verbiest and Yang Guangxian), the missionaries finally succeeded in installing themselves on a long-term basis in the capital. After Schall,

[52]Cf. Dehergne (1), *1973*, p. 255.

[53]Crossley and Lun (1), *1987*, p. 208.

[54]Liu Dun (2'), *1986*.

[55]Contrary to the title, the works of Mei Wending in this collection are not complete. Cf. *Wu'an lisuan shumu* (Catalogue of the mathematical and calendrical works of Mei Wending), Shangsha: Shangwu Yinshuguan, *1939*.

[56]Peng Rita Hsiao-fu (1), *1975*, p. 386.

[57]His works on this subject remind one of the content of books 15 and 16 of Clavius's commentary on Euclid's *Elements*. Cf. Martzloff (2), *1981*, p. 265 ff.

[58]Cf. Li Di and Guo Shirong (1'), *1988*.

Verbiest became President of the Bureau of Astronomy and was succeeded by F. Grimaldi, K. Stumpf and others[59] prior to the suppression of the Society of Jesus promulgated in China in 1775.[60] Thus, the missionaries came into direct contact with the Emperors themselves.

It was in this way that the Belgian Jesuit missionary F. Verbiest (1623–1688) came to teach the young Kangxi, who was Emperor of China from 1661 to 1722, the rudiments of European science.[61]

Kangxi was very open-minded and subsequently showed a constant desire to be initiated into European scientific techniques. For this, he had recourse to the services of Joachim Bouvet[62] (1656–1730), Jean-François Gerbillon[63] (1654–1707), Jean de Fontaney[64] (1643–1710) and other Jesuit mathematicians sent to China by Louis XIVth.[65] These missionaries learnt the Manchu language in order that they might give the Emperor special courses. This period from 1689–1690 saw the publication of new translations (into Manchu, then into Chinese) of Euclid's *Elements*, the old translation then being judged muddled and abstruse.[66] As Bouvet wrote in his journal, one of the oldest of these translations was based on a manual by Ignace Gaston Pardies *SJ* (1636–1673).[67] Pardies' manual is a textbook based on the idea that elementary geometry must be assimilated in a less abrupt way than that originally proposed by Euclid. Thus, the order of the propositions is drastically altered and there is very limited insistence on the formal exactness of the proofs which provides for a direct and directly applicable approach. This book was first issued in 1671; it was then constantly reprinted and translated into several languages including Dutch (1690), Latin (1693) and English (1746) (Plate 1).[68] The importance of these new Chinese translations of Euclid's *Elements* stems from the fact that one of them was included in the *Shuli jingyun*.

Subsequently, the missionaries continued to teach the Emperor the rudiments of arithmetic, algebra and trigonometry. The latter used all his new knowledge about evaluation of distances, calculation of volumes and

---

[59]Complete list in Dehergne (1), *1973*, p. 307.

[60]Cf. Dehergne, op. cit., p. 343.

[61]Cf. Bernard-Maître (4), *1940*, p. 112.

[62]Collani (1), *1985*.

[63]See Dehergne (1), *1973*, p. 359.

[64]Cf. Bernard-Maître (6), *1942*.

[65]On the mathematics training received in France by these Jesuit fathers, see Delattre (1), *1949–1957*, III, p. 1186.

[66]"*Li Madou* [Matteo Ricci] *suo yi, yin wenfa bu ming, xianhou nan jie, gu ling yi*" (Given that Matteo Ricci's translation of (Euclid's *Elements*) is written in confused language and that it is difficult to understand its organisation, we have produced a new translation). Cf. Li Yan (42′), *1937*, p. 219.

[67]Bibliothèque Nationale, Paris, French manuscript no. 17239, fol. 266b: "He [the Emperor] allows us to take any approach we wish to explain these propositions, leaving us free to follow the *Eléments de Géometrie* by Father Pardies which we had proposed as the most appropriate for his Majesty." See also Du Halde (1), *1735*, IV, p. 228. On the Manchu versions of the *Elements* see Chen Yinke (1′), *1931*.

[68]Cf. Ziggelaar (1), *1971*, pp. 47–68.

ELEMENS
DE
GEOMETRIE,
OU

PAR UNE METHODE COURTE
& aisée l'on peut apprendre ce qu'il faut
sçavoir d'Euclide, d'Archimede, d'Ap-
pollonius, & les plus belles inventions
des anciens & des nouveaux Geometres.

Par le P. IGNACE GASTON PARDIES,
de la Compagnie de JESUS.

L.C

A PARIS,
Chez SEBASTIEN MABRE-CRAMOISY,
Imprimeur du Roy.
M. DC. LXXI.
Avec Privilege de sa Majesté.

BEGINSELEN
Der
GEOMETRIE,

Waar in door een korte en ge-
makkelijke methode geleert wort, 't geen
dienstig is geweeten te worden uyt EU-
CLIDES, ARCHIMEDES,
APOLLONIUS, en eenige treffe-
lijke Inventien so van oude als niewe
Wiskonstenaars.

Samengestelt door

P. IGNATIUS GASTON PARDIES.

En nu uyt het Frans vertaalt.

Tot HOORN,
Gedrukt by *Stoffel Jansz. Korting*, in compagnie met *Johannes van Ceulen*, Boek-
verkooper, tot AMSTERDAM, 1690.

J. J.
ELEMENTA
GEOMETRIÆ,
in quibus
Methodo brevi ac facili summe
necessaria ex Euclide, Archimede,
Apollonio, & nobilissima veterum & re-
centiorum Geometrarum in-
venta traduntur
per
P. IGNAT. GASTON PARDIES S.J.
Gallico Idiomate conscripta,
Nunc vero
POST QVARTAM EDITIONEM
IN USUM STUDIOSÆ JUVENTUTIS
latinitate donata.

JENÆ,
SVMTV TOBIÆ OEHRLINGJ,
Bibliopolæ,
A. Æ. S. cIↄ Iↄc XCIII.

Short, but yet Plain
ELEMENTS
OF
GEOMETRY.

SHEWING

How by a Brief and Easy Method, most of what
is Necessary and Useful in EUCLID, ARCHI-
MEDES, APOLLONIUS, and other Excel-
lent *Geometricians*, both Ancient and Modern, may
be Understood.

Written in *French*
By F. IGNAT. GASTON PARDIES.

And render'd into *English*,
By *JOHN HARRIS*, D. D.
And Secretary to the *Royal Society*.

The EIGHTH EDITION.

LONDON,
Printed for A. WARD, at the *King's-Arms* in *Little-
Britain*. MDCCXLVI.

Plate 1. Title pages of various European
editions of I. G. Pardies *SJ*'s manual on
geometry.

measurement of the flow of rivers with enthusiasm.[69] In 1713 he created a college of mathematics *suanxue guan*.[70] At the same time, he ordered the compilation of the *Shuli jingyun*.

The monumental text of the *Shuli jingyun* was incorporated in an even more vast work on calendrical astronomy and music, the *Lüli yuanyan* (Ocean of Calendrical and Musical Computations).[71]

One may wonder about the motives for compiling such a gigantic work which, from the point of view of mathematics, at least, included relatively little original material in comparison with both the European works which were adapted into Chinese during the 17th century and the European works of the same period: for example, analytical geometry, algebraic symbolism in the style of Descartes, and infinitesimal calculus were passed over in silence. Would it not have been sufficient to have published only that which had not already been described? In fact, as the notices of the important catalogue of critical bibliography *Siku quanshu zongmu tiyao* (Critical Notices of the General Catalogue of the Complete Library of the Four Branches of Books) (1782) explained some 50 years later, the existing material could not be used directly:

歐羅巴人自密其學、立說復深隱不可解。〔 … 〕
算法新書往往有雜引之處、讀者未之能詳。

Ouluobaren zi mi qi xue, li shuo fu shen yin bu ke jie. [...] Suanfa xinshu, wangwang you zayin zhi chu, duzhe wei zhi neng xiang.

The Europeans have deliberately confused everything, so that their theories have become obscure and incomprehensible.[72] [...] The new books on computation [i.e. mathematical works of European origin] often contain multiple references [i.e. we do not possess the contexts from which they are drawn] so that they are impossible for the reader to understand.[73]

In particular, in astronomy, it was normal for there to be difficulties, since over the space of a few decades, from Ricci to Schall, Aristotelian concepts (treatise on Sacrobosco's sphere)[74] had been superseded by those due to Tycho Brahe. However, more profoundly, the core of the problem lay elsewhere, since one must acknowledge that the syncretic coexistence of different, possibly contradictory systems, had never appeared to constitute an obstacle in the past, as far as the Chinese were concerned. After all, Muslim and Chinese astronomy had existed side by side without anyone raising the question of determining which was true: finite and infinite universes had met and no one was biased either way.[75] In the first chapter of the *Zhoubi suanjing*, the Earth

---

[69]Cf. Peng Rita Hsiao-Fu (1), *1975*; Spence (1), *1977*, pp. 73–74.

[70]Cf. Bernard-Maître (4), *1940*, pp. 103–104; Porter (1), *1980*.

[71]Cf. Hashimoto, (2'), *1971*.

[72]Notice on the *Lixiang kaocheng* in the *Siku quanshu zongmu*, j. 106, p. 897.

[73]Notice on the *Shuli jingyun*, j. 107, p. 908.

[74]See Ricci's *Qiankun tiyi*.

[75]According to the theory of the "Canopy of Heaven" (*gaitian*) described by the *Zhoubi suanjing*, the universe is finite. On the other hand, according to the theory of "the announcement of the night" (translation of *xuanye* by Ho Peng-Yoke (1), *1966*, p. 49) it

is assumed to be flat, since this assumption matches that which can be done with the gnomon; however, in the second chapter of the same book the Earth is likened to a curved surface since this time, the instrument on which the deductions are based is an armillary sphere.[76] In such cases, the conflict between right and wrong was irrelevant, no one was concerned about possible logical aberrations since, for them, a technical subject had essentially to correspond to a purely pragmatic aim.[77] If astronomical tables became incorrect, it was sufficient to replace them with others, the nature of which was unimportant provided that they were efficacious, since there was a belief that these would have to be changed in any case after a short while: had not Chinese astronomical systems been reformed some 50 times between the Han and the Qing?[78] Thus, they found the need to submit to the strange dogmatic structure of Euclid's *Elements* before they could access the countless practical applications enticingly described to them by Ricci,[79] difficult to accept and apparently futile. "Why," Chinese scholars interested in Western mathematics asked, "does Euclid explain the construction of the golden section,[80] without telling us what it is used for?" (*qie lifenzhongmoxian dan you qiuzuo*[81] *zhi fa er mo zhi suo yong*).[82] But the criticism of geometry was not limited to the question of the strange ordering of the theorems. Significantly, the very influential mathematician Mei Wending, whose work makes a syncretic association between European and Chinese knowledge,[83] also found absurd the fact that the golden section was called the "divine proportion" in the Chinese version of Euclid's *Elements*:

> [After having understood how to make use of the golden section], I began to believe that the different geometrical methods could be understood and that neither their attitude [their = that of the missionaries] of considering this simple technique esoterically as a divine gift [allusion to the "divine proportion"] nor our attitude [our = that of the Chinese] of rejecting it as heresy is correct.[84]

---

is infinite. It is known that infinitist conceptions also existed in ancient Greece. Heraclides Ponticus believed that the cosmos was infinite (Dreyer (1), *1906*, p. 123). But, as A. Koyré wrote "neither should we forget that the infinitist conceptions of the Greek atomists were rejected by the stream or the main streams Greek philosophical and scientific thought [...] although they were never forgotten, they were never accepted by the thinkers of the Middle Ages." (Koyré (1), *1957*). In contrast, we note that in China, no school ever constituted orthodoxy in this area.

[76]Cf. Nakayama (1), *1969*, p. 30.

[77]However, when most Chinese scholars were presented with cosmological systems which conflicted too violently with their dominant conception of the World, the reaction was predictable (think of the Chinese critics of the Aristotelian cosmological system, in the 17th century). See Gernet (3), *1980*.

[78]Yabuuchi (3'), *1969*.

[79]See the preface to the *Jihe yuanben* (Elements of geometry) compiled by Ricci (there is an Italian translation by Pasquale d'Elia). See d'Elia (2), *1956*.

[80]Euclid's *Elements*, book 6, definition 3 (Clavius (2), *1591*, p. 286).

[81]*qiuzuo*: lit. "to seek to do." The technical meaning of this term is "construction."

[82]*Siku quanshu zongmu*, j. 107, p. 908 (notice on the *Shuli jingyun*).

[83]Cf. Hashimoto (1') and (3'), *1970* and *1973*.

[84]Mei Wending, *Jihe tongjie* (Complete explanation of [Euclid's] geometry) in *Meishi congshu jiyao*, edited by Mei Zuangao (1874), j 18, p. 3b. See also Martzloff (3), *1981*.

Mei Wending apart, many other scholars found unacceptable the association of mathematics with divinity when, rightly or wrongly, they perceived theology in Jesuit secular works. For example, the author of one of the prefaces of the *Jihe lunyue* (Summary of [Euclid's] geometrical proofs) (around 1700), a work highly valued by Mei Wending, which was composed by a *juren* (provincial graduate) called Du Zhigeng, felt obliged to state that the geometrical arguments had no affinity with any theology whatever, especially Buddhist theological thinking, which was generally viewed as vain and ineffective by many Chinese of his time. More broadly, many suspected that there was an organic link between Euclidean arguments, and those of Christian theology which they rejected. Thus, they felt that the European contribution should be completely restructured whilst preserving that which they saw to be excellent, namely the written computation (*bisuan*), the trigonometrical and logarithmic tables and the mathematical instruments.

In fact, as explains Clavius in his preface to the *Elements*, the usefulness of geometrical reasoning with respect to theology was great.[85] For the missionaries, there was certainly an affinity between geometric and theological modes of thought and moreover, that affinity was ancient. The neo-Platonist Proclus, an important commentator on Euclid of the beginning of the fifth century, stated explicitly:

From what we have said it is clear that mathematical science makes a contribution of the greatest importance to philosophy and to its particular branches, which we must also mention. For theology, first of all, mathematics prepares our intellectual apprehension. Those truths about the gods that are difficult for imperfect minds to discover and understand, these the science of mathematics, with the help of likenesses, shows to be trustworthy, evident, and irrefutable.[86]

Thus, many Chinese drew a parallel between geometry and theology based on the discursive affinity which they established between religious and scientific texts. This provided them with a good reason for rejecting such texts en bloc.[87] But for Jesuit missionaries the deductive component of "the rigour and precision of mathematics and astronomy could reinforce the authority of religion. By a very simple argument, if what the Western literati say of the visible world were actually proved to be true, the Chinese should also believe what they say about the invisible world [...]. Secular science and religion lent each other a mutual support."[88]

If mathematics became more acceptable once stripped of its theological component, real or imagined, and hence of its proofs, it also had to be sinicised and the same Mei Wending was also behind a powerful sinocentric current which championed that idea that all Western sciences were of Chinese origin. Throughout the 19th century, this idea was to be spread via the *Chouren zhuan*

---

[85] Clavius, *Euclidis Elementorum*, 1591, Prolegomena: "Utilitates variae mathematicarum disciplinarum."

[86] From Morrow (1), *1970*, p.18.

[87] See Martzloff (13), *1993*.

[88] Gernet (3), *1980*, p. 3.

(Bio-bibliographical Notices on Specialists of Calendrical and Mathematical Computations) (1799), which was drawn up under the supervision of the influential Ruan Yuan (1764–1849). In this connection, we note that some European missionaries, the figurists, helped to reinforce this idea.[89]

After the reign of Kangxi, and until the end of the 18th century, the European contribution to mathematics remained as much a tributary of astronomy as previously. The *Lixiang kaocheng houbian* (1742), which is the most important work of mathematical astronomy of this period, includes a number of examples of computation using the properties of the ellipse, with Kepler's second law in mind.[90]

For their part, Chinese mathematicians marked time, which is not particularly surprising, since as the scholar Tan Tai wrote:

中法之絀於歐羅巴也。由於儒者之不知數也。

Zhong fa zhi chu yu Ouluoba ye. youyu ruzhe zhi bu zhi shu ye.

If Chinese methods are inferior to European methods, it is because the literati, for the most part know nothing about mathematics.[91]

What is more, the major works of autochthonous Chinese mathematics were practically inaccessible at that time. Significantly, the monumental encyclopedia *Gujin tushu jicheng* (Complete collection of pictures and writings of ancient and modern times) (1726) includes only minor works of Chinese mathematics.[92]

However, as far as the 18th century is concerned, mention should be made of the particularly original works of the astronomer of Mongol abstraction Minggatu[93] (?–1764) (known as "Ming'antu" in Chinese and "Miangat" in modern Mongolian) on the expansion of circular functions in infinite series, which were inspired by formulae of a completely European origin, which he apparently learnt of outside their theoretical and demonstrative context. These works, which were unpublished during the author's lifetime and were first published some 50 years after his death, were the subject of remarkable extensions in the 19th century (expansions in series, trigonometric and

---

[89]On page 8 of the first chapter (j. 1) of the first part ("Shangbian"), of the *Shuli jingyun*, it is said that European missionaries such as A. Thomas (Anduo) and F. Grimaldi (Min Mingwo) believed that mathematics was invented in China. This is a theory conceived by J. Bouvet. Bouvet believed that the sciences were invented more than 4000 years ago by the mythical Chinese Emperor Fohi [Fuxi], but that this ancient knowledge had then been lost. He thought he had found traces of it in the enigmatic hexagrams of the *Yijing*. This theory led him to believe that the ancient Chinese had been aware of the Gospel. Cf. Gatty (1), *1976*, p. 142.

[90]On this subject, see the notice compiled by Hashimoto Keizo in Itō (1), *1983*, p. 1128. See also Hashimoto (2′), *1971*.

[91]Tan Tai was admitted as a *juren* (provincial graduate (cf. Hucker, notice no. 1682)) in 1786. Cf. *CRZ*, j. 50, p. 666.

[92]Complete list of these works in Li Yan, *Dagang*, II, p. 461 ff.

[93]"Minggatu" (or more exactly "Mingγatu") means "having a thousand, chief of a thousand men, chiliarch" (from F.D. Lessing (ed.), *Mongolian–English Dictionary*, Berkeley and Los Angeles: University of California Press, 1960, p. 539).

logarithmic, apprehended algebraically and inductively without the aid of differential and integral calculus). On the very edge of mathematics itself, we also note the research on perspective by a governor of the province of Guangdong, Nian Xiyao (?–1738), better known as the director of a porcelain factory,[94] who had been a student of Giuseppe Castiglione (1688–1766), the famous painter and architect of the Court of Qianlong. His *Shixue* (The science of vision) (published in 1729 and reprinted in 1735) is a very rare book (only three examples now remain worldwide)[95] inspired by a little known book by Andrea Pozzo *SJ*, the *Perspectiva pictorum et architectorum* (Rome, 1693 and 1700).

From 1772, as a result of the immense collection of ancient works throughout China organised by Qianlong, who was Emperor of China from 1736 to 1795, numerous works which had been completely lost from view for centuries were found and reconstituted by specialists in textual criticism, based essentially on sparse quotations extracted from the *Yongle dadian* encyclopedia of the very beginning of the 15th century.[96] This encyclopedia which, because of its quotations constitutes the most important source for the history of Chinese mathematics before the 14th century, was never circulated. By the 18th century, only a single incomplete example remained (which has since been almost entirely destroyed), the two copies which were made in 1562 disappeared at the time of the fall of the Ming dynasty.

These efforts led to the collection of 79 582 volumes known as the *Siku quanshu* (Complete Library of the Four Branches of Books). An important catalogue, the *Siku quanshu zongmu* (1782), provides bibliographical notices for each book.[97] Mathematics and astronomy are grouped together under the common heading *tianwen suanfa lei*, which, however, is divided into two sections corresponding to each of these two disciplines. As far as mathematics is concerned (*juan* 107) there are 25 notices on:[98]

- Nine of the Ten Computational Canons of the Tang (*SJSS*).[99]

- Three works of the Song–Yuan period, namely the *Shushu jiuzhang* by Qin Jiushao, together with the *Ceyuan haijing* and the *Yigu yanduan* by Li Zhi.

- Four works of the Ming period, including Euclid's Geometry, *Jihe yuanben*.

- Nine works of the Qing period, all minor except for the *Shuli jingyun*.

---

[94]Cf. Hummel, pp. 588–590; *CRZ*, j. 40, p. 505.

[95]Cf. Beurdeley (1), *1971*.

[96]Cf. Taam (1), *1977*; Guy (1), *1987*; Yan Dunjie (30′), *1987*.

[97]This book is also known under the title *Siku quanshu zongmu tiyao*.

[98]Cf. Li Yan, *Dagang*, II, p. 464 ff.

[99]The missing work, the *Zhoubi suanjing* is listed under the heading of "astronomy" (*tianwen*).

Other ancient mathematical works, including, in particular, the *Suanxue qimeng*,[100] the *Siyuan yujian*[101] and the *Yang Hui suanfa*[102], were found a little later. After these became available again, they were the subject of countless historical works throughout the 19th century.[103]

This phenomenon is reminiscent of the extraordinary dynamic energy which resulted from the rediscovery and philological study of scientific texts from classical antiquity during the Renaissance in Europe. As B. A. Elman puts it:

> Recent historians of China who have ridiculed Ch'ing philologists are apparently unaware of how great a role Renaissance bookmen played in recovering ancient Greek and Roman literature and thought for Western civilization.[104]

However, the consequences were not the same in both cases, for Chinese traditional knowledge was based on conceptions very different from those of ancient Greece which were influential from the Renaissance onwards in Europe. Indeed, axiomatico-deductive reasoning and mathematical idealities were not a fundamental component of the Chinese tradition and when discussed were essentially rejected.

From 1850, new translations of Western mathematical works began to appear, although in a different spirit to that of the previous period. These were no longer limited initiatives, as in the 17th and 18th centuries, depending closely on the requirements of imperial astronomy, but instead, constituted a response from the Chinese authorities to the catastrophic pressure of events (opium war, unequal treaties). In fact, the latter had progressively, and without unanimity, reached the view that the basis for the superiority of the West was directly dependent on the level of technical development. This led to the foundation of schools providing a scientific and technical training associated with language courses in Peking (creation of the Tongwen guan (College of Combined Learning) in 1862),[105] Shanghai, Canton and Fuzhou (1866).

But, even before these schools were created, the ground had been prepared by missionaries, Protestant this time, who had become established in the open ports (Shanghai, in particular). On the whole, they did not share the Jesuit vision of the evangelisation of China, in that they did not seek to gain the confidence of the literati or to become astronomers to the Court, but wished, on the other hand, to convert as many Chinese as possible through their teaching at the schools which they founded in Chinese towns.[106] What is more, whilst it is true that, as in the past, translators such as the Protestant missionary A. Wylie (1815–1905) of the London Missionary Society took it upon themselves to translate prestigious works such as Augustus De Morgan's[107] *Elements of*

---

[100]Cf. Lam Lay-Yong, (7), *1979*.

[101]Cf. Hoe (2), *1977*.

[102]Cf. Lam Lay-Yong, (6), *1977*.

[103]QB, *Hist.*, p. 283 ff.

[104]Cf. Elman (1), *1984*, p. 168.

[105]Cf. Biggerstaff (1), *1961*, p. 94 ff.

[106]Cf. Latourette (1), *1929*.

[107]On this mathematician, see the notice of the *DSB*.

Algebra (*Daishuxue*) (1859) or the part of Euclid's *Elements* left uncompleted by Ricci and Xu Guangqi two centuries earlier (books 7 to 15) (*Jihe yuanben*) (1857), by and large, their translations were not only far more numerous than those of their predecessors, but also often of a more elementary level, and above all, more within the reach of their potential interlocutors (sometimes they published in Chinese dialects rather than in a classical language).[108] But in 1859, China for the first time had a textbook of differential and integral calculus, the *Daiweiji shiji*, adapted into the Chinese by the same A.Wylie and Li Shanlan (1811–1882) from a textbook by the American Elias Loomis, *Elements of Analytical Geometry and of the Differential and Integral Calculus*, New York: Harper and Brothers, 1851.[109] This was followed, up to the end of the Manchu dynasty, by works on conics, probability and mechanics.

In the second half of the 19th century, some missionaries offered their services to the government technical schools and improvised as 'all round' translators covering areas ranging from differential calculus to navigational techniques and from medicine to diplomacy. The most astonishing of these was undoubtedly J. Fryer (1839–1928),[110] who translated no less than 29 textbooks of various types, including 15 on mathematics,[111] between 1870 and 1900, and managed to sell over 30 000 examples of his translations between 1870 and 1880 (including 1781 on mathematics alone).[112]

However, it was to take China much longer to resign itself to definitive adoption of a mathematical culture alien to its own. For a long time, few literati (fewer and fewer then remained in China, which was in decline) were interested in these works, and a change of attitude was first perceived only in 1895.[113] The uninterrupted progression of a foreign current, which was as powerful as it was irresistible, was often felt to be an intolerable aggression inasmuch as foreign political aggression was everywhere visible and tangible. Thus, the reaction was quick to follow. Some tried to preserve the national tradition by publishing the major works of the Chinese past[114] for use by the technical schools, others tried to convince themselves that mathematics, and, more generally, all the other sciences were universal, in other words, in spirit, of Chinese origin.[115] In both cases (the conservatives and the modernists), originality was not tolerated unless it was adapted to Chinese usage: the translators had to build up a mathematical symbolism specially adapted to the Chinese language.[116] In fact, it was not until the beginning of the 20th century that Arabic numerals began

---

[108]Wylie (2), *1867*.

[109]This work was reprinted many times up to 1872. It was also translated into Arabic, see the article 'Elias Loomis' in *Dictionary of American Biography*, London, *1933*, vol. 11, pp. 398–399.

[110]See Bennett (1), *1967*.

[111]Ibid., pp. 84–85.

[112]Ibid.

[113]Tsien Tsuen-Hsuin (1), *1954*, p. 313.

[114]Cf. Li Di, *Hist.*, p. 376.

[115]See below, p. 168.

[116]See below, p. 119.

to creep into Chinese textbooks (Fig. 2.6). It would still be some time before they were definitively accepted. As Wang Ling (Chinese historian, collaborator with J. Needham) wrote:

The Arabic numerals were not accepted until our generation. The logarithmic table used by the author's father and printed in 1902 is still in counting-rod numerals.[117]

One of the first Chinese mathematical journals, the *Suanxue bao*, was created in 1899.[118] Only three issues were published. In their editorial, its creators explained that they refused to yield to the supremacy of Western mathematics: "Western methods should not be adulated and Chinese methods despised." Their model mathematicians were Mei Wending, Li Shanlan and Hua Hengfang, the prestigious champions of the selective adaptation of Western knowledge and renewal of ancient Chinese tradition which, they believed, had been distorted by poor editions of ancient texts. Consequently, they directed their journal towards the correction of errors in certain solutions of problems from the *Jiuzhang suanshu* or the *Ceyuan haijing*. As for Western knowledge, they concentrated their articles on explanations concerning elementary properties of conics or Euclidean propositions such as "If two circles intersect one another, they will not have the same centre."

However, it is true that it would not be difficult to find Chinese texts of the same period of a distinctly higher level (works such as those of Li Shanlan on prime numbers, of Xi Gan on conic sections, of Xia Luanxiang on transcendental curves). Fundamentally, however, Chinese mathematicians were clearly 'not exactly playing the same game' as their Western or Japanese equivalents of the same period (infinitesimal contribution to the international mathematical community, methods closer to those of the *Nine Chapters* than to those of Hilbert or Cantor).

During the last decade of the Manchu dynasty, as a result of the impetus provided by the reformist current, a network of elementary schools in the districts and secondary establishments in the provinces was created.[119] New textbooks were published and supplied to these schools. Thus, between 1902 and 1910, the Commercial Press (Shangwu Yinshuguan) published a total of 43 mathematics books.[120]

After the revolution of 1911, the situation evolved progressively. In an article published in 1947, the mathematician Li Zhongheng wrote that the German mathematician K. Knopp, who taught for seven years from 1910 to 1917 at the University of Qingdao, had wasted his time since the level of the students was too low.[121] However, this opinion is perhaps not representative and, in 1917, a Chinese mathematician Minfu Tah Hu (or Hu Da, styled Mingfu)

---

[117] Wang Ling, *Thesis*, chapter 3, p. 91.

[118] According to Hong Zhenhuan (1'), *1986*, p. 36, the very first Chinese mathematical journal was created in 1897 and bore the same title as the *Suanxue bao*; it lasted only one year.

[119] Cf. Bastid (1), *1971*.

[120] Li Di, *Hist.*, p. 500.

[121] Li Zhongheng (1'), *1947*, p. 67.

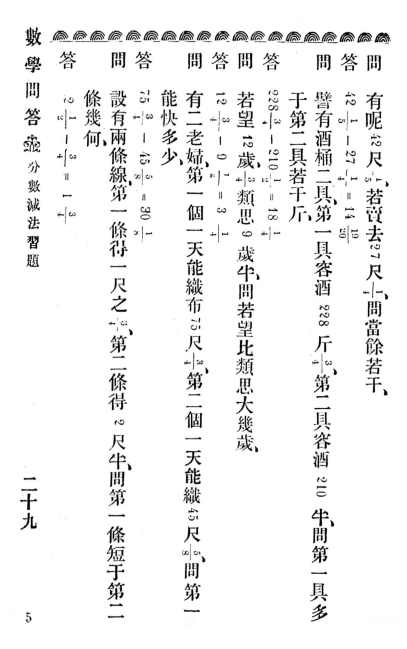

Fig. 2.6. Page 29 of the *Shuxue wenda* (first published in 1901 and cited here from the 1912 reprint). This little primer which was intended for the teaching of arithmetic in Jesuit elementary schools was among the first to use Hindu-Arabic numerals.

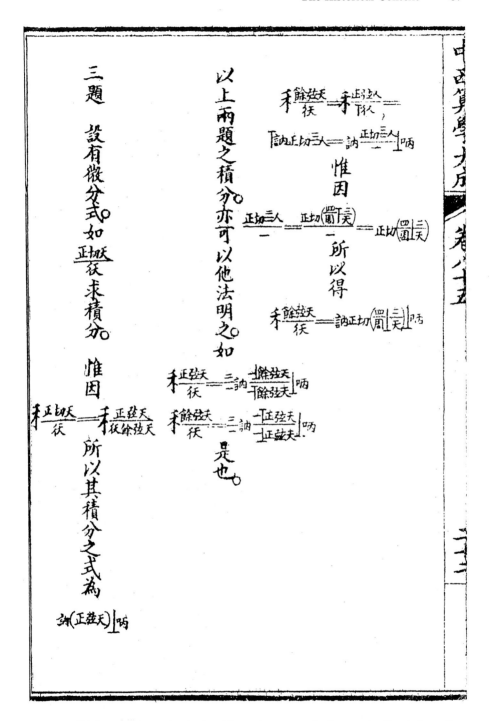

Fig. 2.7. Higher mathematics in the *Zhongxi suanxue dacheng*, a textbook from the Tongwen guan, first published in 1899. (a) Computation of $\int \frac{dx}{\cos x}$, $\int \frac{dx}{\sin x}$, $\int \frac{dx}{\tan x}$

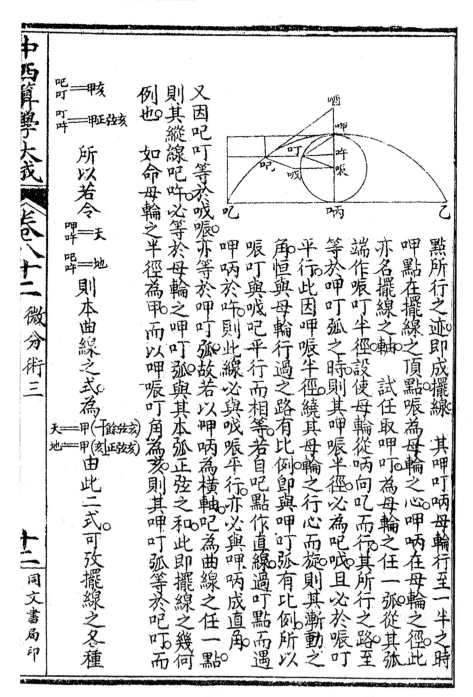

Fig. 2.7 (continued). (b) Explanation of the cycloid.

obtained for the first time a Ph.D. at Harvard[122] which was later published in the *Transactions of the American Mathematical Society.*[123]

In fact, the integration of China into the international mathematical community did not begin to become a reality until Chinese mathematicians ceased learning mathematics through translations and began to learn it directly with the mathematicians of their time, either in China itself (B. Russell, W. Blaschke, Sperner, G. D. Birkhoff, N. Wiener and Hadamard all taught in China) or in Europe, Japan or the U.S.A.

China was represented for the first time at the International Congress of Mathematicians in 1932 and the Chinese Mathematical Society was founded in 1935.[124] In 1946, the Institute of Mathematics of the Academia Sinica was founded under the direction of Professor Shiing-Shen Chern [Chen Xingshen].[125]

The communist takeover of 1949 was followed by ten years of close cooperation with the Soviet Union. During this period, bourgeois mathematics was criticised to the benefit of proletarian mathematics.[126] Later on, during the Cultural Revolution (1966–1976), pure mathematics (especially number theory) was actively developed. Chen Jingrun, born in 1933, obtained his famous result which marks a progression towards the proof of Goldbach's conjecture: every sufficiently large even number is the sum of a prime and a number with at most two prime factors. More generally, the new Chinese school has shown particular brilliance in an area which many considered to be the most difficult of the whole discipline, namely number theory (works of Hua Luogeng,[127] of Wang Yuan on diophantine equations). At the same time, the authorities tried to promote operations research in order to rationalise Chinese economics. Great mathematicians such as Hua Luogeng were associated with the movement.[128]

Finally, from 1911 onwards, solely Western mathematics has been practised in China. At the same time, traditional Chinese mathematics essentially ceased to be practised,[129] and became the subject of purely historical studies, which are still actively pursued to this day under the auspices of the Academia Sinica. Needless to say, since then, the creativity of Chinese mathematicians is equivalent to that of other mathematicians.

For details of the recent history of Chinese mathematics, the reader should consult the following sources:

- Hua Luogeng, Guan Zhaozhi, Duan Xuefu *et al.* (1), *1959.* This book begins with an overall view of the development of Chinese mathematics

---

[122]Yuan Tong-li (1), *1963*, preface, p. vii.

[123]*1918*, vol. 19, pp. 363–407.

[124]Fan Huiguo and Li Di (1'), *1981.*

[125]Yuan Tong-li, ibid., p. vii. On Shiing-Shen Chern see also Chinn and Lewis (1), *1984.*

[126]Hua Luogeng, Guan Zhaozhi, Duan Xuefu *et al.* (1'), *1959.*

[127]On Hua Luogeng, see: Salaff, *1972.*

[128]Ibid.

[129]Except for a very special area of mathematics: computation on the abacus. However, the case of mathematics is radically different from that of traditional Chinese medicine, for example, which is still actively practised and taught.

from 1912 to 1959 and continues with a description of Chinese work in different branches of this science from 1949 to 1959.

It was collectively compiled by mathematicians specialising in the areas involved, the technical descriptions are accompanied by bibliographies classified first by author, then by date of publication. The subjects covered include mathematical logic (pp. 28–54), number theory (pp. 55–75), algebra (pp. 76–106), topology (pp. 107–131), function theory (pp. 132–226), differential equations (pp. 227–260), integral equations (pp. 261–276), functional analysis (pp. 277–301), probability, statistics and operational research (pp. 302–329), numerical analysis (pp. 330–357) and geometry (pp. 358–395).

- Yuan Tong-li (1), *1963*. (Covers the period 1918–1961).

- Fitzgerald Anne and Saunders MacLane, *Pure and Applied Mathematics in the People's Republic of China*, 1977, Committee on Scholarly Communication with the People's Republic of China report no. 3, Washington (National Academy of Sciences).

- Wang Yuan, Yang Chung-chun and Pan Chengbiao ed., *Number Theory and its Applications in China*, 1988, Providence, Rhode Island (American Mathematical Society), p. 170.

- Items no. 2331 to 2351 of: Tsuen-Hsuin Tsien ed., *China, an Annotated Bibliography of Bibliographies*, 1978, Boston, Mass. (G. K. Hall and Co.), pp. 459–463.

We also mention the selected or collected papers of Chinese mathematicians published from 1978 by Springer-Verlag, including, in particular, those of Shiing-Shen Chern [Chen Xingshen], Pao-Lu Hsu [Xu Baolu] and Loo-Keng Hua [Hua Luogeng]. Finally, we mention a recent book on computer-based automated proof of theorems of geometry using algebraic methods, which is based on the ideas of Wu Wenjun, a contemporary Chinese mathematician who views his modern research on algorithmics and automatic theorem proving as a continuation of the ancient tradition of Chinese mathematics: Shang-Ching Chou, *Mechanical Geometry Theorem Proving*, 1988, Dordrecht: D. Reidel.

3. The Notion of Chinese Mathematics

西方算術本出希臘方言謂之瑪底瑪底克譯即致知之義

Xifang suanshu benchu Xila, fangyan wei zhi madimadike,
yi ji zhizhi zhi yi.

Western computational techniques originate in Greece and are called
'madimadike,' a word which means 'the highest form of knowledge' in
the dialect [of this country].   *Suanxue keyi, 1880*, preface, p. 1a.

Unlike other Chinese sciences such as geology or physics which were created
teleologically in the 20th century from scattered elements by universalist
historians,[1] mathematics, as an area of knowledge which the Chinese world
would have explicitly recognised as such, may be easily delimited.

The official biographies recognise mathematics in its own right, generally
distinguishing it from divination, and special treatises have been written on it
since the Han. For example, in the bibliographic part of the *Suishu* (Sui History)
works on mathematics appear in the last part of the section entitled *lishu*
(calendars and numbers) after works on calendrical astronomy and treatises
on the clepsydra;[2] in the *Songshi* (Song History) they are found in the section
entitled *lisuan* (calendars and computations) just after the section on divination
and before that on the art of warfare;[3] in the *Mingshi* (Ming history) they are
found partly in the section entitled *xiaoxue* (lesser learning)[4] and partly in that
entitled *lishu* (calendars and numbers) (in particular, the latter section includes
the *Jihe yuanben* (Euclid's *Elements*)).[5] Finally, in the imperial catalogue, *Siku
quanshu zongmu*, mathematics is classified separately, just after astronomy.[6]

---

[1]In other words, essentially J. Needham and his team. For another approach to the
definition of Chinese sciences, see Sivin (2), *1977*, pp. xi–xxiv. See also Kim Yong-Sik (1),
1982.

[2]Cf. *Suishu*, j. 34, pp. 1025–1026.

[3]Cf. *Songshi*, j. 207, p. 5271–5276.

[4]This section essentially covers works on philology. Cf. Elman (1), *1984*, pp. 165–167.

[5]*Mingshi*. j. 96, p. 2374. Here, the distribution of mathematical works across two sections
is also based on a chronological criterion. Works prior to 1600 are listed in the first section
and others, of Jesuit origin, in the second.

[6]*Siku quanshu zongmu*, j. 107, second part of the section entitled *tianwen suanfa*
(astronomy and mathematics).

Moreover, the authors of this catalogue of the late 18th century wrote that, "mathematics and astronomy go together like a coat and its lining" *suanshu tianwen xiang wei biao li.*[7] This rapid overview clearly demonstrates the durability and solidity of the bond between mathematics and astronomy in China. However, this does not imply that Chinese works on mathematics are concerned exclusively with astronomy. In fact, they handle a very varied range of subjects, within which astronomy as such occupies only a minor position. Instead, the constant association between mathematics and astronomy bears witness to the essentially quantitative nature of the latter science.

The special terms which are used to denote mathematics, namely *suanshu,*[8] *suandao,*[9] *suanfa,*[10] *shushu,*[11] and *shuxue,*[12] indicate that it is concerned with numbers *shu* and with computation *suan* and that it is an art (*shu, fa, dao*). In particular, the suffix *shu* used in the work *suanshu* is the same as that found in terms such as *jishu* (techniques), *fashu* (magic arts), *quanshu* (boxing), *jianshu* (swordsmanship), *wushu* (martial arts), and *yishu* (medicine). In addition to its general meaning of "art," *shu* also means process, method, device, trick, stratagem, artifice or prescription. Thus, the various *shu* are essentially collections of prescriptions which, in the case of mathematics, are more or less similar to our algorithms. In addition, etymologically, *suan* denotes a set of specific objects used for calculation: counting-rods. Thus, at least originally, it relates to the faculty of knowing how to manipulate certain rods with dexterity in order to carry out computations in accordance with fixed prescriptions. But, exactly like general terms such as geometry (which long since lost its meaning of measurement of land), these terms do not have an immutable sense. *Suan* lost all reference to the idea of counting-rods at an early stage in its history and generally means nothing more than computation or mathematics (this is understandable, since available methods of computation were not limited to counting-rods). But conversely, *shu* has almost always retained its initial meaning of method or prescription.

These computational prescriptions are traditionally subdivided into a fixed number of categories governed by the number 9: the "nine types of computation"

---

[7] *Siku quanshu zongmu,* op. cit., j. 106, middle of page 891.

[8] This term is by far the most common. It is already found in the *Hanshu* (lüli zhi – monograph on the calendar and pitchpipes). In modern Chinese, *suanshu* means *arithmetic* (in the sense of elementary arithmetic or logistics rather than number theory); the term which corresponds to mathematics (in general) is *shuxue* (lit. science of numbers) but the term *suanxue* (science of computation, arithmetic) is also found. See Loh Shiu-chang et al. (1), *1976.*

[9] See *DKW* 8-26146: 37. Note that this *suandao* (lit. the way of the rods) has a structure analogous to that of terms such as *judo* (lit. the way of suppleness) or *aikido* which have been borrowed from the Japanese language, which has itself lent them to the Chinese language.

[10] This term commonly appears in the titles of works from the Ming dynasty onwards.

[11] This term has a numerological connotation. Cf. *DKW* 5-13363: 62.

[12] This term was used from the 13th century (cf. the title of a work by Qin Jiushao, the *Shuxue jiuzhang,* more frequently called the *Shushu jiuzhang*).

jiushu, which form the basis for the Nine Chapters (*jiuzhang*),[13] in other words for the archetype of Chinese mathematics. The taste of ancient China for numerical classifications is well known; here, as in many other cases,[14] the number 9 does not relate to subdivisions dictated by a philosophy or logic, but rather to mnemonically organised areas. The titles of the chapters of the *Jiuzhang suanshu* of the Han, all (except one) consist of exactly two written characters, while those of the *Sijuan yujian* (1303) run alongside them with a uniform four characters, which, moreover, is also the case for works such as the *Shushu jiuzhang* (1247). These computational rules are often extremely concise and are occasionally given in verse form, as in the case of the *Suanfa tongzong* (1592), which makes them easy to remember. This partition into nine areas has tended to evolve with time, and even to disappear; however, at the same time, it is often used as a mandatory reference against which to locate new developments.

Thus, for example, for Mei Wending (1633–1721), who was amongst the mathematicians most widely admired by the Chinese of the Qing dynasty (1644–1911), "the Nine Chapters include everything [i.e. all of mathematics] without exception."[15] Clearly, such a declaration should not necessarily be taken literally, since it might have been designed to avoid the criticism of conservatives keen to preserve the purity of the Chinese mathematical tradition. However, the fact is that Mei Wending reinterpreted Euclidean geometry, relating it back to the framework of the ninth chapter of the *Jiuzhang suanshu* on right-angled triangles *gougu*. In addition, in 1867, Li Shanlan proudly explained that his work in the *Duoji bilei*[16] went beyond the Nine Chapters. Again under the Qing dynasty, mathematical treatises were often classified according to their origin (Chinese or Western), distinguishing between Chinese methods *zhong fa* and Western methods *xi fa*. But this amounted more to a pragmatic division than a recognition of the existence of two truly distinct areas. As Mei Wending again explained, only the terminology was different in the two areas, and Western formulations were always reducible to Chinese formulations. In other words, mathematics was universal, since it could be fundamentally interpreted in a purely Chinese framework.[17] Consequently, Chinese authors often provided ingenious explanations to prove that some new question or other related to one of the ancient categories, so that the latter became embedded in other classification systems, partly based around the specific type of application envisaged or determined by the extent to which certain groups of prescriptions resembled one another (for example, false-position rules, rules for extracting

---

[13] According to Zheng Xuan (commentator of the Han), these *jiushu* were identical to the Nine Chapters (*jiuzhang*). See QB, *Hist.*, *1964*, p. 31. But it may be that these *jiushu* were just the "nine numbers," 1,2, ..., 9.

[14] Cf. Granet (1), *1934*, p. 242.

[15] Martzloff (2), *1981*, p. 247.

[16] See the end of the preface to the *Duoji bilei*. On Li Shanlan, see below, p. 173.

[17] Cf. *Jixuetang wenchao*, j. 1, p. 26b (cited from the copy preserved at the Naikaku bunko, Tokyo).

roots, rules for solving linear systems). If, anticipating what is to follow, one also adds the fact that Chinese textbooks could be essentially described as collections of problems and prescriptive rules, it becomes clear that this mathematics was radically different from most of Greek mathematics, not only because the influential part of the latter was based on a very characteristic kind of hypothetico-deductive logical reasoning, but also because Chinese mathematics does not go into questions relating to irrational numbers, ruler and compass constructions, special curves (quadratrix, cissoid, etc.), conic sections, parallelism and infinity. Not that certain of these questions absolutely never appear in Chinese mathematics but rather that, when they do, they do so *à doses homéopathiques*. To be sure, it is possible to interpret a certain laconic Chinese expression composed of four characters as meaning that certain numbers are to be understood as irrational,[18] perhaps also, an astute philological exegesis will show that certain figures are to be interpreted as a testimony of a Chinese theory of parallels. But at any rate, as far as we know, these questions were practically never discussed in China. Save perhaps, if some hitherto utterly unknown manuscript is discovered, nobody will ever write a history of the theory of the infinite, of irrationals, of parallels, or of conic sections in medieval China.

Moreover, neither is there any convenient internal subdivision into categories which are familiar in another context (for example, arithmetic, logistics, algebra, geometry) or into theory and practice, which it links inextricably. It is not that these categories would be a priori inapplicable to Chinese mathematics, since its proportional-sharing and double-false-position rules belong to arithmetic, its calculations of volumes are based on geometry, whilst its manipulation of numbers in two-dimensional arrays resembles algebra. However, often, it makes use of all available means, and procedures of different kinds are used to derive certain results.

Chinese mathematics is visibly similar to the mathematics of Mesopotamia, ancient Egypt, India, the Islamic world and medieval Europe (in part, at least, in other words, to the extent that it does not involve deductive geometry). However, this clear similarity does not imply that, within the framework of problems and prescriptive rules, everything is identical or differs only by small local variations of secondary interest. Certainly, there are many common themes such as the rule of three, Pythagoras' theorem, the extraction of roots etc. However, there are also differences, which are explained both by the diversity of the socio-economic conditions (for example, unlike in the Islamic world, problems relating to inheritance did not arise in China) and by the differences in the stage of historical development (Egyptian and Babylonian mathematics are on the whole less "advanced" than Chinese mathematics, if only because the rules they include are essentially formulated in terms of

---

[18]Or, much more exactly, root numbers or square roots, that is linguistic expressions denoting the square root of certain numbers without any need to compute their approximate value. But in the Chinese context, irrational is not opposed to rational. See below pp. 226–227.

particular numbers,[19] while, in the Chinese case, general formulations are by far predominant). One other, no-less-fundamental difference relates to the structure of the computations: the Egyptians used unit fractions, the Babylonians used sexagesimal fractions, and the Chinese used decimal metrological numbers.

Finally, one should not make the error of thinking that Chinese mathematics always keeps turning over the same themes (those of the Nine Chapters) out of blind respect for tradition, and that it is incapable of evolution. This is not the case, as the observation that many of the subjects which it covers (congruences, polynomial systems) can only be seen as terminal processes resulting from a slow accumulation of knowledge shows.

Moreover, Chinese mathematicians have always pointed out, sometimes vehemently, the imperfections of certain approaches of their predecessors, thereby admitting their desire to surpass them.[20] On the other hand, at various times in its history, Chinese mathematics has experienced difficulties of transmission and long periods of "decline" following upon very brief oases of "progress" at specific times and in specific areas. From its origins to the 19th century, many authors dwelt upon the great difficulty that they had in procuring the works of the past, including the *Jiuzhang suanshu*, in particular.[21] Texts such as Zu Chongzhi's (429–500) *Zhuishu* (apparently on the computation of π)[22] were lost at an early stage, although specialists all recognised their importance. In the 17th century, Mei Wending often acquired essential mathematical texts either by copying them or having them copied, or by borrowing them from bibliophiles.[23] Of course, this phenomenon was not restricted to China. However, it has arisen there so constantly from the

---

[19]The text of *BM* 34568 (Thureau-Dangin (1), *1938*, p. 63) (see bibliography) is one of the rare exceptions to this. It states "Tu multiplieras ... l'un à l'autre. Tu additioneras l'un avec l'autre, puis tu soustrairas (le carré de) l'excès .... [Tu soustrairas] le flanc et la diago[nale], puis tu poseras le front." (Multiply ... the one by the other. Add the one to the other, then subtract (the square of) the excess ... [Subtract] the flank and the diagonal, then set down the front. [...]). Thus, it resembles the Chinese rules.

[20]See, for example, the preface to the *Zhang Qiujian suanjing* (QB, II, p. 329) in which the author criticises Xiahou Yang; the preface to the *Jigu suanjing* (seventh century) in which Wang Xiaotong criticises Zu Xuan (ibid. p. 493); the criticism of the works of Zhang Heng (first century) in Liu Hui's commentary, vol. I (ibid. p. 156); etc. There are many more such examples.

[21]See Liu Hui's preface to the *Jiuzhang suanshu* (QB, I, p. 91); that of Bao Huanzhi, an editor of the same work at the beginning of the 13th century; Wu Jing's preface to the *Jiuzhang suanfa bilei daquan* (1450).

[22]The exact meaning of the title *Zhuishu* is not known. According to the *Suishu*, j. 16, p. 388 (Peking edition, *1975*), Zu Chongzhi is said to have "invented extremely subtle methods for extracting diminished areas and volumes (*kai cha mi, ka cha li*) [these terms of unknown meaning do not occur in any other known Chinese technical text] in order to square the circle. He may be considered as the prince of mathematicians (*suanshi zhi zui zhe*). His work is called the *Zhuishu*. He was excluded (from the textbooks used for teaching) because none of the students at the Imperial College could understand him." However, according to Shen Gua (*Mengqi bitan*, j. 18, *tiao* no. 300) the term *Zhuishu* refers to a method of computation used in astronomy (interpolation?). See also Libbrecht, (2), *1973*, pp. 60 and 275, respectively.

[23]According to the prefaces of the *Meishi congshu jiyao*.

origins to the present day that it seems more appropriate to view Chinese mathematics in a homogeneous manner, within a very long time interval (i.e. within the Braudelian *longue durée*),[24] rather than locate it within shorter, largely arbitrary periods with no real historical or epistemological importance (defined, for example by the political chronology or the succession of dynasties).

---

[24]Cf. Braudel (1), *1969*, p. 44 ff.

4. Applications of Chinese Mathematics

Generally speaking, the idea that Chinese mathematics is divided into two autonomous, naturally interacting areas of theoretical and practical mathematics seems rather misconceived.

Instead, to a first approximation, a majority of Chinese mathematical works may be better represented as pedagogical tools, in other words, didactic aids used to teach numerical computation, together with prescriptive texts (for example, user manuals).

To see this, we note that:

(i) most texts never contrast theory with practice;

(ii) most prefaces, including those to works which appear to have a theoretical basis, stress the economic utility of the art of computation;

(iii) texts intended for those working in a particular domain of application rarely contain references to other works forming the theoretical source from which they draw their substance.[1]

In the case of practical manuals, the mathematical terminology appears strongly dependent on the particular domain of application dealt with. Thus, although the same technique of cubic interpolation is found in both mathematics and astronomy, the terminology is far from the identical in the two areas, since, in each case, the terms used depend on the specific situation concerned. For example, on this subject, the mathematician Zhu Shijie talks of *zhao cha* (recruitment differences) since the problem in which he uses this technique involves the recruitment of soldiers,[2] while the astronomer Guo Shoujing, who lived at the same period, uses other peculiar terms for the same thing, which vary according to the astronomical phenomenon to which the method is applied.[3] In fact, all this is understandable, if one acknowledges that Chinese

---

[1]But exceptions exist: in the introductory part of the *Yingzao fashi* (manual on architecture) by Li Jie (?–1110), the preliminary mathematical explanations end with an explicit quotation from Li Chunfeng's commentary on the *Jiuzhang suanjing* [i.e. *Jiuzhang suanshu*]. (See p. 21 of vol. 1 of the *Wanyou wenku* edition of the *Yingzao fashi*, published in Shanghai in 1933 by Shangwu Yinshuguan).

[2]*Siyuan yujian*, II-10, analysis in *SY*, p. 185 ff.; complete translation of the original texts in Hoe (1), *1976*, p. 95 ff. and 292 ff.

[3]Cf. Li Yan (52'), *1957*. The terminology of interpolation techniques varies considerably from one astronomer to another.

astronomical texts which have reached us were intended not for logicians but above all for technicians, mechanically carrying out repetitive, clearly defined tasks. This agrees with Qiu Jiushao's description of the astronomers of his time, who undoubtedly provided appreciable services, even though they did not understand the whys and wherefores of the computations they carried out: "they applied formulae without understanding them."[4] In the same way, astronomers, masons, musicians, hydrologists and other practitioners also needed some knowledge of ready-made formulae. Consultation of Chinese manuals on these various specialisms such as the *Yingzao fashi* (Manual on architecture, 1103),[5] the *Lülü chengshu* (Treatise on pitch-pipes) (Mongol period),[6] the *Hefang tongyi* (General Discussion of the Protection Works along the Yellow River) (Mongol period)[7] confirms this. Music requires a knowledge of how to extract square roots, architecture requires a knowledge of how to apply ready-made formulae (calculation of the diagonal from the sides of the square, calculation of the sides of various regular inscribed polygons as a function of the diameter of the circle, Pythagoras' theorem),[8] while hydraulics, in turn, requires a number of stereotyped formulae to calculate the volumes of the solids met in excavation work.

The same remarks also apply to economics even though in that case the level of mathematical sophistication appears even more limited. But that was also the case elsewhere outside China. According to Neugebauer:

The mathematical requirements for even the most developed economic structures of antiquity can be satisfied with elementary household arithmetic which no mathematician would call mathematics.[9]

Even so, the case of China, where there is a dominant preoccupation with the planning of work, the rationalisation of economic phenomena and a manifest will to "regulate the political world (rather than the physical world) by weights, numbers and measures", is perhaps different.[10] We shall leave the problem open here, since the history of mathematical economics would require further study.

In any case, all this demonstrates an intelligent adaptation of the means available to desired objectives.

---

[4]Cf. the preface to the *Shushu jiuzhang* (1247): "Only calendar-makers and mathematicians [*ch'ou-jen*] (i.e. *chouren*) were able to manage multiplication and division, but they could not comprehend square-root extraction or indeterminate analysis." (cited in Libbrecht (2), *1973*, p. 58).

[5]The indispensable reference for any study of this work is still Demiéville (1), *1925*.

[6]Cf. *Zengding Siku jianming mulu biaozhu*, reprint, Shanghai: Guji Chubanshe, *1979*, p. 152.

[7]Ibid., p. 299. Now lost, this work of the Mongol period has been reconstituted from quotations recopied in the *Yongle dadian*.

[8]More generally, the same remark is also valid in the case of European architecture during the medieval period: mathematical knowledge necessary for the real practice of architecture was very limited. Cf. Shelby (1), *1972*.

[9]Neugebauer, (3), *1957*.

[10]Gusdorf (2), *1969*, p. 467 (genesis of this type of preoccupation in 17th century Europe).

Whatever the complexity of mathematical knowledge involved in various applications, the problem of the origin of these is left open.

In the case of architecture, music or warfare, for example, the answer is straightforward for practically everything is convincingly explained from a logical point of view in the various commentaries of the *Jiuzhang suanshu*.

Astronomy, however, seems to constitute a case apart for, on the one hand, its numerical techniques are much more sophisticated than those of economics or architecture. They consist of sophisticated formulae for interpolation by polynomials of degree 2, 3 or more, simultaneous congruences (Chinese remainder theorem), tables, and, above all, a plethora of approximation formulae intended, e.g. for the conversion of coordinates (ecliptic to equatorial and conversely), modelling of the equation for the centre of the Sun, Moon and planets in view of the prediction *tuibu* of the positions of the celestial bodies, solstices, eclipses and related phenomena. Actually, this aspect of Chinese astronomy, which was almost wholly unknown until the 1980s, has been brought to light by researchers such as Chen Meidong and a few others.[11] According to research in progress by the author of the present book, the precision attained by the Chinese predictive techniques of astronomy under the Ming dynasty seems equivalent to that offered in Europe by the Alphonsine Tables during the same period.

It would seem that the problem of the reconstitution of the underlying logic of astronomical predictive techniques is hopelessly complicated. However, the overall algebraic character of these might perhaps render their logical origin easier to understand. But, even if this were not the case, it is by no means obvious that the inventor of some new technique would have been able to describe the logic of his invention in all cases. This still occurs nowadays, for example, the computer scientist J. Arsac in one of his popularising works wrote:

I guessed what the program might be [...]. I was only able to show why it works afterwards.[12]

Second, pure and simple borrowing of ready-made, off-the-shelf techniques (case of Muslim astronomy in the 13th century and European astronomy in the 17th century) might also have played an important role. Such 'technology transfers' may have taken place at other periods.

Last but not least, assuming that an inventor had succeeded in creating novel procedures, it is not certain that he would have been inclined ipso facto to reveal the secrets of their creation; in fact, the existence of rivalry between calendarist astronomers is known.[13]

---

[11]Chen Meidong (2′), *1988*.

[12]Arsac (1), *1985*, p. 82.

[13]Prefaces to books often indicate such a rivalry (e.g. see that to the *Jigu suanjing*).

5. The Structure of Mathematical Works

It is difficult to give a synthesis of the structure of Chinese mathematical works. There is no specific invariable form for these in China, any more than there is elsewhere. Even within the Ten Computational Canons (*SJSS*), which themselves form a well-defined historical unit as a collection of manuals brought together in the Tang, the organisation of the material does not follow a unique scheme. In two out of ten cases the text consists of a sequence of dialogues between master and student,[1] another case is a commentary slanted towards philological questions applied to numbers,[2] while a succession of problems forms the framework for the seven remaining cases.

Despite this relative diversity, the fact remains that the works based on collections of problems are by far the most common at all periods.

Just as works merely consisting of raw statements of theorems with no other indications of any kind are not the most common, Chinese works practically never consist solely of sequences of problems in the raw state. In this respect, they differ from Babylonian or Egyptian collections. Their material composition certainly varied ("volumina" *juan* of wooden or bamboo sheets tied together by cords, rolls of paper, xylographed sheets of paper which were then folded and assembled). Most of the texts which we possess are of this last sort; the oldest of these date back to the 13th century, but the majority of the available editions date back to the 18th or 19th centuries. However, we do have four arithmetical manuscripts from Dunhuang, which probably date back to the 10th century or earlier.[3] But however far one goes back in time, Chinese works always have a structured form, including first a precise title by which they may be individually recognised and second an internal organisation based on a precise scheme. They generally begin with one or more prefaces, in order of age, possibly followed by a table of contents which lists the titles of the successive chapters. Finally, each chapter groups together successions of problems, according to various criteria, which are in turn associated with sets of textual or graphical items of various sizes, such as resolutory rules, commentaries, sub-commentaries, numerical tables and figures. We note, in passing, that there exist two types of figures, namely geometric figures, *tu*, and computational diagrams, *suan tu*,

---

[1] Case of the *Zhoubi suanjing* and the *Shushu jiyi*, resp.
[2] Case of the *Wujing suanshu*.
[3] Cf. Libbrecht (5), *1982*.

which are special figures intended to show the state of a computation at a
particular stage of its execution.

Titles

The titles may indicate:

- The fact that a work was composed by a given master (example: *Sunzi
  suanjing*, Sunzi's computational canon).

- The form of the work (example: *Licheng suanjing*, Canon of ready-made
  computations or "ready reckoner"– an allusion to the fact that the text
  consists of ready-made numerical tables).[4]

- The name of a particular mathematical method (example: *fangcheng lun*,
  On the *fangcheng* method).[5]

- The type of mathematical formulae included in the work (example:
  Hushi suanshu, The arc–sagitta computational methods[6] (i.e. methods
  for circular arcs and their versed-sines)).

- The intended application area (example: *Celiang jiyao*, Rudiments of
  surveying).[7]

Prefaces

Prefaces are often composed according to a fairly strict plan which associates
series of almost obligatory stereotyped sequences with portions of text including
information about the work in question.

The introduction to the prefaces usually consists of a discourse on either
the historical or the logical and philosophical origins of the art of computation.

The historical discourses essentially consist of proclamations on the mythical origin of computation, invented by Li Shou,[8] minister under the Yellow
Emperor, the civilising genius,[9] sometimes described as the father of the
tradition of the Nine Chapters,[10] which tradition was handed down imperfectly

---

[4]This is one of the Dunhuang manuscripts. Cf. Libbrecht, op. cit.; Li Yan, *Gudai*, p. 23.

[5]Work by Mei Wending (around 1700). On the *fangcheng* method (linear systems), see
below, p. 249.

[6]Work by Gu Yingxiang, (1483–1565).

[7]Work dating from 1890. Cf. Ding Fubao and Zhou Yunqing (1′), *1957*, p. 90b.

[8]The myth of Li Shou and the Yellow Emperor, the inventors of computation, dates back
to the *Shiben.* cf. Li Yan, *Gudai*, p. 11.

[9]Expression due to Marcel Granet. Cf. Granet (1), *1934*.

[10]Under the Song, there existed a work entitled *Huangdi jiuzhang suanfa* (The Nine
Chapters of the Yellow Emperor). See QB, *Hist.*, p. 145. Numerous other references in Li
Yan, *Gudai*.

as a result of the criminal acts of the First Emperor of the Qin, who was responsible for ordering the burning of books.[11] Other scenarios date computation back to a less ancient past, referring only to the utopian society described in the *Zhouli* (Records of the Rites of the Zhou Dynasty).

The philosophical allusions are most often in laconic form. They frequently stress the privileged role of numbers conceived as a universally dominant principle, in a manner reminiscent of Pythagorism.[12] They often consist of juxtaposed quotations from the *Yijing* and other classics; frequently, these quotations are not echoed anywhere in the body of the text itself. For example, the *Shuli jingyun*, the important encyclopedia of mathematics published in 1723 at the end of the reign of Kangxi,[13] begins with an explanation, according to which mathematics is derived from the eight trigrams by way of the Yellow River Chart, *Hetu*, and the Luo River Diagram, *Luoshu*; however, except at the very beginning of the text, all this is of no importance in what follows. However, similar allusions scattered throughout Liu Hui's preface to the important *Jiuzhang suanshu* (preface apparently written in the third century AD) certainly find echoes in the commentary composed by the same author.[14] Consequently, it appears that the link between the art of computation and the *Yijing* is not due to a confusion between mathematics and divination (which is, essentially, never noticed)[15] but to a resemblance between the operational procedures governing the two systems, which both depend on:

(i) the manipulation of rods;

(ii) diagrams and concrete images;

(iii) verbalisation (comments on the diagrams).

After these brief historical and philosophical incursions, the texts of the prefaces often explain why it is necessary to study the art of computation. The almost unanimous response, which is interminably repeated, consists of an invocation of the utility of a particular piece of knowledge, whether for evaluating the dimensions of the universe or for ensuring the harmonious development of all human activities involving computational ability (economics, great works,

---

[11]See Liu Hui's preface to the *JZSS*.

[12] "References to the modes of divination or comprehension of the universe in antiquity show that (in the *Shushu jiuzhang*) the mysticism of numbers was experienced as strongly in the China of the Song as in Pythagoras's Greece." (C. Diény, *T'oung Pao, 1979,* Vol. LXV, pp. 81–92); see also Libbrecht on the Pelliot 3349 MS.: "The pan-mathematical conception gives at the same time a strange Pythagorean and a striking modern impression – in both of these cases we have also had the basic idea that everything could be reduced to mathematical relations." Cf. Libbrecht (5), *1982*, p. 212.

[13]Cf. Li Yan, *Dagang*, II, p. 401 ff.

[14]In his preface to and his commentary on the *JZSS* Liu Hui repeatedly expresses his conviction that phenomena with a different appearance conceal a common principle and uses for this an expression from the *Yijing*: *tong gui* ("Xici," part 2, section 5).

[15]One of the rare Chinese problems, the solution of which depends on numerological considerations is that at the end of the *Sunzi suanjing*.

the art of warfare, etc.). This is followed by biographical information about the author (this is sometimes the only such information available, whence its importance) together with details of the circumstances which led the author to compose this work. Finally, the whole is completed with a mention of the place, the date and the name of the author.

Problems

The problems of the mathematical works may be divided into the following four main categories,[16] depending on the situation they address:

(i) Real problems which apply to specific situations and are directly usable.

(ii) Pseudo-real problems which pretend to address situations of daily life, but are neither plausible nor directly usable.

(iii) Recreational problems, modelled on riddles, which use data from everyday life in a particularly unrealistic, sometimes grotesque, but always amusing way.

(iv) Speculative or purely mathematical problems.

It goes without saying that the boundaries between these categories are not hermetic; problems of all types may lead to true mathematical questions and, conversely, mathematical techniques may be applied in specific, pseudo-specific or recreational problems.

Real Problems

The themes of these problems relate equally to finance, trade, transport, irrigation, excavation, architecture, warfare, meteorology, etc. There are no available statistics which enable us to assess their quantitative importance, but, according to Libbrecht, real problems are better represented in the Chinese tradition than elsewhere:

> In European medieval mathematical handbooks the greater part of the problems were imaginary, whereas in China they were derived from situations of daily life. An important corollary is that the numbers used are realistic and represent historically valid information, providing an interesting supplement to the socio-economic chapters in the dynastic histories.[17]

---

[16]Vera Sanford identifies three categories of problem: (i) "genuine problems which men have had to solve in real situations;" (ii) "puzzles whose sole aim is to mystify or to amuse;" and (iii) an intermediate class "which deals with real objects acting under unusual conditions – problems which because of their plausibility may be called pseudo-real" (cf. Sanford (1), *1927*, p. 1).

[17]Cf. Libbrecht (2), *1973*, p. 416.

A priori, there is no lack of indicators which lead one to believe that this judgement is broadly correct, even though such problems also exist in the European tradition, namely those which are the subject of political arithmetic.[18]

Certain problems, notably those of the *Shushu jiuzhang* (1247), those of chapter 7 of the *Jiuzhang suanshu*, and those of the *Jigu suanjing*, contain geographical data (names of villages, small towns and prefectures), economic data (precise allusions to monetary systems, to goods distribution networks, to taxes), technical data (architecture, the art of warfare), etc. of a local value.[19] The factual richness of certain results reaches unheard of proportions; the *Shushu jiuzhang* contains a problem which occupies a whole chapter, the statement of which includes no less than 175 unknowns![20]

However, as already noted, the level of mathematical sophistication of these problems is never very high.[21] Arithmetical notions of the most elementary type are generally sufficient to solve them.

Far from stimulating the blossoming of mathematics, the insistence on real problems may have constituted a powerful brake to its development. In fact, since such mathematical problems were too specific, it was difficult to use them to construct a cumulative knowledge; the context only had to change for people to lose sight of its true significance. This could explain why works which were too oriented towards real problems, such as the *Shushu jiuzhang* and the *Jigu suanjing*, had little influence.

This is the vicious circle of Chinese mathematics: either its development is limited by the overpowering complexity of the real world which it claims to represent down to the last detail, or freed from a specific straitjacket, it holds great promise, but becomes useless to the point of being unacceptable and consequently loses its social value.

Pseudo-Real Problems

For essentially pedagogic reasons, this category is well represented, truly real problems (an affirmed objective of Chinese mathematicians) being, respectively, too simple, too complicated or of too limited scope to be used to support teaching. It was better to conceive exercises, which were possibly less real but, on the other hand, permitted more flexibility. This goal was achieved by various procedures, including the borrowing of problems from tried and tested ancient collections the data of which no longer corresponded to anything real, and the fabrication of fanciful results within a specific framework. The roles of data and

---

[18]Cf. Tropfke (3), *1980*, p. 514. This expression, which was commonly used in 18th century Europe was already used by the 14th century Byzantine mathematician N. Rhabdas. See also the notice of Diderot and d'Alembert's *Encyclopedia* and Gusdorf (2), *1969*, p. 461, chapter entitled "political arithmetic and statistics."

[19]See Libbrecht, op. cit., p. 416 ff.

[20]*SSJZ*, j. 9 (Shanghai: Shangwu Yinshuguan, *1936*, vol. 3, p. 209).

[21]Cf. Libbrecht, op. cit., p. 444, note 126: "on the whole, they are not interesting from the mathematical point of view."

unknowns are often inverted: for example, one is often asked to determine the dimensions of a figure given its area or volume,[22] the total quantity of taffetas distributed knowing the overall proportion received by each beneficiary,[23] or the area of a field from the level of the farm rent.[24] In fact, one might think that it was not appropriate to introduce students directly to certain rules in their true context; this seems to have been the case for certain problems relating to bodies moving at speeds varying like the terms of arithmetic progressions which, in reality, hid serious questions of calendrical astronomy.[25]

Recreational Problems

This category of problems of recreational mathematics corresponds to the Chinese category of difficult problems, *nan ti*.

The heading "difficult problems" appears in Chinese arithmetics from the Ming dynasty (1368–1644). As the small introductory notice at the beginning of *juan* 13 of the *Suanfa tongzong* (General Source of Computational Methods) (1592) explains, an anthology of these problems was collected in the fourth year of the Yongle period (1406), on the initiative of the arithmetician Liu Shilong, who was associated with the compilation of the great thematic encyclopedia of the start of the Ming, the *Yongle dadian* (Yongle Reign Period Great Encyclopedia) (1407). These problems, were classified according to the ancient categories of the Nine Chapters and compiled in a pleasing, but intentionally disconcerting and surprising way, even for specialists in computation *suan shi*, in verse; thus they appeared difficult even though they were not (*si nan er shi fei nan*). They were intended to show off the virtuosity and professional skill of arithmeticians in comparison with the mass of those with no arithmetical competence. By way of confirmation of this interpretation, we note that problems 10 and 19 of chapter 14 of the *Suanfa tongzong* exhort the reader calling him *neng suan zhe* and *gaoming neng suan shi*, in other words, "man with a knowledge of mathematics" and "cultivated and intelligent man, with a knowledge of mathematics," respectively. The same thing can be found in Indian mathematics: "Dear intelligent Līlāvatī, if thou be skilled in addition and subtraction [...]."[26]

---

[22] See for example *JZSS* problems 4-19 to 4-24.

[23] *WCSJ* problem 2-7 (cf. Schrimpf (1), *1963*, p. 460).

[24] *JZSS* 6-24.

[25] *JZSS* 7-19 (problem of the horse which runs an additional 13 *li* each day). Note that arithmetic progressions are used in Chinese astronomy (see the *Zhoubi suanjing*, for example).

[26] Colebrooke (1), *1817*, p. 5. On the mathematical recreations of the Hellenistic and Roman worlds, J. Høyrup (3), *1990*, p. 11, writes that "*the formation of professional identity and pride* is served in particular by so-called 'recreational problems.'" The same author also makes analogous remarks about the sub-scientific mathematics of the Islamic world (Høyrup (1), *1987*). Thus, this is a common phenomenon which goes far beyond the particular case of China.

It has long been noted that recreational problems occur in similar forms in all cultures.[27] Thus, these problems are very important, not only from the point of view of the sociology of mathematics, but also as potential indicators of diffusion.[28] By way of illustration, here are some examples including problems involving moving bodies and filling of reservoirs, problems made up of episodes and counting rhymes:

(i) Moving-body problems. These in general involve people or animals (even vegetables) which move in the same direction (grow) or set out to meet each other following a more or less complicated trajectory defined by a straight line, a circle or a broken line, at speeds which are constant or vary according to an arithmetic or geometric progression. For example, *JZSS* problem 6-14 involves the pursuit of a rabbit by a dog;[29] *JZSS* problem 7-12 involves two rats setting out to meet each other by making a hole in a wall; problem 7-19 involves horses; problem 6-20 involves a duck and a wild goose; the 170 problems of the *Ceyuan haijing* (1248) by Li Zhi all involve men walking around the sides of a town in the shape of a right-angled triangle; *JZSS* problems 7-10 and 7-11 involve the growth of a cucumber and a gourd and that of rushes and sedge, respectively.[30] In *ZQJSJ* problem 1-11, the guards of three camps of a round shape go round them several times before meeting up at the South gate.[31]

(ii) Problems about reservoirs. *JZSS* problem 6-26 asks how long it will take to fill a pond, given the flow rate of five canals feeding into it.[32]

(iii) Problems made up of episodes. *JZSS* problem 7-20 is a good example:

Suppose that a man with many *wen* (coins) enters the land of Shu [Sichuan]. [He trades] and obtains an interest of 3 for 10. First he sends 14 000 *wen* [home] then, a second time, 13 000 *wen*, then a further 12 000 *wen*, then 11 000, then 10 000. After doing this five times, his capital and interest are completely exhausted. How many *wen* did he have originally, and how much did the interest amount to?

Answer: 30 468 *wen* and 84 876/371 293 *wen*; interest: 29 531 *wen* and 286 417/371 293 *wen*.[33]

---

[27] Tropfke (3), *1980*.

[28] Høyrup (3), *1990*, p. 15: "recreational problems are important, not only for the understanding of the cultural sociology of the craft of reckoners but as 'index fossils.'"

[29] A similar scenario is found in Alcuin's *Propositiones ad acuendos juvenes*. See Migne (1), *1851*, vol. 150, col. 1151, problem 26.

[30] Analysis in Wu Wenjun (1'), *1982*, p. 267.

[31] Analogous Indian, Arabian, etc. problems: Tropfke, op. cit., p. 588.

[32] This type of problem is also found in the Heronian *De mensuris*, in Metrodorus's *Greek Anthology* (ca. 500 AD) Example: "I am a brazen lion, my spouts are my two eyes, my mouth and the flat of my right foot. My right eye fills a jar in two days, my left eye in three, and my foot in four. My mouth is capable of filling it in six hours; tell me how long all four together will take to fill it?" (Paton (1), *1918*, p. 31 and Sanford (1), *1927*, p. 69). See also Tropfke, op. cit., p. 578.

[33] QB, I, p. 218.

Or, in elementary algebraic notation[34]:

$$((((1.3x - 14\,000)1.3 - 13\,000)1.3 - 12\,000)1.3 - 11\,000)1.3 - 10\,000 = 0.$$

(iv) *Counting rhymes.* For example, *SZSJ* problem 3-34:

> Suppose that, after going through a [town] gate, you see 9 dykes,
> with 9 trees on each dyke, 9 branches in each tree, 9 nests on each
> branch, and 9 birds in each nest, where each bird has 9 fledglings
> and each fledgling has 9 feathers each of a different colour chosen
> from among 9 possibilities. How many are there of each?
> Answer: 81 trees, 729 branches, 6561 nests, 59\,049 birds, 531\,441
> fledglings, 4\,782\,969 feathers, 43\,046\,721 colours.[35]

Speculative Problems

Problems of an incontestably speculative nature have also absorbed mathematicians in China. These include, for example, questions associated with the
planimetry of the circle (calculation of π) or the stereometry of the sphere. They
are treated in a wealth of detail and from a demonstrative angle in the various
commentaries of the *Jiuzhang suanshu.*

From the 18th century, other speculative questions arose in large numbers.
The most notable of these relate to infinite-series expansions and finite
summation formulae.

But, apart from these obvious examples, problems which seem a priori more
or less real should also be considered as speculative (or at least speculative in
the same way exercises in textbooks are).

Those contained in the *Siyuan yujian* (1303), which relate to solving right-
angled triangles in so artificial a way that real applications are certainly out of
the question, are typical examples. This is not an isolated curiosity, since there
are 101 such problems, or 35 per cent of the total number of 284 problems in
the book.

Resolutory Rules

The term resolutory rule refers to any sequence of unambiguous instructions or
directives which may, in principle, be used to obtain the solution of a given class
of problems in a limited number of elementary stages. Thus, resolutory rules
may be thought of as algorithms, although the latter term should not be taken
in the strict sense, since the Chinese rules are not always completely explicit.
More to the point, resolutory rules are dogmatic "stratagems of action,"[36] which
if followed mechanically should lead automatically to the expected result, but

---

[34]Tropfke, op. cit., p. 582.

[35]Cf. Gay Robins and Charles Shute, *The Rhind Mathematical Papyrus, an Ancient
Egyptian Text*, New York: Dover, *1987*, p. 56.

[36]Libbrecht (2), *1973*, p. 377.

which in practice may permit a certain degree of freedom of action, since, not all the stages of the calculations are rigorously specified, as they should be in the case of true algorithms. For example, the Chinese Ruffini–Horner methods always leave the mathematician to evaluate the order of magnitude of the root. But the same remark also holds for all medieval mathematics, whether Chinese or not.

In Chinese mathematical texts, the resolutory rule rarely occurs in isolation, for example, as an object of a theory, but often in association with a specific problem to which it provides the key. When this pattern is not adhered to, algorithms are stated independently of any specific situation and in a quite general way. None the less, even in such a case, specific problems very often accompany the text of the algorithm itself (but the formulation of the algorithm as such is not related to the specific content of the problems). In such a case the ordering is the following:

Statement of problem + Numerical answer + Resolutory rule

In general, the word *shu* (prescription)[37] is used to introduce the resolutory rule, but other terms including *fa* (rule, method), *jue* (trick), *mijue* (secret trick) (sic), *ge* (song) are also used, with or without a mention of the corresponding musical tune to enable the mathematician to sing his algorithms (the last three terms, which are not of general usage, occur above all in the popular arithmetics of the Ming period). Although it is not obligatory, some rules have precise names, which may indicate their function explicitly; for example, *hefen shu* (rule for combining parts or rule for adding fractions), *kai lifang shu* (rule for extracting the cube root). However, some of these names may only indicate how the rules operate; for example, the Chinese version of Horner's method is often called the rule "add-multiply," *zeng-cheng*. In other cases, rules have names which are unusual in mathematics. For example, the rule *dayan*, which is used to solve residue problems, takes its name from a divinatory method of the *Yijing*. However, this does not imply that the rule *dayan* has a mathematical connection with divination.

The rules may be short or long; their length is not always linked to their complexity. The rule of alternating subtractions which is used to simplify fractions,[38] is stated in 33 written characters, while the rule *dayan*, mentioned above, requires 873.[39]

From the algorithmic point of view, certain rules are described in a linear manner, although iterative processes ("loops") are often also discernible; a complicated rule may itself refer to other rules playing the role of subprograms long before the invention of the term. It goes almost without saying that such characteristics are never the object of any explanation or theorisation in the

---

[37]This term, which is of general usage, is not reserved for mathematics alone, see above, p. 42.

[38]*JZSS* 1-6 (QB, I, p. 94).

[39]Detailed breakdown according to the *Yijiatang* edition (1842) of the text of the *Shushu jiuzhang* (1247).

texts.[40] However, Chinese mathematicians often find it necessary to clarify their rules from various points of view: for example, in order to explain the logical origin of a rule when composing a commentary (*zhu*); in order to trace the application step by step using particular numbers which appear in the text (when the rule is itself formulated in general terms, when giving what the Chinese mathematicians call a rough solution (*cao*) (called "rough" probably because the steps of the resolution rarely contained the detailed working of complex computations such as those of root extraction)). Alternatively the expression *xicao* (lit. "detailed solution") is also used instead of *cao*. According to the *Zhizhai shulu jieti*,[41] it means "the complete detail of the dividends, the divisors, the multiplications and the divisions." Lastly, in order to monitor the correctness of computations calculations are carried out in reverse order. This technique is known as the *huan yuan* rule (lit. "restitution of the initial state")).

---

[40]Generally speaking, everywhere in the world, algorithms began being studied theoretically from the 19th century, not before.

[41]j. 14.

6. Mathematical Terminology

In the Chinese case, it is not easy to clarify the meaning of technical mathematical terms since the texts rarely contain definitions. The problem is even more acute because most dictionaries, however complete they are, are of little help in this matter.[1] Under these conditions, many authors essentially adopt one or other of the following two techniques:

(i) prudent recourse to phonetic transcriptions with no translation at all;

(ii) translation of ancient terms using modern mathematical terms.

The first technique is clearly unsatisfactory, but since it concerns a small number of terms, it may be considered as the lesser of two evils. After all, very few translators dare to translate *yinyang* or *dao*. But, taken to its extreme, this procedure brings to mind that employed by Jean-Baptiste Du Halde in the 18th century who, when translating certain Chinese medicinal recipes, left certain special terms of the Chinese pharmacopeia as they were, with no real explanation. "For the lung which is exhausted by short breath and other deep-rooted discomforts, take a cup of *Po hi* [name of herb] broth with *Teou che* [name of bean] and a little onion [...]."[2]

The second technique is apparently justified by the observation that the modernising conceptualisation of ancient Chinese texts enables us to obtain the same results as the ancients. In other words, it is as though the ancient and the modern procedures were simply two superficially different modes of expression hiding a common profound reality. The danger of this procedure is immediately apparent since, given that modern concepts are more general than the ancient ones, it is easy to attribute characteristics to the ancient terms which they do not have.[3] To convince oneself of this, it is sufficient to observe the gap which exists between reconstitutions of ancient Chinese proofs, which are obtained by algebraic calculations (for example) and the actual proofs as explained by authors such as Liu Hui.

---

[1]Note however that T. Morohashi's *Dai kanwa jiten* (Great Chinese–Japanese dictionary) is largely based on the lexicon of the *Suanfa tongzong* (1592).

[2]Du Halde (1), *1735*, III, p. 469.

[3]Marrou (1), *1954*, pp. 151 and 157, in particular: "Under the pretext of attaining the profound reality, the authentic real is very ingeniously replaced by a set of reified abstractions [...]."

However, the historian is not completely powerless in the face of this type of difficulty because, although they have been little exploited until now, glossaries of mathematical terms do exist. For example:

- The *Jiuzhang suanshu yinyi* (Meanings and Pronunciations of [terms] occurring in the *Jiuzhang suanshu*) by Li Ji, imperial librarian of the Song,[4] responsible for the revision of the computational canons (*suanjing*).

- The *Yong zi fanli* (Terminological foreword) which is found at the beginning of the *Suanfa tongzong* (1592).[5]

- The *Jigu suanjing yinyi* (Meanings and Pronunciations of [terms] occurring in the *Jigu suanjing* (by Wang Xiaotong), seventh century) (1823) by a certain Chen Jie.[6]

In addition, certain texts such as the commentaries of Liu Hui and Li Chunfeng on the *Jiuzhang suanshu* contain glosses of special terms.

Certainly, even given the above, many difficulties still remain, if only because these glossaries and glosses only bear witness to the Chinese understanding of certain texts at a given time, which may sometimes be separated by several centuries from the supposed period of the text. Insofar as, in every period, reinterpretation of the past as a function of the presuppositions of the time is the rule rather than the exception, the same difficulties recur. For example: in his work, Zhang Dunren (1754–1834) uses the *tianyuan* algebra to reinterpret Wang Xiaotong, although this algebra did not yet exist in the seventh century,[7] to explain the *Ceyuan haijing* (1248); in the 19th century Li Shanlan uses an algebra like that of Descartes. There are numerous other examples of such an approach, and the 20th century is far from immune. The situation is aggravated by the fact that the known original texts, and in particular the oldest ones, consist in their current state of heterogeneous strata separated from one another by a distance of centuries, which appear difficult to dissociate and, moreover, are of differing conceptual levels.

Is it nevertheless possible to speak of technical terms? One could of course decree that only words or expressions which clearly cannot be understood in their usual sense, but acquire a special meaning when used in a mathematical text count as such. This could include terms such as *shi*, *yu*, or *lü*. But at what level should this be implemented? From the arithmetical point of view, *shi*, which normally means "the fruit" could be said to have the technical meaning of the number which is operated on, the operand. From the algebraic point of view, *shi* would become the constant term of a polynomial equation. *lü* (the rule, the norm) would be associated with pairs of numbers varying proportionally,

---

[4]Dates unclear, there are two possibilities: either under the Tang dynasty (618–907) or under the Northern Song dynasty (960–1127).

[5]*SFTZ*, introductory chapter.

[6]Li Yan, *ZSSLC-T*, IV-2, p. 537.

[7]Cf. *CRZ*, j. 52, p. 695.

and *yu* (the corner) would become the coefficient of the term of highest degree in a polynomial.

Interpreted in this way, Chinese texts do not become contradictory; on the contrary, they often acquire an excessive consistency which they did not have at the beginning, or which they did have, but not necessarily for the reasons one might suppose. This mirrors the way in which Euclid's *Elements* may be read teleologically using Hilbert's geometry as a reference. In a word, this leads back to the difficulties referred to above. There is a risk that history will disappear. Perhaps it is better to acknowledge that we do not know what led to terms such as *lü* and *fa* being used as we have described? This is often not because there are no explanatory hypotheses, but because, for lack of philological and chronological reference points, it is very difficult to choose between these.

In fact, neither the glossaries nor the commentaries restrict their explanations to mathematical terms, since they also include biographical details, historical or geographical points, philological information, all apparently arranged in a random order. Like dictionaries, they continually jump from one subject to another.

Moreover, one must always take particular account of the context. The Chinese terms are not defined *in abstracto*, at the conclusion of Platonist procedures, but form part of an incessant dynamism in which their meaning is continually negotiated.

For Mei Wending (1633–1721), the term *lü*, meant either a *constant* coefficient (in fact a binomial coefficient extracted from Pascal's triangle) or a *variable* term of a proportion.[8]

In addition, within the same work, what appears to be a single mathematical notion may be referred to in different ways, depending on the context in which it is used. For example, Li Shanlan (1811–1882) refers to the sequence $1^2, 2^2, 3^2, \ldots n^2$ in two different ways at two different parts of the same text.[9] There are many other examples.

Ultimately, it seems preferable to recognise that Chinese mathematics has no technical terms, as such, but rather a vast range of technical expressions[10] which it would be futile to wish to imprison a priori in a rigid framework. A case by case study should enable us to delineate the problem more clearly.

Let us consider, for example, the lexicon of the *Suanfa tongzong* (1592), mentioned above. It contains glosses for 72 technical expressions, including 35 verbs, 35 nouns and 2 adverbs.

The verbs, which alone account for half of the explanations, are always *active* and include, for example: *zhéban* (to divide by two); *biancheng* (to multiply all the numbers in a column of numbers by the same factor); *gui* (to divide a number by a number with a single digit); *ru* "to make," as in the expression "two threes make six."

---

[8] *Meishi congshu jiyao*, j. 10 and 20, resp.

[9] See chapters 1 and 2 of the *Duoji bilei* (1867) by Li Shanlan.

[10] On the distinction between "technical terms" and "technical expressions," see Hominal (1), *1980*, pp. 10–11.

The nouns may be subdivided into several categories, as follows:

(i) Terms which denote parts of the abacus. Example: *ji* (the transverse bar).

(ii) Terms which indicate the relative position: *zuo* (to the left of); *you* (to the right of); *shang* (above the bar).

(iii) Dimensions: *guang* (width); *gao* (height).

(iv) Assorted words: *xian* (hypotenuse); *jing* (diameter); *mian* (side); *ji* (result of one or more multiplications — *cheng cheng zhi shu ye*).

The adverbs are *ruogan* and *jihe* (several, how much); these two words have the same meaning but *ruogan* is less literary.

If the word "definition" is taken to have the meaning it has in Greek mathematics,[11] the above cannot be said to be definitions. Euclid, to mention but one, never defines verbs or adverbs, or anything to do with calculating instruments or terms indicating the relative position.[12] In fact in this case, it is a question of ensuring that the student acquires a certain reflex behaviour. When the author of the *SFTZ* writes that *fazhe suo qiu zhi jia ye* (that which we call *fa* is the price we seek), he clearly hopes that the student will react by thinking of *fa* when he sees the word "price" in a problem; thereafter, that with which *fa* is associated from the point of view of the operations will be acquired by impregnation/imitation from individual cases.

One possible objection might be that the *SFTZ* only represents the state of Chinese mathematics at the time of its decline (Ming dynasty). However, most of the above observations may often be transposed. In Liu Hui's commentary on the *Jiuzhang suanshu* of the Han, the definition of area also uses inductive words: *fan guang-cong xiang-cheng wei zhi mi* ("In all cases, we call area the product of width by length").[13] This assumes that, when the Chinese author asks how large the area is in a problem, he wishes his student to begin by identifying the words "width" and "length" in the statement or perhaps to let him discover to what rectangle such and such a figure is equivalent in area. Still in the same work, active verbs are also key terms (*suan zhi gangji*);[14] for example, *qi* "to make level" (action which involves multiplying the numerators by the corresponding denominators when reducing fractions to the same denominator), *tong* "to equalise" (multiply the denominators together in the same operation). All this gives the impression that this mathematics aims at action rather than contemplation. Moreover it is highly context-dependent. For example, rather than uniformly using a specialised term which would provide an abstract definition of the concept of cylinder, the Ten Computational Canons (*SJSS*) of the Tang use a multitude of concrete descriptions such as: the fort *baodao*, the round silo *yuanqun*, the round cave *yuanjiao*, the pillar *zhu*, the round basket

---

[11] Heath (3), *1926*, p. 157; Morrow (1), 1970.
[12] Mueller (2), *1981*, p. 39.
[13] *JZSS* 1-2 (QB, I, p. 93).
[14] *JZSS* 1-9 (QB, I, p. 96).

yuanchuan.[15] In the *JZSS*, the sphere *liyuan* (lit. the "upright ring") is not defined discursively by the fact that:

When, the diameter of a semicircle remaining fixed, the semicircle is carried round and restored again to the same position from which it began to be moved, the figure so comprehended is a *sphere*.[16]

but analogically as a ball *wan*.[17] As L. Vandermeersch writes:

Phenomena are not represented mentally by concepts which would imply their essential characteristics, but only by artificial symbols, which are the names of things *ming*; this is why the normative science of judgement is defined as the method of rectifying names, *zhengming*.[18]

These names may be thought of as proper nouns, "quanta of meaning, below which there is nothing else to show."[19]

Moreover, Liu Hui himself found it hard to make a judgement (*jue*) on a certain method[20] without "empty words" *kong yan*, i.e. "abstract words" — words independent of context. However, it is not true that Chinese mathematics does not use abstract terms; otherwise, how can one explain the presence of terms such as *li* (principle),[21] *shi* (situation, tendency),[22] *qing* (reality of facts)?[23] But, we note that these are terms of general usage, the large number of uses and connotations of which goes far beyond the readily imaginable framework of a strictly defined, unambiguous mathematical corpus. It is as though, with an acute sense of the dynamic nature of thought, Liu Hui and his successors refused to establish compartments, as though reality could be better described by intuitive, ambivalent, multiform, variable concepts, which apparently conveyed greater meaning and possibilities for action. In this, the Chinese attitude agrees with that of Bergson:

At first sight, it may seem prudent to leave the consideration of facts to positive science [...]. But it is impossible not to see that this would-be division of the work (into science and philosophy) amounts to muddling and confusing everything [...].[24]

as E. Nagel notes:

According to Bergson, for example, there is something "artificial in the mathematical form of a physical law and consequently in our scientific knowledge of things;" and he assigns to philosophy the task of transcending "the geometrizing pure intellect" by reintroducing us "into the more vast something out of which our understanding is

---

[15]See resp.: *JZSS* 5-9 (QB, I, p. 163 – cf. also Schrimpf (1), *1963*, p. 283); *JZSS* 5-28 (QB, I, p. 177); *SZSJ* 2-10 (QB, II, p. 298); *XHYSJ* 1-5 (QB, II, p. 563); *XHYSJ* 1-2 (QB, II, p. 562).

[16]Euclid's *Elements*, book 11, def. no. 14, Heath (3), *1908*, III, p. 261.

[17]*JZSS* 4-24 (QB, I, p. 155).

[18]Vandermeersch (1), *1980*, II, p. 270.

[19]Lévi-Strauss (1), *1962*, p. 258.

[20]The method in question was the *fangcheng* method, *JZSS* 8-1 (QB, p. 221).

[21]Cf. QB, I, pp. 93, 95, 96, 106, in particular. See also Chan Wing-tsit (2), *1969*, pp. 45–87.

[22]*JZSS* 7-6 (QB, II, p. 208); *JZSS* 4-24 (QB, I, p. 158). Cf. Bai Shangshu (5′), *1986*.

[23]Ibid.

[24]Bergson, *Oeuvres*, Paris: PUF, *1963*, p. 660.

is cut off and from which it has detached itself through its symbolic rendering of natural processes. One motivation of Bergson's philosophy thus seems to be rooted in his felt disparity between the vague discourse of daily life and the relatively more precise language of science.

But unless one adopts the absurd conception of knowledge, according to which the goal of thought is an intuitive, empathic identification with the things to be known, there is very little substance in Bergson's critique of scientific knowledge as "abstract," and of language and symbolism as impediments to knowledge.[25]

If Liu Hui readily quotes Zhuangzi, could it not be because the mathematician of the Kingdom of Wei shared a common view with the antirationalist philosopher of ancient China, who was "known for his attachment to intuition and spontaneity and his distrust of words and logic"?[26]

This situation is not necessarily disadvantageous however. In Liu Hui's commentary, for example, terms such as *qi* and *tong* (mentioned above) are used in various contexts, even though fractions are no longer involved; the commentator was considering pairs of numbers which formally behave like fractions and thus deemed it legitimate to use these terms. This leads to similarities between situations which are not a priori interrelated, but are in fact isomorphic, in the sense that the same type of numerical manipulations may be applied to them.[27] This surprising characteristic is reminiscent of that which now prevails in mathematics, but which it has been so difficult to master; formalism allows one to establish unthinkable relationships between objects which are not visibly connected.

What then is the basis for this analogy which everything until now seems to make impossible? It is certainly not the result of axiomatising theorisation, on the contrary it is precisely due to the operational nature of Chinese mathematics. In fact, we know that, however complicated the problems were, mathematicians were able to apply their thoughts to situations stripped of their concrete straitjacket, insofar as, for their calculations, they had to transform the latter into sober configurations of counting-rods. They must have noticed that they might have had to carry out the same structural manipulations on the counting-rods in response to different initial verbal problems. The calculations to be carried out became visible. However, it may be that mathematicians would have noticed none of this if they had had recourse to discursive arguments alone. After all, the most abstract terms of Chinese mathematics seem to be associated with numerical operations. In other words, the Chinese abstraction referred to material objects and not to essences.

In this context, it is easy to see that the translation of mathematical works belonging to other traditions may have posed serious problems. But the question does not seem to have arisen until a late stage. In fact, whatever the extent of the contacts between China and other civilisations throughout history, it

---

[25]Nagel (1), *1979*, p. 54.

[26]Review of Graham (5), *1981* by Yves Hervouet in *T'oung Pao*, vol. 69, *1983*, p. 126.

[27]Note also that in the *Siyuan yujian* the terminology of fractions is used although the calculations actually involve polynomials.

was only from the end of the 16th century that European mathematical works were first adapted into Chinese in large numbers. The vocabulary of geometry, in particular, which until then consisted of mono- or dissyllabic terms, was reworked in an attempt to reflect the substance/accident contrast which is not normally marked in Chinese.[28] For example, instead of *gougu* (lit. BASE-LEG),[29] which probably corresponds semantically to the "front" and "flank" of Babylonian mathematics,[30] the expression *zhijiao sanjiaoxing* (lit. triangular figure with a right-angle) was invented; the *guitian* (lit. field shaped liked a *gui* tablet)[31] became a *liangbiandeng sanjiaoxing* (lit. triangular figure with two equal sides), and so on. The hope was that, in this way, the meaning of the compounds could easily be guessed from each of the components. From the 17th to the 19th century, the terminology of geometry was subject to numerous amendments which, however, did not go so far as to question this new mode of derivation of new terms. The same happened in areas previously unknown in China, for example trigonometry and logarithms. But Chinese arithmetic and algebra were less easy to change in this drastic way. In fact, this then led to the syncretic coexistence of various terminological systems, even though some (not necessarily those of European origin) might advantageously have replaced others. For example, in the important mathematical encyclopedia of the end of the reign of Kangxi, the *Shuli jingyun* (1723),[32] the ancient Chinese terms to denote positive and negative (*zheng* and *fu*, respectively) are used solely in the context of the ancient *fangcheng*[33] method independently of the terms *duo* and *shao*, which play exactly the same role, but only within another algebraic technique of European origin, the *jiegenfang*,[34] and no globalising design unifies the common concept carried by these terms.

---

[28]Gernet (4), *1982*, p. 322 ff.

[29]Etymologically, *gou* = "base," *gu* = "leg," or perhaps "gnomon." Cf. *DKW*-9-29284, p. 96.

[30]Thureau-Dangin (1), *1938*, pp. 224 and 226, resp.

[31]*JZSS* 1-25 (QB, I, p. 101).

[32]Li Yan, *Dagang*, II, p. 401 ff.

[33]*SLYJ*, 2nd part (*xia bian*), j. 10, p. 535.

[34]Ibid., j. 31, p. 1233 (the term *jiegenfang* means literally "the borrowing of roots and powers [of the unknown]").

7. Modes of Reasoning

For anyone claiming to write the history of a science of which reasoning forms the very essence the question of the logic is of paramount importance. However, despite this obvious fact, general histories of Chinese mathematics rarely show concern for this question. They insist above all on presenting results, the initial *raison d'être* of which is unclear, even though they incidentally provide the reader with hypothetical reconstitutions of proofs. While this approach to the history of mathematics is naturally a result of various causes, one which probably plays an essential role is the fact that most Chinese mathematical works contain no justifications. However, there is one major exception, namely a set of Chinese argumentative discourses which has been handed down to us from the first millennium AD. We are essentially referring to the commentaries and sub-commentaries of the *Jiuzhang Suanshu*, the key work which inaugurated Chinese mathematics and served as a reference for it over a long period of its history. This fact, which was long unrecognised, means that we are now in a position to know a lot more about the logical construction of mathematics in China than, for example, in Egypt, Mesopotamia or India.

Thus, the understanding of the oldest Chinese mathematical reasoning involves an analysis of a particular literary genre, the commentary. In what follows, we shall not stray far beyond the commentaries of Liu Hui and Li Chunfeng on the *Jiuzhang suanshu*, since few texts provide such substantial information.

Just like any other Chinese literary commentary, the commentaries of Liu Hui and Li Chunfeng are primarily textual explanations. Consequently, philology occupies an important place in them and literary allusions and metaphors abound.

However, everything which touches upon the mathematics itself, either from the point of view of the teaching methods or, above all, from the point of view of the reasoning, is in the foreground. In his preface to the *Jiuzhang suanshu*, the famous mathematician of the Kingdom of Wei, actually explains that his commentary is intended to

析理以辭，解體用圖，庶亦約而能周，通而不黷，覽之者思過半矣。

xi li yi ci, jie ti yong tu, shu yi yue er neng zhou, tong er bu du, lanzhizhe si guoban yi.

Analyse the principles[1] by virtue of verbal formulations,[2] explain the substance of things using figures in the hope of achieving simplicity[3] while remaining complete and general but not obscure, so that the reader [of the commentary] will be able to grasp more than half[4]. [5]

These laconic words indicate Liu Hui's own theoretical conception of mathematics; they also underline the didactic vocation of his commentary. Of course, other motivations do come into play in some cases.

For Euclid, for example, what mattered most was formal rather than pedagogic consistency.[6] Certainly, the quotation is too elliptical for this to be anything other than an impression but, in the body of the commentary itself, Liu Hui also associates in the same sentence two famous passages from the *Lunyu* (Confucian Analects) which both suggest an idea of the same order:

〔 … 〕告往而知來，舉一隅而三隅反者也。

[...] *gao wang er zhi lai, ju yi yu er san fan zhe ye.*

I told him what had gone before and he understood what followed: [I] showed him a corner [i.e. an aspect of a question] and he replied with the other three.[7]

Here, the author indicates his wish not to disclose all the details of his reasoning to the student; it is assumed that the efforts of the latter are minimal. Consequently, instead of giving the details of his own thought processes, he often merely indicates that the solution of a given problem is analogous to that of some other problem,[8] or that "the remainder follows in the same way" *ta jie*

---

[1]*xi li.* This expression seems to occur for the first time in the *Zhuangzi*, "Tian xia": *xi wan wu zhi li.* In his translation of the *Zhuangzi* (op. cit., p. 275), Graham (5), *1981*, writes "chop up the pattern of the myriad things." Liou Kia-hway, in his translation *L'oeuvre complète de Tchouang-tseu*, Paris: Gallimard, *1969*, p. 265, translates it as "split up the structure of beings." Thus, it seems that *xi li* could be translated as "analyse the principles."

[2]This term is difficult to translate. For certain Chinese authors, it was a term of mathematical logic signifying "proposition." This translation is not completely out of court, but it could certainly be misleading. Chinese texts are not divided into "propositional phrases," since they are not punctuated (and even if they were, that would not provide a decomposition into phrases (think of the rhythmical rather than logical Chinese punctuation)). If one considers the various ways in which the word *ci* was used in ancient China, it seems that it should be translated by "phrase(s)," "verbal formulation(s)." Cf. Graham (3), *1978*, p. 207 ff.

[3]*yue.* Depending on the context, the simplicity referred to here is that of the computations and the computational methods.

[4]*si guo ban*: expression taken from the *Yijing*, "Xici," 2nd part: "If the sage considers the verbal formulations *ci* attached to the commentaries on the hexagrams, he will be able to understand more than half."

[5]Liu Hui's preface to the *JZSS* as cited in QB, I, p. 91.

[6]Cf. Heath (3), *1926*, p. 1; Morrow (1), *1970*, pp. 56–57.

[7]From Liu Hui's commentary on the beginning of the second chapter of the *JZSS* (QB, I, p. 114). These two expressions are borrowed from the *Lunyu*, chapters "Xue er" and "Shu er."

[8]For example, in *JZSS* 2-10 (QB, I, p. 117).

fang ci.[9] In fact, he probably attempts to convince rather than to elaborate a self-sufficient theory. This is also shown by the fact that commentary often explains that the objective is to persuade the interlocutor using all available means *bian*, to impart an understanding *xiao* and to verify a fact *yan*. From this angle it is natural that the arguments should not be constrained by fixed forms, which would force them to respect a rigidly defined rhetorical architecture in the Euclidean manner (think of the divisions of the proposition in Euclid: *protasis, ekthesis, diorismos, kataskeuē, apodeixis, sumperasma*).[10] Instead, these arguments use whatever appears the most appropriate in a given situation. This is true for both Liu Hui and Li Chunfeng and also seems to apply generally to most Chinese mathematicians. There is no fixed method of reasoning. As Ji Kang explained in the same period:

No one has the good method *liang fa*. In this world there are no naturally correct ways, and among methods, no solely good techniques.[11]

Thus, the argumentation inevitably depends on methods. For example:

- Passage from the particular to the general, based on a specific, well-chosen example. Commenting on the resolutory rule of *JZSS* problem 8-1,[12] Liu Hui explains that this is a general algorithm *dou shu*, but since it is difficult to impart an understanding of it with empty words *kong yan*, he would give his verdict on it by associating it with a problem relating to cereals, in other words, using the specific numbers of the particular problem stated in *JZSS* 8-1.[13]

- Reasoning by comparison. Confronted with a situation with which the student is unfamiliar, the commentator transposes the discussion to a more familiar framework which, however, is semantically different from that in question. For example, in the commentary on *JZSS* problem 1-21,[14] which concerns the multiplication of fractions (to calculate the area of a field), the explanation consists of an evocation of another problem concerning men buying horses. As H. Maspero notes "this procedure is well-known in the school of Mo-Tseu and is frequently found in the anecdotes and speeches attributed to the master."[15]

---

[9] *JZSS* 2-5 (QB, I, p. 116).

[10] Mueller (1), *1974*, pp. 37–38. See also Heath, op. cit., p. 129.

[11] Henricks (1), *1983*, p. 156.

[12] QB, I, p. 221.

[13] QB, I, p. 221. On this point, see Gillings (1), 1972, p. 233. "A non-symbolic argument or proof can be quite rigorous when given for a particular value of the variable; the conditions for rigor are that the particular value of the variable should be typical, and that a further generalisation to any value should be immediate."

[14] QB, I, p. 100.

[15] Maspero (1), *1928*, p. 19.

- Use of analogy. For example, the technique for computation of the cube root (and, more generally, other higher roots) follows closely the pattern for the square root.[16]

- Use of empirical methods. To show that the side of a regular hexagon inscribed in a circle has the same length as the radius, six small equilateral triangles are assembled and the result is determined *de visu*.[17] One proof technique for determination of the volume of the sphere involves weighing it.[18]

- Recourse to heuristic methods. The commentator advises his reader to proceed by dissection (in geometry) or to try to reduce the problem to the rule of three,[19] since "everything reduces to that."[20]

- Recourse to non-linguistic means of communication.[21] This is necessary because, according to the adage of the *Yijing* cited by the commentator, "not all thoughts can be adequately expressed in words" *yan bu jin yi*.[22] In place of a discourse, the reader is asked to put together jigsaw pieces, to look at a figure or to undertake calculations which themselves constitute the sole justification of the matter in hand. In each of these cases, the language is purely auxiliary to such procedures. For example, the explanation of Pythagoras' theorem may only suggest how to set about it and since the commentator's excessively laconic text is clearly, on its own, not sufficient to reconstitute the details of the process, it follows that it is not only what the student will have read or heard that is important but the manipulation which he will have seen the master undertake. The fact that these two- or three-dimensional figures of Chinese geometry often refer to actual concrete objects reinforces this interpretation. If one has to speak of "proofs," it might be said that, from this point of view, the whole of the mathematician's art consists of making visible those mathematical phenomena which are apparent not in Platonic essences but in tangible things (in the commentary on the *JZSS*, numbers themselves refer to things).[23] All this relates to the historical problem of the idea of the origin of the proof. The Greek technical term meaning "to prove" is the verb δείκνυμι. Euclid uses this at the end of each of his proofs. Originally, this verb had the precise meaning of "to point out," "to show" or "to

---

[16]QB, I, p. 153.

[17]QB, I, p. 106.

[18]See below, the section on the volume of the sphere, p. 286.

[19]This rule is called *jin you* (we now have) i.e. "suppose we have". See *JZSS*, j. 2.

[20]Cf. QB, I, p. 114.

[21]Cf. Mounin (1), *1970*, p. 25.

[22]Cf. Liu Hui's commentary on *JZSS* problem 4-22 (QB, I, p. 154). The exact quotation from the *Yijing* is the following: *shu bu jin yan, yan bu jin yi* "The written characters are not the full exponent of speech, and speech is not the full expression of ideas" ("Xici," I, p. 302, in Z. D. Sung (1), *1935*, p. 302).

[23]In other words, counting-rods.

make visible." Thus, it appears that the Chinese proofs of Liu Hui and Li Chunfeng were similar in nature to the first known historical proofs, an example of which was given by Plato (well-known dialogue in which Socrates asks a slave how to double the area of a square);[24] moreover, visual elements remained an essential component of proofs in China for a long time, while in Greece these were abandoned at an early stage although figurative references were retained. In the words of A. Szabó:

So it is true that Euclid routinely observes the convention of illustrating numbers by line segments, but his use of the verb $\delta\varepsilon\ddot{\imath}\xi\alpha\iota$ (to show, to point out) is merely figurative. He talks of pointing things out, but his arguments are essentially abstracts, their steps cannot be seen.[25]

---

[24]Heath (1), *1921*, I, p. 297 ff.
[25]Szabó (1), *1974*, p. 288.

8. Chinese Mathematicians

A problem of definition arises: how do we know that a particular person was a mathematician?

To answer this question, we shall use the biographical notices of the ancient and medieval Chinese mathematicians, assembled by Li Yan,[1] since, at the time of writing, no other work on this subject is as complete.

These notices essentially consist of scattered quotations from ancient documents (official biographies, annals, prefaces of books, etc.). The immediate striking feature of these texts is the thoroughly administrative precision and abundance of mentions of titles and official functions. However, as far as the mathematics itself is concerned, vagueness dominates, so that it is very difficult to distinguish between specialists on logical reasoning, accountants, cashiers, financiers, intendants, scribes who had mastered computation as well as writing (the texts often associate these two areas), diviners, numerologists, and astronomical technicians who were specialists in computation. In fact, the original sources do not have a special term for mathematicians, but frequently use circumlocution such as "he was good at computation" (*shan wei suan*),[2] or "he knew all about ingenious mechanisms, and was particularly strong on *yinyang*, astrological astronomy, the calendar and computation" (*shan qiao ji, you zhili yinyang, tianwen, lisuan*).[3]

From all the evidence, the class of mathematicians defined in this way, on the basis of vague allusions to computation, is much too large. Thus, one might contemplate a more appropriate criterion; for example, anyone who wrote mathematical works might be deemed to be a mathematician. But, even with this definition, would it not be necessary to give the same importance to everything and thus confuse formularies with theoretical works or with elementary arithmetics? However, unless external criteria, which are rather arbitrary from a historical point of view, are used the raw texts do not necessarily reveal their nature on their own. The context must therefore be taken into account. In particular, it would be useful to enquire as to the social groups to which those presumed to be mathematicians belonged. Unfortunately, except in rare cases, very little information is available, so that often the names of the authors of ancient mathematical works are only illusory labels as far

---

[1] Li Yan, *Dagang*, I, p. 30 ff.; Li Yan, *Gudai*, pp. 44, 62, 61.
[2] Li Yan, *Dagang*, p. 31.
[3] Ibid., p. 34.

as we are concerned, with which it is difficult to associate anything, even the period. Where should we situate Zhao Shuang, commentator on the *Zhoubi suanjing*? In the first century AD? In the third? Not to mention Sunzi, Zhang Qiujian and Xiahou Yang, authors of manuals incorporated in the collection of the Ten Computational Canons (*SJSS*). Where should we situate Xie Chawei, whom the *CRZ4B* places in the Tang,[4] but who, according to the *Songshi*, belonged to the Song?[5] What of Liu Hui, unanimously considered to be the most eminent mathematician of ancient China, about whom we know nothing except that he probably lived in the Kingdom of Wei towards the end of the third century? Certainly, one might think that this has something to do with the fact that all these people belong to antiquity rather than that they are Chinese. For that period, the rarity of information is a uniformly verified rule rather than an exception. However, the same difficulty arises for less ancient periods. For example, in the almost exhaustive bio-bibliographical notice about the mathematicians of the Qing dynasty, compiled by Li Yan,[6] some 75 per cent of the people remain obscure, out of a total of 586 names. All this confirms Wang Ling's statement:

Of all the outstanding mathematicians [...], a few have their independent biography in the twenty four official dynastic histories, with the single exception of Shen Kua [Shen Gua]. The *I-Shu Chuan* [*Yishu zhuan*] chapters giving some short account of mathematicians are invariably placed at the very end of the official histories, perhaps a little before the chapter on barbarians, it is not a coincidence that of the ten officially recognised mathematical classics five are anonymous, two of the five authors are unrecorded in the literature and, of the remaining three, there are only incidental references in chapters on the calendar and occasional miscellaneous records.[7]

Moreover, ancient Chinese sources such as the *Yanshi jiaxun* (Family Instructions for the Yan Clan (sixth century)) abound in the same sense:

Mathematics is an important subject in the six arts.[8] Through the ages all scholars who have participated in discussions on astronomy and calendars have to master it. However, it is a minor occupation, not a major one.[9]

Given that this text explains that calendarists and astronomers had to know mathematics, one might wonder about the relationships which existed between mathematicians and astronomers.

Information on this subject is often based on a famous work by Ruan Yuan (1764–1849), the *Chouren zhuan*, since this contains considerable documentation on the subject.

---

[4]*CRZ4B*, j. 4, p. 42.

[5]Quoted from Li Yan (42′), *1937*, p. 171.

[6]Li Yan, *ZSSLC-T*, IV-2, pp. 419–638.

[7]Wang Ling, *Thesis*, p. 36.

[8]Ritual, music, archery, the driving of carts, calligraphy and mathematics.

[9]Teng Ssu-Yü (1), *1968*, p. 205 (cited by Libbrecht (2), *1973*, p. 4).

If one is to believe Ruan Yuan, one might think that there was effectively a privileged relationship between astronomy and mathematics and even that it was difficult to separate these two areas.[10] However, before the 18th century, the term *chouren* rarely qualifies mathematicians, even if they were also astronomers.[11]

Although in practice Ruan Yuan uses the term *chouren* in the sense of calendarist-mathematician, it is clear that, for him, it is an extremely vague term which equally well denotes engineers and geographers, economists, administrators responsible for military matters, and even ritualists, musicians and poets.[12]

Consequently, the idea that it might be possible and desirable to begin with the notion of *chouren* in order to obtain information about Chinese mathematicians does not appear utterly firm, even though it is incontestable that most mathematicians were also calendarist-astronomers.

In fact, in ancient China, specialists in any area had to be polyvalent. Thus, rather than viewing Chinese mathematicians as specialists devoting the essential part of their activity to mathematics alone, we should instead undoubtedly view them as polymaths or amateurs.

In any case, analysis of a previously mentioned work by Li Yan[13] provides some evidence in support of this. If we consider Li Yan's candidate mathematicians, about whom we have biographical information (25 per cent of cases, or slightly more than 140 individuals), we see that there is not a single one who could not be viewed as an amateur mathematician; it is possible to check that all were declared to be mathematicians by their own superior. For example, Gu Guanguang (1799–1862), an acknowledged admirer of Euclid's *Elements* was a doctor;[14] Liu E (1857–1909), known as the author of a work on spherical trigonometry, was a novelist;[15] Wu Jiashan (ca. 1860) was an ambassador;[16] Xu Jianyin[17] (1845–1901) and Hua Hengfang[18] (1833–1902) were translators. There are many more examples, covering all periods of Chinese history. However, we note that, over the centuries, the percentage of amateur mathematicians amongst the cultivated elite increased steadily. From the Tang to the Qing, the proportion of *jinshi* amongst the *chouren* increased regularly, passing from 5 per cent under the Tang to approximately 19 per cent under the Qing.[19]

---

[10]See the statistical study by Wang Ping (3′) *1976*, based on the *Chouren zhuan.*

[11]The term *chouren* is not found in any of the Ten Computational Canons. As far as mathematical works prior to the 16th century are concerned, it apparently only occurs in Qin Jiushao's preface to the *Shushu jiuzhang* (Cf. Libbrecht, op. cit., p. 58).

[12]End of the foreword to the *Chouren zhuan* entitled *Chouren jie* (explanation of the term *chouren*).

[13]See note 6, above.

[14]*CRZ3B*, j. 5, p. 799.

[15]Li Yan, *ZSSLC-T*, IV-2, p. 590. See also J. Reclus's foreword in the French translation of *Lao Can youji* (The Odyssey of Lao Ts'an), Paris: Gallimard, 1964.

[16]Fairbank (1), *1954*, p. 93.

[17]Quoted in Bennett (1), *1967*, p. 84; see also Li Di, *Hist.*, pp. 365–366.

[18]QB, *Hist.*, pp. 335–337; Li Di, *Hist.*, p. 358 ff.

[19]Wang Ping, op. cit.

None of the above is radically different from the prevailing situation in Europe up to the 15th century. In a certain sense, it can even be said that during the 16th and 17th centuries, European mathematics was mainly practised by amateurs. The mathematicians of that period were not necessarily academic professionals but formed a relatively homogeneous group. Viète (1540–1603) was a barrister until 1564, Cardan (1501–1576) was a doctor and Fermat (1601–1665) was 'Conseiller' to the parliament of Toulouse.

According to R. Hall

One can never predict the social circumstances or personal history of a seventeenth-century scientist. Given the taste, the ability, and freedom from the immediate necessities of the struggle for subsistence, any man who could read and write might become such. Latin was no longer essential [...] nor wide knowledge of books, nor a professional chair. Publication in journals, even membership in scientific societies, was open to all; no man's work needed the stamp of academic approval.[20]

These brief remarks show that one should proceed prudently when attempting to evaluate the consequences of the social structure on the development of mathematics. Thus, in order to understand why Chinese mathematics scarcely progressed over the last three centuries, while Western mathematics witnessed exponential growth over the same period, factors other than those linked with social status must be brought to the fore or, at least, these factors should not be considered in isolation.

---

[20]Hall (1), *1962*.

9. The Transmission of Knowledge

抑人之傳不傳與夫書之存不存殆有數焉〔…〕
此豈非斯人之不辛也歟

yi ren zhi zhuan bu zhuan yu shu zhi cun bu cun dai youshu yan [...]
ci qi fei si ren zhi bu xin yeyu.

Cases of biographies or of works not handed down to posterity might well be numerous [...]. Aren't these cases of misfortune?   *CRZ*, j. 52, p. 711.

We only possess extremely fragmentary data about the most ancient period, from the origins to the beginning of the Christian era. Apart from a number of legends, we can however pick out here and there a number of laconic phrases about the teaching of computation, such as: "at six, children were taught to count and to name the cardinal points."[1]

Given that in China mathematics was generally considered to be a subject of minor importance, marginal in comparison with the preoccupations of the literati, it is not surprising that it is difficult to assemble information about the way in which it was transmitted. However, we are able to advance a number of remarks:

(a) It appears that there existed "independent mathematicians,"[2] who practised their activity outside any institutional framework. According to Libbrecht, this situation only occurred from the Song dynasty.[3] These include the unknown hermit of whom Qin Jiushao was the student (around 1220),[4] Zhu Shijie who peddled his knowledge throughout China (around 1300),[5] the anonymous masters of the abacus from whom Cheng Dawei (author of a famous arithmetic, the *Suanfa tongzong*) gained his knowledge (around 1580),[6] and Mei Wending (1633–1721) a member of a famous

---

[1]*Li Ji*, "Nei ze," (cited from Couvreur, *Mémoires sur les bienséances et les cérémonies*, Paris, *1950*, vol. 1, 2nd part, p. 673). See also Li Yan, *Gudai*, p. 21.

[2]The expression is due to Libbrecht (2), *1973*, p. 5: "We have reason to believe that the 'independent mathematician' appears for the first time in the Sung, judging from what we can deduce from biographical data."

[3]Ibid.

[4]Libbrecht, op. cit., p. 62.

[5]See below, p. 152.

[6]See below, p. 159.

family who preferred to remain independent rather than join the Manchu administration[7] and whose fame spread far beyond the boundaries of his province (cf. the case of one Liu Xiangkui who "sold all his possessions and travelled a thousand *li* to study with Mei Wending").[8]

(b) The mathematician's art could be handed down from father to son. One possible etymology of the term *chouren* which applies to such persons (but also to other specialists), namely "specialists transmitting their knowledge from father to son,"[9] implies this. The best known case is that of the family of the calendrical mechanician Zu Chongzhi, inventor of the approximation of π by 355/113, whose son Zu Xuan is famous for his calculation of the volume of the sphere[10] and whose grandson Zu Hao was also a calendarist.[11] Other examples of such families are known, above all for the Manchu dynasty (1644–1911); that of Mei Wending is particularly typical. Mei Wenmi, Mei Wending's youngest brother is known as the author of a star catalogue;[12] Mei Wennai, another brother, was well versed in the same subject as well as in Chinese traditional computational astronomy;[13] his son Yiyan, who died young, helped him in his scientific work;[14] his grandson Juecheng[15] became famous as the compiler of the vast mathematical encyclopedia commissioned by the Emperor Kangxi, the *Shuli jingyun* (1723);[16] the latter's sons, Fen and Fang also practised mathematics.[17] But, as J. Porter remarks, while family associations were important in a few cases "the great majority of associations recorded were not based on family relationship (157 associations)."[18]

(c) The Academies (*shuyuan*), which were originally private and provided centres for free discussion,[19] also played a role. Hu Yuan[20] (993–1059) who was the master of the philosopher Cheng Yi[21] taught mathematics (*suan*) in his Academy. Qian Daxin (1728–1804)[22] taught mathematics when he was the director of the Ziyang Academy at Suzhou.[23] Li Zhaoluo (1769–

---

[7]Yabuuchi (8′), *1978*, p. 243 (chapter on Mei Wending); Liu Dun, (2′), *1986*.

[8]*CRZ*, j. 40, p. 506; on Liu Xiangkui see also Hummel, p. 867.

[9]Cf. *CRZ*, p. 1 ff.; *Chouren jie* (explanation of the term *chouren*).

[10]Cf. Li Yan, *Dagang*, I, p. 61.

[11]*CRZ4B*, j. 3, p. 31.

[12]*CRZ*, j. 39, p. 491; *SCC*, III, p. 185; Hummel, p. 570.

[13]*CRZ*, j. 39, p. 490.

[14]*CRZ*, j. 39, p. 485.

[15]Idem.

[16]Li Yan, *Dagang*, II, p. 401 ff.

[17]*CRZ*, j. 39, p. 490.

[18]Porter (2), *1982*, p. 539.

[19]Yan Dunjie (24′), *1965*.

[20]*DKW*, 9-29400: 33; *Songshi*, j. 432; see also the preface to the *Jiuzhang suanfa bilei daquan* (1450) by Wu Jing.

[21]Chan Wing-tsit (1), *1963*, pp. 544–546.

[22]*CRZ*, j. 49, p. 639.

[23]Elman (1), *1984*, p. 122.

1841), a famous scholar, known as a geographer[24] became interested in mathematics during his studies at the Longcheng Academy at Changzhou (Jiangsu), under the tutelage of Lu Wenchao.[25]

(d) Sometimes the only way to study mathematics was to teach oneself. When pleading at the Bureau of Appointments,[26] Gu Yingxiang[27] (1483–1565) took advantage of his plentiful spare time to learn mathematics, however he was only able to do this by immersing himself in books on his own.[28] Three centuries later, Ding Quzhong[29] complained that he was unable to find a master to teach him mathematics and that he was reduced to learning this science from old books. Such cases may be found in all periods of Chinese history.[30]

(e) There is scarcely any need to stress the fact that high-level mathematics (for the period) was officially taught from the Sui dynasty (589–618) and then under the Tang dynasty (618-907). Most histories of Chinese mathematics actually mention only that, as far as the teaching of mathematics in China is concerned. Let us briefly remind ourselves of the facts. Students of mathematics were recruited from amongst ordinary people and commoners at the bottom of the hierarchy.[31] They were divided into two classes of fifteen students each and the teaching was supplied by doctors of mathematics (*suanxue boshi*) supported by assistants. The studies, which lasted seven years, revolved around a set of texts specially collated for this purpose and later known, somewhat improperly, as the Ten Computational Canons (*SJSS*).[32] The examinations took place once per year and involved both oral and written tests. The candidates had to recite the text of rules by heart and show that they were capable of understanding the mathematical principles involved (*xiang ming shu li*).[33] With the division of the Tang Empire into five kingdoms, which became independent at the end of the ninth century, the official educational institutions for mathematics which had held their own as well as could be expected[34] disappeared completely for two centuries. A hundred years

---

[24]Ibid., p. 448.

[25]Cf. Elman, op. cit., p. 121. On Lu Wenchao, see Hummel, p. 549.

[26]*nancao*. Cf. Hucker (1), *1985* article no. 4127.

[27]*CRZ*, j. 30, p. 359.

[28]See the preface composed by this author for his *Hushi suanshu* (1552).

[29]Ding Quzhong is principally known as the editor of the collection of ancient and modern mathematical works entitled *Baifutang suanxue congshu* published in 1874. See *CRZ3B*, j. 6, p. 819.

[30]A systematic perusal of the prefaces of mathematics books would undoubtedly be very useful in this respect.

[31]Des Rotours (1), *1932*, p. 154.

[32]Ibid.

[33]Ibid. "The candidates had to write essays *tiao* about the meaning of the material studied in answer to questions; they had to find numerical solutions and have a detailed knowledge of the rules of the problems."

[34]Li Yan, *ZSSLC-T*, IV-1, p. 255 ff.

after the reunification of the empire by the Song, the official examinations in mathematics were re-established, albeit intermittently (in 1104, 1106, 1109, 1113, respectively).[35] This renewal of interest is undoubtedly not unconnected with Wang Anshi's reforms of the imperial examinations.[36]

On this subject, it seems that there are two particular mistakes which one should avoid:

(i) to take everything in the Chinese historical annals as an incontestable fact;

(ii) to accord too great an importance to a phenomenon which was in fact marginal.

According to des Rotours, the sources "do not give an exact description of the administrative organisation which existed under the dynasty, but do give us a picture of what the institutions of the dynasty should have been like, according to the reveries of the Confucian literati."[37] Even though this criticism was excessive, we should nevertheless take into account the fact that, even under the Tang, the number of students of mathematics, which was already small to start with (thirty students of this discipline, compared with a total of a thousand for all others in 656), decreased steadily.[38] After 1113, although the system of literary examinations was deeply and firmly entrenched in Chinese society, the teaching of mathematics disappeared; mathematics did not then figure on the state examination programmes until 1887:

Under continuous pressure, the throne finally conceded and decreed in 1887 that mathematics questions be included in the state examination. The triennial examination at Peking in 1888 saw the first instance of this new policy. Of the sixty candidates present, thirty-two were allowed to take the test and one emerged successful.[39]

---

[35]Li Yan, ibid., pp. 274–278.

[36]Cf. Libbrecht (2), *1973*, op. cit., p. 15, note 7, which cites the work of H. Franke and Kuo Ping-wên on this problem. See also *SCC*, vol. 3, p. 40, which explains that Cai Jing, successor to Wang Anshi, "is specifically stated to have encouraged the study of mathematics for the imperial examinations" (in 1108).

[37]Des Rotours, (2), *1975*, p. 199.

[38]QB, *Hist.*, p. 99. Six years after the foundation of the College of Mathematics, *suanxue guan*, the number of students had fallen sharply (from thirty to ten, between the years 656 and 662, resp.). In 807 there was a total of 12 students of calligraphy *and* mathematics (*shu-suan*) – see Li Yan (41'), *1978*, p. 46. According to Yushkevich (1), 1964, introductory chapter, "At this period China had a large staff of graduate mathematicians, the number of which under the Emperor Taizong (627–649) was estimated to be 3260." In fact, 3260 becomes 300 or even 200 depending on the original sources of the Tang to which one refers (see Li Yan, op. cit., p. 46).

[39]Swetz (3), *1974*, p. 46.

We also note that, very exceptionally, the provincial examinations included computational tests.[40] But the civil service examinations played an unexpected role in the diffusion of mathematics, since the examination centres became privileged places where the candidates could easily establish direct or indirect contact with those interested in mathematics, often through the intermediary of great scholars.[41]

At the level of elementary education, E. Rawski notes that the regulations of Chinese schools passed over the teaching of arithmetic in silence. Thus, it seems that, in general, this subject was not taught.[42] The same author notes that in the 19th century, in a school associated with a Jiangsu clan, arithmetic was taught to the students who were not outstanding enough to prepare for the imperial examinations.[43] In fact, arithmetic was reserved for merchants, artisans and tax collectors who often seem to have acquired their know-how on the job rather than at school.[44]

(f) Under the Manchu dynasty (1644–1911), in liaison with the needs of astronomy, cartography and other applied sciences, the Emperor Kangxi (who reigned from 1662 to 1722) founded (in 1713) a College of Mathematics *suanxue guan* for the youths of the families of eight banners (troops of the Manchu army). Having acquired the rudiments of mathematics from special lessons which he had asked the French Jesuits to give him,[45] he indicated the example to follow by personally becoming a teacher (his students included Chen Houyao,[46] Mei Juecheng[47] and Minggatu).[48] He also called upon the services of his family. Initially, the teaching staff included princes of the royal family together with high dignitaries.[49] In 1723, there were 16 teachers (*jiaoxi*) for eight classes of 30 pupils. In 1818, there were only two teachers of the Han ethnic group for 12 Manchu pupils, 6 Mongols, and 6 Han. These students studied two- and three-dimensional geometry for three years then astronomy for two years.[50]

---

[40]Li Yan, op. cit., p. 284 cites an example from the 14th century. See also Mo Ruo's preface to the *Siyuan yujian* (1303), which states that "mathematics degree courses have recently been reintroduced."

[41]Extensive details on this subject may be found in the notices of the *Chouren zhuan*.

[42]Rawski (1), *1979*, p. 52.

[43]Ibid., p. 52.

[44]Ibid., p. 125.

[45]Peng Rita Hsiao-fu (1), 1975.

[46]*CRZ*, j. 41, p. 509.

[47]Ibid. j. 39, p. 485.

[48]Ibid. j. 48, p. 623.

[49]Li Yan (42′) *1978*, p. 224.

[50]Ibid.

(g) In the second half of the 19th century, after the Opium War, a reform movement advocating the modernisation of China using foreign countries as a model led to the creation of scientific and technical schools. The most famous of these, the Tongwen guan (College of Combined Learning) opened in Peking in 1863 after the acceptance by the Chinese government of a memorial supporting the idea that the learning of foreign languages was the key to Western knowledge.[51] In December 1866, the *zongli yamen* (Foreign Office) submitted another memorial to the Throne. The memorandum declared that "the manufacture of machinery, armament and the movement of ships and armies were based upon astronomy and mathematics and that China must master these subjects if it was to be strong."[52] Opponents contended that it was disgraceful for China to learn from foreigners[53] but the teaching of mathematics was inaugurated the next year after 30 applicants (out of 72) had passed an entrance examination in June 1867.[54] Initially, Baron Johannes von Gumpach was appointed professor of astronomy and mathematics but he considered the level of teaching expected of him beneath his dignity and refused to teach during the two years he lived in Peking.[55] Finally, in 1869, Li Shanlan, the celebrated co-translator of Euclid's *Elements* was appointed as a professor there, although he was entering his 58th year.[56] In 1869, the college had recourse to the services of 13 teachers, four Chinese and three Europeans.[57] There was a total of 120 students who could choose between an eight-year degree course (including foreign languages and foreign sciences) and a five-year degree course principally oriented towards mathematics and its applications. In the latter case, the programme combined traditional Chinese and European mathematics, as follows[58]:

- **First year**. Basic principles of mathematics, the Nine Chapters, algebra.

- **Second year**. Explanation of the "four origins" (*siyuan*, the Chinese algebra of the Song and the Yuan), Euclid's *Elements*, plane trigonometry.

---

[51] Biggerstaff (1), *1961*, p. 103; Swetz (2), *1974*, p. 17.

[52] Biggerstaff, ibid., p. 109.

[53] Ibid., p. 110.

[54] Cf. Su Jing (1′), *1978*; Swetz (2), *1974*, pp. 39–45; Biggerstaff (1), ibid., p. 121.

[55] Ibid., p. 120, note 50.

[56] After Li Shanlan left, one of his pupils, Xi Gan (1845–1917) was appointed to the chair of mathematics. Cf. Su Jing, op. cit., p. 64. See also Li Yan, *ZSSLC-T*, IV-2, p. 498.

[57] Swetz, op. cit., p. 43. No foreign teachers taught mathematics. Except for language teachers, who were all foreign, few Chinese taught the sciences, but, unlike the case of mathematics, no Chinese taught astronomy. The director of the Tongwen guan was the American Protestant missionary William Alexander Parsons Martin (1827–1916). Cf. Duus (1), *1966*, pp. 11–14.

[58] Su Jing, op. cit., p. 44.

- **Third year**. Elements of the sciences, chemistry, engineering.

- **Fourth year**. Differential and integral calculus, navigation, astronomy.

- **Fifth year**. International law, astronomy, mineralogy.

It is possible to obtain a somewhat more precise idea of the contents of this teaching by analysing the contents of the mathematics examination papers of the Tongwen guan. A collection of the best mathematics papers over the ten years from the creation of the Tongwen guan was in fact published in 1880 by teachers at this establishment, Xi Gan and Gui Rong (the text was reviewed by Li Shanlan). This anthology, which includes approximately 200 papers, was entitled *Suanxue keyi* (Fig 9.1). Slightly more than a quarter of this book relates to Western applied mathematics (astronomy (Fig. 9.2), navigation, engineering, ballistics), the themes and the terminology of the rest of the problems are essentially taken from ancient and medieval Chinese mathematics, although some problems relating to tangent circles may have been taken from traditional Japanese mathematics (the *wasan*). The solutions of the problems are fundamentally algorithmic and algebraic and derive their symbolism from innovations brought into China during the second half of the 19th century (Fig 9.3). We also note that one of the problems (j. 2, p. 43) is solved "in the ancient way" using the notation of numbers on rods, in the style of Zhu Shijie; thus, the students may have had the choice between the options of medieval and modern algebra.

This type of education attracted very few students: in 1879 for the first time there were 10 students enrolled, there were again 10 in 1887, four in 1893 and the section disappeared in 1902.[59]

Indeed, from the beginning, the recruitment of promising students was a serious problem. Many of them were more attracted by unusually high stipends and an easy access to official positions (for those who were unable to succeed via the system of literary examinations) than by scientific studies (many Tongwen guan graduates were sent abroad as officers in Chinese legations in Europe and the United States).

In 1898, the Tongwen guan dropped its five-year curriculum[60] and in 1902, it was transferred to the department of translation (*fanyi ke*) of the Peking University created in 1898 as a result of the Hundred Days of Reform.[61]

---

[59]Ibid., p. 45.
[60]Biggerstaff, ibid., p. 138.
[61]Ibid., pp. 137 and 139.

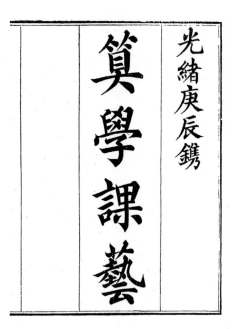

Fig. 9.1. Title page of the *Suanxue keyi* (1880).

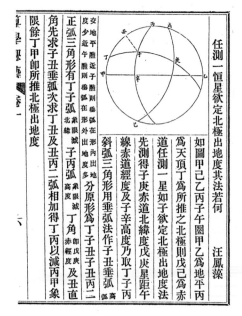

Fig. 9.2. A problem on the determination of the latitude of a place from the *Suanxue keyi* (j. 1, p. 6a).

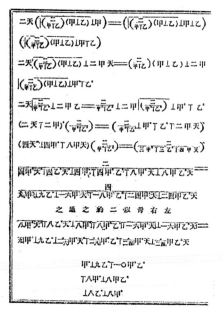

Fig. 9.3. A page of algebra from the *Suanxue keyi* (j. 3, p. 29b).

After the creation of the Tongwen guan, a similar establishment opened in Shanghai, on the initiative of Li Hongzhang, the famous statesman and viceroy of the province of Zhili.[62] Schools created by Protestant missionaries devoted part of their activity to the teaching of mathematics in the provinces[63] and, in liaison with the arsenals, work to translate science manuals of all types intensified.[64]

(h) It is also appropriate to mention another important aspect of the question associated with the dissemination of the written culture. While the printed book played an important role, the circulation of the information also depended, until a very late stage, on the copying of mathematical texts. This is well illustrated by the way in which the first Chinese research work on infinite series was disseminated. The oldest Chinese text dealing with these problems was not published until 1839, almost 80 years after the death of its author, but the manuscript was copied several times and gave rise, in the interim, to a number of original publications which founded a tradition of research on the subject.[65]

---

[62]Hummel, p. 464 ff.
[63]Latourette (1), *1929*.
[64]Bennett (1), *1967*.
[65]See below, p. 356.

10. Influences and Transmission

In works on the Chinese sciences, no question has been touched on more often than that of the circulation of ideas. However, we still know very little about the subject. It is certainly easier to identify the dissemination of material inventions than that of ideas, since in the medieval period, authors seldom mention their sources.

In view of the difficulty of providing direct proofs of transmission in one direction or another, the question was first approached indirectly by establishing similarities between apparently analogous ideas, techniques or mathematical problems found both in China and elsewhere. This led to the establishment of lists of ideas which were liable to have been transmitted. That raised by J. Needham (1959) is still the most representative list;[1] many others have been inspired by it.[2]

In *Science and Civilisation in China*, J. Needham writes: "when [. . .] we ask what mathematical ideas seem to have radiated from China to the southwards and westwards, we find a quite considerable list"[3] and he justifies this idea by the following arguments:

1. The decimal notation (14th century BC in China) and the zero as a blank space (fourth century BC in China) are only found in India and in Europe after a lapse in time of between one and two millennia.

2. The procedures for extracting square and cube roots (first century BC in China) are similar to those of Brahmagupta (ca. 630 AD). The method for extracting roots of high degree (Chinese version of Horner's method; Horner lived in the 19th century) which has its genesis in the aforementioned Chinese techniques and which was developed by Jia Xian in the 11th century seems to have influenced al-Kāshī (15th century) and traces of these advanced methods are shortly afterwards found in Europe.

3. The rule of three first appears in the Chinese Nine Chapters before it is found in Indian texts; the terminology is the same in both cases.

4. The Indian vertical notation for fractions was inspired by the practice described in the Nine Chapters (*JZSS*).

5. Negative numbers (first century BC in China) appear much later in India.

---

[1] *SCC*, III, pp. 146–148.
[2] In particular: Yushkevich (1), *1964*; Liang Zongju (1'), *1981*.
[3] *SCC*, III, p. 146.

6. The visual proof of Pythagoras' theorem (commentary on the *Zhoubi suanjing*, third century AD) appears for the first time outside China in the work of Bhāskara, in the 12th century, in India.

7. The problems of practical geometry of the Nine Chapters (such as that of the broken bamboo *JZSS* 9-13, or that of the two walkers walking around the sides of a town in the shape of a right-angled triangle *JZSS* 9-14) are reproduced by Mahāvīra (ninth century).

8. The same Mahāvīra uses an erroneous ancient Chinese formula to calculate the area of a segment of a circle (*JZSS* 1-35).

9. The fundamental identicalness of algebraic and geometric relationships in Chinese mathematics is found later in the works of al-Khwārizmī.

10. The double-false-position rule, called the "regola elchataym" ("à la Chinese,"! since the term "elchataym" is said to come from "Kitan" – Cathay), which has its origins in the Nine Chapters, is said to have reached Europe via the Islamic world.

11. The problem of residues (Chinese remainder theorem) first occurs in the works of Sunzi (China, fourth century) and appears later in India in the works of Brahmagupta and then, later, in Europe.

12. The same applies to the problem of indeterminate analysis known as the problem of the hundred fowls (*ZQJSJ*, 5th century AD).

13. The numerical solution of equations of degree three, as practised by Wang Xiaotong (seventh century AD) is said to have been transmitted to Leonardo of Pisa (13th century).

14. Pascal's triangle (China, 1100 AD) first spread to India before reaching Europe in the 16th century.

As far as the flow of ideas in the opposite direction is concerned, J. Needham reconstitutes the transmission to China of:

(i) trigonometric tables of Indian origin (Tang dynasty),

(ii) Euclidean geometry (Mongol period),[4]

(iii) Arabic spherical trigonometry (Mongol period),

(iv) the Arabic notation for numbers without position markers,

(v) the European "gelosia method of multiplication" possibly introduced to China by the Portuguese around 1590.[5]

---

[4]It seems unlikely that a Chinese translation of Euclid's *Elements* would have been produced in China in the Mongol period. Anyway, there is not the slightest trace of Euclidean influence in later Chinese mathematical texts until the end of the 16th century. Cf. D'Elia, (2), *1956*.

[5]Cf. Smith (1), *1925*, II, p. 115.

J. Needham concludes that between 250 BC and 1250 AD, far more inventions would have left China for the external world than would have entered it in the opposite direction, not to mention the fact that China was scarcely receptive to anything reaching it from the outside: "India was the more receptive of the two cultures [*sic*]."[6]

This type of argumentation is controversial because of its chronological and methodological imprecision; there is nothing to prove that concomitant developments could not have taken place. Precisely which algebraic and geometric relationships are referred to in point 9? Does Needham mean those which may be used to deduce algebraic formulae from geometric figures? But this approach dates back to Babylonian mathematics,[7] far earlier than the oldest known Chinese mathematics. Which "Pascal triangle" is referred to in point 14? What form did it have? What was it used for? Was it used to solve combinatorial problems, as described by historians of Indian mathematics? Or was it used to extract roots, as in China? Or, again, was it used, as Pascal used it, to calculate probabilities? Furthermore, the idea that China may have known of the zero in the form of a blank space in written texts from remote antiquity is a hypothesis which the known documentation does not confirm.[8] The association of the methods of extracting square and cube roots used by the Chinese of the Han with "Horner's method" does not appear well founded.[9] The rule of three is already found in the Rhind papyrus.[10] We do not know exactly how the Chinese of the period of the Nine Chapters (*JZSS*) denoted fractions. The negative numbers of the Chinese are not the same as those of the Indians (since the *JZSS* is unaware of the rules for multiplying these numbers). As far as the approximation formula of point 8 is concerned, Roman land surveyors are also known to have used it and, moreover, the Heronian corpus refers to it as an "ancient method."[11] But all this dates to approximately the same period as that of the Nine Chapters. "Pythagoras' theorem" is already found in Babylonian mathematics long before it occurs in China.[12] As far as the false position rule is concerned, J. Needham refutes his own argument in a note.[13] In fact, the Arab term "al-ḫaṭa'ayn" (Europeanised to the form "elchataym") means the rule of two errors.[14] This term has no connection whatsoever with the word "Kitan" which originally denoted the Liao Empire (947–1125) and which ultimately was extended to denote the whole of China, "Cathay."

---

[6] *SCC*, op. cit., p. 148.
[7] Cf. Høyrup (4), *1990*.
[8] See below, on Chinese numeration.
[9] See below, on Horner's method.
[10] Cf. Problem no. 69 of the Rhind papyrus (Tropfke (3), *1980*, p. 359).
[11] Bruins (1), *1964*.
[12] Neugebauer (3), *1957*, p. 37.
[13] *SCC*, III, p. 118, note (b).
[14] Cf. Yushkevich (2), *1976*, p. 46.

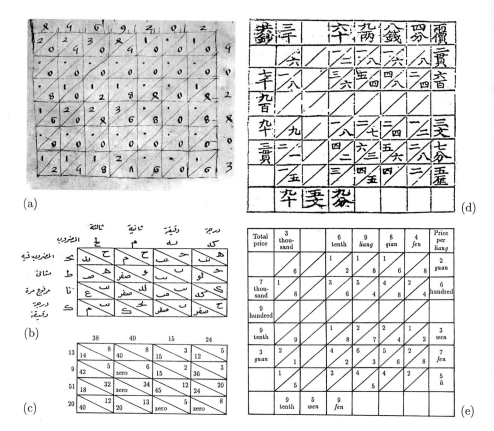

(a)

(b)

(c)

(d)

(e)

Plate 1. Three examples of multiplications performed using the gelosia method: (a) $4\,569\,202 \times 502\,403$ (from a Latin manuscript (ca. 1300 A D) preserved in Oxford. Cf. Murray (1), *1978*, plate 6. (b) $13, 09, 51, 20 \times 38, 40, 15, 24$ (sexagesimal notation). From al-Kāshī's *Miftāḥ al-ḥisāb* (Calculator's key), (1427). Reproduced from the edition of the text by A. D. Damirdash and M. H. al-Hafni, Cairo, *1967*, p. 110. (I am indebted to Ahmed Djebbar for this reference). (c) Translation of (b). See also Berggren (1), *1986*, p. 47. (d) $306\,984 \times 260\,375$ (with units). From Wu Jing's *Jiuzhang suanfa bilei daquan* (1450), introductory part. Cf. Martzloff (2), *1981*, p. 54, note 13. (e) Translation of (d).

Furthermore, the Chinese knew the gelosia rule (Plate 1) for multiplication as soon as 1450. Thus, there is no need to suppose that they learnt it from the Portuguese. As far as the problems referred to in point 7 are concerned, the same have recently been discovered on engraved tablets in cuneiform script.[15]

Moreover, it would not be difficult to show that China was outstripped in various areas which some see as typifying its mathematical originality.[16]

---

[15]Cf. van der Waerden (3), *1983*, pp. 175–181; Sesiano (4), *1987*.

[16]On this, see, for example, Wu Wenjun (1'), *1982*.

Nonlinear interpolation techniques already occur in Babylonian mathematics.[17]
Indivisibles and Cavalieri's principle existed in embryo in ancient Greece five
centuries earlier than in China.[18] The idea that China was at the origin
of part of medieval mathematics, in turn suggests the idea of precursors
from which the former would have drawn its knowledge, which precursors
would themselves have been fatally outstripped. This type of reasoning leads
diffusionist historians[19] to the view that, in order to prove that two ideas
inevitably had a common origin, it is sufficient to show that they are the same.

Usually, the degree of sameness of two ideas is proven not by comparing
these in their historical context but by means of a sort of "reduction of ideas
to the same denominator" which often involves translating everything into
modern (or rather, more-or-less "modern") mathematical notation. Moreover,
the conclusion so obtained is generally that the direction of the diffusion of
ideas is a one-way process: from China to other parts of the world in the case
of J. Needham, but from Indo-European territories to anywhere else in the case
of van der Waerden who follows just this approach to construct, by infinite
regression, his hypothetical pre-Babylonian mathematics, the common source
of Egyptian, Chinese, Indian, Babylonian and Greek mathematics.[20]

However, the fact that the answers to the problem of the diffusion of ideas
are not satisfactory, clearly does not mean that this problem does not arise,
less still that it is unimportant; for, as long as it remains unsolved, it will not
be clear how to judge what is original among Chinese scientific ideas. What
is certain, in any case, is that China was in contact with the outside world
at different periods of its history, both over land and by sea. Material relics
of Roman origin are scattered along the Silk Road. The French archaeologist
Malleret discovered the remains of a port on the coast of Cochin China, where
he unearthed Roman coins corresponding to the reigns of Antoninus Pius and
Marcus Aurelius.[21] Silk from China reached Rome from the first century AD.[22]
Much has already been written about the diffusion of Indian Buddhism to China
during the first millennium AD.[23] The same is true of the history of Chinese
Islam.[24] We must also mention the expansion of the Mongol Empire and, more
recently, the European penetration into China from the end of the 16th century.

Thus, we should not really be surprised that the mathematics are analogous
in many ways, to a close approximation, the problems are similar everywhere.
What can this be attributed to? Is this a result of diffusion phenomena with

---

[17]See below, p. 94 ff.

[18]See below, p. 292 ff.

[19]This term appears to have been used for the first time in connection with the history of
mathematics in Wilder (1), *1952*, p. 263.

[20]Cf. van der Waerden, op. cit. See also the important review of van der Waerden's book
in Knorr (2), *1985*.

[21]Etiemble (1), *1988*, I, p. 52.

[22]Ibid. p. 52.

[23]Cf. Zurcher (1), *1959*.

[24]Leslie (1), *1986*.

adaptation and local assimilation of exogenous borrowed elements? Or is it the result of parallel stages of development?

The latter hypothesis is perhaps correct in certain cases but it would be absurd to believe that the Chinese developed voluntarily in isolation by systematically rejecting everything from abroad. As a rule, the cultures of the most isolated peoples have remained the most primitive known. What happened in the case of China was probably a multi-sided diffusion followed by an adaptation of foreign cultural traits limited by geographical, social, political and linguistic factors.

However, in the present state of our knowledge, not a single indisputable proof of influence of Chinese mathematics on Western mathematics is really convincing (yet in domains other than mathematics proofs of Chinese influences exist).

In what follows, rather than establishing genealogies, we shall endeavour to make a number of comparisons between Chinese and non-Chinese mathematical techniques and to highlight (without claiming to be exhaustive) a number of salient facts about the interactions between China and other cultural areas.

Possible Contacts with the Seleucids

The Chinese seem to have known the Seleucid Empire (312–64 BC).[25] It turns out that certain problems of the *Jiuzhang suanshu* are similar to mathematical problems of this civilisation,[26] both from the point of view of the question asked in the statement and the rule which is used to solve them. A parallel may be drawn, for example, between the problems *BM* 34568 no. 2 and 12 and those of *JZSS* no. 9-3 and 9-8. Here is a translation of the texts:

- **BM 14568 no. 2** [The fl] ank [is 4] and the diagonal 5 what is the front? Since you do not know, 4 times 4: [1]6; 5 times 5: 25. Subtract 16 from 25: remainder 9. What should I multiply by what to make 9? 3 times 3: 9. The front is 3.[27]

- *JZSS* **9-3** Suppose LEG: 4 *chi* [feet], hypotenuse: 5 *chi*; question: BASE? Answer: 3 *chi*.
  "BASE–LEG" rule[28]: [...][29]: LEG automultiplied, by subtracting [the result] from the hypotenuse automultiplied [then] by extracting the square root of the remainder gives BASE.[30]

---

[25]Cf. Leslie and Gardiner (1), *1982*, p. 254. In this article, the authors identify the Seleucid Empire with the Ligan/Tiaozhi, mentioned in the Chinese sources.

[26]Cf. van der Waerden (3), *1983*, p. 58.

[27]Thureau-Dangin, op. cit., p. 58.

[28]BASE = small side; LEG = large side (of the right angle of a right-angled triangle).

[29]This part of the text, which we omit, gives the general statement of Pythagoras' theorem.

[30]QB, I, p. 241.

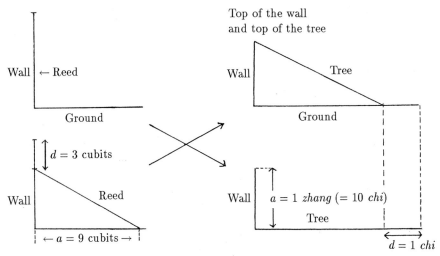

Fig. 10.1. Figures (not in the original texts) for the problems *BM* 34568 no. 12 and *JZSS* 9-8.

- **BM 34568 no. 12** A reed is placed vertically against a wall. If it comes down by 3 cubits, it moves away by 9 cubits. What is the reed, what is the wall? Since you do not know, 3 times 3: 9 $[d^2]$; 9 times 9: $1'21^{31}$ $[a^2]$. [Add] 9 to $1'21$: $(1'30)$ $[d^2 + a^2]$. [Multiply] $1'30$ [by] $30'$: 45 $[(d^2 + a^2)/2]$. The inverse of 3 is $20'$ $[1/d]$. [Multiply] $20'$ [by 45] $[(d^2 + a^2)/2d]$: 15, the reed [...].[32]

- **JZSS 9-8** Suppose a wall is 1 *zhang* [10 *chi*] high. A tree [or a wooden pole] is rested against it so that its end coincides with the top of the wall. If one steps backwards a distance of one *chi* [1 foot] pulling the tree, this tree falls to the ground. How big is the tree?

  Rule: multiply 10 *chi* by itself $[a^2]$, divide by the step back $[\frac{a^2}{d}]$, add the step back to what is obtained and divide the result by 2 $[\frac{\frac{a^2}{d}+d}{2}]$, that is the height of the tree.[33]

The first two problems both depend directly on Pythagoras' theorem, which dates back to the Old Babylonian period (1800 to 1600 BC),[34] while the last two, in fact, require one to solve a right-angled triangle, knowing one side (denoted by *a*) and the difference between the hypotenuse and the other side (denoted by *d*), where the hypotenuse is represented by the reed and the tree, respectively (Fig. 10.1). The differences in the structure of the calculations are self evident,

---

[31]$9 \times 9 = 1 \times 60 + 21$.

[32]Thureau-Dangin, op. cit., p. 60.

[33]QB, p. 244.

[34]Cf. Neugebauer's well-known study of the Plimpton tablet no. 322 (Neugebauer (3), *1957*, p. 37).

but the scenarios of the two statements are closely akin to one another, except that the ground plays the role of the wall in the second problem. Furthermore, the two resolutory formulae are almost identical:

$$\text{reed} = \frac{a^2 + d^2}{2d}; \qquad \text{tree} = \frac{1}{2}\left(\frac{a^2}{d} + d\right) .$$

It is clearly impossible to deduce anything from this. However, many other similar comparisons may be made.[35] Thus, the possibility of influences cannot be eliminated; the Chinese of the Han may only have conserved the framework of the problems and the solutions while adapting the resolutory formulae to their usual computational customs.

One could make other comparisons of the same type, particularly with the mathematical texts of the ancient Babylonian age, above all if one takes into account the recent research of Jens Høyrup, which has underscored the fact that the conceptualisation within the Babylonian texts seems fundamentally of a geometric rather than an arithmetic or algebraic nature.[36] In fact, this type of geometric conceptualisation also runs constantly through Chinese texts, and the *Jiuzhang suanshu* of the Han in particular, but the traditional arithmetic (or algebraic) interpretation of the Babylonian texts has masked the similarity of the two approaches.

Contacts with India

Although it is difficult to establish the precise chronology of the diffusion in China of mathematical knowledge of Indian origin following the introduction of Buddhism from the first century AD onwards, a number of sparse but indisputable facts may be gleaned.

For example, we know that the calendarist He Chengtian[37] (370–447) tried to find out about Indian astronomy by questioning his contemporary, the famous monk, Hui Yan.[38] We also know that the bibliography of the *Suishu* cites two mathematical (or perhaps numerological) works irremediably lost but obviously of Indian origin, the *Poluomen suanfa* (Computational Methods of the Brahmans) and the *Poluomen suanjing* (Computational Canon of the Brahmans),[39] both in three *juan*. In addition, the *Shushu jiyi*, one of the Ten Computational Canons of the Tang, uses Chinese terms transliterated (semantically or phonetically) from the Sanskrit, such as *daqian* and *chana*.

Daqian, the literal meaning of which is "great thousand", is an abbreviation from the Sanskrit *tri-sāhasra-mahā-sāhasraḥ loka-dhātavaḥ* which may be

---

[35]Cf. van der Waerden (2), *1980*.

[36]Cf. Høyrup (4), 1990.

[37]Li Yan, *Dagang*, I, p. 59.

[38]According to Li Yan, *Gudai*, p. 161, which cites an extract from j. 7 of the *Gao seng zhuan* (Biographies of Eminent Monks).

[39]*Suishu*, j. 34, p. 1026.

rendered, approximately, by "a major chiliocosm or universe of 3000 great chiliocosms;"[40] *chana* (Sanskrit: *kṣaṇa*) is a purely phonetic transliteration which denotes the shortest measure of time.

It has also been shown that various Chinese sources of the Tang period contain technical terms which are incontestably of Indian origin. These relate to systems of notation for large numbers, elements of mathematical astronomy, and Indian numbers for the first nine integers and the zero represented by a dot.[41]

Systems of Notation for Large Numbers

Indeed, Chinese adaptations of Indian Buddhist works from the fifth–sixth centuries contain various representations of large numbers. One of these representations uses terms such as *juzhi, ayuduo* and *nayouta* which were all transliterated phonetically from the Sanskrit *koṭī, ayuta* and *niyuta*.[42] (Note here that the phonetic rendering of Chinese characters used here follows modern usage and is therefore irrelevant. Of course, Chinese transcriptions should be based on ancient usage. However, since the correspondences have been well established by specialists in Chinese Buddhism there is no need to go into further details). According to Li Yan:[43]

- 1 *koṭī* = one hundred thousand
- 100 *koṭī* = 1 *ayuta*
- 100 *ayuta* = 1 *niyuta*.

Numbers so expressed can be fairly large but in Chinese classics special terms (such as *yi* and *zhao*) for numbers greater than 10000 are also found episodically. According to ancient Chinese philologists *yi* was sometimes worth 10 myriads (10^5), sometimes a myriad myriad (10^8), whereas *zhao* was equivalent to 10 *yi* or to a myriad myriad *yi*, respectively. But other systems, built upon the invention of other special terms for still greater numbers were perhaps also used in China towards the beginning of our era or before.[44] The oldest of these extends the ordinary system for denoting small numbers using a system of position markers, composed of purely Chinese characters, which play the same role as normal markers in a decimal series associated with ten, one hundred, one thousand and one myriad (ten thousand), respectively. An extant example of such a system is found in the Dunhuang manuscripts.[45]

Similar systems are also mentioned explicitly in three of the Ten Computational Canons (*SJSS*), the *Sunzi suanjing, Wujing suanshu* and *Shushu*

---

[40]Cf. Soothill and Hodous (1), *1976* (reprint), p. 61 and Mochizuki Shinkō, *Bukkyō daijiten*, Kyoto, *1954*, vol. 2, p. 1598.

[41]See later, *the zero* (p. 204).

[42]Li Yan, *Gudai*, pp. 182–183.

[43]Li Yan, ibid.

[44]Li Yan, ibid., p. 181.

[45]Cf. Libbrecht (5), *1982*, p. 228.

jiyi, respectively. The presence of these systems of numbers which are called "lower degree" *xiadeng*, "medium degree" *zhongdeng* and "higher degree" *shangdeng*, respectively in the *WJSS* and *SSJY*, might or might not betray an Indian influence but, at any rate, these seem of fictitious importance from the point of view of mathematics because texts which cite them rarely use them.

The same remark also applies to words of Sanskrit origin. In all likelihood, the fact that Sanskrit words for uselessly large numbers are mentioned at the beginning of some mathematical texts probably indicates that the authors of mathematical manuals aimed not only at mathematical knowledge *per se* but also at the diffusion of an elementary knowledge of the terminology of numbers typical of Buddhism. In fact, the same phenomenon persisted and lists of mathematical terms borrowed from Sanskrit and denoting very large (or very small) numbers are also found at the beginning of Zhu Shijie's *Suanxue qimeng* and Cheng Dawei's *Suanfa tongzong*.

Even if the need to use very large or very small numbers in mathematical contexts was not felt, it is interesting to note that, exceptionally in the notice (*tiao*) no. 304 of chapter 18 of his *Mengqi bitan* (ca. 1086), Shen Gua states and solves an interesting combinatorial problem, the solution of which leads to an immense number. The problem is to calculate and explicitly write down the number of possible configurations of a *weiqi*-board (that is, of a *go*-board) of size 19×19 ($= 361$ points of intersection) in which each point may be unoccupied, occupied by a white pawn, or occupied by a black pawn.[46] The answer is 3^{361}. Since this number is approximately equal to 1.74×10^{172}, it may be expressed, at least approximately, in terms of the "higher degree" system, which permits even higher powers (Table 10.1). However, curiously, Shen Gua does not use this system to transcribe the final answer[47] but uses instead a multiplicative string consisting of a certain number of characters "*wan*" ($= 1$ myriad) merely juxtaposed: *wan wan wan wan wan ... wan* (the precise number of juxtaposed *wan* given in the present editions of the *Mengqi bitan* is erroneous).[48] Yet, this particular technique of notation (or rather, of approximate notation) for very large numbers is not attested in Chinese mathematical texts, although certain Chinese Buddhist sources of Indian origin do use the same technique.[49] Thus, Shen Gua's notation is perhaps of Indian origin. Moreover, his problem evokes another famous recreational problem, namely that relating to the number ($2^{64} - 1$) of grains that can be placed on a chessboard, one grain being placed on the first square, two on the second, four on the third, and so on, doubling the number of grains each time.[50]

---

[46]For a complete translation of Shen Gua's text and a full analysis of this problem cf. Brenier (1), *1994*.

[47]But Shen Gua relies on the "medium-degree" system to transcribe certain intermediate results.

[48]Cf. *SY*, pp. 266–269.

[49]Cf. Soothill and Hodous, op. cit., p. 285, which mentions a number composed of eight characters *wan* concatenated; see also Mineshima (1'), *1984*, p. 9 (examples of strings of three characters *yi* 億億億).

[50]Cf. Smith (1), *1925*, II, p. 549; *SCC*, III, p. 139; Tropfke (3), *1980*, p. 630 ff.

Table 10.1. Nomenclature for large numbers as found in the *Shushu jiyi* (QB, II, p. 540). Translation of the text of the *Shushu jiyi*:

"In the method produced by the Yellow Emperor, numbers have ten degrees [*shi deng*]. In practice [these ten degrees] are used in three ways. The ten degrees are *yi*, *zhao*, *jing*, *gai*, *zi*, *rang*, *gou*, *jian*, *zheng* and *zai*. The "three degrees" [i.e. the three ways according to which the "ten degrees" are manipulated] are the higher, the medium and the lower. According to the lower [degree of] numbers, numbers are transformed progressively by 10; for example, ten myriads are called *yi*, ten *yi* are called *zhao* and ten *zhao* are called *jing*. According to the medium degree, numbers are transformed progressively by a myriad; for example, a myriad myriad is called *yi*, a myriad myriad *yi* is called *zhao* and a myriad myriad *zhao* is called *jing*. According to the higher degree, numbers are modified when available numbers are exhausted; for example [in this system] a myriad myriad is called *yi*, *yi yi* is called *zhao* and *zhao zhao* is called *jing*."

Markers		Lower degree	Medium degree	Higher degree
wan	萬	10^4	10^4	10^4
yi	億	10^5	10^8	10^8
zhao	兆	10^6	10^{16}	10^{16}
jing	京	10^7	10^{24}	10^{32}
gai	該 〔垓，陔〕	10^8	10^{32}	10^{64}
zi	梓 〔秭〕	10^9	10^{40}	10^{128}
rang	讓 〔壤〕	10^{10}	10^{48}	10^{256}
gou	溝	10^{11}	10^{56}	10^{512}
jian	間	10^{12}	10^{64}	10^{1024}
zheng	政	10^{13}	10^{72}	10^{2048}
zai[a]	載	10^{14}	10^{80}	10^{4096}
Modern formulae $n = 1, 2, 3, \ldots$		10^{n+3}	10^4 and 10^{8n}	$(10^4)^{2^{n-1}}$

[a] After *zai*, Libbrecht (op. cit., p. 219) gives *ji* 極 (but this marker only appears in the Dunhuang manuscripts).

Note that in *SCC*, III, p. 87, J. Needham gives a different interpretation of numerical values corresponding to the medium degree. The problem might come from the particular version of the *Shushu jiyi* used by Needham, namely QB (13′), *1932*, p. 76. But in fact, Qian Baocong's text is the same as the one used above and I see no way of interpreting it as Needham does. Note also that the present interpretation coincides with that of Lam Lay-Yong and Ang Tian-Se (4), *1992*, p. 19.

Elements of Mathematical Astronomy

The development of Indian astronomy from the end of the fifth century, based on Greek mathematical astronomy, had repercussions in the China of the Tang. We know that, during this period, Indian astronomers of the schools of Kāśyapa (*Jiaye*), Kumāra (*Jumoluo*) and Gautama (*Qutan*) collaborated with the Chinese Imperial astronomers.[51] One of these Indian astronomers, a certain Qutan Xida (the Chinese transcription corresponds to the Indian name Gautama Siddhārta) adapted various Indian works into Chinese. At the current state of research, it is not generally possible to determine the precise correspondence between the Chinese texts and the Sanskrit texts which served as their source. However, it is known that the *Jiuzhi li* (Navagrāha Calendar)[52] is partly based on the Pañcasiddhāntikā.[53]

This *Jiuzhi li* is not a theoretical treatise; it contains only the procedure to be followed to obtain the value of astronomical parameters such as the longitude of the Sun or the Moon, the apparent diameter of the Moon, or the duration of lunar eclipses, by computation and using tables. It also contains an astronomical table called *yue jianliang ming* (word for word: *ming*[54] of the lunar intervals). In modern terminology, we would say that this table gives[55] $3438 \sin x$ for angles beginning with $3°45'$ and increasing in steps of $3°45'$ (Fig. 10.2, see next page). The origin of this table seems to be an ancient table of chords due to Hipparchus.[56]

Bibliographic Comment. See also R.C. Gupta, "Sino-Indian Interaction and the Great Chinese Buddhist Astronomer–Mathematician I-Hsing [Yi Xing] (683–727 AD)," *Gaṇita Bhāratī*, 1989, vol. 11, nos. 1–4, p. 38–49.

---

[51]The information in this paragraph is essentially taken from Yabuuchi (4), *1979*.

[52]Navagrāha = "the 9 planets" i.e. the Sun, the Moon, two fictitious planets and five other true planets (on the connection between Indian and Hellenistic astrology induced by this notion of "9 planets" see Yano (1), *1992*). The *Jiuzhi li* forms chapter 104 of the *Kaiyuan zhanjing* (Treatise of astrological astronomy of the Kaiyuan reign-period (713–741)), which contains no other elements of Indian origin. This work was classified as secret and does not appear to have been divulged; having been ignored for centuries, it was rediscovered in 1616 in a Buddhist statue. Cf. Bo Shuren's foreword in *Tang Kaiyuan zhanjing*, 1989, Peking: Zhongguo Shudian, p. 15.

[53]Yabuuchi, op. cit., p. 10.

[54]This *ming* corresponds exactly to the Sanskrit "*jīva*." As a technical term it denotes the half-chord (= sine) of an arc.

[55]The constant 3438 is an approximation to the radius (in minutes). Cf. Mazars (1), *1974*: "Indian astronomers usually divided the circumference into 21600 *kala*; this angular unit corresponds to our minutes, and rounding to the nearest minute we have $R = 21600/2\pi \simeq 3438$."

[56]Toomer (1), *1974*.

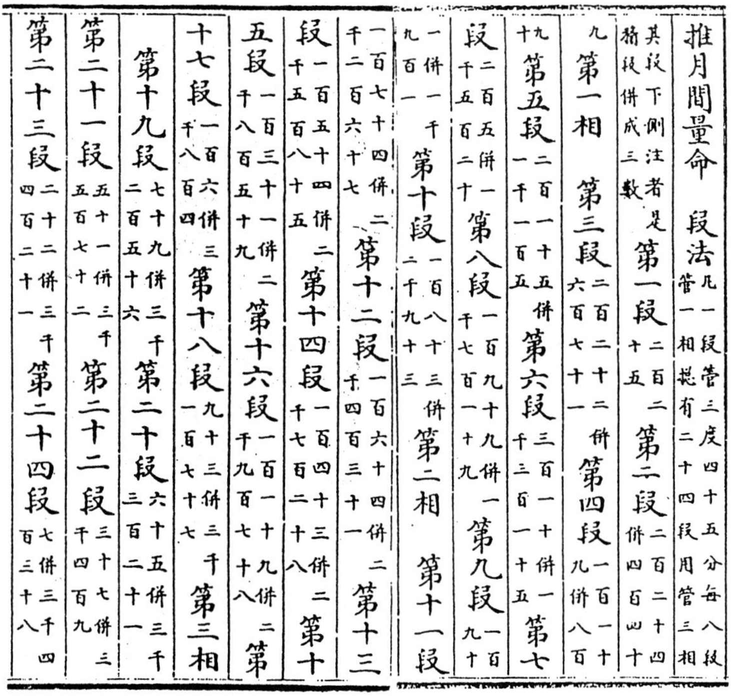

推月間量命段法（凡一段管三度四十五分每八段管一相總有二十四段用管三相）
（其段下側注者是積段併成三數）第一段（二百二十五）第二段（二百二十四併四百四十
九）第一相 第三段（二百二十二併六百七十一）第四段（一百一十九併八百
九十）第五段（二百一十五併一千一百五）第六段（三百一十併一千三百一十五）第七
段（二百五併一千五百二十）第八段（一百九十九併一千七百一十九）第九段（一百九十
一併一千九百一）第十段（一百八十三併二千九十三）第二相 第十一段
（一百七十四併二千二百六十七）第十二段（一百六十四併二千四百三十一）第十三
段（一百五十四併二千五百八十五）第十四段（一百四十三併二千七百二十八）第十
五段（一百三十一併二千八百五十九）第十六段（一百一十九併二千九百七十八）第
十七段（一百六併三千八百四）第十八段（九十三併三千一百七十七）第三相
第十九段（七十九併三千二百五十六）第二十段（六十五併三千三百二十一）
第二十一段（五十一併三千五百七十二）第二十二段（三十七併三千四百九）
第二十三段（二十二併三千四百二十一）第二十四段（七併三千四百三十八）

Fig. 10.2. The table of sines reproduced in the section of the *Kaiyuan zhanjing* devoted to the *Jiuzhi* calendar (*Jiuzhi li*), a calendar of Indian origin. Translation: "determination of the *yue jianliang ming*. Intervals: each interval controls 3 degrees 45 minutes; 8 intervals control one sign (*xiang*). The total number of intervals, 24, controls 3 signs [...]. First interval: 225; second interval: 224, total: 449; third interval: 222, total: 671, fourth interval: 219, total: 890 [...]." The first listed value corresponds to $3438 \sin k\alpha$ where $\alpha = 3°45'$ and $k = 1, 2 \ldots 24$, the other values correspond to the the sum of the preceding first values. Source: *Kaiyuan zhanjing*, j. 104, pp. 12b-13a in *Wenyuange Siku quanshu*, vol. 807, p. 939 (Taipei: Shangwu Yinshuguan, *1986*).

Contacts with Islamic Countries

Seek knowledge even unto China.
Prophet Muhammad

There is ample evidence of the contacts between China and the Islamic countries from the Tang dynasty (618–907),[57] but scientific exchanges are generally only thought of as going back to the Mongol period. This is probably a mistake, since, as the Taiwanese historian Luo Xianglin showed, these contacts go back to at least the beginning of the Northern Song dynasty (960–1127).

[57]Cf. Chang Jih-ming (1), *1980*.

According to Lo Hsiang-lin [Luo Xianglin],[58] the family register of sinicised Muslims of Huaining (Shenxi) having the typical patronymic name of Chinese Muslims, Ma, mentions that the first ancestor of this family, one Ma Yize, originated in the "land of Lumu" (now Ramithan, Samarkand). Recruited as an astronomer in 961, this ancestor then compiled a calendar on behalf of the Chinese, which was presented to the Emperor by Wang Chune, under the name of *Yingtian li* and used officially from 964 to 982.[59] Later:

> Instead of massacring the Muslims, Chinggis [Khan] wished to use their skills and expertise. He ordered his troops to spare Muslim craftsmen. Having few artisans among his own people, Chinggis depended upon foreigners [...] Chinggis [...] deported a large number of Muslims from Central Asia to the east. Juvaini notes that thirty thousand Muslim craftsmen from Samarkand were distributed among Chinggis's relatives and nobles. Many of these artisans were eventually settled in Mongolia or in North China.[60]

In 1267, a certain Muslim astronomer, called Zhamaluding in Chinese transcription (Jamāl al-Dīn?) presented the Mongol Emperor Qubilai with a perpetual calendar, the *Wannian li* (lit. "Myriad Years Calendar") and a number of astronomical instruments, certain of which were similar to those then used in Maragha.[61]

Many historians, such as J. Needham,[62] suppose that this Zhamaluding may have been the Jamāl al-Dīn whom Hulagu Khan had asked to found the Maragha Observatory in 1258. However, according to K. Yabuuchi and K. Miyajima, this is probably incorrect.[63]

Anyway, Zhamaluding was then appointed director of the Bureau of Astronomy, founded in Peking, on the orders of Qubilai in 1271. Two years later, in 1273, the *bishujian* (Directorate of the Palace Library),[64] a government service intended to administer the imperial library, was created. This service was then temporarily responsible for supervising the Islamic Bureau of Astronomy, newly created at the same time as the Bureau of traditional Chinese astronomy.[65]

Chapter 7 of the *Bishujian zhi* (Monography of the Palace Library)[66] (ca. 1350) includes a laconic list of the titles of 23 books and three instruments destined for the "Northern Observatory" *bei sitian tai*, or, in all likelihood, the Muslim observatory.[67] The brief introduction accompanying this text says that the list in question refers to the tenth year of the Zhiyuan era (1273). The titles and the names of instruments in the list are given not in Chinese,

---

[58] *RBS*, 14–15th year (1968–70), notice no. 18.

[59] Chen Zungui (1'), *1984*, pp. 1404 and 1470; *CRZ*, j. 19, p. 221; *Songshi*, j. 68, p. 1498.

[60] Rossabi (1), *1981*, p. 263.

[61] The problem of the identification of these instruments is not settled (cf. Miyajima, (1'), *1982*).

[62] *SCC*, III, p. 372.

[63] Yabuuchi (5), *1987*, p. 548; Miyajima (1'), *1982*, p. 407 ff.

[64] Cf. Hucker (1), *1985*, notice no. 4588.

[65] Yabuuchi (5), *1987*, p. 548.

[66] *Siku quanshu zongmu*, j. 79, p. 684.

[67] Tasaka (1), *1957*, p. 101.

but as phonetic transliterations of Arabic/Persian terms. This previous list has been studied in great depth from the philological point of view by the Japanese orientalist Tasaka Kōdō.[68] From this research it appears that the list probably contains, amongst other things, the title of some version of Euclid's *Elements* in 15 books, Ptolemy's *Almagest*, astronomical tables (*zīj*), a star map, a treatise on astrology and a perfect compass. Many historians have wondered if these texts were then translated into Chinese.[69] It seems that they were not; indeed, it is by no means certain that such translations would have been necessary, since the non-Chinese specialists on Muslim astronomy who were in the service of the Mongols, must have been able to understand Persian or Arabic texts. As D. Leslie notes, Chinese was not the only language in current use in China in the Mongol period. In particular, "Persian was a lingua franca during this period, one of the languages used by the Mongols for their government."[70]

When the Mongols came to power in China, they were interested not only in Islamic astronomy, but also in traditional Chinese astronomy. They then subjected the latter to a reform, based not on Islamic astronomical techniques but on typically Chinese methods. The resulting astronomico-calendrical system was called the *Shoushi li* (Season-Granting Calendar) and promulgated in 1281. It remained in service throughout the Mongol dynasty, in parallel with the Muslim system of Astronomy *Huihui li*. In 1368, the new dynasty, that of the Ming, symbolically founded its power on the promulgation of a new calendar, the *Datong li* (Imperial Calendar).[71] This calendar, which was not very different from the *Shoushi li* remained in service until the fall of the Ming dynasty (1644). However, during this long period of more than three and a half centuries, Muslim calendrical techniques were still used at the same time as Chinese techniques. They were to remain in use until 1656.[72]

When the first Ming emperor, Zhu Yuanzhang, came to the throne in 1368, he inherited two Bureaus of Astronomy, the Chinese Bureau and the Muslim Bureau.

In Autumn 1382, the Emperor had various works of Islamic astronomy and astrology translated into Chinese:

In the 15th year of the Hongwu era [1382], Wu Bozong and Li Chong of the Hanlin Academy were commissioned to translate the whole of various texts on astronomy and the degrees of longitude and latitude of the Muslim calendar. When the work was completed, Wu Bozong was asked to write a preface.[73] Here is what he wrote: "When,

---

[68] Tasaka, op. cit.

[69] For example, Needham in *SCC*, III. p. 105; Yan Dunjie (10′), *1943*; D'Elia (2), *1956*, p. 162; *SY*, p. 263.

[70] Cf. Leslie (1), 1986, p. 95; Huang Shijian (1), *1986*. Yet, as M. Rossabi remarks "[...] the lingua franca of the Mongol empire, at least in its eastern portion was almost certainly not Persian, as it is sometimes assumed, but Turkish." Cf. Rossabi (2), *1983*, p. 308, note 76.

[71] The expression '*Datong*' denotes the Emperor. Cf. *DKW*-3-5831: *1769*, p. 2707.

[72] Yabuuchi (5), *1987*, p. 550.

[73] This is the preface to the *Tianwen shu* (treatise on astrology, apparently adapted from Ptolemy's *Tetrabiblos* by Wu Bozong, based on a text by Kushyār ibn Labbān). On this

the Great General[74] pacified the capital of the Yuan (now Peking) at the beginning of the Hongwu era, he took possession of myriads of documents [...] He had these transported to the capital [Nanking] and placed them in the imperial library [...]. This mass of documents included hundreds of volumes from the Western Regions; however, since these texts were compiled in various languages, using many different scripts, no one was able to read them.[75]

According to this quotation, knowledge of foreign languages at the beginning of the Ming dynasty seems to have been uncommon, whence the need for translations. Perhaps too the Chinese authorities felt it undesirable that certain technical texts (astrology, astronomy, etc.) should only be understood by specialists of foreign ascendancy, who could have monopolised these to the detriment of the Chinese.

The original translations of Wu Bozong et al. have only survived through the intermediary of later Chinese and Korean editions.[76] The most complete of these editions is that known as the *Qizheng tuibu* (Computational techniques of the "seven governors," i.e. the Sun, the Moon and the five planets). This is a revision carried out between 1470 and 1477 by a certain Bei Lin, then vice-director of the Muslim Bureau of Astronomy at Nanking. The work consists of seven chapters (*juan*). The first chapter consists of computational prescriptions relating to the Muslim calendar, to the movement of the Sun, the Moon and the planets, to eclipses and to the duration of the day, etc. The six other chapters contain nothing but tables and a star catalogue. The terminology used betrays a Ptolemaic type of astronomy, but there is no theory and no explicative figures. Some of the technical terms in the work were later reused as such in translations of astronomical texts of European origin carried out after 1630.

All this shows that Chinese interest in Islamic Astronomy continued well beyond the Mongol dynasty, having started well before it. But what were the consequences for the development of Chinese mathematics?

Firstly, we note that the mathematics practised in China at the end of the Song and the beginning of the Yuan included techniques for numerical computation (Horner's method, decimal fractions) very similar to those found in the works of certain Arabic authors such as al-Samaw'al (ca. 1172).[77] While, in China, the idea of decimal fractions predates the Tang (618–907), the same is not true of Horner's method. In China, the latter method only appears towards the 11th century and, contrary to some assumptions,[78] there is no evidence to support the fact that its roots lie in Han mathematics. However, generally speaking, the originality of Chinese algebra (*tianyuan*) is such that even though this algebra were the result of a Chinese adaptation of foreign techniques, Chinese mathematicians of this period were so creative that their

---

subject see Pelliot (1), *1948*, p. 234; Yabuuchi (5), *1987*, p. 551; Yano and Viladrich (1), *1991*.

[74]Probably refers to Xu Da. Cf. *Mingshi*, j. 125.

[75]*CRZ*, j. 29, pp. 347–348.

[76]Cf. Yabuuchi, op. cit., p. 551, and Yabuuchi (3'), *1969*, pp. 202–234.

[77]Cf. Rashed (1), *1984*, p. 110.

[78]Cf. See below, p. 247 ff.

algebra far surpasses its supposed model. Moreover, the Chinese spherical trigonometry of the period is also highly original and quite unrelated to Greco-Islamic techniques, since it is intrinsically based on approximation formulae and not on exact trigonometry.[79] However, Chinese mathematics of the 13th century undoubtedly includes a number of non-native elements. On the one hand, magic squares of order greater than three are first found in China in the *Yang Hui suanfa* (1275) and, on the other hand, there is accurate information that magic squares of Islamic origin were introduced into Northern China at the same period (Fig. 20.2, p. 363).

If we now turn to the question of the Chinese influences on the Islamic countries, we note that when the highly qualified medieval Moslem authors al-Mas'ūdī, al-Bīrūnī and Ṣa'id refer to Chinese science, unlike in their treatment of Indian science, they talk only of technology: "Al-Mas'ūdī (*d.* 956) writes about China and refers to other Moslems who before him visited it and wrote about it. Like other early Arabic authors, he writes much about Hindu wisdom and learning, and refers to Chinese technology and social life, but never to Chinese science. The learned al-Bīrūnī does not refer to Chinese science in his *Chronology* nor does Ṣa'id in his *Ṭabaqāt.*"[80]

Despite this, some authors still hypothesise about a transmission of Chinese ideas (for example, about decimal fractions, extraction of roots (Horner's method), simultaneous congruences (Chinese remainder theorem)) to Islam. In this respect, Saidan observes that: "Al-Ṭūsī who worked in the Marāgha observatory with the Chinese shows no acquaintance with decimal fractions. But al-Kāshī's likeness [*sic*] for tabulation may be the outcome of the Chinese line-abaci, and his idea of decimal fractions may be suggested by Chinese patterns."[81]

This conclusion is perhaps correct but we do not know what Saidan's "Chinese line abacus" exactly refers to: the Chinese abacus is a later innovation from the 13th or 14th century; moreover, no Chinese "line-abacus" has ever been found, either materially or abstractly (in the form of some textual description).

Transmission of Chinese Mathematics to Korea and Japan

If traditional historiography is to be believed, Korean and Japanese mathematics began as a pure reflection of Chinese mathematics. In both cases, the development was apparently conditioned by university institutions exactly copying those of the Tang.[82] In the kingdom of Silla, as in the Japan of the Asuka (552–645) and Nara periods (710–794), a teaching system revolving around the Ten Computational Canons and involving examinations is thought to have

---

[79]See below, p. 328 ff.

[80]Saidan (1), *1978*, p. 455.

[81]Saidan, op. cit., p. 485.

[82]Sugimoto and Swain (1), *1978*, p. 33 ff.; Kim Yong-Woon (1), *1973*; *Meijizen*, I, p. 3 ff.; Kim Yong-Woon and Kim Yong-Guk (1'), *1978*.

existed. However, as far as we know, none of these systems led to autochthonous mathematical traditions and the influence of Chinese mathematics did not begin to make itself felt in a lasting way in these regions until much later.

In Korea, during the reign of the King Se-djong (1419–1450), fourth monarch of the Yi dynasty (1392–1910), two important Chinese manuals of the end of the 13th century, the *Yang Hui suanfa* and the *Suanxue qimeng*,[83] together with an arithmetic from the start of the Ming, the *Xiangming suanfa*,[84] were reprinted and used in the civil service examinations.[85] These works, and later works such as the *Suanfa tongzong*, were then imported to Japan in circumstances which are still unclear (possibly during Japanese military expeditions to Korea, 1592–1598).

In Japan, the mathematical tradition, which had always been insignificant, began to develop at a late stage. Initially (first half of the 17th century) the influence of the popular Chinese tradition of the abacus was felt. The famous *Suanfa tongzong* was then used in small part as a source for the widely distributed manual, the *Jinkōki* (1627).[86] Later, from 1650, the Chinese instrumental algebra of the Song–Yuan period spread via the *Suanxue qimeng*. Rival schools of amateur algebraists were then constituted, largely outside official circles. These algebraists essentially concerned themselves with the solution of gratuitous problems (often on a geometrical pretext) using only computational techniques; the problems were often chosen more for their supposed intrinsic beauty than for their utility.

Paradoxically, while Japan and, to a lesser extent, Korea developed a mathematics which was very marked by Chinese algebra, both as far as its written form and its principles were concerned, China underwent a period of decline during which its mathematicians were unaware of the algebraic concepts of the Song and the Yuan. The *Suanxue qimeng*, which was reprinted numerous times in Korea and Japan (in 1658, 1660, 1672, 1690, 1715, 1750 and 1810)[87] became completely unobtainable in China itself and did not reappear in its country of origin until the beginning of the 19th century, thanks to the Korean edition of 1660 found by Ruan Yuan.[88] However, it was not until 1839 that Zhu Shijie's elementary manual, which had been forgotten for five centuries, appeared for a second time in a Chinese reprint.[89]

With its purely Chinese basis, but supplemented with Japanese innovations (in particular, literal algebraic notation) and probably European elements too, brought there first by Jesuit missionaries[90] and later by Dutch merchants, Japanese mathematics led to non-trivial results which compare well with work in Europe in the same areas at the same period (infinite series, determinants,

---

[83]Kodama, (1'), *1966*.

[84]Ibid., p. 18 ff.

[85]Kim Yong-Woon, op. cit.

[86]Original text reprinted in 1978 (Tokyo: Iwanami Shoten (Iwanami Bunko collection)).

[87]Cf. Hirayama (1'), *1981*, p. 241.

[88]Cf. Yan Dunjie (14'), *1945*.

[89]Idem.

[90]Hirayama (3'), *1993*.

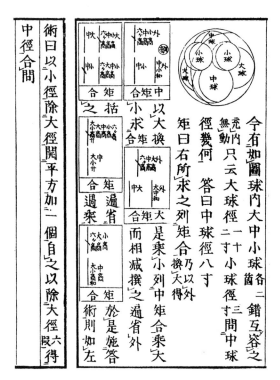

Fig. 10.3. A geometrical problem from the *Sanpō tenseihō shinan* (from Hirayama Akira and Matsuoka Motohisa (eds.), *Aida Sanzaemon Yasuaki, 1982* (private publication, no place name) ch. 4, pp. 5b–6a), an algebraical work by Aida Yasuaki (1747–1817) which was incorporated, in 1898, by Liu Duo, in a Chinese mathematical encyclopedia, the *Gujin suanxue congshu* (Collection of mathematical works, ancient and modern). Aida Yasuaki is the founder of the school of Japanese mathematics known under the name of *Saijō-ryū*, an ambiguous expression which means both "The very-best school" and "The style of Mogami," the name of a river in Aida's native district (see Sugimoto and Swain (1), *1978*, p. 363). Liu Duo was a *zhongshu sheren* (Secretarial Drafter) from the *zongli yamen* (Foreign Office).

Translation of the problem: "Let two little spheres, two medium spheres and two large spheres be packed in a big sphere in such a way that any motion is impossible. It is only said that the diameters of the large and little sphere are 12 *sun* and 3 *sun*, resp. What is the diameter of the medium sphere? Answer: 8 *sun*." After the statement, the remainder of the text shows some intermediate equations (inside small rectangles) together with verbal indications referring to computations to be carried out. Lastly, the following resolutory formula (which is correct) is given: $m = 6D/\left(\frac{D}{d} + 1\right)^2$ (*D*, *d* and *m* = diameter of the large, small and medium sphere, respectively). This sort of problem is fairly typical of Japanese traditional mathematics whose manuals frequently consist of more-or-less of successions of elegant but unrelated and artificial problems. The text is written in classical Chinese but a few marks (*kunten*) such as $\sqrt{}$, carrying various linguistic information useful for Japanese readers, have been added beside certain Chinese characters. In spite of their uselessness for Chinese readers, these marks have been kept intact in the Chinese edition of Aida's work.

complex diophantine equations, combinatorics (Bernoulli numbers), geometrical problems such as that of Malfatti,[91] (Fig. 10.3)). The Japanese example is a magnificent instance of an elaborate development of a mathematics essentially based on the Chinese tradition and outside the Euclidean sphere.

From 1868, the Meiji reformers undertook an unprecedented overhaul of the Japanese educational system. Almost overnight, traditional Japanese mathematics gave way to Western mathematics in the teaching programmes.

As far as scientific exchanges are concerned, if we consider the nature of the relationships which China maintained with its neighbours within its own cultural sphere, we note a type of circulation of ideas apparently different from that which prevailed in the Mediterranean West.[92]

Ideas seem essentially to have radiated from China towards its sinicised periphery rather than in the opposite direction. We know of no examples of Chinese travellers learning Mongolian, Tibetan, Korean or Japanese and being influenced in their work by their local discoveries (except, of course, at the end of the 19th century). Neither do we know of examples of commercial exchanges of mathematical works, other than in a single direction from China and towards the sinicised regions.

The vastness of the Chinese territory and the relative distance between the countries involved may have played a determining role in this respect, but this does not explain everything (think of the case of the development of "Dutch Learning" (*rangaku*) in 18th century Japan). It is likely that the following factors also played an important role:

(i) In China, as in Korea and Japan, mathematics has always been viewed as a minor art.

(ii) In Japan, mathematics was often transmitted secretly by direct initiation of the disciple by the master, by the copying of manuscripts rather than via printed books with a large circulation.

Bibliographic comments. No recent book covering the whole history of Japanese mathematics exists in Western languages. Apart from the book by Smith and Mikami (1), *1914*, which is already old, the reader may also consult the article by Nakayama, "Japanese Scientific thought," in the last volume of the *DSB* (pp. 728–758), together with Sugimoto and Swain (1), *1978* and Horiuchi (2), *1994*. On the works of certain mathematicians, see also the copious English summaries in Japanese books such as *Seki Takakazu zenshū* (Complete works of Seki Takakazu (?–1708)), edited by A. Hirayama. K. Shimodaira et al., Osaka: Osaka Kyōiku Tosho, *1974*; see also Martzloff (12), *1990*.

---

[91] Malfatti's problem is the following: inscribe three circles in a given triangle so that each of these is tangent to the two others and to two sides of the triangle. It was first solved in 1789 by Ajima Naonobu (1732–1798). Cf. Hirayama and Matsuoka (1′), *1966*, p. 37.

[92] Cf. Riché (1), *1979*, p. 131 (chapter entitled "Relations culturelles entre l'Occident et le monde oriental").

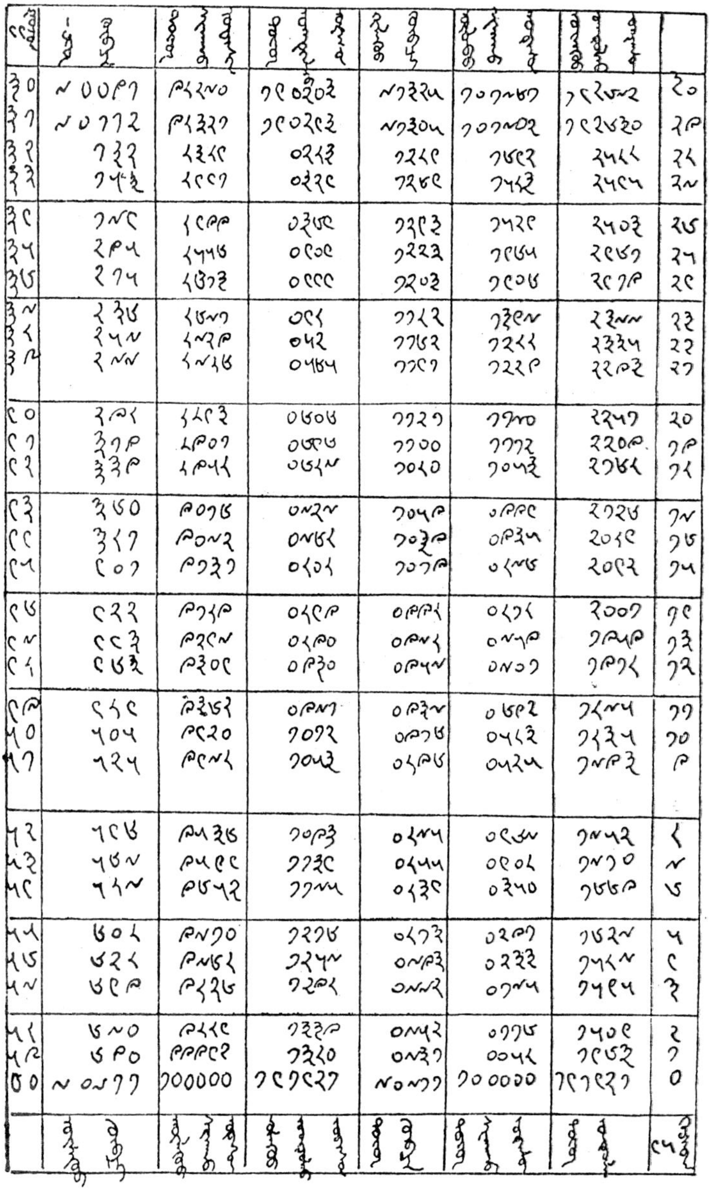

Fig. 10.4. A page, from a Mongol mathematical manuscript, showing the values of trigonometrical lines of angles lying between 44°30′ and 44°60′. According to Yan Dunjie this manuscript dates back to 1712 and was based on a late edition of the *Chongzhen lishu*, a Jesuit treatise of mathematical astronomy compiled by Giacomo Rho, Adam Schall and others eighty years earlier (From Yan Dunjie (27′), *1982*, p. 39).

Contacts with Mongolia

According to Yan Dunjie, European mathematical techniques were translated into Mongolian from the *Chongzhen lishu* in 1712 (Fig. 10.4).[93]

Contacts with Tibet

The Chinese influence on Tibet seems limited to calendrical astronomy from the Qing period.[94] In 1989, two Chinese historians published some original sources in the form of a Sino-Tibetan edition composed of research papers together with Tibetan texts edited in their original language and translated into Chinese.[95]

Contacts with Vietnam

Vũ-Văn-Lập's *Nam sử tập biên* (1896) mentions two Vietnamese mathematical works entitled *Cửu chương toán pháp* (Chinese: *Jiuzhang suanfa*, (mathematics in nine chapters) and *Lập thánh toán pháp* (Chinese: *Licheng suanfa*, (ready reckoner)). This title is almost the same as that of the Stein 930 manuscript from Dunhuang.[96] Both were composed in 1463 by two successful candidates in the civil service recruitment examination.[97]

Two centuries later, the famous *Suanfa tongzong* and other elementary arithmetical books reached Vietnam.[98] From the 17th to the 19th centuries, Chinese adaptations of European works initiated by Jesuit missionaries for the reform of Chinese astronomy made their way into Vietnam.[99]

For up-to-date information on Vietnamese mathematical books, the reader should, above all, consult the catalogue of Vietnamese ancient books recently published in Hanoi in a bilingual (French and Vietnamese) edition.[100]

---

[93]Cf. Yan Dunjie (27'), *1982*, p. 25. For further details of Mongolian translations of Chinese works and a bibliography in the Mongolian language cf. Shagdarsüren (1), *1989*.

[94]Cf. Joseph, *1994*.

[95]Huang Mingxin and Chen Jiujin (1'), *1989*.

[96]On the *Licheng suanjing*, cf. Libbrecht (5), *1982*, p. 205.

[97]From Trần Văn Giàp (1), *1937*, p. 93 (cited by Schrimpf (1), *1963*, pp. 354 and 373, note B 18).

[98]Cf. Han Qi (2'), *1991*.

[99]Bernard-Maître (8), *1950*, p. 154.

[100]Trần Nghĩa and Gros (1), *1993*.

Contacts with Europe

> *But no sooner had we got outside the door than a shot whizzed past us*
> *[...]. We then decided that we must keep watch in turn that night. Never*
> *before had I found my knowledge of Euclid serve me so well. I went over*
> *in memory the first book, proposition by proposition, and I was able to*
> *keep awake.*
>
> Timothy Richard, *Forty-Five Years in China.*
> London: T. Fisher Unwin Ltd., 1916, p. 42.

We have already noted that prior to the 16th century, at the current stage of research, it is often impossible to determine exactly which parts of Chinese mathematics are borrowed and which are original developments.

However, from the end of the 16th century, this is no longer the case. Thanks to the documents left by the missionaries and thanks also to the original work of historians, it is much less difficult to analyse what went on, even though there are still large shady areas.

In the past, it has often been written that the period from the beginning of the 16th century was when European sciences were introduced into China, and that the universalisation of Chinese mathematics dates back precisely to the beginning of the 17th century. As J. Needham put it:

The year 1600 is a turning point, for after that time there cease to be any essential distinctions between world science and specifically Chinese science.[101]

It is difficult to adhere to such a conclusion. In the first place, the scientific contacts between China and Europe at that time were unilateral and subordinate to the propagation of the Christian religion. Secondly, the transmission process was solely a result of the translation and adaptation of European works. None of the open-minded Chinese mathematicians who were interested in new developments (and there were many of these) ever worried about having access to the original texts themselves.[102] Yet the need to learn the languages of peoples from outside the Chinese world was ancient in China: at least as early as the Tang dynasty China had an official bureau of interpreters and a Russian language school is said to have been established in 1757.[103] But, as far as we know, these institutions did not concern mathematics during the 17th and 18th centuries and no Chinese mathematician from that period ever studied this science in Europe.

Inevitably, however much effort the missionaries and their autochthonous collaborators were able to contribute, they could only translate at best a small part of existing European material into Chinese.

The list of mathematical works translated into Chinese between the beginning of the 16th century and the end of the 19th century, given at the end

---

[101] Needham (1), *1958*, p. 1.

[102] See Li Yan's bibliographies in *ZSSLC-T*, IV-2.

[103] Biggerstaff (1), *1961*, p. 96.

of this book, clearly indicates the nature of the gaps in transmission and the belatedness of some contributions. For example, this literature does not contain echoes of the philosophical controversies about infinitesimal calculus, imaginary numbers and parallelism. It does not contain any information whatsoever about the theories of Cauchy, Gauss, Riemann or Dirichlet and algebraic symbolism similar to that of Descartes did not appear in China until the second half of the 19th Century. Similarly, differential equations, analytic geometry and mechanics are all absent from Chinese mathematics before this period.

However, the theory of an "instant universalisation" of Chinese science inevitably triggered by a Chinese access to some Western developments from 1600 is not merely dismissed by a question of documentary limitation. More fundamentally, the question hinges on the assessment, by influential Chinese, of what was or was not relevant, acceptable or borrowable. Whence the enhancement or, conversely, the rejection of specific aspects of Western mathematics by the majority of Chinese mathematicians of the time.

Above all, and especially during the 17th and 18th centuries, it turns out that European science, or rather Jesuit science, was perceived by Chinese scholars as a body of knowledge founded on Euclid's *Elements*.[104] For Jesuit missionaries, the *Elements* had indeed the double advantage of uniting the certainty of mathematics with its applicability to the study of nature,[105] especially astronomy, a particularly strategic science in the Chinese context.

But whereas Jesuits and Chinese scholars both agreed that the *Elements* were the most fundamental Western science, the latter did not admit that the totality of geometry was equally important. Indeed, most Chinese mathematicians were convinced of the fact that the particular mathematical results contained in the *Elements* had nothing to do with the underlying axiomatico-deductive structure. In other words, they tended to believe that the form of the *Elements* was independent of its content and that the two could and should be dissociated.

For the influential proponents of the evidential research movement (*kaozheng xue*), knowledge such as that displayed in the *Elements* amounted to an instrumental compilation of computational methods which were to be conveyed by means of prescriptions readily understandable and usable; by this very fact, such a knowledge excluded literary ornaments and excessively long passages such as those found in Euclidean theorems and proofs. In particular, syllogisms and other logical forms were especially unacceptable for they involved numerous repetitions and redundancies: they were contradictory with the canons of Chinese literary redaction which, in the case of technical subjects, valued conciseness above all.[106] For similar reasons, the separation between Euclidean propositions expressing ideas semantically close 'but 'deductively

---

[104] *Siku quanshu zongmu*, j. 107, p. 907 (end of notice on the *Jihe yuanben*); *CRZ*, j. 45, p. 594.

[105] Cf. Cosentino (1), *1970*; Krayer (1), *1991* among others.

[106] Martzloff (13), *1993*.

distant' the one from the other (that is, expounded the one and the other in different books of the *Elements*) was no longer acceptable.

Thus, Chinese authors endeavoured to rid the geometry of its useless ornaments and retained only what they considered to be the essential, namely the calculable results, either because they believed they could compare them with those which existed in their tradition or because these results provided them with the means of carrying out new calculations.[107] For example, in his reinterpretation of the *Elements, Jihe tongjie* (Complete explanation of geometry, end of the 17th century), out of Euclid's whole treatise, the influential Mei Wending (1633–1721), only retained a number of the propositions from books 2 and 6 (precisely those which historians class as geometric algebra), which he included as computational rules for solving right-angled triangles, like in the *Jiuzhang suanshu*.[108] In his *Jihe bubian* (Complements of Geometry), the same author made much use of the division into mean and extreme ratio,[109] which he saw as a novelty, to calculate the volumes and relative dimensions of regular and semi-regular polyhedra inscribed within one another[110] (most probably, Mei Wending knew that the untranslated part of Clavius's *Elements* dealt with this subject).[111] In so doing, he only retained the parts of the demonstrative discourses which Euclid termed *protasis* and *diorismos* and which do not exactly form part of the proof proper (*apodeixis*).[112] He modified the figures to make them immediately readable, although Euclid operated in the opposite direction thus making it necessary to resort to discursive reasoning.[113] He regrouped the theorems on a semantic basis, thus breaking the Euclidean framework. Even more radically, in his *Jihe yue* (Summary of geometry, around 1721)[114] Fang Zhongtong, son of the famous scholar Fang Yizhi (?–1671)[115] suppressed practically all the proofs to retain only the theorems; in his *Jihe lunyue* (Summary of geometric proofs, end of the 17th century)[116] Du Zhigeng, who was considered by Mei Wending as the best Chinese mathematician of his generation, identified the theorems with their converses since they amounted to "the same thing said backwards" (*li tong ben ti fan yan zhi*).[117] The same Du Zhigeng also thought of a point as "the place at which there is a mark on a string" (*xian zhi suo zhi chu yue dian*)[118] so that, as far as he was concerned, geometry practically merged with a branch of drawing.

---

[107]Martzloff, (3), *1981*.

[108]Martzloff, (2), *1981*.

[109]Clavius, *Euclidis Elementorum* (op. cit.), book 6, def. 3.

[110]*Meishi congshu jiyao*, j. 25 to 28.

[111]Clavius, op. cit., books 15–16, pp. 305–355.

[112]On these divisions of the Euclidean propositions, see Mueller (1), *1974*, pp. 37–38.

[113]Martzloff (3), *1981*, p. 47.

[114]Bibliographic references in Li Yan, *ZSSLC-T*, IV-2, p. 439.

[115]Hummel, p. 233.

[116]Ibid., p. 484.

[117]See Martzloff (12), *1993*, p. 169 (*Jihe lunyue*, j. 1, corollary to proposition I-15 of the *Elements*).

[118]Ibid., p. 170 (*Jihe lunyue*, j.1, commentary on propositions I-28 and I-29 of the *Elements*).

But even after the geometry had been so accommodated, other criticisms of a different order were levelled against the very possibility of relying on geometry in order to understand natural phenomena. Fang Yizhi (1611–1671), for example, criticised the Jesuits' optical geometry applied to eclipses, on the grounds that the Sun's rays cannot be assimilated into straight lines:

> They do not realise that the Sun's rays are always fat. The Earth's image of itself thin, cannot be obtained by acute angles and straight lines. Why? Material things being obstructed by form, their image is easily made to vanish. Sound and light rays are always more subtle than the "number" (shu) of things.[119]

Wang Xishan (1628–1682),[120] whom, for a long time after his death, the Chinese still considered to be one of their greatest astronomers, voiced similar criticisms against geometrical representations of celestial phenomena[121] and deemed the recording of geometric proofs inessential:

> On the whole, when the ancients established a mathematical method (fa) they necessarily relied on organising principles (li) and yet they expounded their methods accurately without stating the underlying principles. In fact, organising principles are contained in mathematical methods and those who are inclined towards study and thorough reflection might find inwardly the necessary energy to grasp organising principles.[122]

Much later, Li Shanlan (1811–1882), the translator of Euclid's *Elements*, found the Euclidean definition "a point is that which has no part" absurd, under the pretext that an ink stain, however small, necessarily has some dimension.[123] For him, the very idea of an idealised point with no counterpart in reality was irrelevant.

Moreover, according to the dominant Chinese conception, technical knowledge could not be limited a priori by dogmatic constraints such as those announced in a foreword to Matteo Ricci and Xu Guangqi's Chinese translation of the first six books of the *Elements*:

此書有四不必。不必疑、不必揣、不必試、不必改。有四不可得。
欲脫之不可得、欲駁之不可得、欲減之不可得、欲前後更置不可得。

ci shu you si bubi. bubi yi, bubi chuai, bubi shi, bubi gai. you si bukede. yu tuo zhi bukede, yu jian zhi bukede, yu qianhou gengzhi bukede.

---

[119] *Wuli xiaozhi*, j. 1, p. 25 (cited from Peterson (2), *1975*, p. 391).

[120] Cf. N. Sivin's notice in the *DSB*.

[121] *CRZ*, j. 35, p. 436.

[122] Wang Xishan, "Zazhu" (various works), p. 3a (quoted from the *Muxixuan congshu* edition of Wang Xishan's collected works).

[123] *Fangyuan chanyou* (The mysteries of the circle and the square explained), j. 1, p. 1a (cited from the edition of this text reproduced in Li Shanlan's collected works *Zeguxizhai suanxue*).

Four things in this book are not necessary[124]: it is not necessary to doubt, to assume [new conjectures], to put to the test, to modify. In addition, four things in this book are impossible. It is impossible to remove any particular passage, to refute it, to shorten it or to place it before that which precedes it, or vice versa.[125]

Yet, among certain of those who were closely associated with Jesuits on the occasion of translation activities, in a very limited number of cases, certain personalities, and not the least, for example, Xu Guangqi (1562–1633) readily accepted everything Western, religion included, and reasoned *more geometrico* in their own mathematical works. Still, the vast majority of Chinese scholars did not consider that the totality of Western mathematics was acceptable and Xu Guangqi's ideas remained quite marginal; for them there was no compelling reason to reject what the Chinese cultural context had taught them, no particular motive to interpret the Chinese version of Euclid's *Elements* in an Euclidean sense inasmuch as even the Chinese translation of the title of Euclid's book, namely *Jihe yuanben* "Elementary book on geometry" invited an immediate interpretation on the part of Chinese readers for whom the word *jihe* was naturally associated with the ritual question at the end of problems in Chinese arithmetics which simply meant "how much is that?" Thus, highly prosaically, Euclid's *Elements* became the "Elementary book of 'how much is that?'", in other words, a banal work on practical computation. It is true that, according to definition no. 1 of book 5 of the *Jihe yuanben*, *jihe* also meant "quantity."[126] Perhaps the trap could have been avoided, but Ricci's preface which ceaselessly referred to specific applications (calendar, music, technology) lent itself naturally to such an interpretation.

There were many other translation problems. On the one hand, it was impossible to make broad use of the stock of the vocabulary of traditional Chinese mathematics, since the latter was highly indexical; for example, geometrical figures were considered as all sorts of fields *tian*.[127] On the other hand, the normal approach which one could have contemplated in the case of translation between European languages, namely transcription using the common basis of Greek or Latin roots, would only have produced incomprehensible barbarisms (this is not surprising – think of the inverse situation; however, think also of the case of modern Japanese which runs counter to this observation). In addition, it is quite clear that the works in which this approach was used, such as the *Mingli tan* (Investigations of the principles of names – that is, the principles of logic – in traditional China, the attribution

---

[124]The usual translation of *bubi* is "it is not necessary to" or "there is no need to." However, the notice of the imperial catalogue *Siku quanshu zongmu*, j. 107, p. 908, about a Chinese work on geometry of the end of the 17th century, the *Jihe lunyue* (Summary of geometrical proofs) induces a different interpretation of this expression: "it is forbidden to."

[125]Xu Guangqi, "Zayi" (various reflexions) – a foreword to the *Elements*.

[126]Clavius (2), *1591* book 5, def. 1.

[127]Instead of "rectangle" or "circle," one finds expressions such as *fang tian* (rectangular field) and *yuan tian* (round field) where *fang* (square) and *yuan* (round) are not defined.

of correct names and denominations played the role of logic), were not very successful.[128]

There were many other difficulties of the same order. For example, Ricci had translated Clavius' Latin *definitio* (definition) by *jie* (limit).[129] In itself, this translation is quite correct, but the Chinese reader could certainly not find in the Chinese context any further explanation corresponding to what "definition" means in the mathematical tradition based on Greek mathematics.[130] Significantly, in the translation, the term *jie* was glossed by *mingmu*, that is "nomenclature." Moreover, in traditional Chinese mathematics, the term *fa* meant rule or computational prescription but in Euclid's *Elements*, Ricci used the term in the sense of "construction", which is not the same thing.[131] The problems, theorems and propositions all became *ti* in Chinese, as though they were arithmetical problems (in Chinese traditional arithmetics, such as the *Suanfa tongzong*, *ti* means "arithmetical problem").[132] In the 19th century, the translator A.Wylie added to the confusion by translating "theorem" by *su* [*shu*] (prescription, rule, recipe) (Fig. 10.5).[133] Moreover, axioms and proofs were both rendered by the same word *lun* (discourse, commentary).[134]

In addition to the terminology, the even more formidable problem of the difference between the Chinese syntax and that of European languages had to be faced. The main difficulty was the absence of the verb 'to be' in classical Chinese.[135] The translators were unable to find better substitutes for it than demonstratives or transitive verbs such as *you*, *wu* and *wei*. For Graham *wei* can hardly be called a copula; it has the flavour of an active verb. The same author also states: "In Chinese, the word *you* is used primarily of concrete things, the English word nothing implies the absence of any entity; the Chinese word *wu* only the absence of concrete things. This phenomenon might explain the difference in status of subject and predicate in the Chinese and Western

---

[128]This is a treatise of Aristotelian logic adapted from the *Logica Conimbricences* (translated by F. Furtado in 1631). This work was not included in the *Siku quanshu* collection. From the very beginning, Aristotle had no success in China.

[129]Heath (3), *1926*, I, p. 143; see also Granger (1), *1976*, pp. 38, 85, 174, 235, 341.

[130]On this, cf. Morrow (1), *1970*, p. 85 ff.

[131]*kataskeuē*; Latin: *constructio*; cf. Heath, op. cit., p. 129.

[132]One might think, a priori, that, in this context, *ti* would be understood as an abreviation for *mingti* (proposition). However, it is not certain that this term existed in the 17th century. According to the *DKW* 2-3473: 56, p. 2046, *mingti* came from the English "proposition" and thus was introduced in the 19th century.

[133]Cf. Doolittle (1), *1872*, II, p. 363 (Alexander Wylie is the author of section no. XXXII *Mathematical and astronomical terms* of Doolittle's dictionary).

[134]Cf. Martzloff (13), *1993*, pp. 163–164.

[135]Graham (1), *1959*; Gernet (4), *1982*, p, 322 ff; Granger (1), *1976*, p. 58 ff. The last author cites Trendelenburg's hypothesis about the influence of the structures of the Greek language on Greek philosophical thought: since Trendelenburg, a seductive hypothesis has been developed, according to which the multiplicity of categories as *typical contents of predication*, would in some way reflect the structures of the Greek language. Benveniste used this hypothesis. (Trendelenburg (1802–1872), German philologist, author of the *Elementa logicae Aristotelicae*, Berlin, 1836).

Mathematical and Astronomical Terms. OF THE CHINESE LANGUAGE. 363

日 'tai yang jih; *radiation*, 日光射 jih kuang shê.
Solid, 體 'ti; *mensuration*, 商功 shang kung; *number*, 體數 'ti so.
Solidity, 體積 'ti chi.
Solstitial colure, 二至經圈 êrh chih 'ching 'chüan.
Space, 有界之形 yu chieh chih hsing.
Sphere, 球 'chiu, 球體 'chiu 'ti, 立圜體 li 'huan 'ti.
Spiral, 螺線 lo hsiang; *of Archimedes*, 亞奇默德螺線 ya-'chi-mo-tê lo hsien.
Spirit level, 酒平 chiu 'ping.
Spot (on sun), 斑 pan.
Spurious disc, 假體 chia 'ti.
Square, 方 fang, 正方 chêng fang, 平方 'ping fang, 冪 mo; *number*, 平方數 'ping fang so; *root*, 平方根 'ping fang kên.
Star, 星 hsing.
Stationary point (of a planet), 留 liu.
Stereographic projection, 渾蓋通憲法 'hun kai 'tung hsien fa.
Straight line, 直線 chih hsien.
Subnormal, 次法線 'tzŭ fa hsien.
Subtangent, 次切線 'tzŭ 'chich hsien.
Subtract, 減 chien.
Subtraction, 減法 chien fa.
Sum, 和 'huo.
Summer solstice, 夏至 hsia chih.
Sun, 日 jih, 太陽 'tai yang.
Sunrise, 日出 jih 'chu.
Sunset, 日入 jih ju.
Superficies, 面積 mien chi.
Superior conjunction, 上合 shang 'ho.
Supplement, 外度 wai tu.
Supplementary chord, 餘通弦 yü 'tung hsien.
Surface, 面 mien; *of revolution*, 曲面積 'chü mien chi.
Symbol of quantity, 元 yüan.
Synodic month, 朔望月 so wang yüeh.
Syzigies, 合衝 'ho 'chung.
Table, 表 piao.
Tangent, 切線 'chieh hsien, 正切 chêng 'chieh.
Tangential force, 切力 'chieh li.
Telescope, 遠鏡 yüan ching.
Temperature, 寒暑率 'han shu shuai.
Temporary stars, 客星 'ko hsing.
Term, 界 chich; *of an algebraic expression*, 項 hsiang; *of ratio*, 率 shuai.
Tetrahedron, 四面體 ssŭ mien 'ti.
Theodolite, 地平尺 ti 'ping 'chih.
Theorem, 術 su.
Thermometer, 寒暑表 'han shu piao.
Tides, 潮汐 'chao hsi.
Time, 時 shih.
Total differential, 全微分 'chüan wei fên.
Trade winds, 貿易風 mao i fêng.
Transcendental *curve*, 越曲線 yüeh 'chü hsien; *expression*, 越式 yüeh shih; *function* 越函數 yüeh 'han so.
Transform, 易 i.
Transversal force, 橫力 'hêng li.
Transverse axis, 橫軸 'hêng chou, 橫徑 'hêng ching.
Trapezium, 無法四邊形 wu fa ssŭ pien hsing.
Trapezoid, 二平行邊四邊形 êrh 'ping hsing pien ssŭ pien hsing.
Triangle, 三角形 san chiao hsing.
Trident, 三齒線 san 'chih hsien.
Trigonometrical survey, 三角法測地 san chiao fa 'tsê ti.
Trigonometry, 三角法 san chiao fa, 句股 chü ku.
Trilateral figure, 三邊形 san pien hsing.
Trinomial, 三項式 san hsiang shih.
Triple star, 三合星 san 'ho hsing.
Triplicate, 三倍 san pei; *ratio*, 三次比例 san 'tzŭ pi li.
Trisection, 三等分 san têng fên.
Tropic *of Cancer*, 晝長圈 'hua 'chang 'chüan; *of Capricorn*, 晝短圈 'hua tuan 'chüan.
Tropical year, 太陽年 'tai yang nien.
True place, 眞位 chên wei.
Twilight, 朦朧 mêng lung.
Umbra in eclipses, 闇虛 an hsü.
Unequal, 不等 pu têng.
Unit, 一 yi.
Unity, 一 yi.
Unknown, 未知 wei chih.

Fig. 10.5. The English–Chinese glossary of mathematical and astronomical terms elaborated by A. Wylie. Note the translation of "theorem." Source: Doolittle (1), *1872*, II, p. 363.

linguistic systems, respectively."[136] But often, the verb "to be" disappeared altogether, as in the following case:

圜者。一形於 平地居一界之間。自界至中心作直線。俱等。

yuanzhe. yi xing yu pingdi ju yi jie zhi jian. zi jie zhi zhongxin zuo zhixian. ju deng.

[The] circle: [a] shape situated on flat ground (*ping di*) [sic] within [a] limit. [The] straight strings (*xian*) constructed from [the] limit to [the] centre: all equal.[137]

Of course, other (more elegant, more grammatical) renderings of this phrase would be possible. However, English grammaticality tends to obliterate the structure of the Chinese and the connotation of the specialised terms and, at least in the present case, it is probably better to change the structure of the Chinese sentence as little as possible. Hence the rendering of *xian* by "string" and not by "line," since in most Chinese geometrical texts one finds only line-like objects but no idealised geometrical lines as such.

Compare with Clavius' original:

Circulus, est figura plana sub una linea comprehensa, quae peripheria appelatur, ad quam ab uno puncto eorum, quae intra figuram sunt posita, cadentes omnes rectae linae, inter se sunt aequales.[138]

One might think that this type of phenomenon contributed to a masking of the conception, according to which geometric objects possess inherent properties, the existence or non-existence of which is objectifiable.

It would of course be possible to find characteristics of the same type and rejections of deductive reasoning in many books published in 17th century Europe.[139] To be sure, certain mathematicians such as Pierre de la Ramée (1515–1572) rejected Euclid.[140] However, while these mathematicians were favourably received by specialists in mathematical education, they never succeeded in reorienting the development of European mathematics towards a rejection of axiomatico-deductive reasoning.[141] Clearly, the Chinese situation was exactly the opposite.

Outside geometry (which, despite the title of Euclid's book, was not at all elementary), many innovations were however accepted. This, even though they may have been seen as forming part of numerical computation expressed in the form of algorithmic rules, either on their own or (most frequently) together with their proofs, provided the latter depended on properties already existing in traditional Chinese mathematics (similarity of right-angled triangles, Pythagoras' theorem, elementary computation of proportions, to name only the most important). Many Chinese mathematicians expressed their admiration for logarithms, because of the convenience and ease they provided in numerical

---

[136]Graham, op. cit., p. 98.

[137]*Jihe yuanben*, book I, def. 15.

[138]Clavius, *Euclidis Elementorum*, book 1, def 7.

[139]Kokomoor (1), 1928.

[140]Cf. M.S. Mahoney's article on him in the *DSB*.

[141]Cf. Montucla (1), *1768*, vol. I, p. 205 ff.

computations,[142] not to mention those who, from the 17th to the 19th centuries composed treatises on plane or spherical trigonometry and calculating instruments and, to a lesser extent, on infinite series.

However, European symbolic algebra was received much less easily than purely numerical computation. The first attempts to introduce it were made at the beginning of the 18th century within the framework of the special lessons given by the French missionaries to the Emperor Kangxi. At first, the enlightened monarch found the subject difficult to assimilate.[143] Could this have had something to do with the fact that the notation used was strange to him? The type of algebra which he was taught used special Chinese characters in the same way as Roman letters were then used for algebraic computations in Europe, both for the unknowns and for the known parameters of a problem, thus permitting purely literal calculations. Subsequently, the missionaries devised a much simpler algebraic system which only permitted calculations using polynomials in one unknown with numerical coefficients, which they called *jiegenfang* (lit. the borrowing of roots and powers, where *gen* (the root) denotes the unknown and *fang* any other power of that). This technique bears a greater resemblance to the algebra of the Cossist mathematicians of the Italian renaissance than to that of Descartes or even Viète. It is in no way superior to the methods of the Chinese algebraists of the 13th century in the sense that, unlike the Chinese methods, it can only handle polynomials of low degree and in a single variable. Thus, when, at the beginning of the 18th century, Chinese mathematicians began to rediscover the methods of their ancestors, they had no objective reasons for preferring the archaic European technique. This probably explains why practically none of them were interested in the *jiegenfang*.

Thus, until the mid 19th century, Chinese mathematics essentially continued to ignore European algebraic symbolism (or rather the primitive variety of symbolism missionaries had introduced them to).

This situation only began to change with the new translations of works on algebra and differential and integral calculus undertaken after 1850. The fundamental concern of the translators (A. Wylie, J. Edkins, Y. J. Allen, and others)[144] was to respect the principles of European algebra, while at the same time, they invented hybrid, half-Chinese, half-European notations. For example, they decided that $y = f(x)$ should become 地 = 函（天）(lit. "earth = contains (heaven)" where "earth" and "heaven" are the names of unknowns in the *tian-yuan* algebra) and that $\int 3x^2dx = x^3$ should become 禾 三天 { 天 = 天 (in other words, sum (of) 3 heaven² small heaven = heaven³, where 禾 and { are ideograms specially created by simplifying written characters of the standard vocabulary standing for "sum" or "total" and "small" or "minute," respectively.[145] In order to make available the equivalent of European letters

---

[142]See the notices of the *CRZ*.

[143]Peng Rita Hsiao-fu (1), *1975*, p. 401.

[144]Bennett (1), *1967*.

[145] 禾 and { are abbreviations for 積 and 微 , respectively.

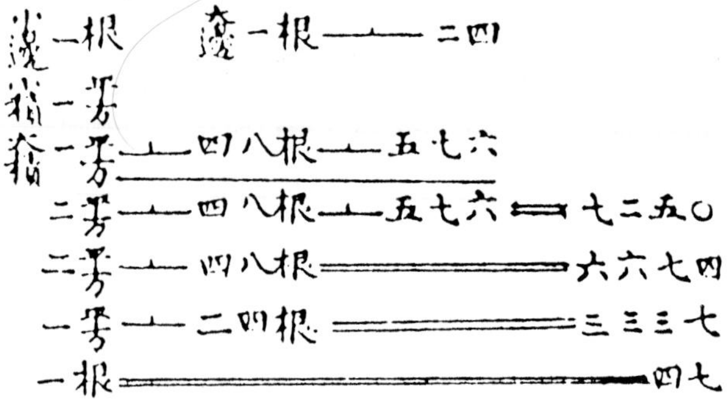

Source: *Shuli jingyun* (1723), j. 35, p. 17a (reprinted, Shanghai, 1888).

Notes:

1. The equals sign is represented by two extended parallel lines. Is this borrowed from Thomas Harriot?

2. The plus sign is denoted by ———. This is undoubtedly in order to avoid confusion with the number ten, which is denoted by ╋.

Translation:

small side	1 root		large side 1 root + 24				
small area	1 square						
large area	1 square	+	48 roots	+	576		
	2 squares	+	48 roots	+	576	=	7250
	2 squares	+	48 roots			=	6674
	1 square	+	24 roots			=	3337
	1 root					=	47

Fig. 10.6. The *jiegenfang* (the borrowing of roots and powers).

亥亥 ⬜ 甲丙 ⊕ 乙亥	means	$x^2 + ac = bx$
亥亥亥 □ □ 甲亥 ⊕ 庚	means	$x^3 - ax = g$
二戌 ⬜ 四亥 ⊕ 乙	means	$2y + 4x = b$

Fig. 10.7. Algebraic formulae in the manuscript *A'erribala xinfa* (New method of algebra) ca. 1710. From QB, *Hist.*, p. 278. See also Jami (1), *1986*. Note the curious signs ⊕, ⬜ and □ □ whose meaning is =, + and −, resp.

vii

SYMBOLS.

a 甲 *Kĕă*	A 呷 *Kĕă*	α 角 *Kĕŏ*	Λ 暃 *Kĕŏ*
b 乙 *Yĭh*	B 叿 *Yĭh*	β 亢 *K'ang*	B 吭 *K'ang*
c 丙 *Ping*	C 陃 *Ping*	γ 氐 *Tĕ*	Γ 呧 *Tĕ*
d 丁 *Ting*	D 叮 *Ting*	δ 房 *Fáng*	Δ 塲 *Fáng*
e 戊 *Mow*	E 哦 *Mow*	ζ 尾 *Wei*	E 呬 *Sin*
f 己 *Kĕ*	F 阣 *Kĕ*	η 箕 *Kê*	Z 喔 *Wei*
g 庚 *Kăng*	G 庚 *Kăng*	θ 斗 *Tòw*	H 噗 *Kê*
h 辛 *Sin*	H 崪 *Sin*	ι 牛 *Nêw*	Θ 叫 *Tòw*
i 壬 *Jĭn*	I 旺 *Jĭn*	κ 女 *Neù*	I 哶 *Nêw*
j 癸 *Kwei*	J 喫 *Kwei*	λ 盧 *Heu*	K 收 *Neù*
k 子 *Tszĕ*	K 呼 *Tszĕ*	μ 危 *Wei*	Λ 嘘 *Heu*
l 丑 *Chŏw*	L 吼 *Chŏw*	ν 室 *Shĭh*	M 帽 *Wei*
m 寅 *Yin*	M 嗔 *Yin*	ξ 壁 *Pĕĭh*	N 窒 *Shĭh*
n 卯 *Maŏu*	N 哪 *Maŏu*	o 奎 *K'wei*	Ξ 噼 *Pĕĭh*
o 辰 *Shĭn*	O 脤 *Shĭn*	ρ 胃 *Wei*	O 崒 *K'wei*
p 巳 *Szĕ*	P 吧 *Szĕ*	σ 鼎 *Maòu*	Π 嚶 *Lòw*
q 午 *Woó*	Q 呼 *Woó*	τ 畢 *Pĕĭh*	P 脆 *Wei*
r 未 *Wĕ*	R 味 *Wĕ*	υ 觜 *Tsuy*	Σ 噪 *Maòu*
s 申 *Shin*	S 呻 *Shin*	χ 井 *Tsĭng*	T 呷 *Pĕĭh*
t 酉 *Yĕw*	T 酉 *Yĕw*	ω 柳 *Lèw*	Υ 嘴 *Tsuy*
u 戌 *Sĕŭh*	U 哦 *Sĕŭh*	F 咂 *Hân*	Φ 嘰 *San*
v 亥 *Haé*	V 咳 *Haé*	ſ 函 *Hân*	X 呷 *Tsĭng*
w 物 *Wŭh*	W 吻 *Wŭh*	φ 楅 *Hân*	Ψ 塊 *Kwei*
x 天 *T'ĕen*	X 吙 *T'ĕen*	ψ 涵 *Hân*	Ω 瞅 *Lèw*
y 地 *T'ĕ*	Y 嗤 *T'ĕ*	Μ 根 *Kăn*	ε 訥 *Nŭh*
z 人 *Jin*	Z 㕥 *Jin*	π 周 *Chŏw*	d 彳 *Wĕ*
			ſ 禾 *Tsĕĭh*

A. WYLIE.

SHANGHAE,

July, 1859.

Fig. 10.8. A list of Chinese characters given as equivalents of Roman and Greek letters used in Western mathematical works. Certain of these characters are wholly new and have been especially created for that purpose. Source: *Daiweiji shiji, 1859*, p. vii.

used as algebraic symbols, the translators also created new Chinese characters by adding the mouth radical 口 to the left of cyclic characters. This led to series such as 呷 , 叹 , 呴   (Fig. 10.8).

However, they were not all in agreement on the page layout. Should the mathematical formulae be written horizontally or vertically? Each solution had its disadvantages; in the first case, incompatibility with the Chinese script made for a poor layout, while in the second case, the whole was difficult to read.

Finally, after 1920, Chinese mathematicians resolved the problem first by unrestrictedly adopting the Western symbolism and second by writing their language in horizontal rather than vertical lines.

11. Main Works and Main Authors (from the Origins to 1600)

The Ten Computational Canons

The Ten Computational Canons (*SJSS*) is the name commonly given to the collection of mathematical manuals compiled officially at the beginning of the Tang dynasty, from ancient or modern texts for the imperial examinations in mathematics.

> The *jianhou* (Astronomical Observer) named Wang Sibian had presented a memorial to the Throne reporting that ten computational canons [such as the] *Wucao* [*suanjing*] or the *Sunzi* [*suanjing*] were riddled with mistakes and contradictions. [Consequently] Li Chunfeng together with Liang Shu, a *suanxue boshi* (Erudite of Mathematics) from the *guozijian* (Directorate of Education), and Wang Zhenru, a *taixue zhujiao* (Instructor from the National University) and others were ordered by imperial decree to annotate ten computational canons [such as the] *Wucao* or the *Sunzi*. Once their task was completed, the Emperor Gaozu [ruled from 618 to 627] ordered that these books be used at the *guoxue* (National University).[1]

In fact, on the one hand (unless one adopts the belated point of view of scholars at the end of the 18th century), it is not very logical to call all the works which comprise this collection canonical: some of them were not ancient when they were incorporated into the *SJSS*. On the other hand, the number of these canons varies somewhat, depending on the source consulted.

What form did the collection of the Ten Computational Canons take originally? Was there a unique manuscript or were there several? Were students able to consult it readily or was it reserved for the masters? These questions deserve further investigation. Nevertheless, we do know that it was first xylographed in 1084 by the imperial administration.[2] According to the bibliography at the end of the *Suanfa tongzong* (1592), this collection consisted of the following works: the *Huangdi jiuzhang* (The Nine Chapters of the Yellow Emperor – the *JZSS* (?)); the *Zhoubi suanjing*; the *Wujing suanfa* (The Computational methods of the Five Classics); the *Haidao suanfa*; the *Sunzi suanfa* (Sunzi's Computational methods); the *Zhang Qiujian suanfa*; the *Wucao*

---

[1] *Jiu Tangshu* (Old Tang history), j. 79 (biography of Li Chunfeng), p. 2717.
[2] Cf. Li Yan, *Dagang*, I, p. 151.

Table 11.1. The transmission of the Ten Computational Canons from the origins to the end of the 18th century (summary). On this subject, see also: Kogelschatz (1), *1981*, op. cit.; Schrimpf (1), *1963*; Li Yan, *Dagang*, II, p. 470 ff.; Qian Baocong (25′), *1981*, p. 282 ff.

Title	Authors	Commentators	Period of compilation
(1) *Zhoubi suanjing* (Zhou Dynasty Canon of Gnomonic Computations)	Unknown	Zhao Shuang (third century?) Zhen Luan (sixth century) Li Chunfeng (seventh century)	100 BC (?)–600 AD
(2) *Jiuzhang suanshu* (Computational Prescriptions in Nine Chapters)	Unknown	Liu Hui (end third century) Xu Yue (start third century?) Zhen Luan Zu Xuan (sixth century) Li Chunfeng (602–670)	200 BC–300 AD and later if commentaries are taken into account
(3) *Haidao suanjing* (Sea Island Computational Canon)	Liu Hui		End third century
(4) *Sunzi suanjing* (Sunzi's Computational Canon)	Unknown	Li Chunfeng	fifth century very approximately
(5) *Wucao suanjing* (Computational Canon of the Five Administrative Sections)	Unknown	Li Chunfeng	fifth century? very approximately
(6) *Xiahou Yang suanjing* (Xiahou Yang's Computational Canon)	Unknown		?
(7) *Zhang Qiujian suanjing* (Zhang Qiujian's Computational Canon)	Unknown	Zhen Luan Li Chunfeng Liu Xiaosun	ca. 466–485 (Northern Wei)
(8) *Wujing suanshu* (Computational Prescriptions of the Five Classics)	Unknown	Li Chunfeng Zhen Luan	ca. 566 ?

Table 11.1 (continued)

Title	Authors	Commentators	Period of compilation
(9) *Jigu suanjing* (Computational Canon of the Continuation of Ancient)	Wang Xiaotong (ca. 650–750)	Li Chunfeng	seventh century
(10) *Shushu jiyi* (Notes on the Traditions of Arithmo-Numerological Processes)	Xu Yue (start third century)	Zhen Luan	?
(11) *Zhuishu* (The exact meaning of this title is unknown)	Zu Chongzhi (429–500)	Li Chunfeng	fifth century
(12) *Sandeng shu* (The art of the Three degrees; notations for large numbers)	Dong Quan[3]	Zhen Luan	sixth–seventh century

644–648	Li Chunfeng[4] and his team gather together and collate the ancient mathematical texts with a view to the official mathematics examinations (*ming suan*).
1084	First xylograph of the Ten Canons (none of these texts has survived).
1213	Second xylograph of the Ten Canons.[5]
1403–1407	Some of the Ten Canons are copied into the *Yongle dadian* (Great Encyclopedia of the Yongle reign-period).[6]
1573–1620	Publication of the *Zhoubi suanjing* and the *Shushu jiyi* in the *Bice huihan* collection.
1728	Publication of the *Zhoubi suanjing* and the *Shushu jiyi* in the encyclopedia *Gujin tushu jicheng.*
1773	Gathering together of reconstituted copies and manuscripts of the Ten Canons for the official compilation of the *Siku quanshu.*
1773	*Weiboxie* edition.[7]
1775–1794	*Wuying dian* edition of the Ten Canons.[8]

---

[3] *Jiu Tangshu*, j. 49, p. 2039.

[4] He was director of the astronomical observation service. Cf. *CRZ*, j. 13, p. 157.

[5] This edition contains all of the Ten Canons except the *Zhuishu* and the *Xiahou Yang suanjing* (cf. Kogelschatz (1), *1981*, p. 47).

[6] Those in question are: the *Jiuzhang suanshu* (9 j.); the *Sunzi suanjing* (in 2 j. rather than 3 j.); the *Haidao suanjing*; the *Wucao suanjing*; the *Xiahou Yang suanjing*; the *Wujing suanshu* and the *Zhoubi suanjing* (according to Li Yan, *Dagang*, II, p. 297).

[7] Cf. Kogelschatz, op. cit., p. 37.

[8] Ibid.

suanfa, the *Jigu suanfa*; the *Xiahou Yang suanfa* and the *Suanshu qiayi* (The *Shushu jiyi* (?)). However, this first edition has not survived.

In 1126, the siege of the capital of the Song (Kaifeng) by the Jin led to the dispersal of the imperial archives; the Song then took refuge in Southern China. Around 1213, with the re-establishment of the examinations in computation, the authorities decided to reprint the Ten Canons. The scholar responsible for supervising the operations, one Bao Huanzhi,[9] who was then a *dali pingshi*,[10] managed, albeit with great difficulty, to gather the works of the collection (which had become almost unobtainable) together again.

Under the following dynasties the collection again sank into oblivion from which it did not wholly reappear until the end of the 18th century,[11] thanks to the efforts of scholars and specialists in textual criticism including, in particular, Dai Zhen[12] (1724–1777) and Kong Jihan[13] (1739–1784).

Since then, the term Ten Computational Canons has been taken to refer to the works listed in the preceding Table.[14]

The *Zhoubi Suanjing*

The *Zhoubi suanjing* is above all important for the history of Chinese astronomy. This famous treatise (which dates back approximately to the Han dynasty),[15] contains cosmological speculations on the dimensions of the universe, based on both observations and mathematics.

As far as the history of mathematics is concerned, the *Zhoubi suanjing* is referred to by Chinese authors for two main reasons:

(i) It contains the "hypotenuse figure" *xian tu*, which provides an immediate visual proof of "Pythagoras's theorem," without words.

(ii) One of the commentaries of the *Zhoubi suanjing* by a certain Zhao Shuang (third century ?)[16] contains a list of 15 ready-made formulae for solving right-angled triangles.

The hypotenuse figure is frequently referred to because of the importance of Pythagoras's theorem and there is little need to dwell on the matter here.

The formulae for solving right-angled triangles formed the basis of a Chinese tradition which lived on until the end of the 19th century. Given that the

---

[9]Cf. *CRZ* j. 22, p. 269.

[10](*dali*: Chamberlain for law enforcement; *pingshi*: Case reviewer) (cf. Hucker (1), *1985*, notices nos. 5984 and 4712).

[11]Certain copies of books of the collection were in the possession of collectors of rare books. Moreover, certain texts such as the *Zhoubi suanjing* or the *Shushu jiyi* were reprinted during the Ming dynasty, in particular.

[12]Hummel, p. 695 ff.

[13]Ibid., p. 637.

[14]The order of the first ten texts in this list is that used by Qian Baocong in his edition of the Ten Canons (cf. Qian Baocong (25′), *1963*). Texts (11) and (12) have not survived.

[15]Cf. Qian Baocong (9′), *1929* and (23′), *1958*.

[16]Cf. Li Yan, *Dagang*, I, p. 37.

complete translation of the text containing these formulae is easily accessible,[17] we shall only cite one of the formulae here, namely that which may be used to solve a right-angled triangle given the two differences $(c-a)$ and $(c-b)$ between the hypotenuse (c) and the two sides (a and b, respectively) of the triangle:

兩差相乘，倍而開之，所得，以股弦差增之，爲勾。

以勾弦差增之，爲股。兩差增之，爲弦。

liang cha xiang cheng, bei er kai zhi, suo de, yi gu-xian-cha zeng zhi, wei gou. yi gou-xian-cha zeng zhi, wei gu. liang cha zeng zhi, wei xian.

Multiply [the] two DIF together; double and extract the [square] root; increase [the] result by LEG-HYP-DIF: makes BASE; increase this by BASE-HYP-DIF: makes LEG; increase this by two DIF: makes HYP.[18]

With BASE-HYP-DIF $=$ $c - a$ and LEG-HYP-DIF $=$ $c - b$, these prescriptions correspond to the following computations:

$$\sqrt{2(c-a)(c-b)} = a+b-c$$
$$(a+b-c)+(c-b) = a$$
$$(a+b-c)+(c-a)+(c-b) = c$$

The *Jiuzhang Suanshu* (*JZSS*)

The Importance of the *JZSS*. Most historians consider the *Jiuzhang suanshu* (Computational Prescriptions in Nine Chapters) to be a sort of Chinese mathematical Bible or (in a very different vein, the *JZSS* does not contain any proofs) as the equivalent of Euclid's *Elements*,[19] in other words, as the classic par excellence. As Wang Ling explains:

It is [...] a remarkable fact that with all these later improvements and extensions, the *Chiu chang* continues to hold, as it were, a nuclear position. Its influence is reflected in all subsequent Chinese mathematical works. Its problems stimulated the creation of a score of new topics of study. It set up a model for mathematical language and a pattern for computation. Calculators and accountants followed its path; calendar experts and astronomers copied and borrowed its technical terms. Written two thousand years ago, the *Chiu chang* opens the first chapter in the history of Chinese mathematics and has remained a kind of 'mathematical bible' ever since. [...] Not until the generation represented by the author of this thesis did the name of the *Chiu chang* cease to be constantly heard.[20]

---

[17]Brendan (1), *1977*.
[18]Original text in QB, I, p. 18, col. 8.
[19]Cf. Ogura (3') *1978*, p. 168.
[20]Wang Ling, *Thesis, 1956*, p. 16.

The *JZSS* is in fact at the origin of not only the Chinese mathematical tradition, but also those of Korea, Japan and Vietnam.[21] Moreover, its level is considerably higher than that of many other later works. According to K. Vogel:

[The *JZSS*], which was compiled in the first century AD, is a work of the very first order; because of its influence, it is probably also the most significant of all Chinese mathematical works. It is the oldest known arithmetic and its 246 problems make it incomparably richer than any surviving collection of Egyptian or Babylonian texts. In fact, the known Greek anthologies of arithmetical problems do not date back beyond the end of the Hellenistic and Byzantine periods.[22]

Indeed, at all periods, countless works have been inspired by the classification of mathematics in nine chapters or have borrowed their vocabulary and their resolutory methods from the *JZSS*. Innumerable authors refer to it as the example of the model to be followed. For a long time, no discordant voices cast doubt on the scientific importance of the *Jiuzhang suanshu*. Apparently, the first such doubts were only voiced in 1865, when Zeng Guofan wrote that:

Our Chinese Mathematical Work [i.e. the *JZSS*] arranged under nine heads founds its terminology throughout on concrete subject matters; treating each by separate method. The student sticking blindly to the track (marked out for him) pursues his search; and after a lifetime spent in practical mathematics knows his rules indeed, but knows nothing of the reason for them. So that mathematics are thought by some an impossible study owing to the wearisome multiplicity (of the rules); (but) simply they look vaguely at the methods and have not the sense to enquire after principles.[23]

Thus, the importance of the *JZSS* in the Chinese mathematical tradition is universally recognised, whether for its 'high level' (as stressed by Vogel) or for its dogmatism. In all cases, these value judgements seem to depend on the tacit idea that throughout its existence the *JZSS* has always been the same book we know today, having been composed and completed at a well-determined date (under the Han, for example at the beginning of the first century AD). However, one should not forget that the *JZSS* was primarily a textbook. It would have been normal for such a book to have been adapted from edition to edition in accordance with its particular function (addition of new problems and techniques, deletion of data which has become meaningless in a new context, etc.). Many chronological uncertainties remain and the process of transmission of the text is not well known.[24]

Textual Problems. The oldest edition of the *JZSS* known at the present time dates back to the 13th century.[25] But this is an incomplete edition limited to the first five chapters of the *JZSS* (or slightly more than half of

---

[21]Cf. Kim Yong-Woon (1), *1973*; Kim Yong-Woon and Kim Yong-Guk (1'), *1978*; Sugimoto and Swain, (1), *1978*.

[22]Vogel (1), *1968*, p. 1.

[23]Zeng Guofan's preface to the translation of Euclid's *Elements*, 1865 (quoted from Moule (1), *1873*, p. 150).

[24]Wu Wenjun (1'), *1982*, p. 28 ff.

[25]Ho Peng-Yoke (6), *1973*, p. 419.

the original text which is assumed to have had nine chapters). The only known example of this edition belonged to the bibliophile Huang Yuji (1629–1691); Mei Wending (1633–1721) is known to have consulted it in 1678. After Huang Yuji, it belonged to Kong Jihan (1739–1784), then to Zhang Dunren (1754–1834). It is now in the Shanghai library. In 1932 it was reproduced in the *Tianlu linlang* collection. The other known editions of the text of the *JZSS* date back to the end of the 18th century and all depend on reconstitutions based on quotations from the text (assumed to be from the venerable classic) recovered from the *Yongle dadian*, the great encyclopedia of the beginning of the 15th century which, unfortunately, is no longer accessible today in its entirety (it was almost completely destroyed at the end of the 19th century). Consequently, the oldest known part of the text of the *JZSS* is approximately 12 centuries later than the original while a further half a millennium separates the complete text. This situation is not a priori necessarily different from that which exists in the case of mathematical works from Ancient Greece even though, in the latter instance, the oldest known sources are much more numerous and varied (Arabian translations) than in the Chinese case. But a recent discovery could lead to a complete renewal of our knowledge of the origins of Chinese mathematics. In 1983–1984 a set of texts inscribed on bamboo strips was unearthed from three tombs, estimated to have been constructed between 187 and 157 BC, at Zhangjiashan near Jiangling in Hubei province. These texts included an arithmetic predating the *Jiuzhang suanshu*, the *Suanshu shu*, about which nothing at all (not even the title) was known until then.[26]

The part of this text which has now been published and translated is very similar to the present-day text of the *Jiuzhang suanshu* in its elementary part.[27]

In any case, it seems difficult to believe that the present-day text of the *JZSS* is not the result of a long evolution. Even Liu Hui's preface (263 AD)[28] states that:

> In the past, the tyrant Qin burnt written documents, which led to the destruction of classical knowledge. Later, Zhang Cang, Marquis of Peiping[29] and Geng Shouchang, Vice-President of the Ministry of Agriculture[30] both became famous through their talent for computation (*yi shan suan ming shi*). Because of the state of deterioration of the ancient texts, Zhang Cang and his team produced a new version removing [the bad parts] and filling in [what was missing]. Thus, they revised some parts, with the consequence that certain of these were then different from the old parts, the parts in question being essentially those which were discussed at that period (*er suo lun zhe duo jin yu ye*).[31]

---

[26] This discovery was announced in the newspapers *Renmin ribao* (Peking) on 17/01/85 and in the *Yomiuri shinbun* and *Mainichi shinbun* (Tokyo) on 18/01/85.

[27] Crossley and Lun (1), *1987*, pp. 56–59.

[28] Liu Hui's preface is undated; the date of 263 AD comes from the *Jinshu*, according to which this commentary on the *JZSS* was written by Liu Hui in the fourth year of the era of Jingyuan of the Wei (263 AD). Cf. *Jinshu*, j. 16, p. 492 and Wagner (2), *1978*.

[29] Zhang Cang (before 202 BC. 125 BC). See Li Yan, *Gudai*, p. 44.

[30] Geng Shouchang, (fl. ca. 61–48 BC), ibid., p. 44.

[31] QB, I, p. 91.

(i.e. the content of the problems was reworked using data including, in particular, economic data, readily understandable by contemporary readers).

Thus, some parts of the *JZSS* may have dated back to the earlier Han or may have been even older. Is it possible to determine which? This is certainly possible to some extent, since the socio-economic information contained in the text provides valuable information. For example:

(i) The sixth chapter of the *JZSS*, called *junshu*, refers to a system for distributing goods created under the Emperor Wu of the Han in 110 BC: "goods that were cheap in one part of the country were transported [...] to regions where they were scarce for resale. The purpose was ostensibly to provide better distribution of goods throughout the nation, though it would seem that the actual object was to make money for the government."[32]

(ii) In *JZSS* problem 3-5 the term *suan* denotes a tax, which is said to have been collected under the Han, in return for exemption from military service.[33]

(iii) *JZSS* problems 3-1, 3-6 and 3-8 mention five titles of staff responsible for road building.[34]

(iv) *JZSS* problems 6-21 and 7-19 both mention Chang'an, which was the capital of the earlier Han.[35]

(v) *JZSS* problem 6-9 mentions the royal park Shanglin which was built under the Qin, then abandoned. At the start of the later Han another park, also called Shanglin, was built near the new capital Luoyang. Animals and birds for the royal hunt were kept in this park. *JZSS* problem 6-9 speaks of the moving of cages.[36]

(vi) *JZSS* problem 1-4 uses the unit of area called the *qing*, which is equal to 100 *mu*, and which was not in everyday use before the Qin; the equivalence 1 *mu* = 240 *bu*, which appears constantly in the *JZSS*, dates back to 375 BC.[37]

In order to obtain other chronological reference points, attempts have also been made to compare the text of the *JZSS* with that on wooden strips discovered

---

[32]Watson, *1961*, II, p. 91. See also *Shiji*, j. 30, p. 1440; *RBS*, I, *1983*, p. 34: review of an article by Kageyama Tsuyoshi.

[33]This is an interpretation by Kusuyama Shūsaku in the *Tōhōgaku* no. 64, 1982, p. 15–29 (summary in *RBS*, I, 1983, no. 33). See also Wang Ling, *Thesis*, p. 61, which explains: "the technical word *suan*, for poll taxation, seems to be also a Han term started at the year 203 BC. In Wen-Ti's time, adults were under this name regularly taxed every three years. Thus, the second century BC would be a fair guess of the date of the third chapter of the *Chiu chang* [*JZSS*]."

[34]E. Chavannes, *Les mémoires historiques de Sse-ma Ts'ien*, Paris, 1895-1905, II, p. 528; Wang Ling, ibid., p. 386, note 63.

[35]On Chang'an see Wang Zhongshu (1), *1982*.

[36]Wang Ling, *Thesis*, *1956*, p. 64.

[37]Ibid., p. 56.

in Juyan (Etsingol) in Inner Mongolia in 1899 and 1930. On this subject, the sinologist Michael Loewe remarks that:

Further material regarding the dating of the book is provided by one of the fragments found at Chüyen [Juyan] [...] which reads: 囗 斗 二 升 廿 七 分 升 卅 六 術 曰 囗 上 下 ...” ... *tou* [*dou*] 2 and 26/27ths *sheng*. Method: ‘ ... upper and lower ... ’.” This is clearly a fragment of an answer to a mathematical exercise followed by a statement of the rule or method (*shu*). This exact pattern is seen in a number of the exercises of *CCSS* [*JZSS*]; although the text of this fragment is not seen in the present text of *CCSS* [*JZSS*], the similarity is striking enough to confirm that the present compilation probably includes authentic material drawn from the Han school books.[38]

M. Loewe's conclusion is all the more acceptable since it is corroborated by other examples.[39] It would then remain to establish the history of the texts associated with the *JZSS*. However, given the current state of the available documentation, that would be extremely difficult. The present-day *JZSS* contains areas of text which cannot be dated since they contain no historical, linguistic or geographical reference points, etc. (for example, chapters 4 and 8); not to mention the fact that we do not know how many chapters of text there were at any given period,[40] the number of problems per chapter, the order of the problems within each chapter[41] or the relationship between the countless works known as *Nine Chapters* and the *JZSS*.[42]

The Structure of the *JZSS*. The *JZSS* consists of a collection of 246 tripartite sequences which always comprise: the statement of a problem, the numerical answer and the ready-made prescription which may be used to calculate the solution from the data. Each sequence follows an invariable scheme and contains neither definitions nor logical explanations. Broadly speaking, the material is arranged in an order which depends on the degree of mathematical complexity (the calculation of plane areas precedes that of curvilinear areas; the simple rule of three precedes the composite rule of three, etc.). However, on the one hand, the *JZSS* does not include the beginnings of mathematics (it says nothing about how the four operations are carried out) and, on the other hand, the way in which the problems are grouped together is not necessarily based on any evident logical considerations, for the simple reason that, within each

---

[38]Loewe (1), *1961*, p. 70.

[39]Wu Wenjun (1'), *1982* (article by Li Di, p. 109–110).

[40]According to Liu Hui's preface (QB, I, p. 91) it would be appropriate to identify the "nine chapters" *jiuzhang* with the "nine types of computation" *jiushu*. See Li Yan, *Dagang*, I, p. 24 and QB, *Hist.*, 1964, p. 31. Wang Ling argues that the *jiushu* have nothing whatever to do with the nine chapters. "Liu Hui prejudiced by the usual commentary on "Chiu Shu" [*jiushu*], thought that the *Chiu Chang* [*Jiuzhang*] must have been written in 1000 B.C. and in order to explain why such an early book contained so many geographical names and terms of Han origin, he probably suggested some of the famous mathematicians in the Han period as its commentators to defend the authenticity of the book. We have so far no reliable source either to discount or to support him." (*Thesis*, p. 49).

[41]Wu Jing's *Jiuzhang suanfa bilei daquan* (1450) includes *all* the problems of the *JZSS*, but in an order not usually attributed to the problems of the *JZSS*. It is not known why.

[42]See for example the bibliography of the *Suishu*, j. 34, p. 1025.

chapter, the problems are, in approximately 60 per cent of the cases, grouped together according to the specific nature of the subject in question.

The Content of the Nine Chapters of the *JZSS*. Chapter 1 of the *JZSS* (38 problems) is entitled *fang tian* (square field). According to its title and Liu Hui's commentary on it, its general subject was the calculation of the area of fields of various different shapes: 'in the shape of a tablet' *gui* (actually an isosceles triangle), oblique (trapezium), in the shape of a tumulus (a non-plane shape, the exact nature of which is unknown), circular, in the shape of an arc or a ring. However, the logic behind the ordering of the problems in this chapter is not evident. The text effectively begins with calculations of area (problems 1 to 5) but continues (problems 5 to 18) with problems about fractions (simplification using Euclid's algorithm (in other words by alternating subtractions), addition, subtraction, multiplication, division, comparison of fractions with one another) and ends with other area calculations (problems 19 to 38) which do not depend on all the operations on fractions described in the preceding problems.

According to Liu Hui's commentary, Chapter 2 (46 problems), which is entitled *su mi* (decorticated and non-decorticated grains)[43] concerns the exchange of various goods (tiles, taffetas, cloth, silk, bamboo, etc.). It is not just about the exchange of grain, as the title might suggest. Thus, as we saw for chapter 1, the title does not refer to the whole chapter. But this may be only a simple 'empty' label with no particular logical value: in China, the chapters of books are often named in a purely mnemotechnical manner according to the first words of their text.

As far as the mathematical content itself is concerned, this chapter begins with what Western arithmetics refer to as the "rule of three" (37 problems), it then ends with other problems, concerning the exchange of goods, the solutions of which do not involve the rules of proportionality (problems 2-38 to 2-46). This may be because the present-day text of the *JZSS* is a late compilation of heterogeneous material. Anyway, it seems that the authors of the manual (or those who compiled latterly) found it difficult to order the problems of the chapter according to two classifying criteria at the same time; they seem to have oscillated between the temptation to retain either the purely mathematical criterion (proportionality) or the criterion of the actual subject of the problems (exchange). However, it is certainly true that the unity of certain mathematical criteria may still have been unclear in their day.

Chapter 3 (20 problems), which is entitled *shuai fen* (decreasing shares) is just as heterogeneous as the previous chapters, since the problems of proportional sharing (partnership) are similar to problems involving inverse proportions or the calculation of interest. As before, the pretexts of the problems concern everyday life; however, one wonders if these are more like exercises than

---

[43]Translation of *su mi* according to the explanations of Li Ji (commentator of the Song period) in his glossary of the *JZSS* entitled *Jiuzhang suanshu yinyi*, in: *Suanjing shi shu*, Taipei: Shangwu Yinshuguan, *1978*, I, p. 95 ff.

truly real situations since, for example, problem 3-5 asks how many men from different villages must contribute to the corvée, and gives a non-integral number of men as the answer:

Suppose that a village in the North has 8758 inhabitants, another in the West has 7236 inhabitants and another in the South has 8356 inhabitants. The 3 villages have to supply a total of 378 men to the corvée. If the number of men contributed by each village is determined by a decreasing function of the number of inhabitants, how many men must each village provide?

Answer: Northern village 135 and 11637/12175 men
　　　　　Western village 112 and 4004/12175 men
　　　　　Southern village 129 and 8709/12175 men.[44]

These calculations may be less gratuitous than they appear; they may be actual calculations carried out by the accounts of a finicky administration.

Chapter 4 (24 problems) is entitled *shao guang* (decrease [in length] to the benefit of the width) since its first 11 problems involve the determination of the constantly decreasing length of a rectangular field of fixed area equal to 1 *mu* with a variable width which increases from problem to problem. Remarkably, the statements are not arbitrary, but adhere to the following uniform prototype which covers them all:

Suppose a field has width $1 + 1/2 + 1/3 + \ldots + 1/n$ $(n = 2, 3, \ldots, 12)$. What must its length be if it [is rectangular and] has area 1 *mu*?[45]

These problems are clearly the pretexts for quite complicated calculations using fractions.

The other problems of the chapter are concerned, respectively, with the determination of the side of a square from its area, the edge of a cube from the volume, the length of the circumference of a circle from the area of the latter and the diameter of a sphere from its volume. In all cases, these problems depend on the extraction of square or cube roots. However, the unity of the chapter stems from the fact that the unknowns of the problems are all lengths.

Chapter 5 (28 problems), is entitled *shang gong* (estimation of works, i.e. civil engineering). It concerns the workforce needed to carry out various construction works, given the volumes to be excavated or erected (dykes, canals, ditches, basins, ramparts, etc.), the daily output of labourers depending on the nature of the earth (soft or compact) and also depending on the season and the nature of the material to be carted, the distances to be covered taking into account possible slowing down due to sloping ground, etc. This time, the computational rules include formulae for evaluating volumes which we would now call prisms, pyramids, tetrahedra and truncated cones.

Chapter 6 (28 problems) is entitled *junshu* (fair distribution of goods). It is principally concerned with the transport of goods, the assignment of soldiers to places far from their homes, sharing etc., taking into account constraints such as

---

[44]QB, I, p. 134.
[45]QB, I , p. 143 ff.

the numerical strength of the populations involved, the distances to be covered and the price of hiring carts. Its approach is apparently quite realistic.

The next two chapters, unlike the first six, are not characterised by a more-or-less pronounced uncertainty concerning the criteria used to classify the problems, since in all cases highly specific computational techniques are involved.

Chapter 7 (20 problems) expounds the double-false-position rules. It is entitled *ying bu zu* (too much and not enough), by allusion to the nature of certain values which result from the assumptions made (also called "false assumptions") about the value of the unknowns in a problem. (If a precise, but arbitrary value is attributed to one of the unknowns and the calculations laid down in the statement are carried out for this false value, it follows that a certain quantity is too large, not enough or "exactly right" (*shi zu*: just enough). Carrying out the same operations for a second time, a second value which is too large *ying* or not enough *bu zu* is obtained. Finally, an appropriate formula involving these excessively high or excessively small values is used to provide the exact value of the desired unknowns).[46]

This time, the ordering of the problems is visibly determined by a concern for pedagogic progression; firstly the statements indicate which false assumptions it is most appropriate to choose (in a real situation this choice would be the responsibility of the mathematician) and, secondly, the resolutory formulae are presented in increasing order of complexity. The simplest case is that in which the two false assumptions lead to errors of opposite types (one too large, one too small) – the formula only involves addition, multiplication and division and not subtraction (problems 7-1 to 7-4). This is followed by the case of errors of the same type and then by the random case in which the exact solution (not too large, not too small, but just right) is determined by chance.[47]

Chapter 8 (18 problems), which is entitled *fangcheng* (square arrays) is certainly by far the most original of all. It shows how to solve problems by manipulating numbers arranged in tabular fashion in parallel columns, possibly using negative quantities in the intermediate calculations. In modern terminology, this corresponds to the solution of linear systems in n unknowns (with $n \leq 6$ in the examples, although the rules are general). One of the procedures for solving these systems is none other than Gauss's method![48]

Chapter 9 (24 problems) is entitled *gougu* (base–leg) where, according to Liu Hui's commentary, the "base" is the "small" side of the triangle and the "leg" is the "large" side.

All the problems in this chapter have in common the fact that they introduce the right-angled triangle to apply Pythagoras's theorem (problems 1 to 4), to use the rules derived from this theorem to solve right-angled triangles (problems 5 to 14), to determine the side (resp. diameter) of a square (resp. circle) inscribed

---

[46]The double-false-position method is constantly used in medieval mathematics. (Analysis and numerous examples in Tropfke (3), *1980*, p. 371 ff.).

[47]Cf. The detailed analysis of chapter 7 of the *JZSS* in Lam Lay-Yong, (5), *1974*.

[48]See below, p. 254-255.

in this triangle (problems 15 and 16) or to measure distances indirectly. The calculations involved are based on similarity properties.

The Commentaries of the *JZSS*. The *JZSS* has many commentaries and sub-commentaries by Xu Yue (second–third century), Liu Hui (third century), Li Zunyi (fourth or fifth century), Zhen Luan (fl. 570), Zu Xuan (sixth century), Li Chunfeng (602–670), Yang Hui (second half of the 13th century), Li Huang (?–1812), and others.

Not all these commentaries have survived. Of those dating back to the first millennium AD, we now only have access to those by Liu Hui, Li Chunfeng and Zu Xuan (in the last case we only have a fragment of the original).

In all probability, these are all the result of a mixture of different commentaries at periods several centuries (even a millennium or more) apart. In fact, it is easy to find many discontinuities in them, which are otherwise inexplicable. For example, the commentary jumps from one subject to another and suddenly broaches a subject which is clearly outside the framework of the problem on which it is supposed to be commenting,[49] the system of units is inconsistent,[50] and the chronology of the historical events mentioned is not without contradictions.[51]

Thus, it is clear that although the texts of the commentaries which we possess are supposed to be by Liu Hui, Li Chunfeng or Zu Xuan, we have reasons to doubt their authenticity. There is another similar difficulty. In all the known editions of the *JZSS*, these three commentaries form a single compact block, so that it is not always easy to determine what belongs to which author. Certainly, we in principle know what is due to Li Chunfeng, since the text states this explicitly (in all cases, the formula *Chen Chunfeng deng qin an*, which means "Your servant, Chunfeng and his collaborators comment respectfully (on the given problem of the *JZSS*)" announces that Li Chunfeng is the author of what is to follow). But in Liu Hui's case, there is nothing like this. Assuming the transcribers who transmitted the text made mistakes (how can we avoid assuming this when we consider the poor condition of the present-day *JZSS*), it follows that the two commentaries are irreparably muddled at numerous places.

Thus, there is a need to dream up appropriate ways of attempting to physically separate the commentaries from one another. This has led to analysis of the style of the text, the quotations which it contains and the subjects with which it is preoccupied.

Thus, Guo Shuchun has recently advanced the plausible conjecture that Liu Hui was influenced by the neo-Taoists[52] Ji Kang (223–263), Wang Bi (226–249), He Yan (?–249) and Guo Xiang (?–312) who left their mark in history as the

---

[49]This is the case, for example, of Liu Hui's commentary on *JZSS* problem 1-32 (calculation of π).

[50]Reifler (1) *1965*.

[51]Wagner (2), *1978*.

[52]On this, see for example, Chan Wing-tsit (1), *1963*, p. 314.

architects of a revival of interest in the thinkers of the period of the Warring States – Laozi, Zhuangzi[53] and Mozi.

The fact that Li Chunfeng was the author of other commentaries apart from that on the *JZSS* has also been used to advantage. Generally speaking, this author tends to rely on numerical proofs and empirical procedures rather than advancing logical and theoretical arguments as Liu Hui does. This provides a way, albeit far from perfect, of attempting to determine what is due to Li Chunfeng.

The *Haidao Suanjing*[54]

The Sea Island Computational Canon by Liu Hui concerns the determination of inaccessible points.

The first of the nine problems of the work involves the determination of the highest point of an island in the sea, whence the title. The other eight problems involve the calculation of the height of a fir tree growing on a mountain side, the distance to and the length of the side of a square town, the depth of a gorge, the height of a tower on a hill given that the observer is in the plain, the width of a river estuary, the depth of an abyss full of clear water, the width of a ford viewed from a hill and the dimensions of a town viewed from a mountain side. Analogous questions are often found in European arithmetics of the Middle Ages.[55]

From the mathematical point of view, the text contains only computational prescriptions. These were probably obtained using the similarity properties of right-angled (rather than arbitrary) triangles (Liu Hui uses only right-angled triangles in his works).

The *Sunzi Suanjing*

Sunzi's Computational Canon is an arithmetic textbook by an unknown author; whilst its period is also unknown, it has long been thought to date back to Early Antiquity. The poet and bibliophile Zhu Yizun[56] (1629–1709) thought that Sunzi was none other than Sun Wu, the well-known strategist and author of the *Sunzi bingfa* (Master Sun's Art of War). In his *Chouren zhuan*, Ruan Yuan[57] included Sunzi at the end of the chapter on the Zhou dynasty.[58] However, Ruan Yuan was well aware of the existence of a chronological problem concerning Sunzi and, among other things, he carefully noted that the mention of a Buddhist sutra in the statement of the fourth problem of the third chapter

---

[53]Note that Liu Hui explicitly cites Zhuangzi (cf. QB, I, p. 237).

[54]Complete translation of the original text in Schrimpf (1), *1963*; see also Ang Tian-Se and Swetz (1), *1986*, pp. 99–117.

[55]Cf. Vogel (2), *1983*.

[56]Biography in Hummel, p. 182. See also QB, II, p. 275.

[57]Hummel, p. 399.

[58]j. 1, p. 8.

of the *Sunzi suanjing* was chronologically incompatible with Sun Wu; at the same time, he also explained that the question had to be studied later and that the listing of Sunzi in the chapter on the Zhou dynasty was provisional.[59]

More recently, authors such as A. Wylie and L.E. Dickson have placed Sunzi at less remote dates, but without justifying their beliefs.[60] Based on the fact that the text mentions very special terms about certain systems of taxes, Wang Ling thought he was able to place the complete work between two very precise dates: 280 and 473 AD:

> The *Sun Tzu Suan Ching* mentions the *mien* as an item of taxation, and the *hu tiao* system. These two were first established in 280 AD. So the book could not have been written before this date. [...] A new scale between *chih* and *tuan* was established in 474 AD, the *Sun Tzu*, still using the old scale by Wu Ch'en-Shih's emendation cannot be older than 473 AD.[61]

However, the bibliography of the *Suishu*,[62] which was compiled in 636,[63] refers to a *Sunzi suanjing* with two chapters although the present-day version actually has three.[64] Thus, it seems possible that the present-day text of this Sunzi is a compilation from various sources.

As far as content is concerned, the three chapters of the *Sunzi suanjing* do not have a homogeneous structure, since only the last two chapters consist of a series of problems.

The first chapter begins with a description of systems of measures of length, weight and capacity and continues with a short presentation of special Chinese characters denoting large powers of ten (from 10^4 to 10^{80}). Then, the text gives simple explanations of counting-rods and the practice of multiplication, division and extraction of square roots with these. For the first time, these operations are described with a relative wealth of detail which makes it possible to reconstruct them without introducing too many suppositions (for this reason, numerous authors have taken the *Sunzi suanjing* as representative of Chinese arithmetical practices of earlier periods for which no such information is available; needless to say, such explanations of the past by means of the future are subject to caution). In addition to arithmetical operations, the chapter ends with a multiplication table for products of numbers with a single digit, beginning with 9×9 and ending

---

[59] *CRZ*, j. 1, p. 8.

[60] Dickson ((1), *1919*, II, p. 57) writes: "Sun-Tsu in a Chinese work Suan-ching (arithmetic) about the first Century AD [...]"; Wylie ((3), *1897*, p. 165)) writes that: "During the third Century Sun-tsze, an author of considerable note, published his *Swan-king*, 'Arithmetical classic'." Yet, in a text published in 1905, the Japanese historian T. Hayashi still thought that Sunzi was the same person as Sun Wu (Hayashi (1'), *1937*, I, p. 950).

[61] Wang Ling (3), *1964*.

[62] *Suishu*, j. 29, p. 1025.

[63] Wright (1), *1978*, p. 14: "[...] the *Sui History* [...] in eighty-five chapters compiled under imperial auspices early in the succeeding dynasty [...] was completed in 636 by Wei Cheng, the tough Confucian counselor to T'ang T'ai-tsung."

[64] *Suishu*, j. 34, p. 1025.

with 1×1.[65] All this has much in common with the Pelliot 3349 Dunhuang manuscript.[66]

Chapters 2 and 3 consist of 28 and 36 problems, respectively. These relate to fractions, proportions, computation of areas, volumes, loans, proportional sharing, conscription, the *fangcheng* and *ying bu zu* methods and other subjects similar to those of the *JZSS*. On the whole, these problems are simpler than those of the *Jiuzhang suanshu*. However, the *Sunzi suanjing* is noteworthy for its famous remainder problem (*SZSJ* 3-26), which we shall analyse later (below, p. 310). Many versions of this problem may be found in medieval arithmetics of all countries; Sunzi's would be the oldest of these.

While the remainder problem has aroused much admiration, the following problem has given rise to negative comments by certain historians of Chinese mathematics[67] who found it illogical and not worthy of a serious mathematical treatise:

A woman aged 29 has been nine months pregnant. What is the sex of her future baby?
Answer: male.
Method: Set down 49, add the gestation period and subtract the age [of the woman]. From the remainder take away 1 [the number of the] heaven, 2 that of earth, 3 the man, 4 the four seasons, 5 the five phases, 6 the six pitchpipes, 7 the seven stars [of *Ursa Major*], 8 the eight winds and 9 the nine territories [of China under Yu the Great]. If the remainder is odd, the infant will be a male, if even, a female.[68]

(The answer is "a male" because $49 + 9 - 29 - 1 - 2 - \ldots - 7 = 1$; 8 and 9 are not subtracted because the subtraction is "impossible").

In fact, as Yan Dunjie showed, j. 14 of the *Neijing suwen*, a classical book in Chinese medicine, contains a passage similar to that of the above rule and problem 3-36 only makes sense if ancient Chinese medical and numerological considerations are also taken into account.[69]

The *Zhang Qiujian Suanjing*[70]

Zhang Qiujian's[71] Computational Canon is a work in 3 *juan* consisting of 15, 22 and 38 problems, respectively. These include several original questions not found in the venerable *JZSS*, notably: problems on the least common multiple (problems 1-10 and 1-11), problems on arithmetic progressions (problems 1-22,

---

[65]For a complete translation of the text, see Lam Lay-Yong and Ang Tian-Se (1), *1992*, p. 151 ff.

[66]See Libbrecht (5), *1982*, pp. 211–225.

[67]See QB, *Hist*, p. 79.

[68]*SZSJ* 3-36 (QB, II, p. 322).

[69]Yan Dunjie (3'), *1937*, p. 27.

[70]I have not had access to Ang Tian-Se's thesis ((1), *1969*). See however the translation of all the statements of the problems in Schrimpf's thesis ((1), *1963*).

[71]Nothing is known about the life of this author.

-23, -24, 2-1 and 3-36)[72] and a curious indeterminate problem which historians of mathematics usually call the problem of the hundred fowls (problem 3-38). In all the editions of the text currently known, the end of chapter 2 and the beginning of chapter 3 are missing. Some authors (Ho Peng-Yoke in particular)[73] believe that the solution of problem 2-22 was based on a simple numerical rule for solving equations of degree two, known in China from the 13th century under the name of *daicong kaifang fa*.[74]

The *Wucao Suanjing*

The *Wucao suanjing* (Computational Canon of the Five Administrative Sections) is an anonymous, undated work which, as its title indicates, must have been intended for those involved with surveying, the management of troops, tax collection, granaries and money, in other words, the Five Sections *wucao* instituted under the Wei dynasty after 220 and retained until around 589 AD.[75]

The first chapter, *tian cao* (Section for Cultivated Fields) contains 19 formula for calculating the area of fields in the shape of a snake, a wall, a flute, a bull's horn, etc., the geometric nature of which is never defined and can only be understood by analysing the formulae. Yang Hui, a 13th-century arithmetician, became interested in these problems in order to expose their mistakes;[76] strictly speaking, some of them are wrong, although they provide a good enough approximation for current practical needs.

The second chapter (12 problems) *bing cao* (War Section) is concerned with determination of the number of soldiers that a population would be liable to provide (problems 2-1 and 2-2 are solved by simple division); calculation of the payment in kind or in money (problems 2-3 to 2-7); determination of the numbers of men and carts needed to defend a town (problems 2-8 and 2-10) and calculation of the provisions needed for men and animals (problems 2-9, 2-11, 2-12).

The third chapter (14 problems), *ji cao* (Accounts Section) is concerned with trade in cereals, the amount of goods which can be bought for a given sum of money and other questions similar to those of chapter 2 of the *JZSS*.

The fourth chapter (12 problems) *cang cao* (Granary Section) deals with taxes, transport and storage of grain.

The fifth chapter (10 problems) *jin cao* (Treasury Section) deals with the purchase and distribution of silk, brocade, etc. It assumes only the rule of three and the conversion of units.

---

[72]The rules in the text give formulae for calculating: the sum of the terms as a function of the first term and the ratio; the number of terms as a function of the mean value of the terms, the first term and the ratio; the sum as a function of the ratio and the extreme terms.

[73]Ho Peng-Yoke (1), *1965*, pp. 37–53.

[74]Lit. "method for extracting square roots accompanied by a *cong*" (the *cong* is the term which follows the square). This involves extracting generalised square roots of the type "square + *cong* = dividend," i.e. $x^2 + ax = b$, cf. Lam Lay-Yong, (6), *1977*, p. 252.

[75]Li Yan, *Dagang*, I, p. 76.

[76]Cf. Lam Lay-Yong, ibid., p. 110 ff.

The *Wujing Suanshu*

The Computational Prescriptions of the Five Classics (2 *juan*) is a commentary on passages extracted from classical works such as the *Yijing*, the *Shijing*, the *Zhouli*, the *Yili*, the *Liji*, the *Lunyu* and the *Zuozhuan*, which a priori have nothing to do with mathematics. The glosses by Zhen Luan (calendarist astronomer known for his adherence to Buddhism)[77] relate to:

(i) the correct interpretation of certain calendrical data found in these classics (involving, in particular, the determination of intercalary months);

(ii) large numbers;

(iii) the *dayan* divination method of the *Yijing*;

(iv) mythical geography;

(v) weights and measures and musical tubes.

The *Jigu Suanjing*

The Computational Canon of the Continuation of Ancient (Techniques) is the work of the calendarist Wang Xiaotong (who is also known as a mathematician and was a doctor of computational science (*suanxue boshi*)).

The work begins with a pursuit problem (pursuit of a hare by a dog). According to the author's explanations, the problem is a pretext concealing an astronomical question (movement of the Moon across the Chinese solar divisions (*jieqi*)). This is the only problem of its type.

This is then followed without any transition by 13 problems concerning the construction of an astronomical observation tower (problem 2), the building of a dyke (problems 3 and 4), the excavation of a canal bed (problem 6), etc. The following partial quotation from the statement of one of these problems will give the reader an idea of the apparent imbroglio which characterises them:

> Suppose that a dyke is to be built. The difference between the lower and upper widths of the West face [is] 6 *zhang* 8 *chi* 2 *cun* [...], the height of the East face is 3 *zhang* 1 *chi* less than the height of the West face [...]. Sub-prefecture A [numbers] 6724 men, sub-prefecture B 16677 men [...]. Each man of the four sub-prefectures excavates 9 *dan* 9 *dou* 2 *sheng* of earth per day and packs down on average 11 *chi* 4 *cun* and 6/12 *cun* of earth per day [...]. Previously, a man was able to cover a horizontal road of length 192 *bu* 62 times per day carrying 2 *dou* 4 *sheng* and 8 *he* of earth on his back. But now, it is necessary to climb a hill and cross a watercourse in order to fetch the earth. A climb of 3 *bu* is equivalent to 4 *bu* [for a flat road] [...] 1 *bu* for a ford is equivalent to 2 *bu* [for a flat road] [...]. What is the daily task of a man who digs, transports and constructs; what are the heights and the upper and lower widths of the dyke [...].[78]

Clearly, Wang Xiaotong mixes questions which actually arise in reality (those relating to the planning of works) with those which do not (those

---

[77]Zurcher (1), *1959*, p. 302.

[78]*Jigu suanjing*, problem no. 3 (QB, II, p. 502).

concerning the dimensions of the dyke, which are normally known at the beginning). This kind of problem was probably inspired by the colossal construction of long canals.[79]

These problems are then followed, again without transition, by a final six problems, the sobriety of which contrasts strongly with the complication of their predecessors. These involve the solution of right-angled triangles, leading to polynomial equations which are either biquadratic or of degree three.

The *Shushu Jiyi*

Traditionally, the Notes on the Traditions of Arithmo-Numerological Processes is attributed to Xu Yue (fl. later Han). In fact, it is most probably an apocryphal work.[80]

One of the subjects of the work is the written expression of large and very large numbers in different numeration systems using special terms beyond the myriad for certain powers of ten.[81] But these systems were perhaps also divised in order to raise the question of the finiteness of numbers and to argue that any number, however large, is expressible using powers of ten in an astute way which would allow exponents to take arbitrarily large values by means of "cyclical variations." In that way the Chinese character corresponding to '1', for example, would sometimes mean 'one' and sometimes some other power of ten.[82] Yet, interspersed as it is with sybilline theological considerations, borrowed from the Buddhist idea of infinite cycles of reincarnation, the relevant part of the text of the *Shushu jiyi* is really difficult to interpret in a mathematical sense even using Zhen Luan's commentary.

The logic of other numeration systems in the *Shushu jiyi* appears no less difficult to apprehend. For example, "the two principles" *liang yi* (*yin* and *yang*) or "the three powers" *san cai* (Heaven, Earth and Man) systems evidently take their names from the vocabulary of the *Yijing*, but their true meaning can only be conjectured at.

The *Xiahou Yang Suanjing*

Xiahou Yang's Computational Canon is an apocryphal compilation[83] consisting of 19 + 29 + 44 elementary problems spread across three chapters. It has no notable features, except possibly its simplified computational technique which may be applied to carry out multiplication using a single line of counting-rods (instead of three lines as in the method of the *Sunzi suanjing*).

---

[79]Cf. Wright (1), *1978*, p. 179. As Wright notes, the amount of labour necessitated by the construction of such canals was so enormous that in 608 "hundred-odd tens of thousands" workers were mobilised and even women were conscripted.

[80]See the notice of the *Siku quanshu zongmu* on this work (j. 107, p. 903).

[81]See above, pp. 97 ff.

[82]Cf. Volkov (2), *1994*'s tentative explanations.

[83]Cf. QB, II, p. 551.

Jia Xian and Liu Yi

According to Qian Baocong,[84] Jia Xian (fl. 1050) was a *zuoban dianzhi* (Palace Eunuch of the Left Duty Group)[85] known both as the pupil of a famous calendarist Chu Yan[86] and as the author of a work which has not survived, the *Huangdi jiuzhang suanfa xicao* (The Yellow Emperor's Detailed Solutions of Computational Methods in Nine Chapters). According to Bao Huanzhi's explanations included at the beginning of the *Xiangjie jiuzhang suanfa* (Detailed explanations of the computational methods in nine chapters) – a work composed by Yang Hui in 1275 – a popular manual entitled *Huangdi jiuzhang* (The Nine Chapters of the Yellow Emperor) was widespread in China during his lifetime but this was nothing other than a downgraded version of the *JZSS* which did not include Liu Hui and Li Chunfeng's commentaries.[87]

As far as Liu Yi is concerned, little is known about him, apart from the fact that he is said to have written a certain *Yigu genyuan*[88] (the precise meaning of this title is not known).

The importance of these two, whose works we only know through Yang Hui's quotations from them, is due to their contributions in the area of the extraction of square and cube roots or even more general operations corresponding to the numerical computation of a root of certain polynomials of degree 2, 3 or 4.

According to these quotations,[89] two mathematically distinct methods for extracting roots may be attributed to Jia Xian. The first of these, which is called *licheng shisuo*,[90] depends on the binomial expansion of $(a + b)^n$ (Pascal's triangle) and the second, which is suggestively called *zeng–cheng* (i.e. additive–multiplicative method) implements the "Horner" rule. Liu Yi's contribution is of the same order; one of the 22 problems attributed to him by Yang Hui is actually solved using a "Horner" method.[91]

---

[84]QB, *Hist.*, p. 144.

[85]Hucker (1), *1985*, article no. 6983.

[86]Biography in *Songshi*, j. 462, vol. 39, p. 13517 ff.

[87]Cf. *Xiangjie jiuzhang suanfa* (Shanghai: Shangwu Yinshuguan), *1936*, p. 5.

[88]Cf. Lam Lay-Yong, (6), *1977*, p. 83.

[89]Quotations due to Jia Xian: see *YLDD*, j. 16344, p. 5b; see also the *Xiangjie jiuzhang suanfa cuanlei* (Shanghai: Shangwu Yinshuguan), *1936*, p. 36; quotations due to Liu Yi: see Lam Lay-Yong, op. cit., pp. 113–131.

[90]*Licheng*, (lit. "immediately obtained") denotes a pre-calculated numerical table. In this case, it is a Pascal's triangle. The precise meaning of the term *shisuo* (lit. unlocking) is not known.

[91]Lam Lay-Yong, op. cit., pp. 130–131.

Li Zhi (1192–1279) [92]

Li Zhi, whose father Li Yu was a *muliao* (attaché or secretary) to a certain Jurchen officer called Hu Shahu,[93] was born in Daxing (now Peking) in 1192. Having taken his doctorate at Luoyang in 1230, he was appointed Assistant Magistrate (*zhupu*) for the district of Gaoling in what is now Shenxi province; but he was unable to take up his post because of the war and was appointed governor of the prefecture (*zhou*) of Jun in what is now Henan.

When the Mongols took Kaifeng in 1233, he escaped being massacred thanks to the intervention of Yelü Chucai (1190–1244), a former high-ranking Jurchen official who had gone over to serve the Mongols.[94] In 1234, he took refuge in Shanxi and lived as a hermit. However, he entered into contact with great literati of his time such as Wang E[95], Zhang Dehui,[96] and Yuan Haowen.[97] It was during this period that he began to compose his main work, the *Ceyuan haijing*.[98] In 1257, the future Mongol emperor Qubilai sent for him to ask his opinion on how to govern, organise recruitment competitions and interpret earthquakes.[99] In 1259, he completed his second mathematical work, the *Yigu yanduan*.[100] In 1264, the same Qubilai admitted him to the newly created Hanlin Academy, so that he could participate in the editing of the official dynastic annals of the Liao and the Jurchen. However, he resigned several months later, on the pretext of old age and ill health; he died in 1279.

The *Ceyuan Haijing* by Li Zhi (1248)

The *Ceyuan haijing* (Mirror comparable with the ocean[101] reflecting [the heaven] of calculations of [inscribed and circumscribed] circles)[102] is a treatise in 12

---

[92] Cf. *DSB*, VIII, p. 313–320; *CRZ*, j. 24, p. 287–289; Mei Rongzhao in *SY*, p. 104–107; Needham in *SCC*, III, p. 40–41; *Yuanshi*, j. 160, pp. 3759–60. Li Zhi is the real name of the person commonly known as Li Ye (note that the character *zhi* 治 is written with only one more 'dot' than the character *ye* 冶).

[93] *SY*, p. 105.

[94] Ch'ên Yüan (1), *1966*, p. 33.

[95] *DKW*, 7-20823: 160; *Yuanshi*, j. 160, p. 3756.

[96] Ibid., 4-9812: 1418; *Yuanshi*, j. 163, p. 3815.

[97] Ibid., 1-1340: 646.

[98] The author's preface is dated 1248. The original edition has not survived. The oldest existing version of the *Ceyuan haijing* still seems to be that which includes a manuscript composed at an undetermined period which belonged to the great scholar Song Lian (1310–1381), the main compiler of the historical annals of the Yuan (*Yuanshi*). This manuscript is now in the Peking library. Cf. *SY*, p. 111.

[99] *Yuanshi*, j. 160, p. 3760.

[100] This title is difficult to translate (the precise meaning of the term *yanduan* is unknown). Cf. Mikami, (9'), *1937*; Lam Lay-Yong and Ang Tian-Se (1), *1984*.

[101] Translation of this title, based on Li Zhi's explanations in his preface: "This book is entitled *Ceyuan haijing* to recall the idea of the heavens which are reflected in the mirror that is the ocean." This explanation probably refers to numerological ideas, such as those of Shao Yong, which were widespread during Li Zhi's lifetime.

[102] Categorically, Li Zhi does not have the computation of π in mind.

chapters (*juan*) which, after the author's preface, successively includes the following:

(i) a geometrical figure;
(ii) preliminary information (chapter 1);
(iii) 170 problems (chapters 2 to 12).

The geometrical figure at the head of the book is called *yuan cheng tu shi* (figure of the round town) (Fig. 11.1). Since this figure is appropriate for all the problems, it is the only one of the text. When one looks at it for the first time, it appears to be a banal figure analogous to those which decorate our geometry books, except that the geometrical points are clearly represented here by characters of Chinese script rather than letters of the alphabet. However, the resemblance is deceptive, since the Chinese characters do not represent points but geographical or geomantic reference points (note that the figure includes the cardinal points, together with certain hexagrams from the *Yijing*). But Li Zhi's geometrical system is not the same as Euclid's, for one even more essential reason: in his case, three written characters do not define either an angle or a triangle (strictly speaking, these concepts do not exist in Chinese geometry). Instead of speaking about the triangle (in general, the equivalent of this word does not exist in the text) the author only ever mentions the specific elements of certain triangles, namely the lengths of the sides. Instead of speaking about "the base AB of the triangle ABC," he uses an expression of the form "the X base," where X is a qualifying adjective. For example, for him, what other mathematicians would call AB becomes "the general base" or "the minute base" (*tong gou, zhuan gou*, resp.), in other words, the base of the largest (resp. smallest) triangle in the figure, and so on. We know of no other author (Chinese or not) apart from Li Zhi who takes this approach.

It is possibly because he is the only one to use such conventions that Li Zhi feels it necessary to include preliminary information which is grouped together into three separate sections within the first *juan*.

The first section is entitled *zong lü ming hao* (names of all the quantities). In this section, the author explains the meaning of special terms (such as *tong xian*, that is, "the hypotenuse [HYP] of the general [GEN] [triangle]") which are used to name various roads on the figure (or rather on the "town plan"). For example:

天之地爲通弦

tian zhi di wei tong xian.

[From] sky to ground [that] makes GEN–HYP.[103]

The second section is entitled *jin wen zheng shu* (correct numbers used in the problems). This is a list of all the values (in integers) of the segment lengths (or sums or differences of these lengths) used in the problems, where the large

---

[103]Bai Shangshu (6′), *1985*, p. 2.

triangle is of type $(680, 320, 600) = 40 \times (17, 8, 15)$.[104] One may wonder about the *raison d'être* of this list. Certainly, the problems of the work are artificially constructed using numbers which "fall out correctly," but that does not justify such a wealth of detail. Of course, Li Zhi may have wanted to specify right-angled triangles in terms of integers. However, as we shall see later, the *Ceyuan haijing* is concerned with algebra and not with number theory. But one might conjecture that Li Zhi used this list to verify the correctness of the algebraic results which he obtained at various stages of his calculations. He may also have used it to obtain general results based on specific numerical calculations; in fact, this may explain certain mistakes in the text.[105]

The third and final section is entitled *shibie zaji* (Various notes designed to distinguish different things from one another).[106] This is a long list of 692 (six hundred and ninety two!) formulae involving the areas of triangles or the lengths of segments.[107] The author gives these formulae in the raw, with no logical explanation of any kind whatsoever. They may refer to the equality of lengths, as in the following case:

天之於日與日之於心同

tian zhi yu ri yu ri zhi yu xin tong.

From sky to sun and from sun to heart: equal.[108]

But they may also refer to the equality of areas. For example:

明弦明股併與 重弦重勾併相乘得半經冪

ming-xian ming-gu bing yu zhuan-xian zhuan-gou bing xiang cheng de banjing

Sum of LUMINOUS-HYP and LUMINOUS-LEG times sum [of] MINUTE-HYP [and] MINUTE-BASE makes [the] SQUARE of the half-diameter [of the circular town].[109]

These formulae may appear quite anodyne, even though their proofs are not always completely self-evident. However, Li Zhi had no interest whatsoever in the geometrical aspect of the question. The only important thing as far as he was concerned was the role the formulae might play in his computational practices. One might think that he must have needed a vast repertoire of ready-made formulae in order to experiment with a far greater number of computational

---

[104](17, 8, 15) is a Pythagorean triplet, i.e. a set of three integers representing the lengths of the three sides of a right-angled triangle. It has long been known that such triplets already occurred in chapter 9 of the *JZSS* (Cf. K. Vogel, (1), *1968*, for example). This triplet is one such. Note also that $40 = 17 + 8 + 15$.

[105]On these mistakes, cf. Bai Shangshu, (5'), *1985*, p. 113.

[106]On the meaning of the term *shibie*, see: *DKW* 10-35974: 59; *zaji* is the title of one of the chapters of the *Liji* and means "miscellanies," "various notes."

[107]Cf. *SY*, p. 114.

[108]Bai Shangshu, op. cit., p. 17.

[109]Bai Shangshu, op. cit., p. 24; *SY*, p. 115. We do not give here the modern equivalent of this formula for we are only interested in Li Zhi's specific formulation.

situations than those provided for him by the few classical algorithms of the Chinese tradition relating to right-angled triangles.

After these substantial preliminaries, the 170 problems finally begin (chapters 2 to 12). They all have the same story line: two men A and B (*jia* and *yi*) walk along certain roads around a circular town, the plan of which is given at the beginning of the text in the form of a geometric figure (that which we have already discussed) which includes, in particular, a circle inscribed in a right-angled triangle. These men attempt to catch sight of one another or of a given object (such as a tree) which was hidden from them by the town walls. The question is invariably to determine the diameter of the town, given the distances they have walked. The answer is invariably the same: 120 *bu*.

Despite their concrete appearance, these problems are highly artificial and unrealistic and are designed to illustrate a new type of computational technique, the *tianyuan* algebra.

Unlike most Western algebraists, Li Zhi never explains how to solve equations, but only how to construct them. But he does not limit his reflections to equations of degree two or three; for him, the fact that polynomial equations of arbitrarily high degree are involved is of little importance. Moreover, he never explains what he understands by an equation, an unknown, a negative number, etc., but only describes the manipulations which should be carried out in specific problems, without worrying about arranging his text in terms of definitions, rules and theorems. In other words, like many other algebraists, Chinese or not, he demonstrates algebra by using it, much as one demonstrates movement by walking.

This probably explains the very distinctive organisation of his *Ceyuan haijing* both from the point of view of the structure of the chapters and from that of the structure of the calculations.

For example, chapter 12 is explicitly organised around the notion of the fraction (it is entitled *zhifen*) and its solutions involve generalised fractions (we would call them rational fractions). In a very different way, chapter 2 gathers together problems which do not depend on a specific area of mathematics, but which all require the application of an *ad hoc* formula; in all cases it is sufficient to inject the numerical values of the data into the formula in question to obtain the result after a sequence of more-or-less complicated calculations, the skeleton of which is specified in advance. Moreover, some of these formulae are quite remarkable (they may be used to calculate the diameter of a certain inscribed or escribed circle, the diameter of a circle with its centre on a vertex or the diameter of a circle tangent to two sides of a right-angled triangle with its centre on the other side–see later, the illustration of these formulae called the "nine inscriptions" *jiu rong*). But Li Zhi refers to them as being already known in his time[110] and does not attempt to explain them in any way.

In chapters 3, 4 and onwards, the problems are grouped together according to a very different type of criterion, namely the fact that the statement of

---

[110]Li Zhi attributes their invention to one Dong Yuan, about whom nothing is known. See Li Yan, *ZSSLC-P*, IV, p. 24 ff.

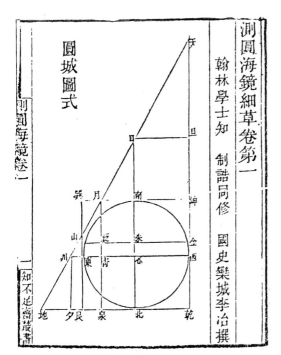

Fig. 11.1. The figure of the round town. Source: *Zhibuzu zhai* edition of the *Ceyuan haijing.*

On this figure note in particular: the cardinal points
North (*bei* 北), East (*dong* 東), South (*nan* 南), West (*xi* 西), the hexagrams of the *Yijing* *qian* 乾, *kun* 坤, *gen* 艮, *xun* 巽, etc.

each problem includes an item of data which is common to all the problems (it may, for example, be the known length of the hypotenuse of a certain triangle). Naturally, this type of criterion implies that solutions of very different degrees of difficulty follow one another within the same chapter. However, when some of his solutions are considered in pairs, appropriately chosen from two different chapters, a remarkable consistency[111] may sometimes be observed. It is as though Li Zhi had applied a series of simple and uniform modifications to the text of one to obtain the text of the other. The technique may be more or less involved but, fundamentally, it hinges on the fact that, in Chinese geometry, as already noted, the two sides forming the right angle of a right-angled triangle have special names: one is called "BASE" *gou*, and the other "LEG" *gu*. Suppose that two triangles are similar (or more generally, analogous from a certain point of view). It is possible that in a given formula the base of one may play the role of the leg in the other (and conversely). This means that any formula about one of the triangles has a "parallel image" for the other. Therefore, to obtain a new formula from an old one which has already been determined, it is sufficient to carry out a mechanical replacement of terms (for example, base may be replaced by leg everywhere, and conversely). It is as though Li Zhi had sensed that he could obtain new results simply by taking the "parallel copy" of calculations

---

[111]Consider (for example) the solutions of problems 3-1 and 4-1 of the *CYHJ.*

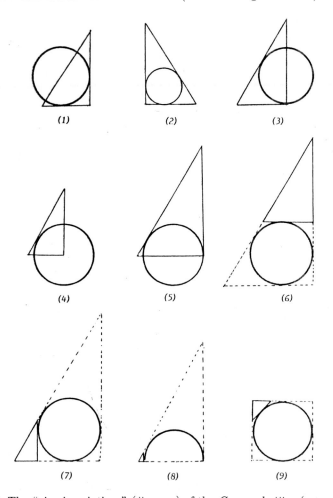

Fig. 11.2. The "nine inscriptions" (*jiu rong*) of the *Ceyuan haijing* (reconstitution).

which had already been made, thereby intuitively applying a duality principle before it was invented.[112]

As far as the text of the *Ceyuan haijing* is concerned, all that is implicit, in the same way someone writing in English generally does not see the necessity to explain the syntax of his prose. But other Chinese mathematical texts use a special terminology to refer to this type of phenomenon. For example, in the *Yang Hui suanfa*, the problems for which the statement (or the solution or both) imitates that of certain typical problems are classified as "imitations" *bilei*.[113]

---

[112]Cf. Chemla, (1), *1990* and other papers by the same author previously published in the journal *Extrême-Orient Extrême-Occident* (Paris).

[113]This term is borrowed from the general literature; one of its possible meanings is "to take something as a model" (from a moral point of view). Cf. S. Couvreur, *1950*, Transl. *Mémoires sur les bienséances et les cérémonies* (French translation of the *Liji*), Paris: Les Belles Lettres, Tome 2, première partie, p. 76.

Thus, for Yang Hui, the typical problem of calculating the area of a rectangular field, solved by simple multiplication, is associated with other problems, which are also solved by a single multiplication but in which the question asks for the total price given the unit price of certain goods. More subtly, the technique for calculating the area of a square field is associated with another technique for calculating the number of arrows contained in a pile arranged contiguously to form a right-angled parallelepiped with a square cross-section. If c denotes the length of the square (in linear units or in terms of the number of arrows), in the first case, the text simply recommends the calculation of c^2, while $(c/4 + 1)^2$ should be calculated in the second case.[114]

If this were literature rather than mathematics, this would be called a case of parallel prose (*pian wen*). One of the salient features of this typically Chinese literary genre was a preponderance of metrical and syntactical matchings between lines of texts.[115] More generally, it is also possible to detect other influences of literary composition on mathematics. For example, as Yan Dunjie explains[116] the term *huan gu* (literally "changing the bones") which occurs in connection with the extraction of roots in the *Shushu jiuzhang* is in fact borrowed from the rhetoric of Chinese poetry. In the latter context it means something like "not changing the meaning (of a former poem) but creating one's own diction"[117] while in mathematics it is used when a constant term changes from negative to positive in the process of transformation of an equation into another one.[118] It may be that there is nothing more here than a vague analogy. Still, research into the common terminology of Chinese poetry and mathematics would probably shed light on hitherto poorly understood aspects of the latter.

Qin Jiushao (ca. 1202–1261)[119]

Born in Anyue in what is now Sichuan, Qin Jiushao's father was admitted as *jinshi* (metropolitan graduate) in 1193 and then occupied various posts in the local administration (prefect of Bazhou (present-day Bazhong) around 1219, Vice-Director of the Palace Library (*bishu shaojian*) in 1224, and prefect of Tongchuan (present-day Santai in Sichuan) in 1225).

In his preface to his *Shushu Jiuzhang*, Qin Jiushao writes:

In my youth I was living in the capital [present day Hangzhou] so that I was enabled to study in the Board of Astronomy; subsequently, I was instructed in mathematics by a recluse scholar.[120]

---

[114]Cf. Lam Lay-Yong (6), *1977*, p. 92 ff.

[115]Nienhauser (1), *1986*, pp. 656.

[116]Yan Dunjie (15'), *1947*, p. 111.

[117]Nienhauser, op. cit., p. 447.

[118]Libbrecht (2), *1973*, p. 195.

[119]See the notice Ch'in Chiu-shao in the *DSB*, vol. 3, p. 249. See also Libbrecht (2), *1973*, pp. 22–34; *SY*, p. 60; *CRZ*, j. 22. p. 277.

[120]Libbrecht, op. cit., p. 62.

From the data given by Libbrecht[121] (date of arrival of Qin Jiushao in Hangzhou), Qin Jiushao was then 22 or 23 years old. According to the same author, from an analysis of the content of the *Shushu jiuzhang*, it follows that the *Jiuzhang suanshu* constituted a central part of Qin Jiushao's studies.[122] It is also known that a fiscal intendant from Chengdu taught him poetry around 1233,[123] at approximately the same time as he became a Commandery Defender (*wei*). In 1234, the Mongol armies invaded Sichuan. In Qiu Jiushao's preface we read:

> At the time of the troubles with the barbarians, I spent several years on the remote frontier; without care for my safety among the arrows and stone missiles, I endured danger and unhappiness for ten years.[124]

In 1244, he reappears as a Court Gentleman for Comprehensive Duty *tongzhi lang* of the prefecture of Jiankang (now Nanking), but in September of the same year, he left in mourning, probably for a period of three years, after his mother died in Huzhou (now Hexian, Anhui). It is thought that he composed his *Shushu jiuzhang* during this period (the preface is dated the ninth month of 1247). In the following years he held posts in various administrations; these frequent moves took him once to the island of Hainan (Zhejiang). He died in 1261.

The *Shushu Jiuzhang* of Qin Jiushao (1247)

The overall structure of the *Shushu jiuzhang* is closely related to that of the *Jiuzhang suanshu* even though the mathematical complexity of the two works is not at all the same: on the whole the problems and algorithms of Qin Jiushao's work are much more sophisticated than those of its illustrious predecessor.

The first section of the *SSJZ* (j. 1 and 2) is abstrusely called *dayan*, a term which calls to mind a divination method of the *Yijing*[125] (*dayan* means "great expansion"). The term appears in section 9 of the first part of the "Great Appendix" of the *Yijing* (Book of Changes): "The number of the great expansion makes 50, of which 49 are used [for divinatory purposes]. [The stalks representing these] are divided into two heaps [...]."[126] Once again, the logic of the *dayan* appellation is not to be sought in mathematics but in mnemotechnics: it represents the first word of the statement of the first problem of the *Shushu jiuzhang*, a problem which concerns a divinatory method analogous to that of the *Yijing* and which is used as a pretext for carrying out calculations involving congruences. It is easy to see that, despite its name, the *dayan* method has nothing irrational about it.[127]

---

[121]Ibid.
[122]Ibid.
[123]Ibid.
[124]Ibid., p. 27.
[125] *Yijing*, "Xici," 1-9.
[126]On this method of divination, cf. Ngo Van Xuyet (1), *1976*, p. 168.
[127]See below, the section on indeterminate problems.

The nine other problems of the same chapter all involve unrealistic situations, except for one, which deals with calendrical cycles.[128] Five centuries later, the famous mathematician Gauss also thought of the calendar to illustrate his mathematical findings about the remainder theorem:

This [i.e. the remainder theorem] applies to the problem of chronology in which it is desired to find the day of the year corresponding to the given indiction, golden number and solar cycle.[129]

The second section (j. 3 and 4) is entitled *tian shi lei* (Category of celestial periods and atmospheric calamities). The expression *tian shi* denotes both "celestial calamities" and "celestial chronology." By using it, Qin Jiushao is alluding both to the *Yijing* and the *Shujing*. In these classics, the expression *tian shi* may have different meanings depending on its context.[130]

Five of the eight problems in the *tian shi lei* deal with calendrical chronology and the three others concern atmospheric precipitation in the form of rain or snow and mention a rain gauge (*tian chi pen*).[131]

The third section (j. 5 and 6) is entitled *tian yu lei* (Category of the limit of fields). As its title indicates, it deals with a banal subject; however, it does so in a very original way. For example, we note the astonishing solution for calculating the area of the "pointed field" *jian tian* (Fig. 11.3),[132] which involves expressing the area of this figure not directly, but in a particularly roundabout way as the root of an equation of degree four, the coefficients of which depend on the dimensions of this figure.[133] This chapter also contains the first occurrence in China of the formula commonly known as Hero's formula,[134] which is used to calculate the area of an arbitrary triangle as a function of the lengths of its three sides.

---

[128]Libbrecht, op. cit., p. 382 ff.

[129]Gauss (1), *1801*.

[130]Cf. *Shujing*, "Pan Geng" in Couvreur, *Chou King*, Paris: Cathasia, 1950, p. 141: "The [people] have rarely foundered when faced with heavenly calamities." Cf. also *Yijing*, commentary on the hexagram *Qian*, "[The saint] may precede the heaven without being opposed by it; he may also follow it according to the celestial chronology."

[131]Cf. Needham, *SCC*, III, p. 471; Libbrecht, op. cit., p. 474. Frisinger (1), *1977*, p. 89–91: "The first known account of a rain gauge occurs in the *Arthaśāstra* by the Indian Kauṭilya ca. 400 BC [...] There is no indication, however, that the Indians of this period thought of rainfall in terms of depth of water. The idea apparently originated around 1000 AD in Palestine." According to the same author, the rain gauge appeared around 1400 in Korea, while in Europe the oldest known specimen is that spoken of by Benedetto Castelli in a letter to Galileo dated June 18th 1639.

[132]Based on the figure of the original text, it follows that this is a quadrangle formed by two isosceles triangles placed side by side along their common base so as to form a convex figure.

[133]Without giving any rational justification, Qin Jiushao states the coefficients of the equation, in general terms (in other words, not based on specific numerical values). We would write this equation as follows:

$$-x^4 + 2(A + B)x^2 - (B - A)^2 = 0$$

where $A = [b^2 - (c/2)^2](c/2)^2$; and $B = [a^2 - (c/2)^2](c/2)^2$ and a, b and c denote certain dimensions of the quadrangle (Fig. 11.3) and x is the unknown area to be determined.

[134]Clagett (1), *1979*, p. 80 ff.

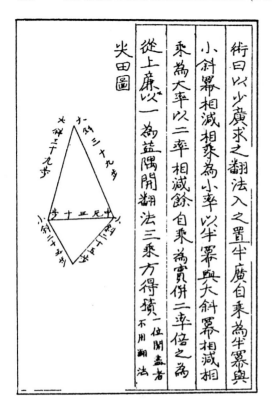

Fig. 11.3. The figure of the "pointed field" *jian tian*. (From the Wenyuange edition of the *Shushu jiuzhang* (republished Taipei: Shangwu Yinshuguan, *1986*, vol. 797, p. 411). We note that: (1) the segments of the figure are not denoted by letters but by special names (large oblique line *da xie*, small oblique line *xiao xie*, etc.); (2) the measurements of each segment are shown on the figure.

In the same vein, the fourth section (j. 7 and 8), entitled *cewang lei* (Category of measurements on the ground of distances of inaccessible points) includes nine problems of a classical type,[135] whose solution depends on equations of high degree (one is of degree ten).[136]

In a much less sophisticated way, the solutions to the problems of sections 5 to 9 depend on straightforward mathematical techniques. They deal with civil and military architecture, commerce and fiscal matters and according to Libbrecht are characterised by the realistic nature of their statements.

Zhu Shijie

Zhu Shijie (fl. end of the 13th century) is the latest 13th-century Chinese algebraist. Everything we know about him is contained in a few lines by a certain Mo Ruo:

---

[135] These problems are analogous to those of the *Haidao suanjing* by Liu Hui.

[136] *SY*, p. 85; Libbrecht, op. cit., p. 136 ff., which gives numerous references on this problem.

Master Songting[137] of Yanshan[138] became famous as a mathematician. He travelled over seas and lakes[139] for more than 20 years and the number of those who came to be taught by him increased each day.[140]

His two works, the *Suanxue qimeng* (1299) and the *Siyuan yujian* (1303) were xylographed by a certain Zhao Cheng who lived in Weizhou[141] (now Yangzhou in Jiangsu), one of the centres of commerce and artisans at that period.[142]

The *Siyuan Yujian* (1303)

Following Mikami Yoshio the title *Sijuan yujian* is often translated as "Precious mirror of the four elements."[143] But J. Hoe[144] has shown that this translation is doubly erroneous because: (i) the allusion to jade has nothing to do with the rarity of this type of stone – it is actually intended to evoke the brightness, the transparency and the fact that the calculations of Zhu Shijie's book are trustworthy; (ii) *yuan* does not mean "element" but "the source from which all the material universe was born"[145] and more precisely from the mathematical point of view, the unknown. Moreover, using the term "mirror" Zhu Shijie implies that his work reflects mathematical phenomena without the least distortion (Chinese historians and philosophers, often use this term). Thus, the true meaning of *Siyuan yujian* is "Mirror [trustworthy as] jade [relative to the] four origins [unknowns]."

The original edition of the *Siyuan yujian* by Zhu Shijie has been lost. The versions of this work available today are all derived from a certain version of the text which Ruan Yuan (1764–1849) found by chance at the beginning of the 19th Century in Zhejiang when he was governor of that province.[146] Unfortunately, not all are accessible to research; in practice, researchers have to make do with those which have been edited.[147] One of these was published in Shanghai in 1937 under the title *Siyuan yujian xicao* [The *Siyuan yujian* with detailed solutions] by the Commercial Press (Shangwu Yinshuguan). Composed of 1394 pages in three volumes, it contains not only Zhu Shijie's text but also Luo Shilin's commentary (which was originally published in 1834) and two appendices (100

---

[137]Songting is a *hao* (literary name) of Zhu Shijie.

[138]Near the present-day Peking.

[139]We would say "up hill and down dale."

[140]Preface to the *Siyuan yujian*, dated the 15th day of the first moon of the year *guimao* of the era *Da de* (1303).

[141]According to the preface by Zu Yi.

[142]Cf. Yabuuchi (4′), *1967*, p. 63.

[143]Mikami (4), *1913*, p. 89.

[144]Hoe (2), *1977*, p. 41 ff.

[145]Ibid., p. 45.

[146]Cf. QB, *Hist.*, p. 297; *SY*, p. 166 ff.; Yan Dunjie (14′), *1945*; article in the *DSB* by Ho Peng-Yoke on Chu Shih-chieh [Zhu Shijie].

[147]Details of these editions in Ding Fubao and Zhou Yunqing (1′), *1957*.

pages each) devoted to the reconstitution of arithmetical operations (division, root extractions, etc.).[148] We shall use this version here.

In this edition, the text consists of four clearly distinct parts comprising, respectively:

(i) a set of seven prefaces, postfaces or notices;

(ii) four figures;

(iii) four preliminary problems for which the author describes his algebraic mathematical techniques;

(iv) 288 problems divided into three parts *men* and into 24 chapters *juan*.

The prefaces are classified according to age. Except for the first two which are both dated 1303,[149] all date back to the 19th century. The first two prefaces are both due to scholars (one of them presents himself as a *jinshi* (metropolitan graduate) but is otherwise utterly unknown). The other prefaces are composed by 18th century scholars, notably Ruan Yuan. Luo Shilin's postface is of much interest because the copious critical notes which it contains give us an idea of the version of the *Siyuan yujian* found by Ruan Yuan: indeed this version was extremely corrupted and teeming with errors.[150]

The figures (*tu*) are of two sorts:

(a) The first figure, which is in fact composed of two schemata, bears the following legend: *jingu kaifang huiyao zhi tu* (Schema including everything that is essential [to know] to extract roots in the ancient or the modern way) (Fig. 11.5)

The first of these two schemata shows a rectangle with a grid pattern of 36 small squares (9×4), where each of these small squares contains information formulated uniformly using three written characters only. Because of the conciseness of the whole and the absence of context the precise meaning of all this is unclear. Could it be a representation of a counting-board?[151]

The second schema is just Pascal's triangle, which shows the coefficients of the successive binomial powers up to eight and which is called (*gu fa qichenfang tu*, i.e. table of the powers up to the eighth, according to the ancient method). (Fig. 14.4, p. 231)

(b) The three remaining figures all look similar. Each shows a large square decomposed into a certain number of squares or rectangles which themselves have a uniform grid pattern with a variable number of small squares. According to the legends accompanying these figures,

---

[148]On Luo Shilin cf. Hummel, p. 538 ff.

[149]Partial translation of one of these in Hoe, op. cit., pp. 115–116.

[150]pp. 23–42.

[151]A reconstitution of the calculations can be found in the *Kaifang guyi* (The ancient meaning of the extraction of roots), published at the end of the last century by Hua Hengfang (cf. Ding Fubao and Zhou Yunqing, op. cit., notice no. 362).

the aim in each case is the visualisation of algebraic identities of type $(a + b + c + \ldots)^2$. More precisely, the corresponding identity in the first figure is $(x + y + z + d)^2$ where x, y and z denote the lengths of the three sides of a right-angled triangle and d is the diameter of the circle inscribed in the same triangle (Fig. 11.4); the last two cases illustrate (i) the square of the "five sums"[152] and (ii) the square of the "five differences."[153]

The four preliminary problems are grouped together in the section entitled *si xiang xicao jialing zhi tu* ("table [of rod-numerals] for the "let us suppose that"[154] with detailed solutions using the four symbols"). In fact, these are typical problems in 1, 2, 3 and 4 unknowns, for which the author describes the manipulations needed to construct a resolutory equation. This succinct part of the book (22 pages, overall, the book runs to more than 1000 pages) is the only one which contains theoretical information; whence its importance (for technical details we refer the reader to the chapter of this book on Chinese algebra and to Jock Hoe's remarkable study).[155]

The text itself begins after these preliminaries. The structure is simple. Like many other Chinese manuals, it comprises a succession of statements of problems accompanied by their answers and some very brief information about how these are obtained.

As always, the influence of the Ten Canons is not negligible; thus, we again find the usual practical problems (architecture, finance, military logistics, etc.).

However, Zhu Shijie is not a narrow-minded sycophant of the tradition of the Nine Chapters. The well-worn paths are certainly not an inexhaustible attraction as far as he is concerned: he combines non-homogeneous quantities and unrestrictedly adds areas to volumes and prices to lengths in a manner reminiscent of Babylonian problems.[156] He performs handstands and starts from the sum of a series to find the number of terms or assumes that the volume of a figure is known before asking about its linear dimensions. However, he rallies and invents (or borrows, but from whom?) astonishing problems about the division of figures (division of a trapezium into equal parts using parallels to the base, division of a disc using parallel chords).[157] Finally, did he not annoy the military administration by suggesting that the numbers of soldiers recruited vary as the squares or the cubes of successive terms of an arithmetic progression?[158]

Does this mean that the *Siyuan yujian* is a fanciful work which should be relegated to the category of mathematical eccentricities? This may have been

---

[152]See below, p. 266.

[153]Ibid.

[154]The "let use suppose that" are the statement of problems.

[155]Cf. Hoe, (2), *1977*.

[156]On Babylonian mathematics, cf. Høyrup (4), *1994* and Ritter (1), *1989*.

[157]Cf. *SYYJ*, II-9 (pp. 521–590 in vol. 2 of Luo Shilin's *Sijuan yujian xicao* quoted above). Here again, these appear to be similar to Babylonian problems such as those in Thureau-Dangin (1), *1938*, p. 88. The similarity with certain Euclidean problems is much less evident. Cf. Archibald (1), *1915*.

[158]See below, p. 340.

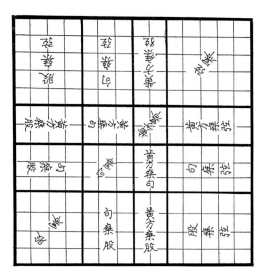

Fig. 11.4. The square of the sum of the four unknowns $(x + y + z + d)^2$. Source: *Siyuan yujian xicao* (1303) (Shanghai: Shangwu Yinshuguan 1937, p. 46).

x = base (*gou*), y = leg (*gu*), z = hypotenuse (*xian*), d = side of the "yellow square" (*huang fang*) = length of the diameter of the circle inscribed in a right-angled triangle.

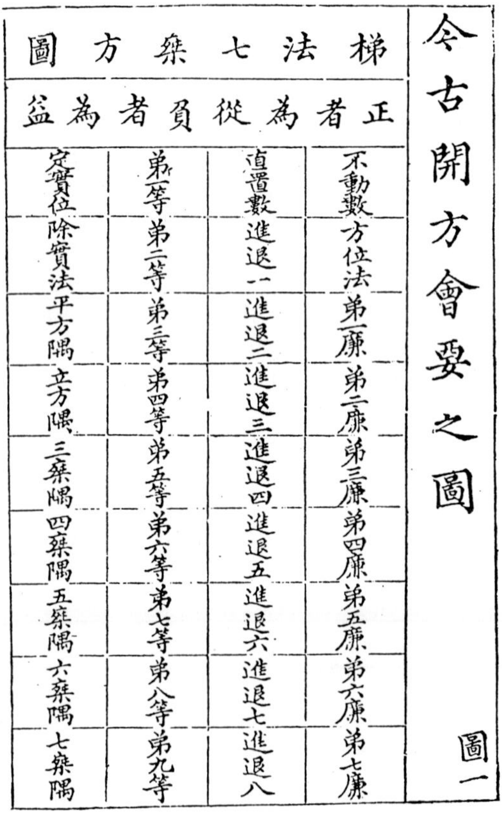

Fig. 11.5. Table including everything that it is essential to know how to extract roots in the ancient and the modern way. Source: *Siyuan yujian xicao* (1303) (same source as Fig. 11.4).

the perception of Zhu Shijie's contemporaries and is possibly the reason why it was four centuries before it excited the interest of literati who were stupefied to discover the existence of a Chinese algebra predating European algebra.

Indeed, the originality of the *Siyuan yujian* is most striking. By freeing himself as best he can from the lead weight of the realism of traditional mathematics, Zhu Shijie at the same time recoups the dividends of a mathematical freedom unknown to his predecessors. His equations are no longer limited to degree two or three (but go up to degree 14) and his series are no longer bridled by the need to correspond to some concrete application (general terms in the form of figurate numbers or obtained by squaring the terms of arithmetic progressions and more complex series). He also uses "Newton's" famous interpolation formula.[159]

Yang Hui

We know nothing about the life of Yang Hui except that he was born in Qiantang (now Hangzhou) under the Southern Song.[160]

Unlike Li Zhi, Qin Jiushao and Zhu Shijie, who were all algebraists who manipulated complicated algorithms, Yang Hui worked in the area of elementary arithmetic. Even though he did sometimes extract roots using Horner's method, his writings rarely exceed the level of the Ten Computational Canons. His output, which in terms of quantity was greater than that of all other ancient Chinese mathematicians, consists of the following titles:

(a) *Xiangjie jiuzhang suanfa* (detailed explanation of the computational methods in the Nine Chapters [i.e. of the *JZSS*]), 12 j.; 1261.

(b) *Riyong suanfa* (computational methods for daily use), 2j., 1262.[161]

(c) *Chengchu tongbian suanbao* (Computational treasure of the multiplicity of variations of multiplication and division), 3 j., 1274. This work, together with the following two works below ((d) and (e)) constitute part of the collection of the works of Yang Hui known as the *Yang Hui suanfa* (Yang Hui's Methods of Computation), which was published in 1275, reprinted in 1378 and partly copied in the *Yongle dadian* encyclopedia (1407). A Korean edition appeared in 1433. Later, the book made its way into Japan and was partly copied by Seki Takakazu (?–1708), the 'father' of Japanese traditional mathematics. Forgotten in China itself, the book was rediscovered much later by Li Rui, Song Jingchang and other Chinese philologists.[162] Kodama Akihito has reproduced certain of these ancient

---

[159]Hoe, op. cit., p. 300 ff.
[160]Cf. Lam Lay-Yong, (6), *1977*, p. xv.
[161]Cf. Lam Lay-Yong, (4), *1972*.
[162]Kodama (1'), *1966*, pp. 29–31.

editions of Yang Hui's works[163] in facsimile. Lastly, in 1977, the whole of the *Yang Hui suanfa* was translated into English by Lam Lay-Yong.[164]

(d) *Tian mu bilei chengchu jiefa* (Quick methods of multiplication and division for the computation of the number of mu[165] of fields, with analogies) 2 j., 1275.[166]

(e) *Xugu zhaiqi suanfa* (Continuation of the tradition of strange computational methods), 2 j., 1275.

The *Xiangjie jiuzhang suanfa* is a commentary on the *Jiuzhang suanshu* which is very different from previous commentaries on the same work: it is partial. Yang Hui chose only to comment on those problems of the *JZSS* which he thought were representative of the whole (80 problems out of a total of 246).[167] But it contains three more chapters than the *JZSS* (one for the geometrical figures, one to explain the fundamental operating techniques and, lastly, one in which the problems are classified differently from in the initial text);[168] it also includes new sections unknown to Yang Hui's predecessors. In the first of these textual units, called *jieti* (explanation of the problem), Yang Hui explains the logic of the problem while in the second, entitled *cao*, he shows how to obtain the solution using a numerical example. Finally, in the third, *bilei*, he shows how problems which are superficially different may in fact be solved using identical or very similar resolutory algorithms.[169]

Yang Hui also explains how to solve equations of type $ax^2 + bx = c$ numerically and how to extract a fourth root corresponding, for us, to the equation[170]:

$$-5x^4 + 52x^3 + 128x^2 = 4096$$

Each time, he gives a single positive root. Yang Hui uses two types of method: methods based on geometry (for equations of degree two) and a method analogous to Horner's method (for an equation of degree four). According to some authors, Yang Hui explained that the equation of degree two $-x^2 + 60x = 864$ has two distinct roots.[171] In fact, the original text of the *Tian mu bilei chengchu jiefa* effectively contains two problems which lead to this equation (problems 46 and 47 – cf. Lam Lay-Yong, op. cit., pp. 118 and 120). (In problem 46,

---

[163]Ibid.

[164]Lam Lay-Yong (6), *1977*.

[165]The *mu* is an agrarian unit of area.

[166]*bilei*: see *DKW*, 6-16743: 186, p. 805. The expression is taken from the *Liji*. For Yang Hui, two problems are said to be analogous if they can be solved by analogous resolutory formulae (regardless of the specific themes of the statements).

[167]See Yang Hui's preface to his *Xiangjie jiuzhang suanfa*, Shanghai: Shangwu Yinshuguan, *1936*, p. 7).

[168]The complete text has not survived. Thus, we cannot say what sort of classification this was.

[169]See Yan Dunjie's article in *SY*, p. 151.

[170]*Tianmu bilei chengchu jiefa*, problem 60. Cf. Lam Lay-Yong (6), *1977*, p. 130 (transl.) and p. 269 (comments).

[171]Ho Peng-Yoke, article *Yang Hui* in the *DSB*; Lam Lay-Yong, (6), *1977*, p. 265.

Yang Hui determines the width of a rectangle, given that length + width = 60 and that surface area = 864; in problem 47, for the same data, he determines the length). Yang Hui uses three separate methods to solve these two problems: two methods for the first problem and one for the second. Based on this, Yang Hui may effectively be said to have found two numbers corresponding to a single equation (i.e. corresponding to a single sequence of numbers on the counting surface). However, it seems hard to say that he had seen that a single equation of degree two has two distinct roots, because what he did does not amount to displaying a single problem leading to a single resolutory form corresponding, for us, to an equation with two distinct solutions (or roots). Rather, Yang Hui was concerned with two distinct problems, each with a unique solution.

Cheng Dawei

Cheng Dawei[172] (1533–1606) (Fig. 11.6) is the most illustrious Chinese arithmetician. Everything we know about him is summarised by what one of his descendants wrote in a preface to a reprint of his *Suanfa tongzong*:

> In his youth, my ancestor Rusi[173] was academically gifted, but although he was well versed in scholarly matters, he continued to exercise his profession as a sincere Local Agent, without becoming a scholar. He never lagged behind either on the classics or on ancient writing with characters in the form of tadpoles, but he was particularly gifted in arithmetic. In the prime of his life he visited the fairs of Wu and Chu.[174] When he came across books which talked about "square fields," "decorticated grain" [...] he never even looked at the price before purchasing them. He questioned respectable old men who were experienced in the practice of arithmetic and gradually and indefatigably formed his own collections of difficult problems[175] [...].[176]

The *Suanfa Tongzong* (1592)

The *Suanfa tongzong* (General Source of Computational Methods) is essentially a general arithmetic for the abacus. There is nothing particularly original about it (it can be shown to be a compilation of earlier works),[177] but it was republished many times under the Manchu dynasty and thus was widely

---

[172] On Cheng Dawei, see: Hummel, p. 117; Li Yan, *Dagang*, II, p. 308; QB, *Hist.*, pp. 139–141; Takeda (2'), *1954*; Li Di (4'), *1986*; Li Zhaohua (5'), *1990*; *Suanfa tongzong jiaoshi* (a facsimile reprint of the original text of the *Suanfa tongzong* with explanations by several authors (Yan Dunjie, Mei Rongzhao, Li Zhaohua) published in 1990 by Anhui Jiaoyu Chubanshe with no name of collective editor); *CRZ*, j. 31, p. 385. Other references in Libbrecht (2), *1973*, note 109, p. 291.

[173] Rusi is the personal public name (*zi*) of Cheng Dawei.

[174] In the present-day Jiangsu and Hubei/Hunan, respectively.

[175] See above, p. 56.

[176] Cited from Takeda (2'), *1954*, I, p. 8.

[177] Takeda, op. cit.

Fig. 11.6. Portrait of Cheng Dawei reproduced at the beginning of his *Suanfa cuanyao* (Essentials of arithmetic). From Li Peiye (2′), *1986*, p. 22.

disseminated.[178] As a certain Cheng Shisui, descendant of Cheng Dawei noted in 1716:

> A century and several decades have passed since the first edition [of the *SFTZ*], during which period this work has remained in vogue. Practically all those involved in mathematics have a copy and consider it to be a classic comparable with the Four Books[179] and the Five Classics[180] used by those who sit the examinations.[181]

Beyond the limited circle of mathematicians, the *SFTZ* also reached a vast popular audience. Even in the mid-20th century (1964), the well-known historians of Chinese mathematics Li Yan and Du Shiran remarked that:

> Nowadays, various editions of the *SFTZ* can still be found throughout China and some old people still recite the versified formulae and talk to each other about its difficult problems.[182]

---

[178]Ding Fubao and Zhou Yunqing (1′), *1957*, notice no. 169, p. 43b.

[179]The Great Learning *Daxue*, the Confucean Analects *Lunyu*, the Doctrine of the Mean *Zhongyong* and the Works of Mencius *Mengzi*.

[180]The Book of Changes *Yijing*, the Book of Odes *Shijing*, the Book of History *Shijing*, the Book of Rites *Liji* and the Spring and Autumn Annals *Chunqiu*.

[181]Ibid.

[182]Li Yan and Du Shiran (2′), *1964*, p. 35.

This observation leads one to wonder about the nature of the tradition of which the *SFTZ* is part. Takeda Kusuo has shown that the terminology and style of the famous arithmetic are by no means literary, but betray instead the background of merchants, the main guardians of the tradition of the Nine Chapters under the Ming dynasty.[183]

The same author has also observed that certain problems from the *SFTZ* were rewritten in a style more "elevated," adapted to the literati, when these were incorporated in the *Tongwen suanzhi* (1614),[184] an arithmetic compiled from Chinese and European sources (especially Clavius's *Epitome Arithmeticae Practicae*). As we know, Jesuit policy was based on a penetration of the Chinese elite so that everything published by the Jesuits could not but respect the standards of Chinese scholarly culture, and a "mathematisation" of Chinese scholars based on popular culture was out of the question.

Like other Chinese arithmetics, the *SFTZ* is built around a series of problems (595 in all, across 17 chapters) which are classified in the same way as in the *Jiuzhang suanshu*. However, unlike the authors of the venerable classic, Cheng Dawei was not afraid of superfluity or verbosity. His book is an encyclopedic hotch-potch of ideas which contains everything from A to Z relating to the Chinese mystique of numbers (magic squares *Hetu* and *Luoshu*, generation of the eight trigrams, musical tubes), how computation should be taught and studied, the meaning of technical arithmetical terms, computation on the abacus with its tables which must be learnt by heart, the history of Chinese mathematics, mathematical recreations and mathematical curiosities of all types.

The way in which works of arithmetic were composed evolved from the Han to the Ming, although certain aspects were clearly unchanged. The most marked innovation was that some problems, rules and solutions were composed in verse like the rhyming formulae used by doctors and other specialists. The following small poem which was used to memorise Sunzi's solution of the remainder problem will give the reader an idea of this:

三人同行七十稀   五樹梅化廿一枝
七子團圓正半月   除百零五便得知

san ren tong xing qi shi xi   *wu shu mei hua nien yi zhi*
qi zi tuan yuan zheng ban yue   *chu bai ling wu bian de zhi*

Three *septuagenarians* in the same family? Rare!
Five plum trees with *twenty one* branches in flower
Seven brides in ideal union?[185]
[It is] precisely the middle of the month![186]
One hundred and five subtracted? Lo the result appears![187]

---

[183] Takeda, op. cit.

[184] Takeda (3′), *1954*, p. 12.

[185] *tuan yuan*: perfectly round. Idea of a harmonious union. Cf. *DKW* 3-4834: 5, p. 2377.

[186] Allusion to the date of the festival of marriage (*tuan yuan jie*) which took place on the 15th day of the eighth lunar month.

[187] *SFTZ*, j. 5, in *Suanfa tongzong jiaoshi*, op. cit., p. 430; *ZSSLC-T*, I, p. 67; Libbrecht (2), *1973*, pp. 291–292. For more details on the verses of the *SFTZ*, see *Meijizen*, I, p. 403.

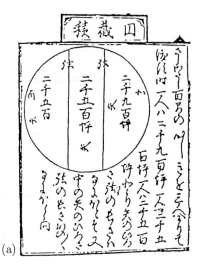

(a)

Fig. 11.7. A problem "bequeathed to posterity" *idai* from the *Jinkōki*. In a reprint of the Jinkōki published in 1641, 14 years after the first edition, the Japanese mathematician Yoshida Mitsukuni inserted 12 "difficult" problems without giving their solution. In the tenth of these, a circular field is divided into three parts whose areas are given and the corresponding chords and sagitta are required (a). This problem was perhaps inspired by the following found in the *Suanfa tongzong*: "A small river cuts right across a circular field whose area is unknown; (b) given the diameter of the field and the breadth of the river find the area of the non-flooded part of the field."

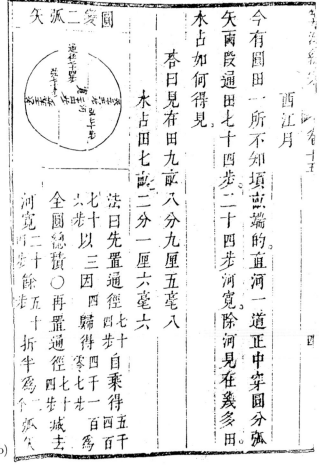

(b)

Sources: (a) *Jinkōki ronbun shū* (Collected papers on the Jinkōki) [no name of editor], Osaka: Kyōiku Tosho, *1977*, p. 33 and (b) *Suanfa tongzong jiaoshi*, op. cit., p. 878.

$((2 \times \mathbf{70}) + (3 \times \mathbf{21}) + (2 \times \mathbf{15})) - \mathbf{105} - \mathbf{105} = 23$. See the statement and solution of Sunzi's problem in Chapter 16 (Indeterminate Problems).

One other interesting aspect of the *SFTZ* is its "difficult problems" (we would call them mathematical recreations). These problems attracted the interest of certain 17th-century Japanese mathematicians. The 1641 edition of a famous arithmetic, the *Jinkōki* by Yoshida Mitsuyoshi (1598–1672), includes a series of 12 curious problems (some of which were taken from the *SFTZ*) such as the following[188]:

- To solve a right-angled triangle given that

$$a = b = 81 \ ken \qquad a + c = 72 \ ken$$

  where a, b and c denote the lengths of its sides with $a > b > c$.

- To solve linear systems with at most 4 unknowns.

- To divide a disc into three parts of given area. (Fig. 11.7)

The solutions of these problems were not given and were left to the reader. This was the start of the Japanese tradition of *idai* (unsolved problems which a mathematician places at the end of a work and hands down to his followers).

The *Shuli Jingyun*

The *Shuli jingyun* (Collected Essential Principles of Mathematics) is a mathematical encyclopedia, in 53 *juan* which covers almost all mathematical knowledge, Chinese and Western, available in China at the time of its publication (1723).

The compilation of this monumental work, commissioned by imperial order, was started in 1713 after the Emperor Kangxi had specifically recruited more than one hundred promising young scholars all over China,[189] regardless of whether these had already received advanced degrees in the civil service examinations.[190] A new Academy was then specially founded inside the Summer Palace in Peking so that these new recruits could study mathematics, astronomy, music and other specialised subjects in contact with both the Emperor himself, who taught them mathematics, and Jesuit specialists.

A large number of instrument makers (mathematics, astronomy) were also recruited to provide for the technical needs of the new Academy.[191]

Although the names of the majority of these scholars have not come down to us, we know at least that Minggatu and Mei Juecheng were among them. The former was made assistant editor and the latter chief editor of

---

[188]Shimodaira (2), *1981*.
[189]Cf. Foucquet (1), *1716* p. 3.
[190]Spence (1), *1974*. p. 51.
[191]Foucquet, op. cit., idem.

the *Shuli jingyun*[192] together with a more classical scholar, He Guozong (?–1766). In addition, a team of 15 calculators responsible for the verification of the computations and as many collators collaborated in the compilation.[193] Of course, Jesuit missionaries had also previously been requested to adapt, into Manchu or Chinese, topics such as elementary geometry, algebra or logarithms.

The *SLJY* is divided into three parts composed of 5, 40 and 8 *juan*, respectively. The first part consists of a presentation of theoretical notions, the second of various mathematical techniques and the third of numerical tables.

On the whole, this particular ordering of mathematical subjects is patterned upon the composition of mathematical textbooks intended for the teaching of mathematics in Jesuit colleges towards the end of the 17th century. This explains why the first part of the *SLJY* reflects the bipartite division of theoretical mathematics into speculative geometry and speculative arithmetic, two domains which are respectively devoted to the study of continuous and discontinuous quantities, as in the geometrical and arithmetical parts of Euclid's *Elements*.[194] However, these Euclidean topics are adapted for school use and essentially consist of a didactic succession of mathematical techniques; proofs are often reduced to their simplest expression and logical consistency is deemed superfluous. In particular, terms such as "axiom," "postulate," "definition" or "theorem" are wholly omitted. Moreover, some notions which are not in Euclid's *Elements* (for example, conics) are freely introduced.

In the same way, the second part of the *SLJY* concerns mathematics oriented towards applications – that is, practical arithmetic and geometry, algebra and logarithms – together with examples of real applications fully worked out (surveying, rough map drawing, use of the proportional compass, construction of a sundial). More precisely, topics are distributed among the following sections which clearly evoke the Euclidean division of figures into lines, surfaces and solids:

1. Preliminaries *juan* 1–2
2. Lines *juan* 4–11
3. Surfaces *juan* 12–22
4. Solids *juan* 24–31
5. Last part *juan* 32–40

The preliminary section introduces metrological units and elementary operations on integers and fractions, not on the abacus but by means of written

---

[192]Hummel, p. 285.

[193]Ibid.

[194]On this conception of mathematics, see the detailed explanations in François Blondel, *Cours de Mathématiques contenant divers traitez composez et enseignez à Monseigneur le Dauphin par François Blondel, Professeur Royal en Mathématique et en Architecture, de l'Académie Royale des Sciences, Maréchal de Camp aux armées du Roy et cy-devant Maître de Mathématiques de Monseigneur le Dauphin.* 2 vols. Paris, *1683*, chez l'auteur et Nicolas Langlois avec privilège du Roy. vol. 1, p. 1 ff. See also Bernard Lamy, *Eléments de Mathématiques*, Paris, *1680*, and Claude François Milliet de Chales, *Cursus seu Mundus mathematicus*, Lyon, *1674* and *1690*.

computations in a characteristic European style which is essentially the same as that still in use in Western countries.

The section on "lines" deals with usual arithmetical topics (proportions direct and inverse, proportional sharing (partnership), double-false-position rule, etc.).

In the following two sections, surfaces and solids are considered from the point of view of effective computations. For example, the section on surfaces explains how to compute the lengths of the sides of regular polygons inscribed or circumscribed in a circle in order to construct trigonometrical tables. The computations begin with the square and the hexagon and are extended to polygons with 4×2^{33} and 6×2^{33} sides respectively. As a result, a value of 2π composed of 40 digits is derived (but only 16 of them are correct). The section on solids follows the same overall plan but with the circle replaced by the sphere and polygons by regular polyhedra circumscribed or inscribed in a sphere. In both cases, various mathematical tools (such as square roots, cube roots, plane trigonometry, division into mean and extreme ratio) are introduced as soon as they are needed.

The last part introduces a sort of algebra called the *jiegenfàng*.[195] There is also a collection of problems said to be difficult concerning, for example, a packing of tangential spheres, which are solved using the *jiegenfang* algebra. The book ends with a detailed description of logarithms to the base ten. As shown recently by a young Chinese historian of mathematics, Han Qi, the section on logarithms is an almost complete translation of the method used by H. Briggs (1556–1630) to compute decimal logarithms.[196] Although logarithms were first introduced in China as early as 1653,[197] long before the publication of the *SLJY*, the translation of Briggs's treatise surpassed what had been previously published in China on the same subject. In this section of the book, everything is explained at length with an almost unimaginable wealth of detail and numerical examples; even arithmetical operations are often fully worked out. This aspect of the compilation of the *SLJY* is moreover characteristic of the whole book and the contrast with Chinese mathematical books published before the 17th century is striking.

Finally, the *Shuli jingyun* also has eight chapters of tables for the sines, cosines, tangents, cotangents, secants and cosecants for every ten seconds up to 90 degrees. A list of prime numbers is also given as well as a table of logarithms of integers from 1 to 100 000 calculated to ten decimal places and perhaps borrowed from Vlacq.[198]

Apart from recent developments such as logarithms or algebra and a full-scale reworking of less novel subjects of European origin, the *SLJY* also includes Chinese autochthonous mathematics. In the introductory part of the encyclopedia, *Yijing*, numerology as well as quotations from the *Zhoubi*

---

[195]Cf. p. 119 above.

[196]Cf. Han Qi (1′), *1992*.

[197]Cf. Li Yan *ZSSLC-T*, I, p. 215.

[198]Wylie (1), *1852*, p. 192.

suanjing are held up as witnesses of the Chinese origin of mathematical sciences. Remarkably, the special terminology of the headings of chapters of the *SLJY* devoted to traditional topics (based on the division of mathematics into nine chapters) is taken from the *Suanfa tongzong*.[199]

These autochthonous topics include, among other things, *fangcheng* methods and *gougu* techniques. While the former are based on Mei Wending's reinterpretation of traditional techniques for solving linear systems[200] the latter is not exactly a mere catalogue of ancient results either. Of its 60 resolutions of right-angled triangles, 17 are solved by "new methods" *xinfa* (second part, *juan* 12). But, unlike techniques mechanically applicable in a variety of mathematical settings (such as algebra or trigonometry), each new method in fact depends on ancient techniques based on *ad hoc* visualisations of problems by means of drawings and geometrical representations. Although these techniques are not easily generalisable the pugnacious reader was not, for all that, left at the end of his resources, for examples of algebraic solutions of *gougu* problems were also given in the section of the *SLJY* on the *jiegenfang* algebra.

Owing not only to imperial patronage but also to its syncretical multiplication of computational techniques made readily usable through numerous numerical examples, the *SLJY* enjoyed a lasting success. Even at the end of the 19th and the beginning of the 20th century the prestige of the great mathematical encyclopedia from the end of Emperor Kangxi's reign was still so great that it was reprinted as late as 1875, 1882, 1888 and 1896 and still later in 1911. Moreover, new publications such as the *Zhongxi suanxue dacheng* (Great Compendium of Mathematics, Chinese and Western) published in 1889 (and reprinted in 1901) by the Tongwen guan Press and containing more recent developments, continued to model explicitly their overall conception of mathematics on that of the *SLJY*.

The *Chouren Zhuan*

The *Chouren Zhuan (CRZ)* is a monumental collection of notices devoted to mathematical astronomers, Chinese and Western, which covers the totality of Chinese and Western history.

The scholar responsible for the collection was Ruan Yuan (1764–1849), the authoritative patron of Chinese scholarship in the provinces of the lower Yangzi region during a large part of the Qianlong and Jiaqing periods.[201]

Ruan Yuan served in 1799 as a director of the mathematics section of the *guozijian* (National University). He viewed the compilation of the *CRZ* not as his own achievement but as a result of a cooperative effort involving those who were then considered as the best specialists of quantitative sciences,[202] Li Rui

---

[199]Cf. Yan Dunjie's explanations in *Suanfa tongzong jiaoshi*, op. cit., p. 4.
[200]Cf. Li Di and Guo Shirong (1'), *1988*, p. 213.
[201]Cf. Elman (1), *1984*, p. 63.
[202]On the notion of quantitative sciences, cf. Sivin (2), *1977*, p. xii.

(1765–1814), Qian Daxin (1728–1804), Jiao Xun (1763–1820), Tan Tai and Zhou Zhiping (dates unknown).

Started in 1795, the compilation of the *CRZ* was completed four years later in 1799 as a collection in 46 *juan* consisting of 316 biographies (275 Chinese and 41 Westerners). It was republished in 1829 (edition limited to the biographies from the Qing dynasty), 1840, 1842, 1882 and, still later, in 1935 and 1955. In 1840, Luo Shilin (1789–1853), a disciple of Ruan Yuan, wrote six additional *juan* comprising 44 new biographies. His supplement was often appended to the initial *CRZ* but it was also known separately as the *Xu Chouren zhuan* (A sequel to the *Chouren zhuan*). In 1886, Zhu Kebao (1845–1903), a *juren* (provincial graduate) of 1867, added a new supplement in seven *juan* consisting of 129 new biographies (113 Chinese, 15 Westerners and 1 Japanese). His work was entitled *Chouren zhuan san bian* (Third supplement to the *Chouren zhuan*). Finally, a substantial last supplement in 11 *juan* and 1 appendix (284 Chinese and 157 Westerners) was published in 1898 by a certain Huang Zhongjun.[203]

Reprinted and augmented during the whole of the 19th century, the *CRZ* provided a large amount of information on astronomy and mathematics to a vast audience which had generally not had access to this through the medium of general education. For this reason, the image of quantitative sciences was cast in the mould of a particular literary genre called *zhuan*.

According to most Chinese–English dictionaries, *zhuan* means "biography." This is not false but the equivalence is rough: the *zhuan* genre reflects the orthodox Confucean conception of "Accounts of Conduct"[204] which tends to describe individuals as exemplary types so that merits and demerits may be assessed. At the end of the 18th century, such a conception was still extraordinarily extant and Ruan Yuan compiled the *CRZ* in keeping with a very ancient tradition whose first lineaments date back to the time of Sima Qian (145–86 BC), the illustrious historian. In accordance with this tradition, the biographical parts of the *zhuan* of the *CRZ* are thus essentially composed of highly impersonal lists of academic or administrative titles conferred upon individuals in the course of their official careers. These dry lists, however, do not usually occupy more than a few lines and are normally filled out by various devices including most frequently the use of stereotyped passages and conventional anecdotes (*topoi*) and the recourse to extensive quotations from various technical works authored by the individual in question. Such a technique frequently results in concatenations of ready-made textual units in which the influence of the historian is generally difficult to perceive except through the special choice and ordering or quotations. (In this respect, it appears that the *CRZ* is characterised by a complete rejection of astrology and other false sciences.) Yet, each *zhuan* is often followed by an autonomous concluding unit, the *lun* ("judgement") in which the author expresses his opinion.

Just as the *zhuan* genre conveys a particular notion of biography, the *chouren* are not merely astronomers and mathematicians but rather "hereditary

---

[203]Cf. the bibliographical notice on the *CRZ* in Ding Fubao and Zhou Yunqing (1'), *1957*.
[204]Cf. Twitchett (1), *1961*. p. 104.

specialists" in general. More importantly, however, the expression refers to the following passage of Sima Qian's *Shiji*:

> After [Kings] You and Li, the House of Zhou weakened. The subsidiary ministers of the great officers held the government in their grasp. Astrologers did not compute the calendar and Princes failed to proclaim neomenies. Consequently, hereditary calendarists *chouren* dispersed. Some remained in China, others went among [Barbarians] *Yi* and *Di* and divinatory methods fell into disuse.[205]

This passage, which seems to refer to nothing but a banal episode of Zhou China, was submitted to an astonishing interpretation: the *chouren* were perceived not only as "hereditary calendarists" but also as mathematical astronomers endowed with a deep knowledge of mathematical astronomy, a knowledge which had supposedly been interrupted in China as a result of political strife and transmitted to Barbarians among whom Chinese specialists had taken refuge. Furthermore, "Barbarians" were identified with ancestors of Western missionaries and the superior scientific knowledge of these was said to originate in the pristine knowledge of ancient *chouren*.

The true meaning of *Chouren zhuan* is therefore not "biographies of mathematical astronomers" but rather "Accounts of conduct of hereditary experts whose ancestors possessed a superior technical knowledge from which Western astronomy and mathematics originate." This explains why throughout the *CRZ*, philological expertise is mobilised in order to establish the Chinese origin of Western novelties introduced into China during the 17th and 18th centuries as a consequence of Jesuit activities.

The idea of the Chinese origin of mathematical astronomy, however, was not particularly original at the time the *CRZ* was composed. It had already been proposed by well-known scholars such as Huang Zongxi, Mei Wending and many others more than a century earlier.[206] The *CRZ*, however, treated the theme more systematically than ever and became so authoritative that criticisms were not levelled before the end of the 19th century, when diversified information on Western countries was becoming more widely available to well-travelled Chinese such as Wang Tao, who wrote that:

> [Ruan Yuan] is superficial and sketchy [...]. He shows himself ignorant of the fact that Western astronomy really originated in Greece [...].[207]

But, beyond the question of the origin of science, Ruan Yuan's aim was also to allow his readers to judge for themselves on the basis of quotations from original documents. To begin with, Ruan Yuan and his collaborators culled their information extensively from the monographs on the calendar *lifa* and astronomy *tianwen* preserved in the 24 dynastic histories and reproduced the corresponding texts almost verbatim. Since similar monographs were not available in the case of the last dynasty, the newly compiled *Siku quanshu* collection which was a wealthy repository of 17th and 18th century sources, both

---

[205] *Shiji*, j. 26, p. 1258.
[206] Jiang Xiaoyuan (1'), *1988*.
[207] Cf. P.A. Cohen (1), *1987*, p. 178.

Chinese and Western, was also extensively used.[208] Later, excerpts from lost mathematical sources from the Song and Yuan dynasties newly rediscovered, such as the *Yang Hui suanfa* or the *Siyuan yujian*, gave material for Luo Shilin's new supplement. Still later, non-technical sources such as *biji* literature or even the works of Mengzi, Zhuangzi or Mozi, were more and more widely exploited. Finally, investigations were so extensive that virtually not a single astronomer or mathematician (real or alleged) was omitted. In each case, the biographies were classified chronologically, in order of successive Chinese dynasties. When uncertainty was such that this turned out to be impossible, dubious cases were arbitrarily inserted somewhere but duly analysed and signalled as such. As for biographies of Westerners, Ruan Yuan and his collaborators sometimes found themselves confronted with contradictory datings and resolved to group foreigners separately, independently of Chinese chronology. The same policy was followed by subsequent editors. In the *CRZ3B*, for example, foreigners were listed apart as usual. But since a small number of Chinese women were also introduced for the first time in the same edition, a new dilemma arose, which was solved by listing these women separately too, together with Westerners, but just before them.

In order to understand what Western astronomy and mathematics refer to in the *CRZ* and its supplements, we note that the Western *chouren* who were attributed a biography are those listed pp. 170–172 (numbers inserted between Chinese names and their translation indicate the approximate length of each biography (*zhuan*) expressed in lines of text; when a *zhuan* relates to several persons the corresponding names are separated by a "/ ". In such as case the first is always the main person and the name of the secondary person is given using small letters exactly as in the Chinese original text); dates of birth and death are given only when they are not well-known. The reader will find all that is needed in this respect in the *DSB* (for example)). Lastly, in the sequel "J" and "P" refer to Jesuit and Protestant missionaries, respectively.

As might be expected, these lists confirm the importance of Jesuit and Protestant missionaries. They also reveal, at various levels, some limitations and obstacles faced by astronomy and mathematics in the process of their transmission from Western countries to China. For example, European names of persons were not always rendered consistently and sometimes two (or more) different Chinese names believed to refer to distinct persons were attributed to the same person. This was the case, for example, of Copernicus who was involuntarily called "Gebaini" and "Nigulao" (from his first name "Nicolaus") at the same time and attributed two unrelated biographies in the same chapter of the *CRZ*.[209] The transliterations "Weiyeda" and "Feiyida" (both for "Vieta" (Viète)) are another example of the same phenomenon. Less anecdotally, the relative importance of Western personalities was often overlooked. Obscure authors of manuals, such as Norie and Landy, were sometimes granted more

---

[208] Cf. *CRZ*, foreword.
[209] *CRZ*, j. 43, pp. 554 and 556, resp.

CRZ, j. 43 (complete list):

Modong	默冬	7	Meton
Yalidage	亞里大各	3	Aristarchus
Dimoqia	地末恰	6	Timocharis
Yibagu	依巴谷	8	Hipparchus
Duolumou	多祿某	33	Ptolemy
Ya'erfengsuo wang	亞而封所王	2	King Alfonso X, el Sabio
Gebaini	歌白尼	8	Nicolaus Copernicus
Ximan	西滿	1	Simon Stevin
Marinuo	麻日諾	1	Giovanni Antonio Magini
Weiyeda	未葉大	3	François Viète
Oujilide /	歐几里得/	5	Euclid /
Ding shi	丁氏		Clavius
Ya'erbade	亞爾罷德	9	al-Battanī
Nigulao	泥谷老	11	Nicolaus Copernicus
Bai'ernawa	白耳那瓦	5	Bernard Walther
Digu	第谷	25	Tycho Brahe
Mojue	默爵	8	Adrianus Metius
Yaqimode	亞奇默德	6	Archimedes
De'aduoxi'ya	德阿多西阿	5	Theodosius (The author of the *Sphaerica* (cf. Heath (1), *1921*, II, p. 245)
Ruowang Nebai'er /	若往訥白爾	13	John Napier
Enlige Balizhisi	恩利格巴里知斯		Henry Briggs

CRZ, j. 44 (complete list):

Li Madou	利瑪竇	74	Matteo Ricci (1552–1610)	(J)
Xiong Sanba	熊三拔	107	Sabatino De Ursis (1575–1620)	(J)
Airulüe	艾儒略	1	Giulio Aleni (1582–1649)	(J)
Pang Di'e /	龐迪莪	1	Diego de Pantoja (1571–1618)	(J)
Long Huamin	龍華民		Niccolò Longobardo (1565–1665)	(J)
Yang Manuo	陽瑪諾	25	Manuel Dias (1574–1649)	(J)
Deng Yuhan	鄧玉函	10	Johann Schreck (1576–1630)	(J)
Luo Yagu	羅雅谷	11	Giacomo Rho (1592–1638)	(J)

CRZ, j. 45 (complete list):

Tang Ruowang	湯若望	121	Adam Schall (1592–1666)	(J)
Nan Huairen	南懷仁	75	Ferdinand Verbiest (1623–1688)	(J)
Jili'an	紀利安	6	Kilian Stumpf (1655–1720)	(J)
Munige	穆尼閣	8	Nicolaus Smogulecki (1610–1656)	(J)

CRZ, j. 46 (complete list):

Naiduan	奈端	11	Newton (Isaac)
Gaxini /	噶西尼	17	Cassini (Which Cassini is intended here, is not clear)
Kebai'er	刻白爾		Kepler
Dai Jinxian /	戴進賢	2	Ignatius Koegler (1680–1746) (J)
Xu Maode	徐懋德		Jackson Pereira (1689?–1743) (J)
Du Demei	杜德美	23	Pierre Jartoux (1669–1720) (J)
Yan Jiale	顏家樂	3	Karl Slaviček (1678–1735) (J)
Jiang Youren	蔣友仁	126	Michel Benoist (1715–1774) (J)

CRZ3B, j. 7 (quasi-complete list):

(In the following list, Hymers, Haswell, Frome, Norie and Lendy are authors of manuals most of which were often reprinted (cf. Bennett (1), *1967*, pp. 84, 92 and 121.)). These authors were not necessarily "mathematicians." Frome, for example, was a British Colonel, Lendy a British Captain and John William Norie, a navigator.)

Huweili	胡威立	6	William Whewell (1794–1866)
Luomishi	羅密士	5	Elias Loomis (1811–1889)
Houshile Yuehan	侯失勒約翰	13	John F. W. Herschel (1738–1822)
Ai Yuese	艾約瑟	21	Joseph Edkins (1823–1905) (P)
Weilie Yali	偉烈亞力	80	Alexander Wylie (1815–1887) (P)
Haimashi	海麻士	6	John Hymers (1803–1887)
Hasiwei	哈司韋	3	Charles Haynes Haswell (fl. ca. 1858)
Fuluma	富路瑪	4	Edward Charles Frome (1802–1890)
Nali	那麗	13	John William Norie (1772–1843)
Lianti /	連提	22	Auguste F. Lendy
Jiayue Zhuanyilang	加悅傳一郎		Kaetsu Denichirō[210]

CRZ4B, j. 9, 10, 11 (incomplete list):

Hema	海馬	3	Homer
Yalisiduo	亞利斯多	3	Aristotle
Bailaduo	百拉多	3	Plato
Talisi	他里斯	2	Thales
Budagela	布大哥拉	7	Pythagoras
Alixi	阿里西	1	Apollonius
Duolini	妥里泥	1	Ptolemy
Diufandou	丟番都	1	Diophantus
Muhanbianmosa	穆罕徧謀撒	1	al-Khwārizmī (Muhammad ibn Mūsā)

---

[210]Obscure Japanese mathematician (fl. ca. 1852), author of a small treatise on geometrical problems solved algebraically which was included in the Chinese mathematical collection *Baifutang suanxue congshu* (1874). See also *Meijizen*, IV, pp. 15 and 75 and V, pp. 176 and 541, resp.

CRZ4B, j. 9, 10, 11 (incomplete list) (continued):

Aboweifa	阿波維法	1	Abu'l-Wafā'
Jiadan	佳但	1	Gerolamo Cardan
Feiyida	肥乙大	5	François Viète (Vieta)
Jialilüe	格里留	25	Galilei Galileo
Daijiade	代迦德	5	René Descartes
Haigengshi	海更士	5	Christiaan Huygens
Lemo'er	勒墨爾	8	Römer
Laibenzhi	來本之	10	Wilhelm Leibniz
Oulou	歐樓	9	Leonhard Euler
Lagelang	拉格朗	2	Joseph Louis Lagrange
Dalangbo	達浪勃	2	D'Alembert
Labailase	拉白拉瑟	10	Laplace
Houshile Weilian	侯失勒維廉	25	William Herschel
Dimogan	棣麼甘	6	Augustus De Morgan
Gaosi	高斯	1	C.F. Gauss

weight than outstanding astronomers and mathematicians like Gauss and Euler. Still worse, Newton was quoted, but not at all in relation to the law of universal gravitation and Kepler appeared as a secondary personage depending on Cassini.

Despite obvious limitations and misrepresentations, Western astronomy and mathematics made accessible to Chinese readers, was none the less, for the most part, based on significant excerpts from texts of primary importance such as Ptolemy's *Almagest*, Copernicus's *De revolutionibus* or Euclid's *Elements* and never on mythical accounts. That was all Ruan Yuan needed to bring to the fore "the strong and weak points" of both sciences in order to prepare a future synthesis.

The very idea of synthesis implies that Western and Chinese astronomy, or rather predictive astronomy *tuibu*, shared common conceptions. Indeed, both the latter and the former, were both engaged in mathematical (but not numerological) predictions of the position of the Sun, Moon and planets, solar and lunar eclipses and related phenomena. As historical records quoted extensively in the *CRZ* clearly show, Chinese predictive astronomy hinged, fundamentally, on observation and mathematics, exactly like its Western counterpart. But, unlike the latter, Chinese astronomers never imposed any a priori restriction whatsoever on the nature of the motion of celestial bodies and on the kind of admissible mathematics. For them, there was no need to distinguish between observable phenomena and their hidden causes and no reason to believe that natural events were caused by immutable laws.[211] In China, mathematics was defined as computational schemes restricted neither by axiomatic constraints such as the uniformity of the motion of celestial bodies along circles nor by deductive reasoning. Furthermore, Chinese astronomers

---

[211]Martzloff (16), *1994*.

deemed the most precise possible agreement between mathematical predictions and observation essential and even recognised the possibility of more and more precise predictions. Consequently, when certain celestial phenomena failed to be predicted, or were not predicted with a sufficient precision, faulty computational techniques were abandoned and replaced by more efficient ones. While predictive failures were often considered as a symptom of inadequate mathematics, they were even more frequently believed to be the result of small celestial perturbations, too minute to be detected in advance, but none the less cumulative and sufficiently important to affect mathematical predictions in the long run. Predictive astronomy was thus necessarily bound to rely on temporary computational techniques. From its origins, Chinese mathematical astronomy had been subject to more than 70 reforms. Naturally, it is true that, historically, certain reforms of Chinese astronomy were prompted by other motives (notably astrology) than the mere desire to readjust mathematical techniques but this observation does not fundamentally change the present analysis since the bulk of Chinese mathematical astronomy is purely mathematical and in itself independent of false sciences.

Seen from that angle, Western predictive astronomy was found very weak for, as noted by Ruan Yuan, "the various principles of astronomy are based on Euclid's *Elements*,"[212] that is, on axiomatic and a priori reasoning (i.e. not necesarily controlled by experience). Significantly, the Chinese mathematician who was judged the greatest (*suanshi zhi zui* "the very best mathematician") by him was not Liu Hui (whose reasoning is praised by modern historians of Chinese mathematics) but Zhao Shuang, the author of a mnemotechnic compilation of ready-made formulae on the right-angled triangle.[213]

Li Shanlan (1811–1882)

Born in a learned family from Zhejiang province, Li Shanlan was educated by Chen Huan[214] (1786–1863), a renowned philologist. According to an often repeated edifying anecdote, when he was eight years old he discovered a copy of the *Jiuzhang suanshu* on the shelves of the library of his private school and mastered the content of the venerable Chinese mathematical Bible without effort.[215] Six years later, he repeated the same feat with the first six books of Euclid's *Elements* in the version translated into Chinese by Matteo Ricci and Xu Guangqi in 1607. A little later, he sat the provincial examination held at Hangshou. Although he did not pass, the unlucky candidate took avail of his sojourn at Hangzhou and bought copies of the *Ceyuan haijing*, the basis of Chinese medieval algebra, and the *Gougu geyuan ji*, a manual of trigonometry by Dai Zhen (1724–1777), the influential member of the evidential research

---

[212]Cf. *CRZ.* j. 45, p. 594.

[213]*CRZ*, j. 4, p. 53. On Zhao Shuang's formulae, see Brendan (1), *1977*.

[214]Wang Ping (1'), *1966*, p. 144. On Chen Huan cf. Hummel, p. 822.

[215]*ZSSLC-T*, II, p. 438.

movement (*kaozheng xue*).[216] These two books orientated his interests in a decisive way towards the study of mathematics which he practised as a self-taught man. Greatly impressed by the *Ceyuan haijing*, Li Shanlan viewed the *tianyuan* algebra as the ultimate technique for mechanical solution of all sorts of mathematical problems. Much later, when lecturing on algebra at the Tongwen guan, he took Li Zhi's masterpiece as the fundamental mathematical reference for his young students. Seriously interested in mathematics, Li Shanlan made contact with scholars having the same inclination as himself and began to write his first mathematical works. At that time, however, mathematics was not a subject to earn a living from and, in 1845, he was compelled to accept a position as a tutor in the family of the late Lufei Chi (?–1790), a famous scholar.[217] It is during this period that Li Shanlan met scholars interested in mathematics such as Zhang Wenhu (1808–1885), Wang Yuezhen (1813–1881) and many others.

In 1852, Li Shanlan took refuge in Shanghai in order to escape the Taiping revolt.[218] There he met Alexander Wylie[219] (1815–1887) from the London Missionary Society. Wylie was fluent in Chinese (he later became an outstanding sinologist) but not particularly versed in scientific subjects. However, like many other Protestant missionaries, he rapidly gained an ability to face the problem of diffusion of fundamental scientific knowledge into the Chinese society which most Protestant missionaries considered essential for the Christianisation of China. But he also learnt about native mathematics and was favourably impressed by a little treatise on logarithms composed by Li Shanlan in 1846:

> Li Shanlan [...] now residing in Shanghai [...] has recently published a small work called *Tuy-soo-tan-yuen* [*Duishu tanyuan*] in which he details an entirely new method for their computation, based upon geometrical formulae [...] [he] has had no better aid than that afforded by the *Leuh-lih-yuen-yuen*[220] and [...] has here given us as a result of four years' thought a theorem which, in the days of Briggs and Napier, would have been sufficient to raise him to distinction.[221]

Consequently, Wylie recruited Li Shanlan as a eclectic co-translator of all sorts of scientific works for the London Missionary Society. Since Li had never been trained in any European language the technique of translation involved teams composed of two persons, one of whom was bilingual.[222] Working in cooperation with the Protestant missionaries Alexander Williamson (1829–1890), Joseph Edkins (1823-1905) and Alexander Wylie, Li Shanlan was involved in the translation of various works on botany, mechanics, astronomy and mathematics. Sometimes he worked on several projects at once. In the

---

[216]On Dai Zhen cf. Hummel, p. 695, on the evidential research movement, see Elman (1), *1984*, p. 17 ff.

[217]Hummel, p. 479.

[218]Wang Ping (1), *1962*, p. 779.

[219]Biography in Wylie (3), *1897*, pp. 1–18.

[220]i.e. *Lüli yuanyuan* – an encyclopedia from the Kangxi period consisting of several specialised treatises, in particular the *Shuli jingyun*.

[221]Ibid., pp. 193–194.

[222]Cf. above, p. 21.

Fig. 11.8. Professor Li Shanlan and his mathematical class. Source: Martin (1), *1896*, p. 312.

1850s he translated John Lindley's *An Introduction to Botany* (*Zhiwuxue*), John F. W. Herschel's *Outlines of Astronomy* (*Tan tian*), William Whewell's *An Elementary Treatise on Mechanics* (*Chongxue*), Augustus De Morgan's *Elements of Algebra* (*Daishuxue*), Elias Loomis's *Elements of Analytical Geometry and of Differential and Integral Calculus* (*Daiweiji shiji*) and the last nine books of Euclid's *Elements* (*Xu Jihe yuanben*, a sequel to the *Jihe yuanben*).

Towards 1859–1860, Li Shanlan joined the staff of the Governor of Jiangsu, Xu Youren (1800–1860) who was a renowned mathematician known for his research into infinite series.[223] Three years later, he was recruited by Zeng Guofan (1811–1872), the famous statesman, general and scholar.

In 1867, Li Shanlan's collected works were published in Nanking under the title *Zeguxizhai suanxue* [The mathematics of the studio taking the ancient as models]. In his preface, Li Shanlan explains that the publication was made possible by a grant from Zeng Guoquan (1824–1890), a famous general who had contributed to the overthrow of the Taiping rebellion and younger brother of Zeng Guofan (1811–1872). It is thus not too strange that the *Zeguxizhai suanxue* contains a treatise on ballistics, the *Huoqi zhen jue* [The true art of firearms]. As

---

[223]Hummel, p. 479. Xu Youren was killed during the Taiping revolt.

shown by Horng Wann-sheng, a Taiwanese historian, this book essentially deals
with the question of the maximum range reached by a projectile by means of a
geometrical study of the parabola (rather than elementary calculus), a subject
which had been previously introduced in the *Chongxue*, a Chinese adaptation
of William Whewell's *Mechanics* on which Li Shanlan had previously worked in
collaboration with Alexander Wylie, as noted above.

Apart from ballistics, more than half of the *Zeguxizhai suanxue* (more
precisely eight sections out of 13) is also devoted to Western mathematics, but
almost exclusively from the point of view of power series dependent on properties
of the ellipse, trigonometric functions (direct and inverse) and logarithms.[224] Li
Shanlan rarely develops these subjects from scratch and many mathematical
techniques he uses can be traced back to 18th century Jesuit treatises such as
the *Lixiang kaocheng houbian* or to Chinese works by his predecessors.

Li Shanlan's treatment of subjects derived from Chinese traditional mathematics (four sections out of 13) is frequently very original. This is especially
true of his *Duoji bilei*, a study of summation formulae stemming from generalised "Pascal's triangles" but based on Zhu Shijie's *Siyuan yujian*. A priori,
the search for originality which is so obvious in the *Zeguxizhai suanxue* might
be accounted for in a variety of ways but, as explicitly stated in Li Shanlan's
preface, its essential motive was the author's desire to beat Western mathematicians at their own game, not by a slavish appropriation of "barbarian"
mathematical techniques but by highlighting traditional Chinese modes of
thought and explicitly taking his inspiration from his medieval predecessors,
especially Zhu Shijie.

The *Zeguxizhai suanxue* collection was reprinted in 1868 and 1882 and
some of its parts were sometimes published separately or in other mathematical
collections.[225]

In July 1869, at the age of 58, Li Shanlan was finally appointed Professor of
Mathematics at the Tongwen guan College[226] which was supervised by William
Alexander Parsons Martin (1827–1916) of the Presbyterian mission;[227] he held
this post until his death in 1882.

---

[224]Cf. Mei Rongzhao (11'), *1990*, p. 334 ff.

[225]Cf. Li Yan, *ZSSLC-T*, IV-2, p. 476. The *Duoji bilei*, for example (which is the fourth
treatise of Li Shanlan's collected works) was reproduced in the *Zhongxi suanxue dacheng*
(Great encyclopedia of mathematics, Chinese and Western), published in 1899 and reprinted
in 1901.

[226]Biggerstaff (1), *1961*, p. 123.

[227]On Martin, cf. Duus (1), *1966*.

Part II

The Content
of
Chinese Mathematics

12. Numbers and Numeration

Knotted Cords (Quipus) and Tallies

Ancient Chinese literature contains numerous allusions to quipus and tallies.

For example, the *Yijing* states that: "in Early Antiquity, knotted cords were used to govern with, later, our saints replaced them with written characters and tallies."[1] The *Zhuangzi* states that: "at the period when the Sovereigns Rong Cheng, Da Ting [...] ruled, people used knotted cords."[2] The *Daodejing* advises readers to "see to it that the use of knotted cords is reintroduced to the people."[3] The *Liezi* reports that "There was a man of Sung who was strolling in the street and picked up a half tally someone had lost. He took it home and stored it away, and secretly counted the indentations of the broken edge. He told a neighbour: 'I shall be rich any day now'."[4] There are many other such quotations.

These practices have continued until today among certain ethnic minorities of Yunnan.[5]

Chinese Numeration

Oracle-Bones and Tortoiseshells (*jiaguwen*)

At the end of the last century, in 1899, an important discovery led to a complete revision of all that was then known about ancient China.

Thousands of inscribed bones and tortoise carapaces were exhumed in Xiaotun, to the North West of what is now the town of Anyang in Henan province, on the site of the last capital of the Shang (14th–11th centuries BC).[6]

Subsequently, throughout the 20th century, new finds, made accidentally during construction or agricultural works, or during systematic excavations, have added tens of thousands more items to the initial collection. Many scholars (Liu E, Dong Zuobin, Luo Zhenyu, Hu Houxuan, Guo Moruo, Kaizuka Shigeki,

---

[1] *Yijing*, "Xici," II-2.
[2] *Zhuangzi*, j. 10 ("wai pian," j. 3).
[3] *Daodejing*, section 80.
[4] Graham (2), *1960*, p. 179.
[5] Chen Liangzuo, (2'), *1978*, p. 268.
[6] Deydier (1), *1976*.

Shima Kunio, J. A. Lefeuvre etc.) have studied these inscriptions and concluded that they were essentially divinatory documents.

Some of these contain numerical information (number of men, animals, shells, days, months, sacrifices, war expeditions, etc.). Two different systems of numeration have been distinguished:

The first system uses a combination of two elementary series of symbols, one consisting of ten elements (the ten celestial trunks, *tiangan*) and the other of twelve elements (the twelve terrestrial branches *dizhi*). Associated in pairs, these symbols give rise to a series of 60 doublets.[7] This series is usually called "the sexagesimal cycle" since it is used to number units of time in blocks of sixty units. Originally, the Chinese counted the days in groups of sixty, later the lunar months and, still later, from the second century BC, the years in the same way. This last system is still in operation today. However, 60 does not correspond to any natural astronomical cycle and this system has never been used for arithmetical calculations.

The second system is decimal and comprises the following 14 symbols[8] consisting of thirteen signs for numbers and an additive conjunction corresponding to "and":

| 1 | 2 | 3 | 4 | 5 | 6 | 7 | 8 | 9 | 10 | 100 | 1000 | 10000 | "and" |

Thus, a simple horizontal stroke is used to represent 1, while repetition of this one-stroke as many times as necessary specifies the numbers 2, 3 and 4 in a highly concrete way. The Sumerians, the Babylonians, the Egyptians, the Mayas and the Romans all used the same procedure to denote small numbers. The other symbols are much more difficult to interpret. In his *Shuowen jiezi* (a famous dictionary), Xu Shen (ca. 100 AD) explains the small numbers using magic correlations of the *yinyang* and the "five phases" (*wuxing*). Under the Tang dynasty, the philologist and exegete of the classics, Yan Shigu (581–645) remained on analogous ground in his attempt to explain that the Chinese symbols for numbers came from the trigrams of the *Yijing*.[9] These explanations are not really convincing, but they at least have the merit of reminding us that in China, numerous scholars believed that numbers and divination often went hand in hand.

More recently, palaeographic study of the *jiaguwen* has led to the formulation of hypotheses of a different order. For Chen Mengjia,[10] the symbols representing 5, 6, 7, 8, 9, 100, 1000 and 10 000 are all "false borrowings" *jiajiezi* (in traditional Chinese philology, this is the name given to any written character

---

[7]Cf. Chang Tai-ping, "The role of the *T'ien-kan Ti-chih* in the Naming System of the Yin," *Early China*, 1978/79, no 4, pp. 45-48.

[8]This "14" does not take into account the graphical variants of these symbols. Cf. Chen Liangzuo (2'), *1978*, p. 272 ff. and QB, *Hist.*, p. 5.

[9]Cheng Te-k'un (1'), *1983*, p. 169.

[10]Chen Liangzuo, op. cit., p. 272.

which is conventionally used in a new sense which it did not have originally; this is one of the six ways of forming Chinese characters). There is certainly some truth in this notion, because, for example, the symbol, which stands for the myriad 𝄐 is simply the pictogram of the scorpion. As Sakai Hiroshi explains,[11] the swarm of colonies of new-born scorpions sticking together along the back of the females of this arachnid is an appropriate evocation of an immense number. In the same vein, we note that the *jiaguwen* uses a single symbol to represent one thousand and a man. To explain this curious telescoping, we may assume that in ancient China, the word "man" was pronounced in the same way as the word "thousand." We may also suppose that there was a link between certain parts of the human body and the symbol for a thousand. For example, just as the number ten may be explained using the fingers of two hands, a thousand could be derived from the ten toes of two feet, if it were true that each toe counted as a hundred. It seems that this was actually the case, since the pictogram for a hundred is precisely the same as that for the toe 𝄐.[12]

Of course, these are not the only possible hypotheses. Like Zhang Bingquan, we may also suppose, for example, that the symbols for certain numbers simply translated gestures of the hands or arms associated with these numbers.[13]

In any case, it is now important to analyse the way in which these symbols are combined to form numbers.

As far as the notation for tens is concerned, we first note that the numbers 20, 30 and 40 are denoted by repetitive addition of the sign for ten. After 40, the following tens are denoted using a multiplicative principle; the same is true for hundreds, thousands and myriads. More precisely, in each case, the elements composing the pairs of symbols which are used to denote tens, hundreds etc. are written either on or above one another (the symbols may then be separate but they may also touch as though one were placed on top of the other)[14]:

[11]Sakai (1'), *1981*, p. 251.
[12]Ibid., p. 249.
[13]Chen Liangzuo, op. cit., p. 272.
[14]Ibid., p. 274 ff.

Intermediate numbers are written using both the additive and the multiplicative principles. For example[15]:

(table of jiaguwen numeral forms: 1, 2, 3, 4, 5, 6, 7; 10, 11, 12, 13, 14 or, 15, 16, 17, 100; 8, 25, 41, 42, 68, 114; 9, 130, 159, 162, 451; 1000, 510, 2656 men)

Sometimes an additive "and" 𝒴 occurs between tens and units, or hundreds and tens, rather akin to the English usage in expressions such as "three hundred and one" and "six hundred and one." For example[16]:

659:    *(jiaguwen glyphs)*   six hundred and five ten and nine

The largest number discovered in this system[17] is 30 000. Could the Chinese of the Shang have denoted arbitrarily large numbers had the need arisen? Strictly speaking, we cannot answer this question, since we do not know if they asked themselves it. If we assume that they could only have done so within the conceptual framework of their system of numerals, we are however forced to conclude that to do so they would probably have had to invent indefinitely many new symbols to denote successive powers of ten.

The Traditional Chinese Numeration

The numeration of the *jiaguwen* has evolved over the centuries. Based on numbers engraved on various types of solid material (pottery, bronze instruments, stelae), we may follow the broad pattern of this evolution.

Apart from 1, 2 and 3 the other digits and the characters marking the powers of ten changed shape before stabilising to a definitive shape around the beginning of the first millennium. This is illustrated by the following Table[18]:

---

[15]Idem.

[16]Example taken from QB, *Hist.*, p. 7. On this point, see also Wang Li (1′), *1980*, II, p. 256.

[17]Li Yan, *Gudai*, p. 1.

[18]Following Cheng Te-k'un, op. cit., p. 174 ff. (inscriptions on bronze).

Shang & Zhou　　　　　　　　　　　Qin & Han

4	亖	亖	亖	𖤍			亖			四				四
5	𐤔	𐤔	𐤔	𐤔			𐤔	𖤍		𐤔		𐤔	五	五
6	介	介	介	介			介					六	六	六
7	十	十					十	十		七	七			七
8	)(	)(	八	)(	八		)(	)(		八	八		八	八
9									九	九			九	九
10	丨	十	十				十	十		十	十			十
100	百	百	百	百	百	百	百	百	百	百	百	百		百
1000	千	千	千	千			千	千	千	千	千	千		千
10000	萬			萬		萬	萬	萬		萬	萬	萬		萬

At the same time, the additive 'and' disappeared, superposition of numbers became a thing of the past and eventually, the fundamental numerical symbols formed a linear sequence. A new symbol *yi* 億 (sometimes denoting ten myriads, sometimes a myriad myriads (one hundred million)) appeared, so that larger numbers could be written down.[19]

Like the old system, the new system provided a very simple way of denoting numbers in words. This comprised a very simple procedure, which involved stating (as in speech) how many hundred millions (respectively hundred thousands), how many myriads, how many thousands, how many hundreds, how many tens and how many units the number contained. For example, in the *Jiuzhang suanshu*, the number 1 644 866 437 500 (which is also the largest number found in the Ten Computational Canons[20] (*SJSS*), is written as[21]:

一萬六千四百四十八億六千六百四十三萬七千五百

(or: one myriad six thousand four hundred [and] forty eight hundred million six thousand six hundred [and] forty three myriad seven thousand five hundred. Modern pronunciation: *yi-wan liu-qian si-bai si-shi ba yi liu-qian liu-bai si-shi-san wan qi-qian wu-bai*; or, word for word: 1/myriad/ 6/thousand/ 4/hundred/ 4/ten/ 8/hundred million/ 6/thousand/ 6/hundred/ 4/ten/myriad/ 7/thousand/ 5/hundred).

In this notation, the myriad plays a distinguished role, since it is used as a reference point for writing large numbers and dispenses with the need for special markers to transcribe one hundred thousand, one million, up to *yi*.

In addition, 20, 30 and 40 may be transcribed in two different ways, using either the regular forms *ershi* 二十, *sanshi* 三十 and *sishi* 四十 (which denote, 2 tens, 3 tens and 4 tens, respectively) or the irregular forms *nian* 廿,

---

[19]Cf. Li Yan, *Gudai*, p. 181.
[20]Schrimpf (1), *1963*, p. 69.
[21]*JZSS* 4-24 (QB, I, p. 155).

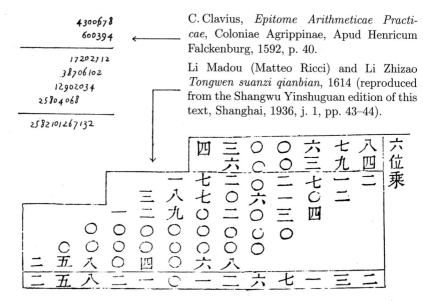

C. Clavius, *Epitome Arithmeticae Practicae*, Coloniae Agrippinae, Apud Henricum Falckenburg, 1592, p. 40.

Li Madou (Matteo Ricci) and Li Zhizao *Tongwen suanzi qianbian*, 1614 (reproduced from the Shangwu Yinshuguan edition of this text, Shanghai, 1936, j. 1, pp. 43–44).

Fig. 12.1. An example of written multiplication in traditional Chinese figures, based on a technique of European origin.

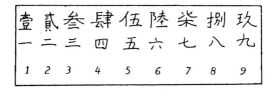

Fig. 12.2. The simple and the complex form of traditional Chinese numbers.

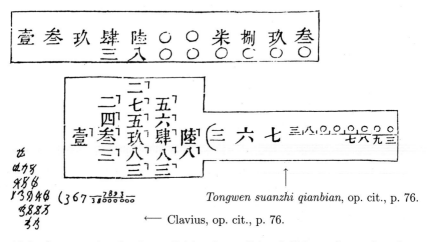

Tongwen suanzhi qianbian, op. cit., p. 76.

⟵ Clavius, op. cit., p. 76.

Fig. 12.3. An example of written division in traditional Chinese figures based on a technique of European origin. (This calls for the division of 13 946 007 893 by 38 000 000. Clavius gives the quotient in the form $367 \frac{7893}{38\,000\,000}$).

sa ⫲, and *xi* ⫿.[22] (which have the global sense of 20, 30, 40). According to J. Needham:

> They [i.e. these symbols] were not used at all in mathematical works nor often (except for poetry and pagination) in general literature.[23]

This judgement is certainly true if one analyses the content of recent editions of ancient Chinese mathematical works. When original manuscripts are consulted, however, a quite different conclusion is obtained. The Dunhuang mathematical manuscripts (which are among the extremely rare Chinese mathematical manuscripts anterior to the 10th century still extant) use these irregular forms for 20, 30 and 40 very frequently for mathematics and not only for poetry and pagination.[24]

After the Han, as previously mentioned, the shape of the figures remained unchanged, but the principles governing how numbers were written were sometimes very slightly modified.[25] For example, during the Tang dynasty, the symbol for "1" *yi* — was optionally written in front of *shi, bai, qian, wan* (ten, hundred, thousand); it later became mandatory.[26] Another more important difference is that from the Ming, Chinese written numeration began to use a particular character, *ling*, which was used as a place holder for skipped units.[27] (a kind of zero)

From the 17th century, Jesuit missionaries introduced computational techniques of European origin into China. They used figures for 1 to 10, exactly like the Hindu–Arabic numerals (Fig. 12.1). Little by little, two sorts of numbers crept into the texts, namely those expressed in words and those based on the principle of the decimal system (with up to ten digits). As far as we can judge from the very late editions of mathematical and astronomical texts which have survived, it seems, however, that such a practice already existed previously, albeit on a limited scale, for denoting numbers within numerical tables.

Finally, we note that Chinese numbers also have more complicated forms, which are difficult to counterfeit, and which are used to make out cheques, invoices and other official documents (Fig. 12.2). Curiously, these are used for purely arithmetical purposes in an arithmetical treatise of 1614 (Fig. 12.3).

The Rod-Numeral Notation and its Derivatives

From the period of the Warring States (403–222 BC) we find coins engraved with numerical symbols associated with small numbers less than ten (Fig. 12.4).

---

[22] According to the *Cihai* dictionary (Shanghai: Shanghai Cishu Chubanshe), *1979*, I, p. 87), this character is pronounced as *xi*.

[23] *SCC*, III, p. 13.

[24] Cf. Li Yan, *Gudai*, pp. 16 and 29, resp.

[25] Brainerd and Peng (1), *1968*, p. 64.

[26] In modern Chinese "1" must obligatorily precede 100, 1000, 10000 but not 10. More generally, for modern usage, see Yuen Ren Chao's [Zhao Yuanren] outstanding *A Grammar of Spoken Chinese*, Berkeley: University of California Press, 1968, pp. 573 ff.

[27] See below, p. 204 ff.

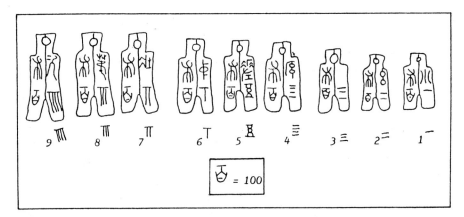

Fig. 12.4. Rod-numerals engraved on coins of the Wang Mang period (9–23 AD) (note that on these coins, the symbols corresponding to '5' and '100' do not consist of rod-numerals). However, rod-symbols for these numbers are also attested well before Wang Mang.

Unlike the figures of the traditional Chinese numeration, these symbols all have a very simple structure: the numbers 1, 2, 3, 4 and 5 consist, respectively of 1, 2, 3, 4, or 5 superimposed strokes while 6, 7, 8 and 9 are decomposed as $5 + 1$, $5 + 2$ $5 + 3$ and $5 + 4$ (where one of the bars denotes 5 and the others each denote a unit)[28]:

$$ \underline{} \quad \underline{\underline{}} \quad \equiv \quad \overline{\overline{\equiv}} \quad \underline{\underline{\underline{\equiv}}} \quad \perp \quad \underline{\perp} \quad \underline{\underline{\perp}} \quad \underline{\underline{\underline{\perp}}} \quad | $$

$$ 1 \quad 2 \quad 3 \quad 4 \quad 5 \quad 6 \quad 7 \quad 8 \quad 9 \quad 10 $$

Another system of "numbers in the form of rods" is also found around the start of the first millennium AD. Here, the figures 1, 2, 3 and 4 are denoted in exactly the same way as above, while 6, 7, 8 and 9 are composed according to the same principle as above but with different orientations. In this case, the bar denoting "5" is horizontal while the units added to this "5" are vertical.[29] Unlike the arbitrary indecomposable representations of the figures to which we are accustomed, these symbols provide a rather concrete image of the numbers they represent. As in the case of the Mayan[30] and Roman systems of numeration units from 1 to 4 are clearly perceptible and 5 plays a distinguished role:

I	II	III	IIII	∧ ∨	VI	VII	VIII	VIIII	X	Ancient roman numerals
•	••	•••	••••	—	—•	••̅	•••̅	••••̅	═	Mayan numerals

---

[28]Chen Liangzuo, op. cit., p. 278.

[29]Following Yan Dunjie (27'), *1982*, Fig. 1, p. 32.

[30]On Mayan mathematics see M. P. Closs's notice in Grattan-Guiness (1), *1994*, I, p. 143 ff.

But in the case of numbers greater than ten, the analogy stops there; the Mayan system uses base 20 and the Roman system uses special symbols for 50 and 500, among others.[31]

How did the Chinese from antiquity denote numbers greater than 10 000 using such a rod-numerals? It is not easy to answer this question since early examples of relatively large small numbers written according to the system of rod-numerals have never been found.

However, based on late accounts (Dunhuang manuscripts, 5th–10th centuries AD, works of the Song and Yuan periods) we can obtain a quite precise idea of the system (or at least of a system using numerical symbols which are graphically analogous to those we have just discussed).

There are two series, each consisting of nine signs representing the numbers from 1 to 10. In each series, the symbols for the numbers in question consist of either vertical or horizontal bars. In the first series, each vertical bar counts for '1' and the horizontal bar counts for '5.' The opposite is true for the second series:

First series	l	ll	lll	llll	lllll	T	ТТ	ТТТ	ТТТТ
Second series	—	=	≡	≣	≣	⊥	⊥	⊥	≛
	1	2	3	4	5	6	7	8	9

In addition, the symbols of the first series are used to mark units, hundreds, myriads (and, more generally, even powers of 10) while the signs of the second system are reserved for tens, thousands and other odd powers of 10.

To represent an arbitrary number in this system, the number of highest-order units is written down first and so on down to the least-order units. For example, 18, 27, 36, 396 and 378, are written as ─Ⅲ, =Т, ≡⊥, Ⅲ≡T, Ⅲ⊥Ⅲ, respectively (after the Stein 930 manuscript).

A priori, these examples give one the impression that it is the position of the numerals within the numbers which determines the power of ten to which a given figure should be attached. In other words, this seems to be a positional notation system. However, this impression turns out to be false; the symbol T on its own may be read as 6, 600, 60 000 etc., while ─Ⅲ may be read as 18, 1800, 1008, etc. Most historians of Chinese mathematics explain that, to get round this difficulty, the Chinese always left an empty space, where necessary, to indicate the missing units.[32]

This explanation is plausible, however, as we shall see in the next section on zero, there are no historical examples which would prove it beyond all doubt.[33]

---

[31] Yan Dunjie (27′), *1982*, p. 32.

[32] *SCC*, III, p. 9: "Before the +8th century, the place where a zero was required was *always* left vacant;" Yushkevich (1), *1964*, p. 14.

[33] See below, p. 205.

In fact, it seems that as far as the rod-numeral notation is concerned, the context played an important role. When one knows what one is talking about in advance the fact that the notation is ambiguous is not necessarily a hindrance. For example, when a merchant says that a piece of furniture costs "deux mille six" (two thousand and six), we generally understand "2600 francs" and not "2006 francs" (in popular French), although in the case of an opinion poll with "two thousand and six positive results," we understand that precisely 2006 individuals have replied "yes."

We also note that it is not difficult to compare this system of rod-numerals with calculating instruments, namely counting-rods. Figures written in the rod-notation merely reflected the figures "on rods" (at least from a certain period).

Counting-rods and rod-numerals existed well before the Han, but we do not know what the relationship between them was. We only know that in the 4th and 5th centuries AD (very approximately) these two systems were related (see the laconic description of numbers on rods in the *Sunzi suanjing*; see also the Dunhuang mathematical manuscripts).[34]

Thus, as many authors suggest, Chinese arithmeticians may possibly have placed their counting-rods on counting-boards including spaces reserved for units, tens, hundreds, etc.[35] If this was the case, arithmeticians could have solved the question of the missing units very easily by placing their rods at the required position. Unfortunately, we know nothing of such counting-boards.[36]

In any case, what is certain is that Chinese mathematicians began to use the zero (in the form of a small circle) from the 12th to 13th centuries; whence, they were able to denote numbers without the slightest ambiguity. For example, to transcribe 22 908, 10 048 and 90 000 they wrote[37] ‖=Ⅲ̅o̅Ⅲ /oo≡Ⅲ Ⅲoooo, respectively. Thus, at this stage the rod-numeral notation became wholly positional.

But simultaneously, the system evolved independently from the rod system and the symbols for some numbers were modified.

Groups of more than three bars were tolerated less and less when writing down figures. Thus, some authors replaced ‖‖ (4) by a cross ✕ [38] and ✕̅ (or ✕̣) (9) by Ⅲ̅ (or ≣).[39] The symbols ō and ȯ (instead of ≣ or ‖‖‖) are found for "5," which probably attests to the influence of the circle-zero (since, in the same way that 6, 7 and 8 are analysed as $5+1$, $5+2$ and $5+3$, ō and ȯ may both be decomposed as $5+0$, where the bar denote 5 units).

[34]QB, p. 282. On this subject see also the material collected by Needham (*SCC*, III, pp. 70ff.), Libbrecht (5), *1982* and Li Yan (22′), *1929*.

[35]See note 1, p. 209.

[36]See below, p. 209.

[37]Examples taken from: *Yang Hui suanfa* (following Kodama (1′), *1966*, p. 61), *Shushu jiuzhang* (1247) (Shanghai: Shangwu Yinshuguan, 1936, p. 165 and p. 119, resp.).

[38]See Kodama (2′), *1970*, p. 22.

[39]Yan Dunjie, op. cit., pp. 31–32.

Under the Ming dynasty, the symbols for 5 and 9 were subject to calligraphic deformation and were transformed into 丂 and 乂 (Ꝺ→丂 ✕→乂). In written expressions for numbers involving several units, the old distinction between the two series of horizontal and vertical bars finally died out and a single series of nine figures came into use, namely the system of "secret marks" (*an ma* 暗碼):

Panzhu suanfa (1573)[40] 盤珠算法	丨	丿丨	丿丨丨	✕	丂	𠄌	二丨	二丨丨	文	丨
Suanfa tongzong (1592)[41] 算法統宗	丨	丨丨	丿丨丨	✕	Ꝺ	𠄌	二丨	二丨丨	文	
	1	2	3	4	5	6	7	8	9	10

From the end of the 16th century, the secret marks were very successful with merchants. The Chinese of today still use them in their everyday life (market, restaurant, etc.), even though they also use other more internationally widespread systems.

Finally, under the Qing dynasty, certain mathematicians invented ingenious techniques for writing down complicated algebraic expressions, which had the astonishing property that they did not use any algebraic symbol but only positional combinations of numbers taken either from the traditional Chinese system or from the system of "secret marks" and written in different sizes. For example, for Xu Youren[42] (1800–1860), the following formula[43]:

$$\tan mx = m \sin x + \frac{(2m^2 + 1)}{3! r^2} \sin^3 x + \frac{(16m^4 + 20m^2 + 9)}{5! r^4} \sin^5 x$$

$$+ \frac{(272m^6 + 560m^4 + 518m^2 + 225)}{7! r^6} \sin^7 x$$

$$+ \frac{(7936m^8 + 22848m^6 + 31584m^4 + 25832m^2 + 11025)}{9! r^8} \sin^9 x + \dots$$

becomes[44]:

---

[40]From Kodama, op. cit., p. 22.

[41]From Yan Dunjie, op. cit.; QB, *Hist.*, p. 141 ff.

[42]Cf. Hummel, p. 479. See also Li Yan, *Dagang*, II, p. 14; Wang Ping, (1′), *1966*, p. 112 ff.

[43]Following Yan Dunjie (27′), *1982*, p. 33.

[44]Ibid. Original text in chapter 4 of Xu Youren's *Geyuan baxian zhuishu* in *Baifutang suanxue congshu*, Shanghai: Hongwen Shuju, 1898.

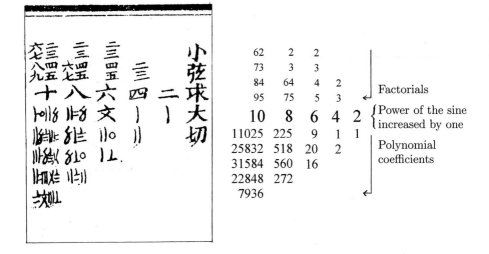

Fig. 12.5. Reproduction of Xu Youren's formula from the original text of his *Geyuan baxian zhuishu*, j. 4, p. 1b in *Baifutang suanxue congshu*, op. cit.

Non-Decimal Systems of Numeration

Although Chinese mathematics is universally dominated by systems of metrology and numeration fundamentally decimal and often positional, real practices were most probably more varied and differentiated than usually believed. We have already mentioned quipus from Chinese antiquity but the *Zuozhuan*, for example, also refers to systems of measures based on quaternary and quinary representations:

> Qi from of old has had four measures, the *dou*, the *ou*, the *fu* and the *zhong*. Four *sheng* make a *dou* and up to the *fu* each measure is four times the preceding and then ten *fu* make a *zhong*. The Chen clan makes each of the [first] three measures once again greater [i.e. each of them is five times the preceding] so that the *zhong* is very large, lending according to their own measure and receiving back again according to public measure [...].[45]

In addition, a system of representation of numbers in the radix 9 was also common among musicians[46] and during the Ming dynasty, for example, the sexagesimal system was used by the Chinese astronomers from the Muslim Bureau of Astronomy.[47]

[45] *Zuozhuan*, "Zhao gong," third year. The present translation is adapted from James Legge, *The Chinese Classics with a translation, critical and exegetical notes, Prolegomena and copious index*. Hong Kong and London, *1872*, V, part 2, p. 589.

[46] Dai Nianzu (1'), *1986*, p. 190 ff.

[47] See Bei Lin's *Qizheng tuibu* which is essentially a set of astronomical tables more or less similar to e.g. the Alphonsine Tables currently used in Europe during the same period.

Units of Measurement

The units of measurements of ancient cultures are often derived from dimensions suggested by the human body. In the case of the Roman empire, for example, we know from Hero of Alexandria (ca. 62 AD) that: "the units of measure are obtained from the human limbs: the finger, the palm, the span [...], the foot, the ell, the step, the fathom."[48]

The same observation also applies in the case of China where the *kui* corresponds to a pace of a man and the *bu* (a very common unit) corresponds to twice the *kui*.[49] In the same way, the Romans also distinguished between one pace and two paces (which they called "pes" and "passus," respectively).[50]

The unit of length called the *zhang* also corresponds to the height of a man. In the same spirit, the length of the gnomon *bi* of the famous *Zhoubi suanjing* was probably defined by the height of a man.[51] But natural objects and phenomena were also made use of to define units of measurement. For example, the *hu* corresponds to the thickness of the thread produced by a silkworm (*can suo tu si wei hu*).[52] According to the *Shuoyuan*, by Liu Xiang (ca. 79–6 BC), the unit corresponding to the smallest possible capacity, called the *su* (millet) is defined by a grain of millet.[53] The units of area are derived directly from the units of length. For example, the *mu* corresponds, in its oldest usage, to the surface area of a square of side 100 *bu*. Later, from the Han, the value of the *mu* curiously became equal to 240 square *bu*.[54]

As Yushkevich notes,[55] this type of relationship does not imply that the values of successive units in the same series are related in a simple way. However, we note that the idea of a decimal series caught on at a very early stage. We know of a ruler from the Zhou period (6th century BC) graduated in tenths and hundredths of a foot (*chi*).[56] At the beginning of the first millennium, Wang Mang, the founder of the Xin dynasty commissioned the fabrication of more than a hundred sets of calibrated standard bronze measures (Fig. 12.6). The corresponding legal measurements, which he promulgated throughout the Empire were, in decreasing order, the *hu* (approximately 20 litres), the *dou* ($= \frac{1}{10}hu$), the *sheng* ($= \frac{1}{100}hu$), the *ge* ($= \frac{1}{1000}hu$) and the *yue* ($= \frac{1}{2000}hu$). More generally, in all the measurement systems used in all Chinese arithmetics from the *Jiuzhang suanshu* of the Han onwards, the units vary mostly in powers of 10.[57]

---

[48] Cf. the commented translation of Hero's *Geometrica* in Bruins (1), *1964*.

[49] Wu Chengluo (1'), *1937*, p. 51.

[50] Tropfke (3), *1980*, p. 74.

[51] Yi Shitong (1'), *1989*, p. 366.

[52] *SZSJ*, j. 1 (QB, II, p. 281).

[53] The *su* is also a unit of weight. See Libbrecht (5), *1982*, p. 223.

[54] *DKW*, 7-21815, p. 8008. See also Wu Chengluo, op. cit., pp. 95–96.

[55] Yushkevich (1), *1964*, p. 19ff.

[56] Ferguson (1), *1941*, p. 357.

[57] Reifler (1), *1965*.

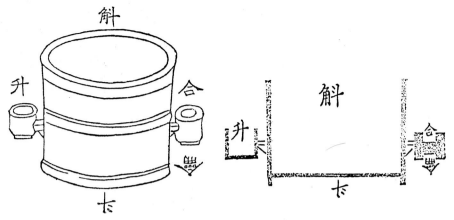

Fig. 12.6. Wang Mang's calibrated standard grain measure with its five cylindrical receptacles (from Reifler, op. cit.).

The decimal regularity of Chinese measurement systems is widely accepted. However, it is uncertain whether the actual usage was as regular or homogeneous. Indeed, there is evidence to suggest that it was not (for example evidence due to F. Swetz relating to the beginning of the 19th century.[58] and to Bernard-Maître relating to the beginning of the 18th century).[59]

However, the legal measurement systems essentially adhered to the decimal system. Various social and political factors probably explain why the systems varied from one province to another.

Fractions, "Models" *Lü*

The heading "fractions" may be taken to include various ideas including sharing, division, ratios, proportionality, rational numbers, etc.

In Chinese mathematics, by far the most common notion of fractions is that which comes from the notion of dividing a whole into an equal number of equal parts (sharing). In particular, the Ten Computational Canons contains many expressions such as

- *san fen lu zhi er*　三分鹿之二　: "2 of 3 parts of a stag" (2/3 of a stag);[60]

- *si fen qian zhi san*　四分錢之三　: "3 of 4 parts of a *qian*" (3/4 of a *qian* (= coin)).[61]

---

[58]Swetz (3), *1974*, p. 32.
[59]Bernard-Maître (1), *1935*, p. 416.
[60]*JZSS* 3-1.
[61]*JZSS* 1-18.

In these expressions, the word *fen* 分 strongly suggests the idea of sharing, since, etymologically, its upper component ㅅ means "to share,"[62] while its lower component represents a knife.[63]

More abstractly, the part of the whole may be given in the form *x fen zhi y* (or: *y* of *x* parts – *y* parts from a whole fragmented into *x* equal parts). The denominator and the numerator are then respectively called *fenmu* (the "mother" of the sharing) and *fenzi* (the "son" of the sharing). According to Wang Ling[64] the inventor of these expressions was thinking of a pregnant mother and her child, thus highlighting both the difference in size and the intimate link between the two terms.

In addition to the above expressions, there also exist special terms which are used solely to denote specific fractions:

- 1/3 is represented by *shao ban* (less than half);[65]

- 2/3 is represented by *tai ban* (more than half);[66]

- 1/2 is represented by *zhong ban* (average half);[67]

- 1/4 is represented by *ruo ban* (weak half);[68]

It seems that these curious expressions, which appear to relate more to approximate quantities than to precise fractions, originally denoted graduations of the clepsydra[69] (the users of this instrument must have known that the measurements of time which they obtained were approximate).

From an operational point of view, Chinese fractions are also associated with division. In fact, when a division did not fall out exactly, the calculator invariably expressed the result in the form

$$A + \frac{a'}{a},$$

where A is an integer *quan* (with respect to a given unit) and $\frac{a'}{a}$ is the remaining fraction (with $a' < a$ because $\frac{a'}{a}$ represents a fraction of the main unit).

When operating on fractions, calculators called the numerator *shi* (dividend) and the denominator *fa* (divisor). *Shi* seems to have been associated with

---

[62] *DKW*, 2-1450, p. 1083.

[63] *DKW*, 2-1853, p. 1277.

[64] Wang Ling, *Thesis*, pp. 133–134.

[65] *JZSS* 6-7, 6-19, 6-23, 6-26. Like Wang Ling (*Thesis*, p. 134) we translate *shao* and *tai* here by "less than" and "more than" rather than "small" and "large."

[66] *JZSS* 3-18.

[67] *XHYSJ* in QB, II, p. 558.

[68] Idem.

[69] Cf. Yang Lien-cheng (1), *1961*.

wealth[70] and *fa* with law, the underlying idea referring to the sharing of a certain wealth according to a legal unit of measurement.[71]

Calculations using fractions sometimes presented the authors of the Ten Computational Canons with great difficulties. (See the preface to the *Zhang Qiujian suanjing*: "anyone who studies mathematics should not be afraid of the difficulty of multiplication and division, but should be afraid of the mysteries of the 'interconnection of parts'" (*tongfen*: reduction of fractions to a common denominator).[72]

For example, while the Ten Computational Canons treats computations on "integers" (i.e. numbers denoting an integral multiple of a certain unit) in an extremely summary manner, computations on fractions are the object of careful expositions. The *Jiuzhang suanshu*, which is the reference text on the subject, contains no fewer than seven computational rules and procedures on this topic. In the order in which they are found, these rules are:

(a)	*yuefen shu*	約分術	rule of simplification of parts;
(b)	*hefen shu*	合分術	rule of combination of parts;
(c)	*jianfen shu*	減分術	rule of diminution of parts;
(d)	*kefen shu*	課分術	rule of comparison of parts;
(e)	*pingfen shu*	平分術	rule of balancing of parts;
(f)	*jingfen shu*	經分術	rule of direct (computation) of parts;
(g)	*chengfen shu*	乘分術	rule of multiplication of parts.[73]

The first of these rules is stated as follows:

Divide that which can be divided by two by two, that which cannot be divided by two [treat it as follows]: set the numbers of the mother [i.e. denominator] and the son [i.e. numerator] of the sharing, decrease the large by the small [and so on], alternately (*geng xiang jian sun*) to determine the "equal" *deng*. Simplify the fraction using this equal [...].[74]

The technique involves an attempt to simplify the numerator and the denominator by two followed by a series of alternating subtractions, as in the classical Euclidean algorithm.[75] Applying this rule, for example to problem *JZSS* 1-6 (simplification of the fraction $\frac{49}{91}$), we successfully obtain:

---

[70]In the *Shuowen jiezi* (Peking: Zhonghua Shuju, reprint, *1965*, 2nd part of j. 7, p. 150) *fu* (wealth) is given as a gloss for *shi*.

[71]Cf. *SCC*, III, p. 65: "If division first arose in connection with land measurement, the divisor would naturally have been the unit measure fixed by law. The dividend would have been the real length of the field."

[72]QB, II, p. 329.

[74]*JZSS* 1-6 (QB, I, p. 94).

[75]Euclid's *Elements*, book 7, proposition 2 (cf. Heath (3), *1908*, II, p. 298).

$$91 - 49 = 42$$
$$49 - 42 = \boxed{7}$$
$$42 - 7 = 35$$
$$35 - 7 = 28$$

$$28 - 7 = 21$$
$$21 - 7 = 14$$
$$14 - 7 = \boxed{7}$$

Since we obtain two equal numbers (the 7 which comes from the subtraction $49 - 42 = 7$ and the 7 which comes from $14 - 7 = 7$), we simplify the fraction by dividing each of its terms by 7. This number "7", which the author of the *Jiuzhang suanshu* calls the "equal" *deng* is, in fact, the highest common factor (hcf) of 49 and 42.

The second rule (b) (addition of fractions) is based on the reduction of the fractions to a common denominator. Because of its relative complexity, it consists of two stages: (i) equalisation (*tong*) of the mothers, in other words, a search for a common denominator by taking the product of the denominators of the fractions to be added (*qun mu xiang cheng wei zhi tong*) – "The multiplication together of the different mothers is called equalisation;"[76] (ii) standardisation (*qi*) of the sons, in other words, multiplication of the numerators by the denominators of the corresponding fractions (*fan mu hu cheng zi wei zhi qi*) – "Generally, the (action of) multiplication of the mothers by the corresponding sons is called standardisation."[77]

We note that these operations of equalisation and standardisation go far beyond simple computation with fractions. For Liu Hui and his emulators, they are fundamental operations which are applied in various situations (for example, rule of three,[78] *fangcheng* method).[79] Such generalisations are possible because, in certain cases, the computations involved have a structure similar to that in the case of fractions.[80]

Rules (d) and (e) both use the procedure of subtraction of parts (rule (c)). In the first case, a subtraction is carried out to determine which is the larger of two given fractions. In the second case, certain fractions are decreased by amounts which are added to others, in order to balance the parts (i.e. to determine the arithmetic mean of several fractions).

The rule of direct computation of parts (f) is used to determine the value of a part corresponding to a sharing into a certain number of equal parts. According to Li Ji's gloss (*Jiuzhang suanshu yinyi*) (Meaning and pronunciation of the terms of the Nine Chapters), *jing* 經 is used instead of *jing* 徑 which means "direct," "in a direct manner." Li Chunfeng's commentary confirms this interpretation: "*jing* (is given) because a man's part is sought for *directly*." (QB, I, p. 99) To understand this, one must remember that there were different ways of sharing: directly, in equal parts (thus carrying out a division) or indirectly in

---

[76] *JZSS* 1-9 (QB, I, p. 96).
[77] Idem.
[78] See Liu Hui's commentary on *JZSS* problem 1-21 (QB, I, p. 100).
[79] *JZSS* 8-1 (QB, I, p. 222, column 4).
[80] Bai Shangshu (3'), *1983*, p. 262.

unequal parts (for example, proportional sharing). Unlike the previous rules, this is intrinsically not a computational procedure on fractions, since the quantity to be shared and the number of parts are not necessarily fractional quantities. Nevertheless, the rule indicates the procedure for the most complicated case possible. For example, the problem *JZSS* 1-18 involves the sharing of 6 *qian*, 1/6 of a *qian* and 3/4 of a *qian* between 3 men and 1/3 of a man (*sic*). The computation begins with the transformation of $6+1/3+3/4$ and $3+1/3$ into two pairs of numbers (namely 85/12 and 10/3) and continues with the "dispersion of the parts" *san fen*, which involves multiplying 85 by 3 and 10 by 12 (85/12 : $10/3 = (85 \times 3)/(10 \times 12)$). The computation ends with the reduction of the numbers found to a canonical form of integer + fraction (whence the result, $2\frac{1}{8}$).

Lastly, rule (g) (multiplication of fractions) simply involves multiplying the numerators and the denominators together.

The variety of techniques associated with computations on fractions is truly impressive. However, in the Ten Computational Canons, this notion of fraction is associated with a notion of much more general importance, that of the "model(s)" *lü*.[81]

Given two quantities A and B which vary in proportion with one another and given two specific values a and b of these quantities, Chinese mathematicians say that the corresponding models of A and B (*xiang yu zhi lü*) are a and b. The corresponding phraseology is "a is the model of A" and "b is the model of B." In the *Jiuzhang suanshu* models arise naturally in connection with the exchange of certain goods. For example, at the beginning of chapter 2 of the *Jiuzhang suanshu*, we find information such as the following: "non-decorticated rice (*dao*) has model 60 in comparison with fermented beans (*chi*) which have model 63," meaning that these two commodities may be exchanged for one other in the proportion 60 to 63.

Chinese mathematicians have used this idea of exchange to structure everything relating to proportionality. For example, in geometry they constantly speak of the "model of the diameter" *jing lü* and the "model of the circumference" *zhou lü* to define the relationship between the diameter of a circle and the length of its circumference, as though the diameter and the circumference were to be exchanged. Similarly, instead of saying that two triangles are similar, they define the "model of the base" *gou lü* and the "model of the leg" *gu lü* of a certain triangle as an archetype for all triangles which are similar to it. As far as linear systems are concerned, they consider the numerical n-tuplet $(a_1, a_2, \ldots, a_n)$ associated with the equation

$$a_1 x_1 + a_2 x_2 + \ldots + a_n x_n = b$$

---

[81] The term *lü* is particularly difficult to translate. Its usual meaning is "rule," "norm." By translating *lü* by model, we wish to highlight the fact that this term relates to certain types of numbers defined for use as a norm, model or standard.

as a generalised model, which may be replaced at will by the model $(ka_1, ka_2, \ldots, ka_n)$; this enables them to carry out computations on integers rather than on fractions.

Decimal Numbers Metrological and Pure

The history of decimal numbers in China is closely related to that of units of measurement. According to Libbrecht "The origin of the decimal fractions can be found in the metrological systems."[82] Indeed, in Chinese mathematics, notation such as

$$9\square \ \ 6\square \ \ 2\square$$

(rather than notation analogous to our 9.62) are constantly found (here the $\square$ represent Chinese characters denoting specific units).

In such a system, one might expect to find as many ways of writing numbers as there are concrete units, but this intuition is not completely borne out. For example, in the Ten Computational Classics of the Tang dynasty, it is apparent that a single series of numerical markers (ten, hundred, thousand ...) is used in conjunction with several types of main unit. For example, in the *Wucao suanjing*[83] and the *Xiahou Yang suanjing*,[84] the units called *li*, *hao*, *si* and *hu* are used episodically as decimal subdivisions of the sapek (*wen*),[85] although originally, in the Nine Chapters (*JZSS*) they were decimal units of length. Indeed, the latter contains an instance where a sum of money such as 5889.216 sapeks is written as:

五貫八百八十九文二分一釐六毫

wu guan ba bai ba shi jiu wen er fen yi li liu hao

or: five *guan* ("ligatures," 1 ligature = 1000 *wen*), eight hundred and eighty nine *wen*, 2 *fen* (tenths of a *wen*), 1 *li* (hundredth), 6 *hao* (thousandth).[86]

Just as the decimal numerical markers in this example are the same as those used for the metrological units, there exist other instances of notation associating the same *fen*, *li* and *hao* with units of surface area.[87]

For example, in problem 1-5 of Yang Hui's *Tianmu bilei chengchu jiefa* (1275) we find the following problem: "A rectangle is 24 *bu* 3 *chi* wide and 36 *bu* 2 *chi* long. What is its area? Answer: 3 *mu* 175 *bu* 4 *fen* 4 *li*."

To solve this problem, it is necessary to know that 1 *bu* = 5 *chi*, 1 *chi* = 10 *cun*, 1 *mu* = 240 [square] *bu* (Yang Hui's text does not distinguish between

---

[82] Libbrecht (2), *1973*, p. 72.

[83] QB, II, p. 409 ff.

[84] Ibid., p. 551 ff.

[85] *SY*, p. 16.

[86] *XHYSJ* 2-8 (QB, II, p. 578).

[87] *SY*, p. 16; Lam Lay-Yong (6), *1977*, pp. 89 and 241.

linear units and units of surface area).[88] The solution involves converting the data of the statement into *chi*

$$24 \ bu \ 3 \ chi = 123 \ chi$$
$$36 \ bu \ 2 \ chi = 182 \ chi$$

so that area = 22386 [square] *chi*. Then 22386 *chi* is converted into *mu* using a division by 6000 (since 1 *mu* = 240 [square] *bu* = 6000 [square] *chi*). The result is 3 *mu* and a remainder of 4386 [square] *chi*. Lastly, this remainder is multiplied "by 4" (in fact by 0.04), since 1 [square] *bu* = 25 [square] *chi* (to divide by 25 is equivalent to multiplying by 0.04), and Yang Hui obtains the final answer

$$3 \ mu \ 175 \ bu \ 4 \ fen \ 4 \ li.$$

This problem clearly illustrates the fact that, for Yang Hui, metrological considerations inextricably permeate the structure of his numbers and computations.

Elsewhere, in astronomy, in the ephemeris of the *Jiyuan* calendar which was promulgated in 1106, the unit called *fen* (used, as we have just seen, as a tenth of a unit) is used to denote one hundredth of a degree (*du*).[89]

All these systems with "joker units" have a hybrid structure which is partly concrete and partly abstract and decimal. However, the concrete aspect of the representation of numbers could possibly recover the lost ground if it was desired to write down a number involving many decimal digits; in view of the limits of the conceptual framework, it would then be necessary to invent new names for decimal units. Qin Jiushao did exactly that in his *Shushu jiuzhang* (1247), when he introduced the decimal subdivisions he called *chen* (dust), *sha* (grain of sand), *miao* (tiny), ... and *yan* (smoke) below the *hao* (in the series *fen, li* and *hao*) to denote ten-thousandths, hundred-thousandths, etc.[90]

All this applies in full to the numbers written using the usual Chinese notation (but not necessarily those transcribed by means of rod numerals which directly reflect the structure of practical computations with counting-rods).

But texts which contain transcriptions of rod numerals are extremely rare. However, Qin Jiushao's *Shushu jiuzhang* contains a large number of diagrams of operations, which lead one to conclude that in the 13th century only a single main unit was used for calculations with rods. In fact, he denoted the number *cun* (= thumb, unit of length), which we would refer to as 0.96644, by:[91]

寸
〇 ≜ ⊤⊥ ⫼ 三

The only sign used, 寸 , denoted the main unit and simultaneously played the same role as our decimal point. This amounts to what we do when we do not use

---

[88]Similarly, the French *perche* of old was at the same time a measure of length, area and even of volume of stone. Many other examples can be cited. Cf. Fowler (1), *1992*, p. 418.

[89]Ibid.

[90]Libbrecht, op. cit., p. 72; *SSJZ*, j. 12, problem no. 5.

[91]Libbrecht, op. cit., p. 74.

the decimal notation but expressions such as, for example, "1 mètre 22." But the system also allows the presence of median zeros (denoted by a small circle, like our present zero) in a way which clearly shows that the idea of decimals was then fully developed.[92]

A level of abstraction similar to that of the *Shushu jiuzhang* is also found in the other mathematical works of the 13th-century algebraist Li Zhi who knew that in a calculation it is always possible to use as many decimal digits as necessary even if one has insufficient special terms with which to reference them.[93] Indeed, this idea is already present in a text attributed to Liu Hui (i.e. more than a thousand years earlier):

微數無名者以爲分子，其一退以十爲母，其再退以百爲母。退之彌下，其分彌細〔…〕

Tiny numbers *without a name* may be viewed as the numerators (of fractions), which would have denominator ten if one moved back a place and one hundred if one moved back two places; thus, the more one moves back, the smaller the fractions are.[94]

Given that the context of this situation is that of a specific operation, namely that of the extraction of the square root, one may wonder, if, more generally in China, the idea of the decimal notation was not suggested by the observations which mathematicians may have made during computations. Indeed:

- The *Xiahou Yang suanjing* states that in order to multiply or divide a number by 10, 100, 1000 or 10000, it is sufficient to move the rods representing it forwards *jin* or backwards *tui* by 1, 2, 3 or 4 decimal places.[95]

- In a problem for which the answer is a non-integral number of men the *Sunzi suanjing* expresses the remainder in the decimalised form 5 *fen* (5 tenths), instead of writing *ban* (a half) as prescribed by the Nine Chapters.[96]

- The Ten Computational Canons sometimes recommends that the data should be converted and expressed in terms of decimal units of measurement before the calculations are carried out.[97]

- Ready-made conversion tables for the units of weight known as *liang* and *jin* (1 *jin* = 16 *liang*) are partially or completely based on decimal expansions. For example, in his *Riyong suanfa* (Computational Methods for Daily Use) Yang Hui quotes the rules:

---

[92]Cf. Libbrecht, op. cit., p. 74. However, it is unfortunate that the sources we are obliged to rely on when discussing such questions only date back to the 19th century (all the studies of Qin Jiushao's *Shushu jiuzhang* are based on the *Yijiatang* edition (*1842*)).

[93]*Jingzhai gu jin tou* (Ancient and new commentaries of Jingzhai [Li Zhi]), j. 1, p. 10 (Peking: Shangwu Yinshuguan, *1935*).

[94]*JZSS* 4-16 (QB, I, p. 150).

[95]QB, II, p. 559.

[96]Ibid., p. 310 (problem *SZSJ* 3-2).

[97]*SY*, p. 17.

- 一兩六釐二毫半 *1 liang* [=] 6 *li* 2 *hao* and a half
- 二兩一分二釐半 *2 liang* [=] 1 *fen* 2 *li* and a half[98]

while in his *Suanxue qimeng* (Introduction to the Computational Science) Zhu Shijie gives the following rules:

- 一退六二五 One? Go back six two five!
- 二留一二五 Two? Keep one two five![99]

which correspond to $\frac{1}{16} = 0.0625$ and $\frac{2}{16} = 0.125$ (since 1 *jin* = 16 *liang*).

Thus, a set of more-or-less highly-decimalised practical systems would have co-existed with the ancient system of the Nine Chapters which is based on mixed (half-integer, half-fractional) numbers.

Negative Numbers and Positive Numbers

Historically, the earliest occurrence of the notion of negative numbers is to be found in Chinese mathematics (towards the beginning of the first millennium AD, under the Han dynasty). This may seem paradoxical insofar as in China perhaps more than anywhere else, the notion of numbers has often revolved around tangible processes. What is more, the texts[100] introduce us directly to the manipulation of negative numbers using "sign rules," as though these numbers were utterly obvious.

Without claiming to clarify this paradox, we would draw the reader's attention to the fact that the Chinese from antiquity often display a predilection for an analysis of all sorts of phenomena in terms of complementary couples, positive and negative. For example, astronomers imagined coupling the planet Jupiter with an anti-Jupiter, whose motion was deduced from the former by inversion;[101] diviners practised a double-sided divination with symmetrically arranged graphics;[102] not to mention also, of course, *yinyang* dualism. At the deepest level, these various phenomena may be dependent on a particular structure of Classical Chinese: parallelism.[103] If so, it may be that the early existence in China of positive and negative numbers is a consequence of a peculiar linguistic structure. Still, positive and negative numbers found in antique Chinese mathematics have special characteristics of their own:

(a) In the *Jiuzhang suanshu*, as in all other Chinese mathematical works, negative numbers are never found in the statements of problems and

---

[98]Lam Lay-Yong (4), *1972*, p. 71.

[99]*SXQM*, introductory chapter.

[100]*JZSS* 8-3.

[101]Vandermeersch (1), *1980*, II, p. 345.

[102]Ibid., p. 300.

[103]Cf. Hightower (1), *1959*; Nienhauser (1), *1986*, p. 656 ff. and no 11, *1989* of the French journal *Extrême-Orient Extrême-Occident* which is entirely devoted to parallelism.

therefore cannot be considered as relative numbers in any situation whatsoever (they only exist via computational procedures). We may assume that this has to do with the fact that for Chinese mathematicians numbers were essentially born of concrete situations (even if, most often, it was a fictitious concrete) – a number always represented a certain quantity of something and, consequently, in the Chinese context, what we consider as two opposite numbers are in fact two complementary aspects of a single number.

Similarly, none of the problems ever have answers which are negative numbers. From all the evidence however, Chinese mathematicians could not have failed to come across problems with thoroughly acceptable data (from the point of view of concrete situations) which led to negative answers. We might then have expected remarks such as those found in medieval European texts in such situations: for example, "this problem is ill-posed" or "the negative number found is fictitious, absurd, wrong, impossible."[104] However, this is not the case.

(b) Positive and negative numbers only occur as computational intermediates during the execution of highly particular algorithms including, for example, square-array algorithms (*fangcheng*) in the *Jiuzhang suanshu* and algorithms for extracting the roots of polynomials in the *Suanxue qimeng* (1299) and the *Shushu jiuzhang* (1247). The notion was not carried over to other areas even though it is generally recognised that this could have simplified the calculations (for example, the double-false-position rules were always subdivided into numerous special cases). In all cases, the context in which these numbers were used is that of the application of dogmatic rules which have been shown to be analogous to our sign rules. Doubtless so that they can be easily remembered, these succinctly formulated rules never use more than ten groups of four, five or six written characters each (list p. 203).

In the *Jiuzhang suanshu* of the Han, these rules concern only addition and subtraction and not multiplication and division. This is not so surprising when one notes that for Liu Hui (commentator on the *Jiuzhang suanshu* who lived in the 3rd century AD) positive and negative numbers have a concrete representation as gains and deficits (*de, shi*)[105].

But the nature of these gains and deficits is not restricted to financial questions. There is far more to the matter, as Liu Hui explains in his commentary on the *JZSS*:

> When a number is said to be negative, it does not necessarily mean that there is a deficit. Similarly, a positive number does not necessarily imply that there is a gain. Therefore, even though there are red and black numerals in each

---

[104]Cf. the article on negative numbers in Diderot and d'Alembert's *Encyclopedia*.
[105]*JZSS* 8-3 (QB, I, p. 225).

column [red = positive, black = negative], a change in their colours resulting from the operations will not jeopardize the calculation.[106]

It is as easy to form an idea of the meaning of addition or subtraction in a given framework as it is difficult to imagine how two deficits may be multiplied together or divided by one another. The fact that one loss is three times larger than another has an obvious significance, but what is the meaning of one loss multiplied by another? This obstacle appears difficult to cross without formal computational manipulations going beyond the concrete framework.

In the *Suanxue qimeng* (1299), the sign rules concern not only addition and subtraction, but also multiplication. This time, the situation is even more mysterious than in the *Jiuzhang suanshu*, because the author does not explain anything about the subject. We find ourselves in the same situation as Stendhal who, after reading Bézout's famous manual (1772), still did not understand the logic of mathematics such as this and wrote that it "has the air of a secret learnt from the good lady next door."[107]

Finally, from the origins to the 17th century, Chinese mathematicians always had recourse to addition and subtraction schemes of the *Jiuzhang suanshu*. After this date, European influences had, of course, to be taken into account.

(c) Chinese negative numbers were given a concrete embodiment by specific objects, namely counting-rods (*suan*). The distinction between positive and negative was then marked by the colour (red for positive numbers, black for negative numbers), the shape (triangular section in one case, square in the other) or the position (upright or crosswise).[108] For written records, a negative number was marked by an oblique bar across one of the digits of this number written in the rod-numeral notation (for example: =⫴ = 24; = ⫴ = −24).[109] Special characters were also used for this, for example *fu* or *yi* for negative numbers and *zheng* or *cong* for positive numbers[110] (*fu* implies the idea of debt; *zheng* means "straight," "correct;" the exact meaning of *cong* and *yi* is unknown).

- Positive–negative rules:

  正負術   *zheng fu shu*

---

[106]Original text in QB, I, p. 226; translation by Lam Lay-Yong and Ang Tian-Se (3), *1986*.
[107]Stendhal (1), *1832*, p. 230.
[108]Cf. QB, I, p. 224; Li Yan (22′), *1929*.
[109]This can be deduced from the works of algebraists of the Song–Yuan period (at least, judging from late editions (the only ones we have) of works by these algebraists).
[110]*fu* and *zheng* are found in the *JZSS* and many other works, but *yi* and *cong* had a much more restricted use.

- Subtraction rules:

  同名相除 *tong ming xiang chu*

  (a) [Rods of the] same name [=sign] [are] mutually reduced.

  異名相益 *yi ming xiang yi*

  (b) [Those with] different names [are] mutually increased.

  正無入負之 *zheng wu ru fu zhi*

  (c) [If a] positive [rod] does not have a vis-à-vis[111] it is made negative (*a positive number subtracted from nothing becomes negative*).

  負無入正之 *fu wu ru zheng zhi*

  (d) [If a] negative [rod] does not have a vis-à-vis, it is made positive (*a negative number subtracted from nothing becomes positive*).

- Addition rules:

  異名相除   *yi ming xiang chu*

  (a′) [Rods] with different names [are] mutually reduced.

  同名相益 *tong ming xiang yi*

  (b′) [Those with the] same name are mutually increased.

  正無入正之   *zheng wu ru zheng zhi*

  (c′) If a positive [rod] does not have a vis-à-vis it is made positive (*a positive number added to nothing is left positive*).

  負無入負之 *fu wu ru fu zhi*

  (d′) If a negative [rod] does not have a vis-à-vis, it is made negative (*a negative number added to nothing is left negative*).

- The sign rules of the *Jiuzhang suanshu* (from QB, I, p. 226).

  同名相乘爲正   *tong ming xiang cheng wei zheng*

  (a) [Rods of the] same name multiplied by one another make positive.

  異名相乘爲負 *yi ming xiang cheng wei fu*

  (b) [Rods] of different names multiplied by one another make negative.

- The sign rule (for multiplication) of the *Suanxue qimeng* (1299) (from the *Guanwoshengshi huigao* edition of the text (1839)).

- Remark: The sign rules for addition and subtraction in the *Suanxue qimeng* are the same as those of the *Jiuzhang suanshu*.

---

[111] *wu ru* "not to have to enter" is an obscure expression. According to Liu Hui, it means *wu dui* "not to have a vis-à-vis." Note that *wu*, which refers here to the absence of all concrete qualities is a very remote ancestor of our zero.

Zero

Whether in relation to China or other civilisations, the question of the origin of zero is particularly difficult because there is essentially a lack of documentation.

The problem arises at various levels. When one speaks of zero, this may mean (for simplicity), zero as a number with the same status as any other number. It may also mean that symbol which when written immediately after the final units of a number enables us to multiply this number by the base (10 in base ten). Yet again, it may mean the special symbol which shows us that certain orders of units are absent.[112]

If the zero was known of in ancient China, it cannot have been of the first type. Chinese problems never have zero for a solution and Chinese mathematics never involves a number zero which is freely subjected to operations like the other numbers.

Symbolic zeros of other types are also unknown in China; before the seventh-eighth centuries AD, there are no known explicitly written Chinese symbols which may be interpreted as zero (unless we include certain terms of the everyday language such as *kong* (empty), *chu*, *ben* or *duan*, which indicate the absence or the beginning of something, respectively, as varieties of zero and which are found in some numerical tables relating to Chinese calendars).[113]

On the other hand, if we include virtual zeros, in other words, representations without a graphical counterpart, then certain Chinese mathematicians may be said to have known the second type of zero. In fact, as we have seen, Xiahou Yang (first half of the sixth century?) knew that in order to multiply a number by 10, 100 or 1000, it was sufficient to move the material representation of this number by one, two or three positions.[114] As far as the third type of zero is concerned, some arguments would tend to prove that it existed in China well before the beginning of the Christian era (still in virtual form). Indeed, in the *Shushu jiuzhang* we read that "in all old books we find empty places."[115] The context shows that these empty places correspond precisely to the kind of zero we are trying to detect in Chinese texts. Nevertheless, Qin Jiushao's masterpiece is not extremely ancient (1247) and his vague allusion to "old books" is not very helpful. Moreover, we do not possess any other written testimony of the existence of empty places in Chinese mathematics prior to 1247. To substantiate the idea that "before the eighth century AD the Chinese always left a gap where a zero was required,"[116] some historians like J. Needham invoke the importance of the notion of position in Chinese notations and the distinguished role of the counting-board. However,

---

[112]Guitel (1), *1975*, p. 675; Smith (1), *1925*, II, p. 69; Tropfke (3), *1980*, p. 141; Yoshida (1'), *1969*.

[113]Cf. Yan Dunjie (15'), *1947* and (27'), *1982*.

[114]See above, p. 199.

[115]See Libbrecht (2), *1973*, p. 69.

[116]*SCC*, vol. III, p. 9. Analogous view in QB, *Hist.*, p. 8; Lam Lay-Yong (6), *1977*, pp. 197 and 240; Yabuuchi (6'), *1974*, p. 19.

the ancient Chinese system of numeration is not entirely positional and the physical existence of the counting-board is still in question, although it is true that the counting-board assumption makes the interpretation of certain texts such as those by Liu Hui on the extraction of roots[117] more evident. Indeed, without a counting-board with well-defined columns for all the different orders of decimal units, these extractions and the logic of many other arithmetic operations are difficult to explain. However, the trouble is that we do not have any early examples of numbers denoted using blank spaces.

Nevertheless, some relatively late texts (fifth–tenth centuries AD, approximately) appear to include such a notation for numbers.[118] In one of these texts, the *Licheng suanjing* (Canon of ready-made[119] computations) the number 405 is written using rod numerals in the form ⫽⫽⫽ ⫽⫽⫽⫽ (we know from the context that this is the number 405). Should the blank between the "4" and the "5" be seen as an intermediate zero, as J. Needham[120] says? Possibly, but one cannot be certain. In fact, it may simply be that its presence is only a result of the pressing need to separate the rods representing "4" and "5" so as to distinguish the corresponding figures.

Thus, it is vital to consider other forms of notation for numbers in this manuscript. We find that 108 and 81 are written as ⊓⫽ and ≜⎮, respectively. Against the logic of the hypothesis of the blank-space zero, the space between the "8" (tens) and the "1" (units) is more important than that between the "1" (hundreds) and the "8" (units) (Fig. 12.7).

Thus, we see that the system of rod-numerals used in this manuscript is strictly speaking not a positional system of numeration but a "dispositional" system (to coin a phrase). How a given figure is written depends on whether this figure relates to an even or odd power of the base (ten); in other words, the *disposition* (and not just the position) of the figures also provides information.[121]

In reality, it is quite difficult to distinguish between the inevitable space between the figures of a number and a space which represents a voluntary blank. How can we distinguish between a script-related space which is indispensable for ease of reading but semantically meaningless, and a symbolic space which is mathematically meaningful and forms a part of the system of numeration?

Before reaching any conclusion, we would like to examine other evidence especially concerning the way numbers with several intermediate zeros or with initial or terminal zeros were represented. There is none. Under these conditions, the hypothesis of the Chinese blank-space zero hypothesis remains as difficult to confirm as to disprove. (However, here, it seems important to distinguish between the writing of numbers and effective computations. It is quite possible

---

[117]On the extraction of roots, see below, pp. 221 ff.

[118]These are arithmetic texts which form part of the Dunhuang manuscripts, which date to before the 10th–11th centuries AD. Cf. Li Yan, *Gudai*, p. 36 ff.; Libbrecht (5), *1982*; Martzloff (5), *1983*.

[119]See the Stein 930 manuscript (cf. the articles cited in the preceding note). 'Ready-made' refers to the fact that the work consists of ready-made numerical tables.

[120]Op. cit., p. 9.

[121]See also Chen Liangzuo (1'), *1977*.

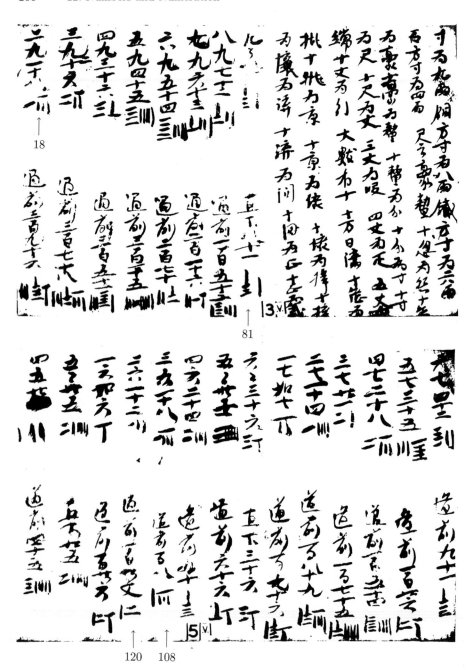

Fig. 12.7. Transcription of the numbers 81, 18, 120 and 108 in rod-numerals. (Photocopy from a microfilm of the Stein 930 manuscript (f. 3v and 5v)).

that the latter fully respected the decimal and positional principle. At least this
is precisely what happens in the case of computations with the Roman abacus.
But, of course, from this, it does not follow that written Roman numerals respect
the positional principle).[122]

In any case, from the eighth century, there is incontestable evidence that
zero was sometimes represented by a particular symbol, namely a dot, a circle
or a square, depending on the period.

In chapter 104 of the *Kaiyuan zhanjing* (Kaiyuan reign-period (713–741)
Astronomico-astrological Canon), which chapter is a Chinese translation of
astronomical texts of Indian origin, we read that:

> On the right[123], one can see the nine written characters[124] which are used to carry
> out multiplication and division using the Indian techniques. Each individual figure
> is written in one piece. After the nine, the ten is written in the next row. *A dot is
> always written in each empty row* and all places are occupied so that it is impossible
> to make a mistake and the calculations are simplified.[125]

Could this dot be the ancestor of the circle-zero (graphically almost identical
with our zero) which appeared five centuries later in the works of 13th-century
algebraists?[126] (Fig. 12.8). It is impossible to say without intermediate texts.

⟵ Zero

Fig. 12.8. The round zero in Zhu Shijie's *Siyuan yujian* (1303).
(From the Commercial Press (Shangwu Yinshuguan) edition,
Shanghai, 1937, vol. 1, p. 65).

---

[122]Cf. Taisbak (1), *1965*; Krenkel (1), *1969*; Fellmann (1), *1983*.

[123]The Chinese text says "on the right" because, as a result of the direction of Chinese
writing, the part of the page which has just been written may be found on the right.'

[124]Unfortunately, these nine Indian figures are missing in the present-day editions of the
text of the *Kaiyuan zhanjing*.

[125]Cited from the *Zhongguo Shudian* edition of the text (Peking, *1989*), j. 104, p. 742.

[126]The oldest Chinese mathematical text containing the symbol for zero in the form of a
circle is the *Shushu jiuzhang* by Qin Jiushao (1247). Cf. Libbrecht (2), *1973*, p. 74. According
to Yan Dunjie, op. cit., this same circle-zero was used a century earlier in certain calendrical
texts. Cf. Yan Dunjie, (15'), *1947*.

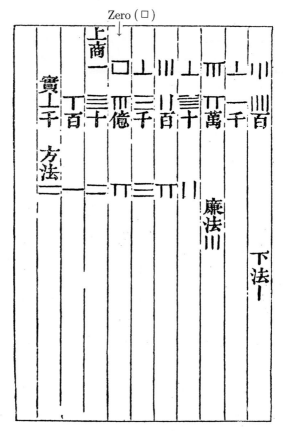

Fig. 12.9. The square zero in the *Lülü chengshu* (Treatise on Pitch-pipes) by Liu Qin (13th century). (From Yan Dunjie, *Zhongguo gudai shuxue de chengjiu* (Realisations of Ancient Chinese Mathematics), Peking, 1956, p. 10).

In any case, this late Chinese circle-zero is identical in shape to that first found simultaneously in Sanskrit inscriptions which appear in Sumatra and in Cambodia in 683 AD.[127] Indeed, the enormous time interval separating the Indian circle-zero from the Chinese circle-zero would militate in favour of an Indian influence. However, we note that Yan Dunjie preferred to think that it was a purely Chinese invention born out of the calligraphic deformation of another symbol for zero, the square-zero (Fig. 12.9).

After the Mongol period, the circle-zero came into general use and progressively, the Chinese used it as an ordinary figure.

Zero was also denoted by a special written character from the Ming; this was the character *ling*, the usual meaning of which is "dewdrop." This *ling* has remained the symbol for zero until today.

---

[127]G. Guitel, op. cit., p. 633.

13. Calculating Instruments

The Counting-Board

Most authors believe that counting-rods were manipulated on a special surface called the counting-board or chessboard,[1] which would have been to rods what the frame and the bars are to the abacus. However there is no proof that such boards existed. As Wang Ling notes:

> It is not necessary to believe that the Han mathematicians always used an elaborate counting-board; they probably worked on any convenient surface. No specific words for such boards occur in the ancient mathematical texts.[2]

It is true that Liu Hui's commentary on the *JZSS* recommends that a certain felt carpet should be spread out before calculations are performed (*yong suan er bu zhan*).[3] It is also true that the mathematician Mei Wending (1633–1721) explains that: "the ancients used rods *chou* laid transversally and longitudinally on a small table" *zong heng lie yu ji'an*.[4] However, we do not have any more precise information. The *JZSS*, to name but one, advises only that the numbers should be "set" *zhi* before an operation, without saying where or how. According to Wang Ling "The table, the ground or any flat surface can be made use of as the board. As the size of counting-rods varies and different problems require different sized boards, it is reasonable not to expect any special reference to the making of a board of a definite size."[5]

---

[1] In *SCC*, III, p. 62, J. Needham writes without providing any evidence that: "Texts earlier than the +3rd century always write out the numbers in full, but it is clear that from the Warring States time the additions must have been made with counting-rod numerals on a counting-board, using a place-value system in which blanks were left where we should put zeros." Writing about Qin Jiushao's *Shushu jiuzhang*, Libbrecht (2), *1973*, p. 398 also affirms that: "all computations are worked out on the counting board, and this must be the reason why such large numbers did not cause difficulties." Later, in a glossary of Chinese mathematical terms, the same author gives a reconstitution of his own of a counting-board, in the form of a square subdivided into 25 equal smaller squares, which he calls *suanpan* (ibid., p. 488). However, there is no proof that such boards existed and the term *suanpan* is normally used to refer to the abacus.

[2] Wang Ling and Needham (1), *1955*, p.365.

[3] *JZSS* 8-18 (QB, I, p. 236):

[4] *Meishi congshu jiyao*, j. 5, p. 7b, col. 9.

[5] Wang Ling, *Thesis*, chapter 3, note 37, p. 42.

Counting-Rods

Counting-rods (*suan* or *chousuan* in the Nine Chapters[6] (*JZSS*)) are also called *ce, chou, chouce, suance,* or *suanzi*[7] according to other sources.[8] They are small objects similar to spillikins, the shape (cylindrical or prismatic, with a square or triangular cross-section), the material (wood, bone, bamboo, ivory, iron and sometimes even jade) and the dimensions of which did not remain unchanged. For example, in the *Qian Han shu* (History of the Former Han Dynasty), we read that: "the computing method uses bamboo stems (*qi suanfa yong zhu*) of diameter 1 *fen* and length 6 *cun*."[9] Based on our knowledge of the units of length under the Han, that corresponds to a diameter of 0.69 cm and a length of 13.8 cm. Later, shorter rods were used; it is known that under the Sui (589–618) they were no longer than 8.8 cm.[10]

Extracts from ancient literary works containing terms such as those mentioned at the beginning of this section are cited to prove the ancient origin of counting-rods. However, the ancient Chinese did not use their rods for counting only. In the hands of diviners, they were a divinatory instrument; for Buddhist monks, they served as monetary tokens, coupons for clothing, etc.[11] and everyone used them to take hold of food.

However, palaeography enables us to formulate a hypothesis about their origin. L. Vandermeersch remarks that, palaeographically, *suan* 算 is written as 𭀩 and that this pictogram represents hands manipulating stems of achillea.[12] Thus, there may be a filiation linking the divinatory instrument and the calculating instrument. In fact, the two usages seem to have co-existed over a long period. In the *Suishu*, the monograph on the calendar refers to rods using terms which refer both to the numerology of the *Yijing* and to mathematics (positive and negative numbers): "[Calendrical] computations use bamboo rods 2 *fen* wide and 3 *cun* long. The positive rods are triangular (*san lian*) and symbolise the positive principle (*qian*). The negative rods (*fu ce*) are square [...] these are the rods of the negative principle (*kun*)."[13]

In any case, counting-rods occupy an important place in Chinese mathematics from the Han to the Yuan, since many techniques of numerical computation in this period involved them. After the Yuan they fell into disuse in China itself; however, in the Japan of the Edo period (1603–1868), after amateur

---

[6] *JZSS* 8-3 (QB, I, p. 225).

[7] In the *Shuihu zhuan* ("Water Margin," a 14th-century Chinese novel which has recently been translated into French (J. Dars (transl.), *Au bord de l'eau*, 2 vols., Paris: Gallimard (Encyclopédie de la Pléiade), *1978*), one of the characters is nicknamed *Shen suanzi*, i.e. "the divine manipulator of counting-rods."

[8] *ZSSLC-T*, III, pp. 29–36.

[9] *Qian Han shu*, j. 21, p. 956.

[10] QB, *Hist*, p. 8.

[11] See the article *Chū* in the *Hōbogirin, dictionnaire encyclopédique du bouddhisme, d'après les sources chinoises et japonaises*, Paris/Tokyo, 5th fasc., p. 431–456.

[12] Vandermeersch (1), *1980*, II, p. 304 and note 53, p. 315.

[13] *Suishu*, j. 16, p. 387, *ZSSLC-T*, III, p. 32.

Fig. 13.1. Mathematicians manipulating counting-rods. From the *Shojutsu sangaku zuye* (1795) (quoted in Smith and Mikami (1), *1914*, p. 29).

mathematicians had come into contact with them via Chinese works such as the *Suanxue qimeng* (1299) or the *Yang Hui suanfa*, they enjoyed a prolonged success (Fig. 13.1) and were only supplanted by written computations from the second half of the 19th century.

The Abacus

The Chinese abacus (*suanpan*) consists of bars set in a rectangular frame (Fig. 13.2–13.6).

In its current version it includes a variable number of bars (there may be 11, 13, 17 or more). On each of these bars two tiers of balls separated by a crossbar may slide freely. The upper tier consistently comprises two balls per bar and the lower tier five balls. By convention, each of the upper balls is worth five units and each of the lower balls one unit. The crossbar serves as a reference; to set the apparatus to rest the balls are pushed towards the frame, while to mark numbers they are slid against the crossbar. According to Yushkevich[14] this particular distribution of the balls corresponds firstly to the five fingers of the hand (lower balls) and secondly to the two hands (upper balls). However, there is another explanation. Originally, the system would have been designed to facilitate the manipulation of units of weight called *jin* (pounds) and *liang* (ounces). In fact, since 1 *jin* = 16 *liang* and only 15 units can be marked per bar, it is as though the inventor of the abacus had wanted to work in base 16. (Of course, the apparatus may be used for calculations other than conversions from *jin* to *liang*).

The notation of numbers on the abacus is based on the principles of decimal numeration, but for the numbers between six and nine, five plays a distinguished role, exactly as in the case of the numeration of counting-rods. Since, moreover,

---

[14]Yushkevich (1), *1964*, p. 19.

1, 2, 3 and 4 are marked on the abacus by 1, 2, 3 or 4 balls, the resemblance between the two systems is striking:

The principle for calculation on the abacus is based on original rules which, like our multiplication tables, must be memorised in full. Applied mechanically, these permit astonishing rates of execution which have amazed all observers. This is often backed up with reference to a surprising competition between a man calculating on the abacus and an electronic machine which turns out unfavourably for the machine.[15]

In principle, these rules may be used not only to carry out the four operations, but also to extract square and cube roots.[16] However, in practice, only addition and subtraction are used, for the simple reason that, unlike counting-rods which were frequently used for advanced computations, the abacus was the basic tool of merchants.

More broadly, the prestige of the abacus in China has spread far beyond the (restricted) category of merchants. It is still taught even now and research of a pedagogical nature is being undertaken to simplify its ancient rules.

A great dynamism is shown by the Chinese Abacus Federation (*Zhongguo zhusuan xiehui*), founded in 1979, which publishes a monthly magazine[17] containing a range of articles on the cultural universe of the abacus and organises "sporting" competitions for the abacus.

The influence of the abacus has long been felt throughout South East Asia, well beyond the borders of China. It became well established in Japan from the end of the 16th century, though not without having undergone minor transformations (balls in the form of a double cone rather than slightly flattened spheres; number of balls per bar equal to 1 + 4 (so as to achieve an overall maximum of 9 per bar, which is sufficient in base 10), instead of 2 + 5 balls (which gives an overall maximum of 15). The latter characteristic should not lead one to think that the Japanese abacus is an improved version of the Chinese abacus. In fact, abacuses of type other than 2/5 were also known in China. One known Chinese abacus dating from the end of the 16th century has one ball in the upper tier and 5 balls in the lower tier (Fig. 13.2); another, from the end of the 19th century, has 3 balls in the upper tier (Fig. 13.4).

---

[15]Cf. *SCC*, III, p. 75. Full details of the competition held on 12th November 1946 in Tokyo in Kojima (1), *1954*, p. 11 ff. (this relates to a Japanese abacus, but the computational principles are not very different from those of the Chinese abacus).

[16]Cf. Li Yan, *Dagang*, II, p. 320 ff.

[17]*Zhongguo zhusuan* (The Abacus), published in Peking.

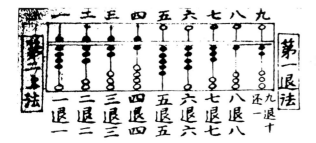

Fig. 13.2. The abacus of the *Panzhu suanfa* (1573). From Kodama (2′), *1970*, p. 23.

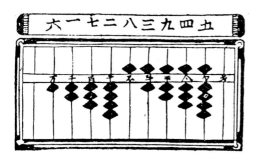

Fig. 13.3. The Japanese abacus of the *Jinkōki* (1627) (chapter 1). This is an abacus of type 1/4. The balls not involved in the calculations are not shown in this schematic diagram.

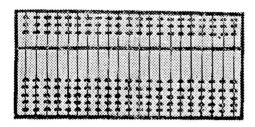

Fig. 13.4. The abacus of the *Suanxue fameng* (1881). From the *Shuzan jiten* (Encyclopedia of the abacus, Tokyo, 1961, p. 5 of the illustrations).

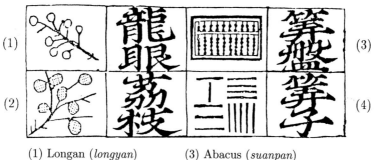

(1) Longan (*longyan*)   (3) Abacus (*suanpan*)
(2) Lychee (*lizhi*)   (4) Counting-rods (*suanzi*)

Fig. 13.5. Picture of an abacus in the reading book entitled *Duixiang siyan zazi* (1337). From the edition of this text published in Tokyo in 1920. Translation of the title, see page 215.

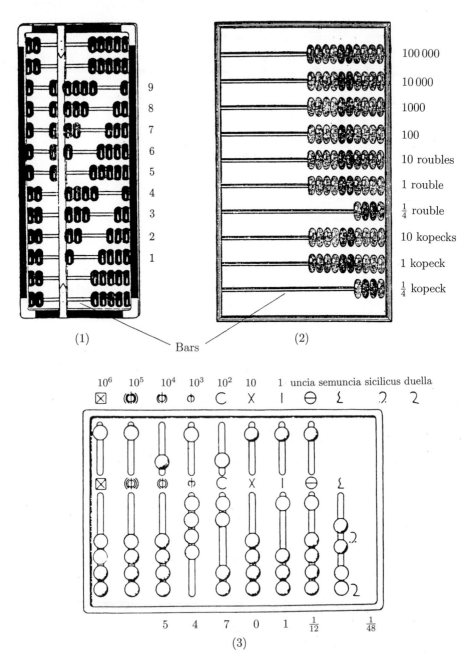

Fig. 13.6. (1) Schematic representation of the usual Chinese abacus (note the representation of the number 123456789). (2) Schematic representation of the Russian abacus. From D. E. Smith, *History of Mathematics*, vol. II, newly reedited. Dover, New York; originally published in 1925. (3) Schematic representation of a Roman abacus (period uncertain). Source: Fellmann (1), *1983*, p. 39. (Note that unlike the other abacuses, the Chinese abacus is purely decimal.)

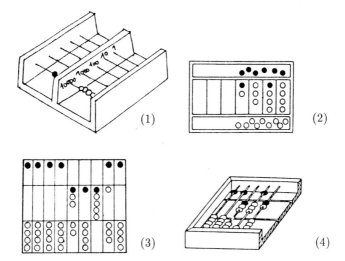

Fig. 13.7. Divergent representations of the abacus discussed in the *Shushu jiyi*. (1) From J. Needham, *SCC*, III, p. 77. (2), (3), (4), from Xu Chunfang, Yu Jieshi, Li Peiye, respectively (see: Toya Seiichi, *Chūgoku shuzan shi* (History of the Chinese abacus), *Sūgaku shi kenkyū*, 1984, no. 101, p. 63).

Was the influence of the Chinese abacus felt in countries outside the Chinese cultural sphere? This seems unlikely. For example, the Russian abacus has little in common with it; the structures of the two instruments are truly different and the operations are carried out in different ways on each.[18]

One may also wonder about the origin of the Chinese abacus. Despite much research into this subject, the question is far from resolved.[19] The only thing that is known for certain is that the abacus only entered into common use in China from the second half of the 16th century. Many illustrated arithmetics xylographed at this period attest to this.[20] However, it is probable that it already existed one or two centuries earlier because:

(i) A book of pictures intended for those learning to read, the *Duixiang siyan zazi* (Various written characters associated with pictures and grouped in fours), the only present-day example of which seems to have been xylographed in 1377, contains an illustration of an abacus (Fig. 13.5).[21]

---

[18]See I.G. Spasskii's article in *Istoriko-matematicheskie issledovaniya*, 1952, 5, pp. 267–420.

[19]Yamazaki (1′), *1962*.

[20]Cf. Kodama (2′), *1970*.

[21]Cf. Endō (1′), *1981*, p. 146. There is another, slightly later example of a work of the same genre. See: *Isis*, 1948, vol. 39, p. 239 (note by Carrington Goodrich): "In 1941, the Columbia University acquired a copy of a rare illustrated primer entitled *Hsin pien tui hsiang ssū yen* first printed in 1436, which includes a picture of an abacus of the type still in use in China among its 308 illustrations of common objects. It is difficult to determine whether or not our copy is a first edition, though paper and printing give that impression."

(ii) The *Lu Ban jing* (The Carpenter's Manual) of the 15th century explains how to make an abacus.[22]

As far as earlier periods are concerned, there are a number of unsatisfactory conjectures. For example, it has been asserted that a picture of an abacus appears in the magnificent painted scroll called *Qingming shang he tu*, which depicts a panoramic instant of daily life in Kaifeng (the capital of the Northern Song) with striking realism.[23] However, the details of the object which is assumed to be an abacus are indistinct. It could be something completely different such as a medicine box, for example.[24] Others, such as Yamazaki Yoemon, have conjectured that the Chinese abacus is descended from the Roman abacus.[25] However, since in this case a central piece of the argument is based on the fact that a sibylline text from a work of uncertain date is said to mention an obscure calculating instrument using balls which, according to some interpretations, vaguely resembles the Roman abacus, ... it is difficult to convince oneself![26]

A more probable conjecture is that the Chinese abacus is derived from counting-rods. Indeed, as we have already mentioned, the representations of numbers on the abacus and with rods are analogous. Moreover, the computational rules are also similar; in particular, the rules for division *jiugui*, which to our certain knowledge were used for counting on rods in the 13th century are also found in the new context of calculation on the abacus.

However, if the abacus did exist in the 13th century, how can we explain the fact that European travellers who visited China never talked about it, although their successors rarely missed a chance to do so?

---

[22] Cf. K. Ruitenbeek *Carpentry and Building in Late Imperial China, A Study of the 15th century Carpenter's Manual* Lu Ban jing. Leiden: E.J. Brill, 1993, pp. 268–270.

[23] Reproduction of this scroll in *A City of Cathay* published by the National Palace Museum, Taipei, 1980.

[24] Cf. Li Peiye (1′), *1984*; Hua Yinchun (2′), *1987*.

[25] Yamazaki (1), *1959*.

[26] Reference to the *Shushu jiyi*. Cf. Yamazaki, op. cit., p. 94 which quotes Mikami: "The method of abacus calculation represented in the *Shushu jiyi* is of the same principle as those practised in Greece and Rome and was probably introduced from the West." In fact, the text in question says the following: "Computation using balls to control the four seasons and govern the three powers [Heaven, Earth and Man]" [commentary by Zhen Luan (sixth century)]: "a plank is carved so as to delimit three compartments, the highest and lowest compartments are reserved for fixed and moving balls, respectively, the central compartment is used to fix the position of the numbers. Each row [?] contains five balls, the value of a ball at the bottom is different from that of a ball at the top [...]." Cf. QB, II, p. 546. This text is difficult to interpret, see Fig. 13.7 which shows reconstitutions of the various abaci to which it gave rise.

14. Techniques for Numerical Computation

Elementary Operations

The *Sunzi suanjing* and the *Xiahou Yang suanjing* are the oldest manuals containing information about how certain operations such as multiplication *cheng zhi fa* "mounting" (which takes its name from the fact that the multiplier is placed above the multiplicand and thus "mounts" it like a horse) or division *chu zhi fa* "the method of reduction" are carried out.

These manuals also tell us that the calculations were carried out using counting-rods. Unlike in modern current practice, the calculations began with the highest-order units. This meant that it was possible to determine the order of magnitude of the result immediately; however, it could lead to complications as a result of the possible effect of carries on figures of the result which have already been determined.

According to the *Sunzi suanjing*, multiplication requires three rows: an upper row *shang wei* for the multiplier; a lower row *xia wei* for the multiplicand, and lastly a central row *zhong wei*.

The calculations require the knowledge of a multiplication table. This is the same as the Pythagorean table, except that it avoids duplication in that it only contains one of the products *ab* and *ba* and begins with the largest possible product (9 × 9) (this is why the Chinese call it "the nine–nine table" *jiu–jiu*) (Fig. 14.1).

Following step-by-step the brief explanations given in the text for calculating the product 81 × 81, we see that the mechanism progresses as follows:

$$
\begin{array}{|c|} \hline
\begin{matrix} \mathbf{8}\ \text{I} \\ 6\ 4 \\ \mathbf{8}\ \text{I} \end{matrix} \\ \hline
\end{array}
\rightarrow
\begin{array}{|c|}\hline
\begin{matrix} \mathbf{8}\ \text{I} \\ 6\ 4\ 8 \\ \mathbf{8}\ \text{I} \end{matrix} \\ \hline
\end{array}
\rightarrow
\begin{array}{|c|}\hline
\begin{matrix} \text{I} \\ 6\ 5\ 6 \\ \mathbf{8}\ \text{I} \end{matrix} \\ \hline
\end{array}
\rightarrow
\begin{array}{|c|}\hline
\begin{matrix} \text{I} \\ 6\ 5\ 6\ \text{I} \\ \mathbf{8}\ \text{I} \end{matrix} \\ \hline
\end{array}
\rightarrow 6\ 5\ 6\ \text{I}
$$

Division is based on the multiplication/subtraction mechanism and also uses three rows. Remarkably, Sunzi regards it as the inverse of multiplication and consequently adopts a scheme which is symmetric with respect to multiplication, as the following illustration shows:

Multiplication	Division	Positions
Multiplier	Quotient	Upper
Product	Dividend	Central
Multiplicand	Divisor	Lower

When the division does not "fall out exactly" the final result is given in the form

Quotient
Remainder
Divisor

in which the resultant fraction is clearly visible.

Naturally, the Ten Computational Canons (*SJSS*) also contains allusions which show that there were special operations for addition, subtraction, multiplication by two and division by two, for example. Later manuals such as the *Suanxue qimeng* (1299) or the *Yang Hui suanfa* (1275) also explain how to multiply a number by 21, 31, 41, etc., how to replace a division by a multiplication using the inverse of the divisor, and how to simplify multiplication by successively multiplying the multiplicand by each factor of the multiplier. Such procedures are found in all medieval arithmetics. Nevertheless, there are also computational techniques which do not occur anywhere other than in China. Rather than listing all these techniques, it seems preferable to linger over the most original ones, namely the *jiugui* rules.

The *jiugui* rules are designed to automate the division of an arbitrary number by a divisor between 1 and 9. They were originally designed for counting-rods[1] but are also suitable for the abacus. In traditional arithmetic, when carrying out an operation, it is usual to begin by setting the operand, the operator and then the successive figures of the result. Under the *jiugui* rules, only the operand is set; when the calculations begin this operand is subjected (as an effect of the rules) to a series of metamorphoses which affect its successive figures. During the calculations, one observes a curious hybrid consisting in part of figures of the definitive result and in part of figures of the initial operand.

Like our multiplication tables, these *jiugui* rules consist of short formulae relating to the divisors 1, 2, ..., 9. For example:

1. *er yi tian zuo wu* 二一添作五 Two and one make five.
2. *san yi sanshi yi* 三一三十一 Three-one: thirty one.
3. *feng ba jin yi* 逢八進一 Eight met: one [unit] upgraded.

---

[1] These rules seem to have been invented around the 11th century. According to Yang Hui (fl. ca. 1275) they were already used in the *Zhinan suanfa* (ca. 1078–1189) (cf. Yamazaki (2), *1962*, p. 137). According to Zhu Shijie "the classical method (of division) is based on the *shangchu* technique (i.e. that which uses three superposed rows); but because it is difficult for beginners it has been replaced by the *jiugui* method which is not orthodox." (From the *Suanxue qimeng*, preliminary chapter).

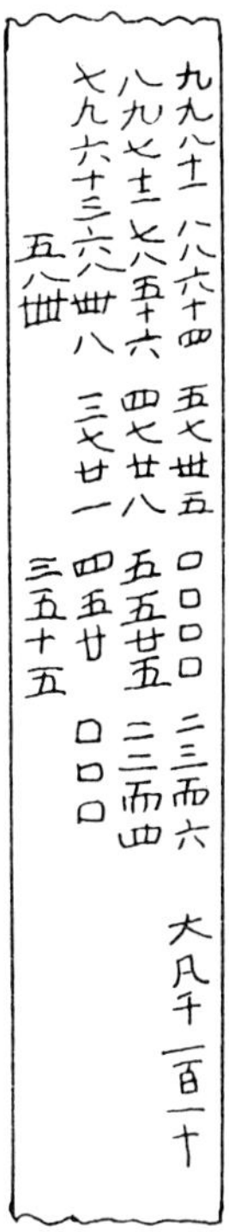

Fig. 14.1. Fragment of a multiplication table inscribed on bamboo strips (ca. 100 BC–100 AD). From Li Yan and Du Shiran, (1′), 1963, p. 20. Translation of the text (beginning at the top of the right-hand column):

1st column: $9 \times 9 = 81$; $8 \times 8 = 64$; $5 \times 7 = 35$; $2 \times 3 = 6$;
  grand total $= 1110$.
2nd column: $8 \times 9 = 72$; $7 \times 8 = 56$; $4 \times 7 = 28$; $5 \times 5 = 25$; $2 \times 2 = 4$.
3rd column: $7 \times 9 = 63$; $6 \times 8 = 48$; $3 \times 7 = 21$; $4 \times 5 = 20$.
4th column: $5 \times 8 = 40$; $3 \times 5 = 15$.

Remarks: (a) the small squares denote illegible characters, (b) 20, 30 and 40 are irregular, (c) the curious expression "grand total $= 1110$" is perhaps nothing other than the result of a computation such as the following $\sum_{m=1}^{9} m\left[\frac{m(m+1)}{2} - 1\right] = 1110$ a computation devised to check the correctness of the multiplication table.

Readers meeting these rules for the first time will undoubtedly be slightly surprised. However, it should not be difficult to follow the logic after a brief moment of reflection:

1. The reason that "two and one make five" is simply because $10 : 2 = 5$ (or again $1 : 2 = 0.5$, etc.).

2. The reason that "three-one" gives "thirty one" is that 10 divided by 3 gives quotient 3 and remainder 1.

3. The last rule is relative to the division by 8 and means that $\frac{80}{8} = 10$; with respect to ordinary units, the "1" which represents tenths is considered as "upgraded."

The *jiugui* rules also exhibit a remarkable sense of economy and simplicity. Unlike the division tables which used to decorate the back of our school exercise books, the Chinese rules take into account a number of rules devised economically. For example, in the case of the rules for division by 3, there are only three rules and not nine, namely[2]:

1. *san yi sanshi yi*   三一三十一   Three-one: thirty one.
2. *san er liushi er*   三二六十二   Three-two: sixty two.
3. *feng san jin yi shi* 逢三進一十   Three-met: ten mounted.

---

[2]These are the rules of the *Suanfa tongzong* (1592).

Thus, the calculator only has to remember three arithmetic slogans.

By way of example, let us consider the division of 1347 by 3 on the abacus:

(a) Firstly the number 1347 is set on the abacus.

(b) Since the first figure one encounters is "1," the rule "three-one: thirty one" applies. Thus, 3 is set in place of the old 1 and a ball is added to the other 3 belonging to the hundreds. The number on the abacus is now **3447**.

(c) The second figure of the number in question is treated as it now appears. Since there is no special rule for dividing 4 by 3, this 4 is treated in two stages. *First stage:* the 4 is replaced by the highest number permitted by the rules. That is 3. The rule "three-met: ten mounted" is then applied. Consequently, 3 of the 4 hundreds balls are taken down and a ten is added to the next higher column. The abacus now shows the number **41**47. *Second stage:* next the remaining 1 in the hundreds column is dealt with. The appropriate rule is clearly "Three-one: thirty one." Exactly as in (b), 3 is set in place of the old 1 and a ball is added to the four tens. The abacus now shows the number 4357.

(d) Since the current number of tens is 5, we again proceed in two stages: (1) 3 of the 5 tens balls are taken down and one ball is added to the hundreds column (rule (3)). The abacus now shows 4427. (2) This **2** is now dealt with using rule (2) and the number 4427 becomes 44**69**.

(e) Since the current number of units is 9, rule (3) must be applied three times in succession. Consequently, the 9 in the units column is cleared and 3 is set in the tens column. This gives the final result of the operation: 449.

Having reached this point some readers will certainly find these rules quite complicated. It is, after all, only division by 3!

However, it should be borne in mind that, in reality, one does not learn to calculate on the abacus by reasoning discursively, but by reproducing mechanically an automatic process. The impression of complexity is quite misleading; the abacus has genuine qualities of its own; which are apparent when one manipulates it effectively.[3]

Of course, the above explanations give only a rough idea of what effectively goes on during calculations on the abacus. For more details, the reader should

---

[3]Even today, outside the Sino-Japanese sphere, some educationalists are in favour of teaching computation using the abacus. Cf. Moon (1), *1971*.

consult, for example, the manuals of Lau Chung Him (1), *1958* or Hua Yinchun (1'), *1979*. In the meantime, we shall answer a number of questions which will undoubtedly arise. How are divisions carried out when the divisor consists of several figures? Given that the figures constantly change position on the abacus, how do we know which figures represent tens, hundreds, etc.? How does one verify the operations?

For division when the divisor lies between 11 and 99, there is a special set of rules called *fei gui jue* (formulae for flying division (= rapid)). For example, 4368 : 78 is calculated in three movements using this technique. One simply applies the rules for the divisor 78:[4]

First the formula *jian si jia yi, xia jia si* 見四加一下加四 applies ("when you see 4 add 1 and also add 1 to the rank below"). Thus, 4368 is transformed into 5468 (add 1 to the 4 of the thousands column and 1 to the 3 of the hundreds column). The same formula is then repeated (since the second figure is again a 4). Thus, 5468 becomes 5578, for the same reason. Finally, the magic formula *qi ba chu yi fei gui* 七八除一飛歸 (Seven-eight: quotient? One! Flying division!) applies (the rule means that 78:78 = 1). Thus, 5578 becomes 56 which is the final result.

The justification for such a specific set of rules that apply only to a specific division is that these allow divisions to be performed extremely rapidly. But general rules also exist and for an arbitrary divisor, the ordinary *jiugui* tables are combined with the subtraction rules so that division is carried out by multiplication/subtraction.

Again as a result of the logic which insists that computation on the abacus be based on the memorisation of tables, the position or metrological value of the figures composing a number is determined using a set of *ad hoc* rules.

Finally, in order to verify that an operation is correct, the calculator must go through all the calculations again in reverse order (the analogue of our checks of seven, nine or eleven does not exist). The Chinese expression for checking operations on the abacus is to restore the initial state (*huan yuan*).

The Extraction of Roots

A priori, when one first considers the question of the extraction of roots in Chinese mathematics, one has the impression that one is dealing with an extremely vast, diffuse and heterogeneous area. The area is vast because the methods proliferate and each author tends to have his own;[5] it is diffuse and heterogeneous because that which applies to the extraction of *n*th roots also applies in the search for the roots of polynomial equations.

---

[4]Example taken from the journal *Zhusuan* (The Abacus), *1983*, no. 3, p. 44.

[5]Cf. chapter 16344 of the *Yongle dadian* encyclopedia (1407) which is devoted to the extraction of roots and contains computational procedures due to numerous arithmeticians.

In reality, the unity of the subject is far greater than one could imagine because:

(i) Generally, the modes of operation follow the same overall scheme. Chinese authors always begin by "setting" the number whose root they wish to extract. They then determine the various decimal figures of the desired result one by one beginning with the figure corresponding to the highest-order units and continuing in decreasing order to the units figure. At this stage they either continue to search for subsequent decimals until the desired accuracy is achieved or apply a formula which gives them an approximate value for the "remainder" in a fractional form.

(ii) Despite an apparent diversity, which is superficially manifest in the names of the methods themselves, the terminology used is very homogeneous. In effect, Chinese mathematicians made up the arithmetical vocabulary for extracting roots using, firstly, the vocabulary from one particular operation, division, and, secondly, that of geometry.

They always refer to the number whose root they wish to extract as the dividend *shi* and to the desired root as the quotient *shang*. In principle, divisors should not exist, however they do use them. For example, in the case of the square root, the author of the *Jiuzhang suanshu* identifies two generalised divisors which he calls the "fixed divisor" *ding fa* and the "fixed total divisor" *zong ding fa*.[6]

The fact that extractions of roots are thought of as special divisions should come as no undue surprise since it is a common idea. For example, in France, as late as 1764, the famous arithmetician N. Barreme wrote that "the square root is very little different from division."[7] But in another respect, roots, at least the simplest ones (square and cube) are also dependent upon particular geometrical representations born of the dissection of a square or a cube. The generic term *kaifang* which denotes these operations literally means "to open" or "to dissect," "to dissociate" or "to disconnect" the square (or the cube). Moreover, the technical expressions used to denote certain coefficients which occur in the computations themselves use two terms *yu* and *lian*, which mean, respectively, corner and edge (Fig. 14.2).[8] None of this is gratuitous. It is the indelible mark of the close relationship which exists between the geometry and the logic of the operations. One can convince oneself of this by reading Liu Hui's logical justifications for the extraction of square and cube roots.[9]

To justify the extraction of the square root, Liu Hui explains in substance that the number A whose root we wish to extract actually represents a square of surface area A and that, in order to find the (unknown) side of this square, it

---

[6]Cf. Wang Ling and Needham (1), *1955*, pp. 359 and 362.

[7]*L'Arithmétique du Sr Barreme ou le Livre Facile pour apprendre l'Arithmétique de soi-même et sans maître*, Paris: chez Musier Fils, 1764, (new edition), p. 216.

[8]Cf. *DKW* 4-9436, p. 3999.

[9]*JZSS* 4-16 and 4-22, resp. (QB, I, pp. 150 and 153).

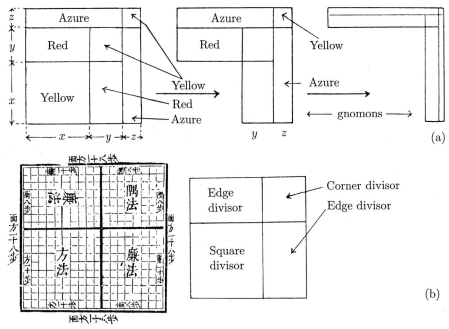

Fig. 14.2. (a) reconstitution of Liu Hui's figures to justify a technique for extracting the square root (b) $(a+b)^2$ according to the *Suanfa tongzong* (1592) (j. 6) .

is necessary to "dissect" it (*kai*) by successively dividing it into yellow (*huang*), red (*zhu*) and azure (*qing*) areas.

In the first stage, he defines the yellow sector of the square A, where this is chosen to have maximum area x^2 corresponding (for example) to the square of the number of hundreds (or thousands) in the root. This leaves an area $(A - x^2)$ in the shape of a gnomon (Fig. 14.2).

In the second stage, he determines the small side y of the rectangles which he calls "red surfaces" (*zhu mi*), the large side of which is known and equals the side x of the previous (yellow) square. According to the figure, the total area of these two rectangles is less than the area of the residual gnomon. Whence:

$$y \le \frac{A - x^2}{2x} \, .$$

He uses this inequality to determine, the second figure of the root by trial and error, by dividing the two numbers $(A - x^2)$ and $2x$ (both known) by one another (note that Liu Hui calls x the "fixed divisor" *ding fa*).

Having done this, he then subtracts the areas of the two red rectangles and that of the yellow square y^2 from that of the gnomon before repeating the same reasoning. In the case of cube roots, the justification is based on analogous ideas, but involving the dissection of a cube. Thus, we see that everything is based on the developments of $(a + b + c\ldots)^2$ and $(a + b + c + \ldots)^3$ which are highlighted geometrically.

All this gives an overall idea of the structure of the calculations which have to be carried out; however it still remains to define the specific operations. The *Jiuzhang suanshu* contains general rules for this; but because of their extreme terseness and unusual terminology, these rules are not very eloquent in themselves even though they are perfectly general. In what follows, we shall only consider the case of the square root. We shall describe the calculations using a numerical example[10]: $\sqrt{71824}$.

Firstly, we shall describe the steps of the calculations using the translation of the corresponding Chinese rule.[11]

Secondly, we shall describe the successive calculations algebraically.

A Prescription for Extracting the Square Root (*kai fang shu*)

"Set the area [of the square] as the dividend *shi*. Borrow a unit rod *jie yi suan* and displace [this rod] by jumping one rank [each time]."

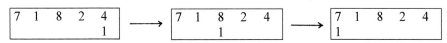

(This is equivalent to dividing the given number into two-figure slices).

"Discuss what is obtained." i.e. determine the first figure of the root. The text does not provide any further information. In the case of 71824 this first figure is 2.

"Multiply this first [figure] by the borrowed rod to make the divisor *fa* and use this to carry out the reduction *chu*."

It can be shown that this "reduction" of the dividend 71824 involves calculating $71824 - 2 \times 20000 = 31824$. Since the *JZSS* does not describe how the calculations are actually arranged, they may be reconstituted in several ways. In the method used here, we place the root above the dividend, the divisor below the dividend and the unit rod below the divisor. This is actually the type of arrangement (up to a number of variants) used by most post-Han mathematicians.

		2							2				Quotient (*shang* 商)
7	1	8	2	4		3	1	8	2	4		Dividend (*shi* 實)
2					→	2						Divisor (*fa* 法)
1					(1)	1					(2)	Unit rod (*jie suan* 借算)

"Once the reduction completed, double the divisor, to make the fixed divisor *ding fa*."

The "fixed divisor" is equal to $2 \times 20000 = 40000$.

"In case of [a further new] reduction diminish the divisor by retrograding it downwards and, again, displace the borrowed rod as at the start so as to discuss [the new figure of the root]."

[10] *JZSS* 4-14 (QB, I, p. 149).

[11] *QB*, I, p. 150.

These condensed prescriptions insist on two aspects of the square root extraction: first, the reduction process should be reiterated and patterned on what has already been done in the same situation; second, it specifies what new manipulations and computations should be performed. More precisely: the divisor is moved back by one position (the text does not say "by one position" but it makes sense if *zhe* (to break) is taken to mean "to retrograde by one position") and the unit rod by two positions.

		2		
3	1	8	2	4
4				
1				

(3)

		2			
3	1	8	2	4	Quotient
	4				Dividend
	1				Divisor

(4)   Quotient / Dividend / Divisor / Unit rod

"Multiply the unit rod by the figure of the root resulting from the new discussion and add what is obtained to the fixed divisor."

The nature of the "new discussion" is not defined specifically but, anyway, the second figure of the root is 6. The bottom rod, which is now worth 100, is multiplied by 6 and the result is added to 4000 so that $4000 + 600 = 4600$.

		2	6	
3	1	8	2	4
	4	6		
		1		

(5)

	2	6	
4	2	2	4
	4	6	
		1	

(6)

"Reduce by means of [the results just obtained]."
Here the reduction corresponds to $31824 - 460 \times 60 = 4224$.

	2	6	8
4	2	2	4
	5	2	8
			1

(7)

"Add what is obtained to the fixed divisor. Reduce and retrograde downwards as before."

The new figure of the root is 8 and the new computation corresponds to $2 \times 260 + 8 = 528$.

After the new reduction $4224 - 528 \times 8 = 0$ and the root of 71824 is precisely 286.

These computations can be justified as follows:

Let us denote the square root of 71824 by x and set $x = 100c + 10d + u$.

We require to solve the equation $x^2 = 71824$. Let us set $x = 100x_1$ where $x_1 = c + y$. Then $10000x_1^2 = 71824$. The coefficient of x_1^2 represents the value of the unit rod after its movement at the beginning of the operation.

Having determined the first figure of the root ($c = 2$) (for example, by finding the largest square contained in 7) we therefore set $x_1 = 2 + y$.

After this change of variable, the equation in x_1^2 becomes

$$10\,000y^2 + 40\,000y = 31824$$

which corresponds to the diagram (3), above.

		2		
3	1	8	2	4
4				
1				

Then, the new change of variable $10y = y_1$ leads to the equation

$$100y_1^2 + 4000y_1 = 31824$$

which corresponds point by point to the diagram (4). We see that the fact of passing from y to y_1 via the relationship $10y = y_1$ enables us to explain the backward movements of the "divisor" and the unit rod described in the rule of the *Jiuzhang suanshu*.

By determining the second figure of the root in one way or another we obtain $d = 6$. We then set $y_1 = 6 + z$ and obtain the new equation

$$100z^2 + (4600 + 600)z = 4224.$$

Setting $10z = z_1$, we find

$$z_1^2 + 520z_1 = 4224.$$

Finally, $z_1 = u = 8$ (units figure of the root) and $\sqrt{71824} = 268$.

When the extraction "does not fall out exactly," Liu Hui explains that it is necessary to have recourse to a certain operation on the side of the square *mian*[12]:

若開之不盡者爲不可開當以面命之

ruo kai zhi bu jin zhe wei bu ke kai dang yi mian ming zhi

In the case of an extraction which does not finish, the root cannot be extracted [exactly] and it is necessary to *ming* it with the side [of the square].

Historians of Chinese mathematics have been puzzled by this. Generally speaking, *ming* has many possible meanings but employed as a verb as is the case here, it can hardly mean anything other than "to name." In the *Suanjing shi shu*, however, *ming* is also used in the context of divisions which do not fall out exactly and means precisely "to name the remaining fraction by taking the remainder of the division as a numerator and the divisor as a denominator."[13] Since, as we have noted, a root extraction is similar to a division, it is tempting to associate *ming* with some fractional remainder. There are several ways of

---

[12]QB, I, p. 150.
[13]See for example QB, II. p. 283.

doing this but the solution generally retained by historians is that when the root extraction is stopped at a certain stage, if one obtains an integral part of the root (with respect to some unit), q and a remainder, r, the approximate value of the root is either $q + \frac{r}{2q+1}$ or $q + \frac{r}{2q}$.

Yet, strictly speaking, the above rule does not explicitly say so. Nevertheless, the two approximations are well attested in the *SJSS* and Liu Hui even writes that the exact value of the root is situated between them.[14]

Recently (1990), a Chinese historian[15] has explained that the expression *yi mian ming zhi* should be taken to the letter and thus interpreted as "to name it [i.e. the root] with the side," that is, to consider that the result of the square root extraction is a new kind of number which should be denoted abstractly by means of the side of the square whose root is being extracted. In other words, this historian considers that the authors of the *Jiuzhang suanshu* had in mind not only numerical computations but also irrational numbers. Insofar as the above prescriptions are primarily about the numerical computation of square roots, this interpretation seems far-fetched. However, the same author has also indicated[16] that in his commentary on the computation of the volume of a sphere[17] Zhang Heng performs computations on roots without trying to determine their exact value. But since in all known Chinese mathematical texts, these roots of a new kind are never proved to be essentially different from fractions which are often non-exactly computable too, it would be better to call them either "non-exactly computable numbers" or perhaps "side numbers" rather than irrational numbers.

Apart from the question of the side numbers, the above prescriptions for the extraction of square roots are also difficult to interpret for many other reasons and have given rise to countless interpretations. Without knowing the exact meaning of all the technical expressions of the Han period it is difficult to clarify certain obscurities (however, readers who would like a translation which is well-argued down to the last detail and approximates as well as possible to the original may always refer to the very full article by Wang Ling and J. Needham which constitutes the authority on the subject).[18]

Still, everybody recognises that an essential ingredient of the above prescriptions is precisely the doubling *bei* of a certain divisor (cf. above, p. 224). But as Liu Hui's geometrical interpretation shows, this doubling comes from a particular dissection of a square, in other words, from the binomial expansion, $(a + b)^2$.

This sort of method is thus not very different from that which was still taught a short time ago before the introduction of pocket calculators. It is also a method of the same family as that used by Theon of Alexandria (ca. 390 AD) in his commentary on Ptolemy's *Almagest*. For him, everything is

---

[14] *JZSS* 4-16 (QB, I, p. 150).

[15] Li Jimin (2'), *1990*, p. 105 ff.

[16] Ibid., p. 108.

[17] *JZSS* 4-24 (QB, I, p. 156).

[18] Wang Ling and Needham (1), *1955*.

based on Proposition II-4 of Euclid's *Elements* which states that: "If a straight line be cut at random, the square on the whole is equal to the squares on the segments and twice the rectangle contained by the segment."[19] From this, Theon explains how to calculate the square root of 144, then that of 4500 degrees.[20] We note the striking resemblance between his figures and those of the Chinese (Fig. 14.3). But, of course, Theon's calculations are in a sexagesimal rather than a decimal system of numeration. What has been said for square roots can be repeated for cube roots: Liu Hui explains this new operation by means of the dissection of a geometrical cube and the corresponding prescriptions involve two multiplications by 3 born of the expansion of $(a+b)^3$.

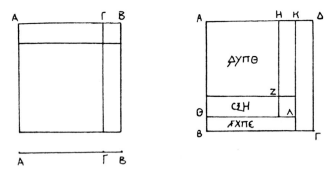

Fig. 14.3. Theon of Alexandria's figures for the extraction of the square root. (From Rome (1), *1936*, pp. 470 and 471).

In exactly the same vein as Liu Hui, Hero of Alexandria explains how to calculate an approximate value for the cube root of 100 using the expansion of[21] $(a+b)^3$.

Thus, the identities $(a+b)^2$ and $(a+b)^3$ are fundamental to the extraction of roots in the ancient mathematics of the Greek and Chinese traditions. They continued to play a crucial role for a long time, notably in the Ten Computational Canons.[22]

Similarly, we also note that the so-called Heronian approximation formulae alluded to above are well attested in the *SJSS* and later.[23] But they are also very common outside China and there is clear evidence of them in a number of demotic papryi from the second–third centuries BC[24] (but unlike the Chinese, Hero transformed the ancient technique into an iterative procedure for calculating square roots).[25] Indeed, Heronian approximations are met so often

---

[19] Heath (3), *1908*, I, p. 379.
[20] Heath (1), *1921*, I, p. 60; Rome (1), *1936*, Tome 2, p. 470 ff.
[21] Bruins (1), *1964*, p. 335 (quotation from Hero's *Metrica*, III-20).
[22] Cf. Schrimpf, (1), *1963*, pp. 124–159 (French translation of the original texts).
[23] Li Yan, *ZSSLC-T*, III, p. 128 ff.
[24] Tropfke (3), *1980*, p. 264 ff.
[25] Bruins, ibid., p. 191.

that it would be tiresome to continue to quote countless Chinese, Indian or Arabic texts which contain these.[26]

However, Chinese techniques for approximating roots are not limited to these. Again in his commentary on the *Jiuzhang suanshu*, after describing his Heronian approximations, Liu Hui also explains that it is always possible to look for the figures for the tenths, hundredths, thousandths, ... of the root right down to "those small numbers for which the units do not have a name."[27] Thus, by using decimal numbers before they were invented, he obtains approximations which are as exact as he wishes. Indeed, his idea seems quite natural since, as far as the operations are concerned, there is no barrier separating integers from decimals. The computational scheme is the same, whether one stops at the units or continues further. This is why the operation of extracting the square root probably acted as a catalyst in the discovery of decimal numbers by the Chinese.

Decimal numbers are only one aspect of square and cube root extractions in China. More importantly, perhaps, the Chinese generalised these techniques not only to nth roots but also to the extraction of roots of polynomials. But remarkably, these various subjects were always viewed in a unified manner.

As stated at the beginning of this chapter, in Chinese mathematics there is no clear-cut distinction between the extraction of nth roots and the extraction of the roots of polynomials (= search for roots of polynomial equations).

In the extraction of ordinary roots, once the calculations needed to determine the second figure of the root have been set up, everything proceeds as though one were dealing with complete equations of degree n. This is easily explained since the calculations implicitly involve underlying changes of variable of the type $x = x_1 + y$, leading to complete equations in y (and not simply to equations of the type $x^n = A$). The numbers involved then correspond to the coefficients of the successive powers of the unknown. For example, for Wang Xiaotong (seventh century) the "operation-equation" of degree two (resp. degree three) corresponds to the specification of numerical coefficients which he calls *lian fa*[28] (lit. "edge coefficient"), *cong fa* (lit. "following coefficient") or, sometimes *fang fa* "side coefficient" and *shi* "dividend." If we denote the desired root by x, all this corresponds algebraically to:

$$x^2 + cong\ fa\ x = shi$$
$$x^3 + lian\ fa\ x^2 + fang\ fa\ x = shi$$

These are only small polynomials, but from the 11th century, we progressively observe the appearance of "operation-equations" of higher degrees. In the 13th century, the terminology used to denote the coefficients stabilised to the following form[29]:

---

[26]Libbrecht, op. cit., p. 196; Saidan, (1), *1978*, p. 454; Li Yan *ZSSLC-T*, III, p. 127 ff.

[27]QB, I, p. 150.

[28]*lian* means "edge" in the sense of "small rectangle (resp. small parallelepiped) situated on the edge of the square (resp. of the cube)" which is used to illustrate the expansion of $(a + b)^2$ (resp. $(a + b)^3$).

[29]Hoe (2), *1977*.

Constant term	*shi*	實	Dividend
Coeff. of x	*fang*	方	Side (or square?)
Coeff. of x^2	*shang lian*	上廉	Upper "edge"
Coeff. of x^3	*er lian*	二廉	Second "edge"
...	...	...	...
Coeff. of x^{n-1}	*xia lian*	下廉	Lower "edge"
Coeff. of x^n	*yu*	隅	Corner

Etymologically, these terms are derived from those used for degree two and three which have a highly specific meaning but of course, here, it would be futile to interpret them in terms of geometrical dissections. In fact, Chinese authors took advantage of the resources provided them by the ancient terms although there was no longer any possibility of a geometrical interpretation.

What form did the generalisation of this transition from dimensions two and three to higher dimensions take? The available documentation does not allow us to follow the details but there is little doubt that Pascal's triangle (as a table of binomial coefficients) played a role here.

At the beginning of his *Siyuan yujian* (1303), Zhu Shijie describes this famous triangle as far as the eighth power (Fig. 14.4) Making no claim to be the inventor, he calls it *Gu fa qi chengfang tu* (Table of the ancient method for powers up to the eighth). A little earlier, the arithmetician Yang Hui in his *Xiangjie jiuzhang suanfa* (1261) had described a table of binomial coefficients up to the sixth power (see below). Again making no claim to originality, he explained that the procedure dated back to Jia Xian (ca. 1050).[30] Exactly like the case of Hero's approximation formula, it is certain that the basic idea of Pascal's triangle had been in the air for a long time.

The *Chandaḥ-sūtra* by Piṅgala (200 BC) mentions a method for determining the number of combinations of n syllables taken p at a time. The commentator Halāyudha[31] (10th century AD) explains this using a version of Pascal's triangle for exponents up to 5. The famous triangle is also found up to the twelfth power in a work by Naṣīr al-Dīn al-Ṭūsī, composed in 1265.[32]

Whether the mathematicians of the Song and the Yuan borrowed Pascal's triangle, or whether they discovered it themselves, one thing is certain: those of their methods for extracting nth roots which use the binomial coefficients have many points in common with the ancient techniques of the Ten Computational Canons for extracting square and cube roots[33] for in both cases, everything depends on the expansion of $(a + b)^m$.

However, the history of Chinese root extractions is more intricate than meets the eye because from the 13th century onwards, and probably earlier,

---

[30]QB, *Hist.*, p. 150.
[31]Bag (1), *1979*.
[32]Yushkevich (2), *1976*, p. 80.
[33]*ZSSLC-T*, II, p. 127 ff.

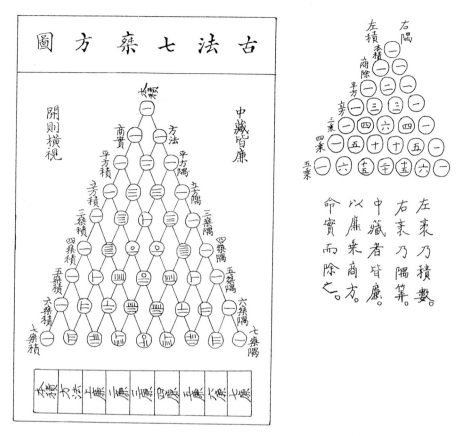

Fig. 14.4. Two Chinese versions of Pascal's triangle. (Source of the illustrations: left: *Siyuan yujian xicao* (cited from the reprint by Shangwu Yinshuguan, Shanghai, *1937*, I, p. 46); right: *Yongle dadian* (1407), j. 16344, p. 6a (reproduction, Peking, 1960)).

other Chinese techniques based not on the binomial expansion but on "Ruffini–Horner" techniques are also attested.

Ruffini and Horner were two mathematicians of the beginning of the 19th century who, independently of one another, invented an ingenious technique for numerical computation of the roots of polynomial equations.[34] Exactly like the above methods, their method is based on the successive determination of the decimal figures of the desired root, digit by digit; it is better than its competitors in the sense that it requires far fewer arithmetical operations to achieve the same result.[35]

For readers who have never met these methods (binomial or Hornerian), we shall now use an example to show what they involve. Suppose we wish to solve

---

[34]Paolo Ruffini (1765–1822) received a gold medal for his discovery, but historians generally believe that W. G. Horner found the same method independently. Cf. Goldstine (1), *1977*, p. 284. Original sources: Ruffini (1), *1804*; Horner (1), *1819*.

[35]Young and Gregory (1), *1973*, I, p. 182.

the equation:
$$x^3 + 2x^2 - 3x - 25 = 0. \qquad (14.1)$$

In such a case, first of all, mathematicians generally ask themselves how many roots of a certain kind (rational, real, complex, etc.) can be expected and, if need be, what their order of magnitude is. But in medieval China, like everywhere else in the same period, nothing of the sort existed and what Chinese mathematicians tried to determine was always a single, specific root, never several roots. Equation (14.1) was considered by them as a sort of arithmetical operation rather than as an equation (the reason for this limitation probably has something to do with the fact that "equations" were always dependent upon specific problems).

Thus, let us suppose that, from the beginning, like our Chinese authors, we knew that the given equation (14.1) has a (real) root x lying between 2 and 3. We may then set $x = 2 + y$, where $0 < y < 1$.

To calculate the second figure of the result, we first need to determine the equation in y resulting from the change of variable $x = 2 + y$. For this, we have to expand the expression

$$(2 + y)^3 + 2(2 + y)^2 - 3(2 + y) - 25. \qquad (14.2)$$

If we use the binomial method, we have to expand $(2 + y)^3$ and $(2 + y)^2$. Thus, we may use the binomial coefficients as they are given in Pascal's triangle. We then obtain:
$$y^3 + 8y^2 + 17y - 15 = 0. \qquad (14.3)$$

If we use a Hornerian method, we simply carry out a series of additions and multiplications such as those shown below.

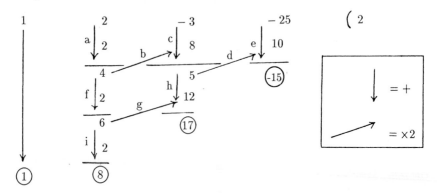

(On the first line, we have written the coefficients of the polynomial (14.1): 1, 2, -3, -25. The 2 at the end of the first line represents the value of the first figure of the result, for the record. An arrow pointing downwards (resp. upwards) indicates an addition (resp. a multiplication). The calculations are carried out in the order shown: a, b, c, d, e, f, etc. We note that, except at the end of the calculations, an addition is always followed by a multiplication, and conversely.

In fact, as a result we obtained the numbers which we have circled, which are the coefficients of equation (14.3), above. Obviously this latter technique needs fewer multiplications and additions (or subtractions) than the former.[36]

By way of example, we shall now show how Qin Jiushao presents one of his Hornerian methods. For this, we shall begin with the sequence of diagrams relating to the numerical solution of the equation

$$-x^4 + 15245x^2 - 6262506.25 = 0$$

which arises in connection with problem no. 2 of *juan* 6 of Qin Jiushao's *Shushu Jiuzhang*, p. 10b ff.[37]:

Diagram 1

	Q			
商	O	⟶	Position reserved for the desired root (called "quotient" (*shang*))	Q
	a			
實	6 2 6 2 5 0 6 2 5	⟶	Constant term (*shi*) lit. "dividend"	a
	O	⟶	Coefficient of x (*fang*) lit. "side"	b
方	b			
	1 5 2 4 5	⟶	Positive (*cong*) coefficient of x^2 (upper *lian*) lit. "upper side area"	c
從上廉	c			
	O	⟶	Coefficient of x^3 (lower *lian*) lit. "lower side area"	d
下廉	d			
	1	⟶	Negative (*yi*) coefficient of x^4 (*yi yu*) lit. "negative corner area"	e
益隅	e			
超三位 二位 益 從上廉超	The *cong shang lian* is shifted by two ranks. The *yi yu* is shifted by three ranks.	⟶	These instructions are taken into account in the next diagram.	

This diagram shows that the coefficients of the "operation-equation," considered in order of increasing powers of the unknown, beginning with the constant term, must first be set on the counting-board (or whatever takes its place), as shown on the figure. We also see that the numbers are transcribed positionally according to the decimal system of rod numerals and that the

---

[36]Full analysis of the question in Knuth (2), *1981*, vol. 2, p. 466 ff.

[37]Cited from the *Yijiatang* edition of the *SSJZ*. On the problem which gives rise to this equation, cf. Libbrecht (2), *1973*, pp. 105–106.

corresponding units are aligned. According to Libbrecht the constant term should be negative;[38] however, the diagram does not show this. This may possibly be because the edition of the *Shushu jiuzhang* which we used is monochrome but in the original manuscript positive and negative numbers were distinguished from one another by their colour. Unfortunately, it is impossible to check this, given that this manuscript is inaccessible and all known editions of the text are monochrome. Did the editors shirk away from the physical and financial problems of printing a work in two colours? Whatever the reason, the author does distinguish the signs of the other coefficients by calling them *cong* (positive) or *yi* (negative). In addition, the verbal explanations at the foot of the diagram indicate the manipulations which must be carried out before passing to the next stage of the calculations.

In what follows, we reproduce integrally all the diagrams of Qin Jiushao's text as they appear in the *Yijiatang* edition of the *Shushu jiuzhang*. We shall also translate them and provide the necessary explanations. The symbol := means that what appears on the right is to be replaced by what is on the left (assignment operation).

The subsequent diagrams begin on page 235.

---

[38]Cf. Libbrecht, op. cit., p. 202: "Ch'in Chiu-shao [Qin Jiushao] always makes the constant term negative".

Diagram 2

商 ○ ○	Q O O	Quotient ⟶ The quotient has 2 digits
實 丁二丁二〣○丁二〣	a 6 2 6 2 5 0 6 2 5	Dividend
○ 方	O b	*fang*
丨〣丨〣〣 上廉	1 5 2 4 5 c	Upper *lian*
○ 下廉	O d	Lower *lian*
丨 益隅	1 e	*yi yu*
二乃 十商 步置	Then set 20 *bu* as quotient	

Diagram 3

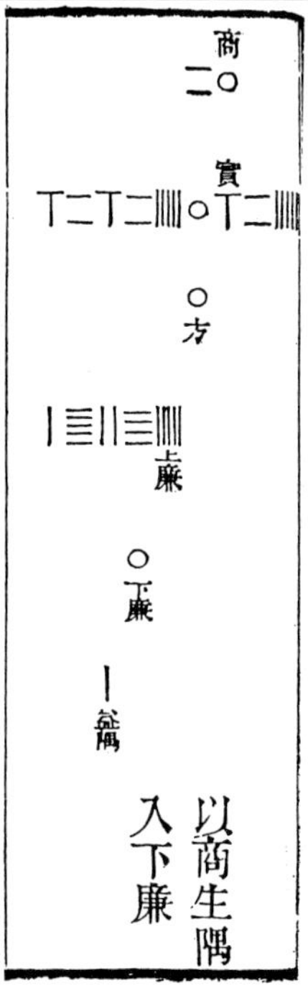

商 二○	Q 2 O	Quotient
實 丁二丁二〣○丁二〣	a 6 2 6 2 5 0 6 2 5	Dividend
○ 方	O b	*fang*
丨〣丨〣〣 廉	1 5 2 4 5 c	Upper *lian*
○ 下廉	O d	Lower *lian*
丨 益隅	1 e	*yi yu*
入以 下商 廉生 隅	Multiply Q by e Add to d	$d := Q \times e + d$ $2 \times 10\,000 + 0$

Diagram 4

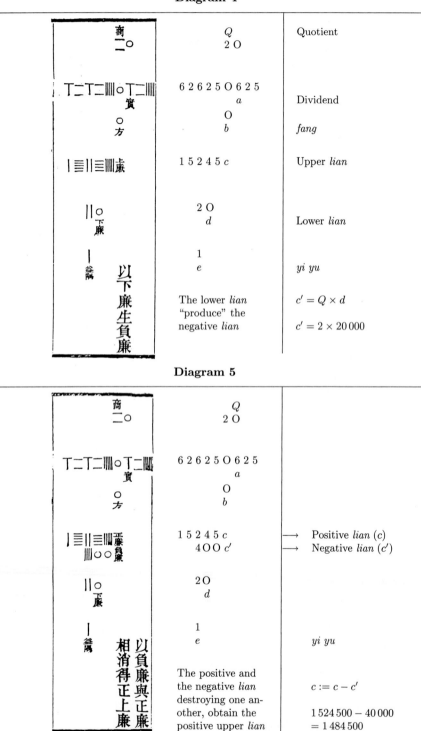

Q		Quotient
2 O		
6 2 6 2 5 0 6 2 5		
a		Dividend
O		
b		*fang*
1 5 2 4 5 c		Upper *lian*
2 O		
d		Lower *lian*
1		
e		*yi yu*
The lower *lian* "produce" the negative *lian*		$c' = Q \times d$
		$c' = 2 \times 20\,000$

Diagram 5

Q		
2 O		
6 2 6 2 5 0 6 2 5		
a		
O		
b		
1 5 2 4 5 c		$\longrightarrow$ Positive *lian* (c)
4 O O c'		$\longrightarrow$ Negative *lian* (c')
2 O		
d		
1		
e		*yi yu*
The positive and the negative *lian* destroying one another, obtain the positive upper *lian*		$c := c - c'$
		$1\,524\,500 - 40\,000$
		$= 1\,484\,500$

Diagram 6

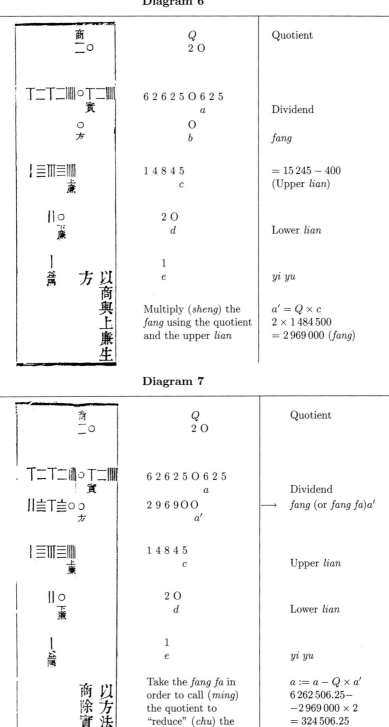

商 二〇	Q 2 0	Quotient
丁二丁二Ⅲ〇丁二Ⅲ 實	6 2 6 2 5 0 6 2 5 a	Dividend
〇 方	0 b	*fang*
｜三Ⅲ三Ⅲ 上廉	1 4 8 4 5 c	$= 15\,245 - 400$ (Upper *lian*)
‖〇 下廉	2 0 d	Lower *lian*
｜ 益隅	1 e	*yi yu*
以 商 與 上 廉 生 方	Multiply (*sheng*) the *fang* using the quotient and the upper *lian*	$a' = Q \times c$ $2 \times 1\,484\,500$ $= 2\,969\,000$ (*fang*)

Diagram 7

商 二〇	Q 2 0	Quotient
丁二丁二Ⅲ〇丁二Ⅲ 實	6 2 6 2 5 0 6 2 5 a	Dividend
‖三丁三〇〇 方	2 9 6 9 0 0 a'	$\longrightarrow$ *fang* (or *fang fa*)a'
｜三Ⅲ三Ⅲ 上廉	1 4 8 4 5 c	Upper *lian*
‖〇 下廉	2 0 d	Lower *lian*
｜ 益隅	1 e	*yi yu*
以 方 法 命 商 除 實	Take the *fang fa* in order to call (*ming*) the quotient to "reduce" (*chu*) the dividend (*shi*)	$a := a - Q \times a'$ $6\,262\,506.25-$ $-2\,969\,000 \times 2$ $= 324\,506.25$

Diagram 8

商 二〇	Q 2 O	Quotient
三॥三‖‖〇丁二‖‖ 餘	3 2 4 5 O 6 2 5 r	Remainder (yu)
‖三丁三〇〇 方	2 9 6 9 O O b	$fang$
┃三‖‖三‖‖‖ 廉	1 4 8 4 5 c	Upper $lian$
‖〇 下廉	2 O d	Lower $lian$
┃ 益隅	1 e	$yi\ yu$
又以商生 隅入下廉	Then multiply the quotient by the yu and add the result to the lower $lian$	$d := Q \times e + d$ $2 \times 10\,000 + 20\,000$ $= 40\,000$

Diagram 9

商 二〇	Q 2 O	Quotient
三॥三‖‖〇丁二‖‖ 餘	3 2 4 5 O 6 2 5 yu	$\longrightarrow$ Remainder
‖三丁三〇〇 方	2 9 6 9 O O b	$fang$
┃三‖‖三‖‖‖ 廉	1 4 8 4 5 c	Upper $lian$
‖‖〇 下廉	4 O d	Lower $lian$
┃ 益隅	1 e	$yi\ yu$
以下廉與商 生負上廉	Multiply ($sheng$) the negative upper $lian$ with the lower $lian$ and the quotient	$c' := d \times Q$ $40\,000 \times 2 = 80\,000$

Diagram 10

商 二〇	Q 2 O	Quotient
餘	3 2 4 5 O 6 2 5 r	Remainder
方	2 9 6 9 O O b	*fang*
正 正廉	1 4 8 4 5 c	Positive (*zheng*) upper *lian*
負 上廉	8 O O c'	Negative (*fu*) upper *lian*
下廉	4 O d	Lower *lian*
益隅	1 e	*yi yu*
負上廉與 正廉相消	The negative upper *lian* and the positive *lian* destroy one another (*xiang xiao*)	$c := c - c'$ $1\,484\,500 - 80\,000$ $= 1\,404\,500$

Diagram 11

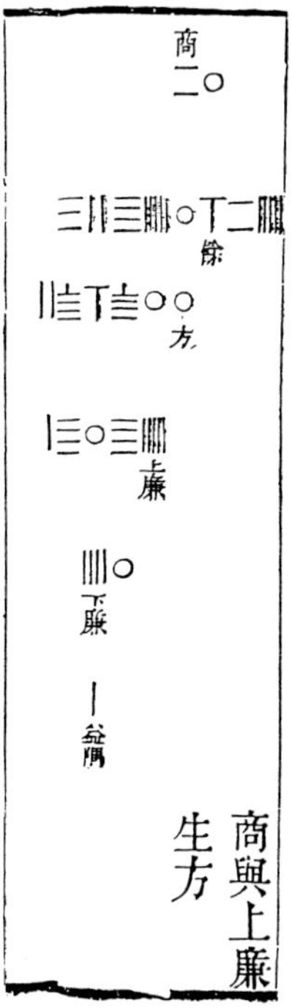

商 二〇	Q 2 O	Quotient
餘	3 2 4 5 O 6 2 5 r	Remainder
方	2 9 6 9 O O b	*fang*
上廉	1 4 O 4 5 c	Upper *lian*
下廉	4 O d	Lower *lian*
益隅	1 e	*yi yu*
商與上廉 生方	The quotient and the upper *lian* "produce" the *fang*	$b := Q \times c + b$ $2 \times 1\,404\,500 = 2\,809\,000$ this product is added to the former *fang* (b) to obtain $5\,778\,000$

Diagram 12

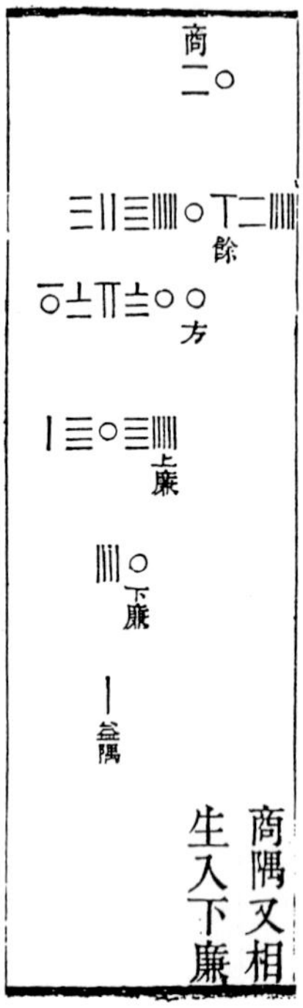

	Q $2\ O$	Quotient
餘	$3\ 2\ 4\ 5\ O\ 6\ 2\ 5$ r	Remainder
方	$5\ 7\ 7\ 8\ O\ O$ b	*fang*
上廉	$1\ 4\ O\ 4\ 5$ c	Upper *lian*
下廉	$4\ O$ d	Lower *lian*
益隅	1 e	*yi yu*
商隅又相生入下廉	The quotient and the *yu* are multiplied by one another [and the result] is incorporated into the lower *lian*	$d := Q \times e + d$ $2 \times 10\,000 + 40\,000$ $= 60\,000$

Diagram 13

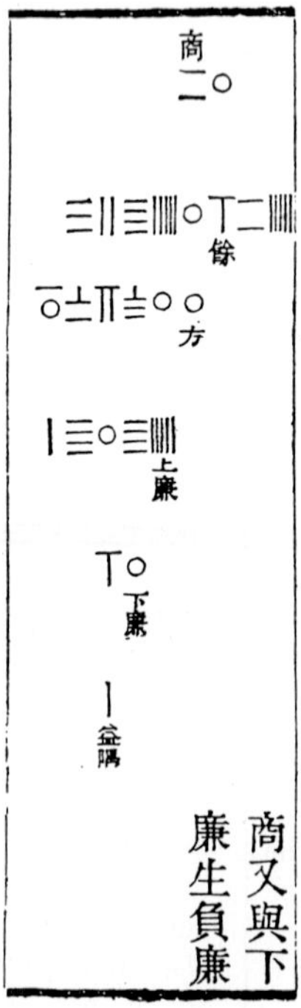

	Q $2\ O$	Quotient
餘	$3\ 2\ 4\ 5\ O\ 6\ 2\ 5$ r	Remainder
方	$5\ 7\ 7\ 8\ O\ O$ b	*fang*
上廉	$1\ 4\ O\ 4\ 5$ c	Upper *lian*
下廉	$6\ O$ d	Lower *lian*
益隅	1 e	*yi yu*
商又與下廉生負廉	The quotient and the lower *lian* produce the negative *lian*	$c' = Q \times 60\,000$ $c' = 2 \times 60\,000$ (taken to be negative)

Diagram 14

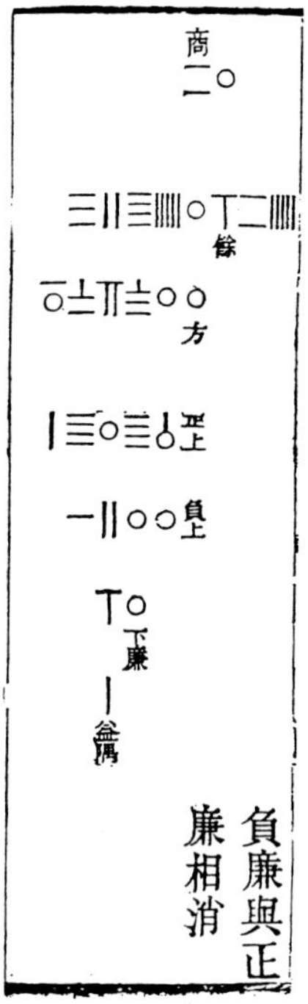

商 二〇	Q 2 O	Quotient
三刂三〸〇丅二帅 餘	3 2 4 5 O 6 2 5 r	Remainder
〇丄帀〸〇〇 方	5 7 7 8 O O b	*fang*
丨三〇三ᒿ�币 正	1 4 O 4 5 c	Upper *lian*
一刂〇〇皀 上	1 2 O O c'	Negative *lian*
丅〇 下廉	6 O d	Lower *lian*
丨 益隅	1 e	*yi yu*
廉負 廉相與 消正	The negative *lian* and the positive *lian* destroy one another	$c := c - c'$ $1\,404\,500 - 120\,000$ $= 1\,284\,500$

Diagram 15

商 二〇	Q 2 O	Quotient
三刂三帅〇丅二帅 餘	3 2 4 5 O 6 2 5 r	Remainder
〇丄帀〸〇c 方	5 7 7 8 O O b	*fang*
丨二帀三帅 上廉	1 2 8 4 5 c	Upper *lian*
丅〇 下廉	6 O d	Lower *lian*
丨 益隅	1 e	*yi yu*
生商 入又 下與 廉隅	The quotient and the *yu* are multiplied [and the result is] incorporated into the lower *lian*	$d := Q \times e + d$ $2 \times 10\,000 + 60\,000$ $= 80\,000$

Diagram 16

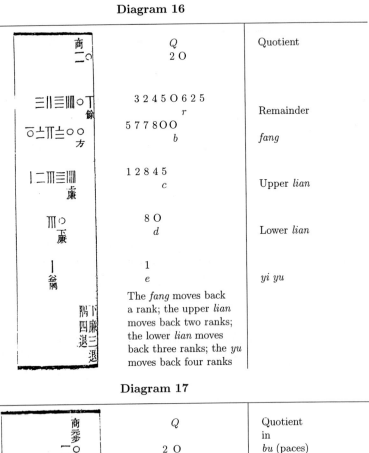

商 二〇	Q 2 O	Quotient
餘	3 2 4 5 O 6 2 5 r	Remainder
方	5 7 7 8 O O b	*fang*
上 廉	1 2 8 4 5 c	Upper *lian*
下 廉	8 O d	Lower *lian*
益 隅	1 e	*yi yu*

The *fang* moves back a rank; the upper *lian* moves back two ranks; the lower *lian* moves back three ranks; the *yu* moves back four ranks

下廉三退
隅四退

Diagram 17

商元步 二〇	Q 2 O	Quotient in *bu* (paces)
餘	3 2 4 5 O 6 2 5 r	Remainder
方	5 7 7 8 O O b	*fang*
上 廉	1 2 8 4 5 c	Upper *lian*
下 廉	8 O d	Lower *lian*
益 隅	1 e	*yi yu*

隅併下廉 廉併入方 無商以上	No quotient (the second figure of the root is zero). Incorporate the upper *lian* into the *fang*; incorporate the *yu* into the lower *lian*

$b + c$; $e + d$

$b + c = 590\,645$

$e + d = \quad\ \ 81$

Diagram 18

商 二〇 步	Q 2 0 bu	Quotient paces
三丨丨三丨丨丨〇丁二丨丨丨 餘	3 2 4 5 0 6 2 5 r	
三丁丨〇丁三丨丨丨 廉方	5 9 0 6 4 5 c,b	$b + c$
〇	O	
二丨 三丨 下廉 隅	8 1 e,d	$d + e$
ᴄ	O	
消命爲母 與正方廉相 益隅併負廉	Add the negative yu to the negative $lian$ and subtract the result from the positive $lian$ and a $fang$. Call the result the mother	$(577\,800 + 12\,845) - $ $-(80 + 1) = 590\,564$ "mother" = denominator

Diagram 19

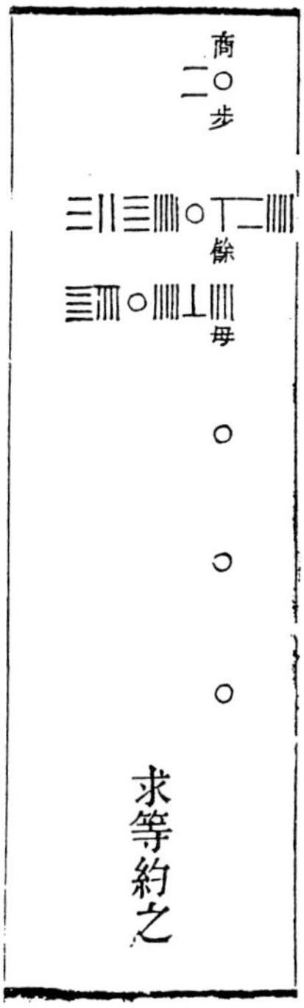

商 二〇 步	Q 2 0 bu	Quotient paces
三丨丨三丨丨丨〇丁二丨丨丨 餘	3 2 4 5 0 6 2 5 r	
三丁丨〇丨丨丨丄丨丨丨 母	5 9 0 5 6 4 m	
〇	O	
ᴐ	O	
〇	O	
求等約之	Find the "equal" ($deng$) and simplify the fraction	i.e. find the highest common factor (hcf) of r and m hcf $(32\,450\,625,\ 59\,056\,400)$ $= 25$ and simplify the fraction $\frac{r}{m}$

Diagram 20

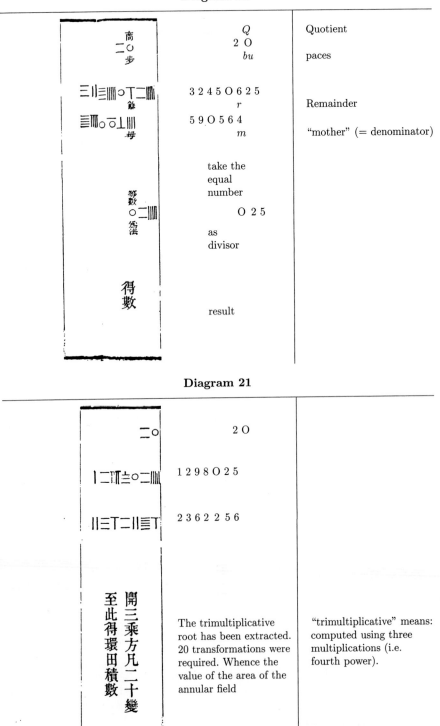

Q 2 O bu	Quotient paces
3 2 4 5 0 6 2 5 r	 Remainder
5 9 0 5 6 4 m	"mother" (= denominator)
take the equal number O 2 5 as divisor	
result	

Diagram 21

2 O	
1 2 9 8 0 2 5	
2 3 6 2 2 5 6	
The trimultiplicative root has been extracted. 20 transformations were required. Whence the value of the area of the annular field	"trimultiplicative" means: computed using three multiplications (i.e. fourth power).

Thus, at the end of these calculations, Qin Jiushao obtains "the" tri-multiplicative root

$$x = 20\frac{1\,298\,025}{2\,362\,256}$$

of the equation

$$-x^4 + 15\,245x^2 - 6\,262\,506.25 = 0. \tag{14.4}$$

We shall explain the current diagrams 1 to 21 using our usual algebraic symbolism.

The instructions of Diagram 1 lead to shifts, in other words, to multiplication of certain coefficients of equation (14.4) by certain powers of 10. It is easy to see that this corresponds to the change of variable

$$x = 10x'.$$

Whence we have the equation (14.5) (corresponding to Diagram 2):

$$-10\,000x'^4 + 1\,524\,500x'^2 - 6\,262\,506.25 = 0. \tag{14.5}$$

Having determined the first figure of the root (the quotient) we see that the calculations of Diagrams 3 to 16 correspond to the transformation of the equation (14.5) which results from the new change of variable

$$x' = 2 + y.$$

Whence we have equation (14.6):

$$-10\,000y^4 - 80\,000y^3 + 1\,284\,500y^2 + 5\,778\,000y - 324\,506.25 = 0. \tag{14.6}$$

Moreover, up to rearrangement of the numbers, the calculations which lead from (14.4) to (14.5) correspond point by point to those of Horner's method (see p. 246).

Next, certain coefficients are moved backwards, corresponding to a new change of variable

$$z = 10y.$$

Whence we have the equation

$$-z^4 - 80z^3 + 12\,845z^2 + 577\,800z - 324\,506.25$$

corresponding to Diagram 17. As far as Diagrams 18 to 21 are concerned, it is as though Qin Jiushao had confused z with its powers so that the remaining fraction is calculated with z as a factor:

$$(-1 - 81 + 12\,845 + 577\,800)z = 324\,506.25$$

or

$$590\,564z = 324\,506.25.$$

$-10\,000$	0	$1\,524\,500$	0	$-6\,262\,506,25$	(2
	$-20\,000$	$-40\,000$	$296\,900$	$5\,938\,000$	
	$-20\,000$	$1\,484\,500$	$2\,969\,000$	$-324\,506,25$	
	$-20\,000$	$-80\,000$	$2\,809\,000$		
	$-40\,000$	$1\,404\,500$	$5\,778\,000$		
	$-20\,000$	$-120\,000$			
	$-60\,000$	$1\,284\,500$			
	$-20\,000$				
	$-80\,000$				
$-10\,000$					

Computational scheme according to the modern version of Horner's method, corresponding to Diagrams 3 to 16 of the medieval method used by Qin Jiushao.

Finally, after the hcf of $59\,056\,400$ and $32\,450\,625$ (namely 25) is determined by successive subtractions, the approximate value of the desired root is found in Diagram 21.

In conclusion, we note that Qin Jiushao's method belongs to the same family as Horner's method in the sense that, among other things, the medieval Chinese algebraist uses a series of alternating additions (or subtractions) which are characteristic of what is generally called "Horner's rule."

Horner's rule is a computational technique which may be used to evaluate the value of the polynomial $f(x) = a_0 x^n + a_1 x^{n-1} + \ldots + a_{n-1} x + a_n$ at x_0 by successively calculating $b_0 = a_0$, $b_1 = a_1 + b_0 x_0$, $b_2 = a_2 + b_1 x_0$, $\ldots b_n = a_n + b_{n-1} x_0 = f(x_0)$. This rule, which uses only additions and multiplications (and does not involve the computation of powers), is very easily explained as soon as one sees that $f(x)$ may be written in the bracketed form

$$f(x) = (\ldots((((a_0 x + a_1)x + a_2)x + a_3)x + \ldots + a_{n-1})x + a_n).$$

This form of calculation, which is an essential ingredient of Horner's method, is far superior to the method of direct calculation of $f(x)$ in the sense that, firstly, it requires far fewer arithmetical operations (only n multiplications and n additions (or subtractions), in general) and, secondly, it immediately reuses the previous calculations.[39]

Qin Jiushao was not the only one to use this procedure. Other Chinese who were adept at extracting polynomial roots using Horner's method include

---

[39] Knuth (2), *1981*, II, p. 467 notes that Newton was already using this rule some 150 years before Horner.

Yang Hui (second half of the 13th century),[40] Zhu Shijie (ca. 1300),[41] Liu Yi (11th–12th centuries?) and Jia Xian (ca. 1050).[42]

Until now, no-one has been able to give a satisfactory explanation of this, because, while it is easy to identify the mechanism of Chinese calculations, it is much more difficult to uncover the logic behind them. The mathematical techniques used to prove Horner's method are not particularly elementary. Horner himself based his justification for his method on Taylor's formula and nth derivatives (given the value of a certain digit of the root x_0, together with the coefficients of the polynomial the roots of which are to be calculated, it is known that:

$$f(x + x_0) = f(x_0) + f'(x_0)x + \frac{f''(x_0)}{2!}x^2 + \ldots + \frac{f^{(n)}(x_0)}{n!}x^n \ .$$

Of course, it is possible to reach the same result by less sophisticated means. In particular, the method of synthetic division ($P(x) = Q(x)(x - a) + R$) springs to mind. But is this really the rational mathematical basis relied upon by the Chinese of the Song? This problem is complicated by the fact that the known texts contain only pure recipes and no explanations.

Certainly, one may conjecture that the Chinese of the Song and the Yuan borrowed this version of Horner's method from the Islamic world, since we know that at the same period, several mathematicians from the Islamic world such as al-Samaw'al (ca. 1172) were using a similar method.[43]

However, this hypothesis raises difficulties (use of sexagesimal rather than decimal numbers in the Islamic world, use of negative numbers by the Chinese only, etc.). Thus, attempts have been made to prove that Horner's method originates in Chinese mathematics alone (with, in the background, the idea that the former method was derived from the Chinese method).[44]

In an article published in 1955, Wang Ling and Joseph Needham tried to explain that this method already existed in the *Jiuzhang suanshu* of the Han.[45]

This idea is made seductive by the fact that there are clear analogies between the techniques of Qin Jiushao and the techniques of numerical computation of the Han (arrangement of numbers in columns in both cases, analogous terminology, shifting and backwards movement of the coefficients in both cases).

If we follow these authors in their comparison of the method for extracting the square root used in the *Jiuzhang suanshu* with Horner's method as used (for example) to extract the root of 71 824, we notice a striking analogy at the level of the series of additions $(400 + 60) + 60 = 520$ (while in the case of a binomial method 520 would come from 260×2):

---

[40]Lam Lay-Yong (2), *1969.*

[41]Lam Lay-Yong (9) *1982.*

[42]*SY.* p. 37 ff.

[43]Rashed (1), *1984.*

[44]Rashed, ibid., p. 100.

[45]Wang Ling and Needham (1), *1955.*

$$
\begin{array}{cccc}
1 & 0 & -71\,824 & (200 \\
 & 200 & 40\,000 & \\
\hline
 & 200 & -31\,824 & \\
\end{array}
$$

400		(60
60	27\,600	
460	−4\,224	
60		
520	−4\,224	(8
8	4\,224	
528		

Thus, the antique method of the Han undoubtedly contains a seed of the Hornerian idea.

However, this is an exception. None of the texts on the extraction of square or cube roots prior to the 10th century develop such an idea. What is more, the particular method in which J. Needham rightly detects a tiny Hornerian passage also uses other principles for the remainder of the calculations.

This is understandable. As previously mentioned, Chinese methods for extracting roots are based on the dissection of the square or the cube; thus, in all cases, the calculations are based on multiplications by 2 or 3 (binomial expansion). In other words, these methods are "binomial" rather than "Hornerian."

But then, how can we explain the "Hornerian" passage in the *Jiuzhang suanshu*? This question is difficult to answer. Perhaps this is due to the confusion of a late transcriber who, knowing Horner's method, may have introduced a foreign element into the method of the Han. The passage may also be a piece of original computational intuition which actually does date back to the Han. However, if this latter supposition were correct, we would still not have enough elements to support the theory of Wang Ling and J. Needham since there is an enormous conceptual gulf between the version of Horner's method used by Qin Jiushao and the microscopic Hornerian fragment on which the authors base their deductions.

However, we could also formulate a conjecture of a completely different type which links Horner's method (or at least, Horner's rule) with a distinctive structural characteristic of the common Chinese numeration. In fact, this numeration frequently abandons the positional principle, to permit the notation of numbers according to what A. Cauty describes as a numeration system of the Hornerian type,[46] that is, a system which admits intermediate factorisations in the notation of large numbers, for example:

---

[46]A. Cauty, 1987, *L'énoncé mathématique et les numérations parlées*, Thesis in mathematics, Nantes, p. 321 and 323.

二千三百六十三萬九千四十

er *qian*   san *bai*   liu *shi*   san *wan*   jiu *qian*   si *shi*

2 *Thousand*   3 *Hundred*   6 *Ten*   3 *Myriad*   9 *Thousand*   4 *Ten*[47]

which corresponds to

$$(2 \times 10^3 + 3 \times 10^2 + 6 \times 10 + 3)10^4 + 9 \times 10^3 + 4 \times 10$$

where symbols above the myriad, *wan*, are not used even though the corresponding number is much larger than a myriad (23 639 040).

A generalisation of this principle would then lead to a system of Hornerian numeration, in other words, to a numeration based on the principle of Horner's rule. We have perhaps here a fine example of the influence of certain linguistic structures on mathematics.

Systems of Equations of the First Degree in Several Unknowns (*Fangcheng* Method)

In the mathematics of ancient peoples one frequently finds computational rules and problems which may be analysed in terms of systems of linear equations in several unknowns.

In the case of the Babylonians, these are essentially small systems in two unknowns only.[48]

In the case of the Greeks, these involve more complicated problems. The Byzantine collection of small arithmetical poems assembled by the grammarian Metrodorus (fifth–sixth centuries AD), known as *The Greek Anthology*, includes problems the solution of which depends on more-or-less complicated linear systems in up to three or four unknowns.[49] However, most remarkable of all is the famous problem known as the "Epanthema[50] of Thymaridas"[51] referred to by the neo-Platonist Iamblicus (ca. 300 AD). In fact, apparently for the first time in history,[52] this problem introduces a whole class of equations in an arbitrary number of unknowns, the solution of which depends on a general algorithm which is described explicitly.[53]

---

[47] *CRZ*, j. 2, p. 21, cols. 5 and 6. Numerous other examples in Brainerd and Peng (1), *1968*, p. 64.

[48] Cf. van der Waerden (1), *1961*, p. 80; Tropfke (3), *1980*, p. 388 ff.; Thureau-Dangin (1), *1938*, p. xix ff.

[49] Cf. Paton (1), *1979*, V, p. 25–109.

[50] The word "epanthema" means "over-blossoming." Cf. Michel (1), *1950*, p. 284.

[51] Ibid. p. 285.

[52] Thymaridas is thought to have been earlier than Euclid. Cf. Michel, op. cit., p. 285.

[53] Ibid., pp. 284–285.

In modern notation this may be translated by the system:

$$x + x_1 + x_2 + \ldots + x_n = S$$
$$x + x_1 = a_1$$
$$x + x_2 = a_2$$
$$\ldots$$
$$x + x_n = a_n$$

which is solved using the formula

$$x = \frac{(a_1 + a_2 + \ldots + a_n) - S}{n - 1}.$$

But this area of mathematics was far more developed in China than elsewhere. In the case of the Chinese, one finds not only isolated problems based on *ad hoc* rules, but a set of general prescriptions (*shu*). Moreover, for the Chinese of the Han, all these methods belonged to a single mathematical category since they regrouped them all under the same generic hat of *fangcheng shu* (*fangcheng* prescriptions), within a complete chapter of the *Jiuzhang suanshu* (the eighth chapter).

This term *fangcheng* is not easy to translate exactly.

Since the end of the 19th century, *fangcheng* means precisely "equation" in modern mathematical Chinese. However, the traditional meaning of this term is very different from "equation" and the reason for choice of such a term to render the notion of equation is not clear.

In general, *fang* means "square." For this reason, Yang Hui explains that *fang* refers to the "shape of numbers."[54] Indeed, the *fangcheng* techniques are characterised by the fact that they involve numbers arranged in the form of a square (or more precisely, a rectangle — *fang* means "square" as well as "rectangle"). But for Li Ji,[55] a philologist of the Song (or perhaps of the Tang), *fangcheng* means "to the right and to the left" because the corresponding computations involve certain groups of numbers set to the right (or to the left) of other groups of numbers. The context shows that, in fact, numbers were laid out in parallel columns. Yet, according to another explanation, the meaning of *fang* was "to compare," that is, to perform certain arithmetical operations between columns of numbers leading to a sort of confrontation between them.[56]

Cheng is also difficult to understand. According to Liu Hui,[57] *cheng* is equivalent to *kecheng*. This *kecheng* exists in classical Chinese but is not a technical term. It means "to determine the value (of certain goods), to assign, to place a quota on or to distribute."[58]

---

[54]From the *Xiangjie jiuzhang suanfa* (detailed explanation of the terms of the *JZSS*) by Yang Hui (around 1270), quoted by Bai Shangshu (3′), *1983*, p. 258.

[55]Li Ji is the author of the *Jiuzhang suanshu yinyi* (end of the 11th century). Cf. *CRZ4B*, j. 5, p. 58.

[56]Martzloff (2), *1981*, p. 166 ff.

[57]*JZSS* 8-1 (QB, I, p. 221).

[58]*DKW*, 10-35589: 26.

While bearing in mind these various explanations, we shall translate *fancheng shu* as "prescriptions which involve distributing numbers in parallel columns so as to form a square" or more simply, "square arrays." But many other equally acceptable translations are possible.[59]

While these philological explanations leave room for doubt, the course of the calculations may be interpreted without ambiguity.

According to Liu Hui,[60] in order to handle an arithmetical problem using a *fangcheng* method the numbers involved in the problem must first be arranged in columns (*hang*).

The term column is taken to mean a sequence of numbers $a_1, a_2, \ldots, a_n, b$ arranged successively one below the other and related by *linear equations* of the type

$$a_1 x_1 + a_2 x_2 + \ldots + a_i x_i + \ldots + a_n x_n = b$$

where the x_i are the unknowns of the problem in question.

According to the vocabulary of Liu Hui's commentary, the right-hand term of the column is called the "total" (*zong*) or the "dividend" (*shi*). The term "total" is self-explanatory; the term "dividend" is explained by the fact that the *fangcheng* methods involve computational procedures which are intended to reduce the solution of linear systems to a simple division in which a certain term b (or rather that which corresponds to it after transformations) plays the role of a dividend.

More specifically, to arrange the numbers in columns, one begins by identifying the first condition of the problem, then the numbers involved are arranged using counting-rods in an initial column which is placed on the right. One then uses the second condition to form a second column immediately to the left of the first, and so on, where the position of the successive columns accords with the right-to-left flow of Chinese writing. All but one of the problems considered are those in which there are generally as many unknowns as equations (or, in Liu Hui's formulation "as many things (*wu*) as columns"):

If there are two things, one makes a new distribution, three things, three distributions and in general, one makes as many distributions as there are things.[61]

Liu Hui's commentary is purely verbal and contains no diagrams. However, as he describes the procedures in great detail, it is easy to reconstitute the appearance of the *fangcheng* diagrams on the counting-board (or on that which took its place). It is clear that, having formed the column on the right, according to the commentary, the calculator must then form column no. 2, then no. 3, $\ldots$, then no. $(n - 1)$, and finally the column on the left.[62] In this way, any problem expressed in words is associated with a counting-rod diagram as follows:

---

[59]Cf. Schrimpf, *1963*, p. 295; Martzloff (2), *1981*, p. 166.

[60]QB, I, p. 221.

[61]*JZSS* 8-1 (QB, I, p. 221).

[62]If there are only three unknowns, Liu Hui prefers to talk of the columns on the right, on the left and in the middle, resp.

$$
\begin{array}{ccccccl}
a_{n1} & \ldots & a_{i1} & \ldots & a_{11} & & \text{First thing} \\
a_{n2} & \ldots & a_{i2} & \ldots & a_{12} & & \text{Second thing} \\
\ldots & \ldots & \ldots & \ldots & \ldots & & \\
a_{nn} & \ldots & a_{in} & \ldots & a_{1n} & & n\text{th thing} \\
b_n & \ldots & b_i & \ldots & b_1 & & \text{Dividends}
\end{array}
$$

This corresponds to the matrix of the following linear system (with the right-hand side added):

$$
\begin{aligned}
a_{11}x_1 + a_{12}x_2 + \ldots + a_{1n}x_n &= b_1 \\
a_{21}x_1 + a_{22}x_2 + \ldots + a_{2n}x_n &= b_2 \\
&\ldots \\
a_{n1}x_1 + a_{n2}x_2 + \ldots + a_{nn}x_n &= b_n .
\end{aligned}
$$

Having formed his diagram (or, if you will, his matrix) of "column-equations" Liu Hui explains that:

此都術也，以空言難曉，故特繫之禾以決之。

ci dou shu ye, yi kong yan nan xiao, gu te xi zhi he yi jue zhi.

It is a general method, but since it is difficult to to explain it with empty words *kong yan* [i.e. with abstract words], we shall give a verdict on it [*jue*] using a particular case of a problem associated with cereals.[63]

In order to remain faithful to the spirit of his method, we shall thus explain this here without separating it from *JZSS* problem 8-1 which may be used to illustrate it:

Suppose we have 3 bundles of high-quality cereals, 2 bundles of medium-quality cereals and one box of poor-quality cereals, amounting to 39 *dou* of grain; [suppose we also have] 2 bundles of high-quality cereals, 3 of medium-quality and one of poor-quality, amounting to 34 *dou* of grain; one bundle of high-quality cereals, 2 of medium-quality and 3 of poor-quality, amounting to 26 *dou* of grain. Question: how many *dou* of grain in 1 bundle of high-, medium- and poor-quality cereals, respectively?

Answer: 1 bundle of high-quality cereals: 9 *dou* 1/4;
1 bundle of medium-quality cereals: 4 *dou* 1/4;
1 bundle of poor-quality cereals: 2 *dou* 3/4.[64]

According to the text of the *Jiuzhang suanshu*, this problem, which translates algebraically to

$$
\begin{aligned}
3x + 2y + z &= 39 \\
2x + 3y + z &= 34 \\
x + 2y + 3z &= 26
\end{aligned}
$$

---

[63] *JZSS* 8-1 (QB, I, p. 221).
[64] Ibid.

should be represented by the configuration of rods denoted by (I) in what follows. However, to make the text easier to read, we shall subsequently use Arabic figures (as in (II)).

I	II	III	
II	III	II	
III	I	I	
=⊥	≡‖‖	≡≛	

(I)

Left	Centre	Right	
1	2	3	High-quality cereals
2	3	2	Medium-quality cereals
3	1	1	Poor-quality cereals
26	34	39	Grain

(II)

Once this is done, the calculations proper may begin. Liu Hui explains that one should:

以右行上禾徧乘中行，而以直除。

use the cereals at the top of the right column to "multiply throughout" the central column in order to proceed to the "direct reduction."

Here, the expression "to multiply throughout" *biancheng* is an instruction to multiply all the terms of the central column by the number at the top of the right column.

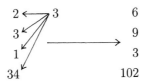

Multiplication throughout

The direct reduction *zhichu* involves carrying out a series of term-by-term subtractions of the right column from the central column ending in the elimination *jin* of the number at the top of the central column

Centre	Right			
6	3	6–3	(6–3)–3	Elimination
9	2	9–2	(9–2)–2	5
3	1	3–1	(3–1)–1	1
102	39	102–39	(102–39)–39	24

Thus, diagram (I) above takes the following successive forms

1	2	3
2	3	2
3	1	1
26	34	39

(1)

1	6	3
2	9	2
3	3	1
26	102	39

(2)

1	3	3
2	7	2
3	2	1
26	63	39

(3)

1		3
2	5	2
3	1	1
26	24	39

(4)

The next step, which is intended to eliminate the number at the top of the left column is based on the same type of manipulation:

1		3
2	5	2
3	1	1
26	24	39

(4)

3		3
6	5	2
9	1	1
78	24	39

(5)

		3
4	5	2
8	1	1
39	24	39

(6)

The same is true of the following step, which is used to eliminate the number at the top of the *new* left column.

		3
4	5	2
8	1	1
39	24	39

(6)

		3
20	5	2
40	1	1
195	24	39

(7)

		3
15	5	2
39	1	1
171	24	39

(8)

		3
10	5	2
38	1	1
147	24	39

(9)

		3
5	5	2
37	1	1
123	24	39

(10)

		3
	5	2
36	1	1
99	24	39

(11)

Finally, in (11), the matrix of the system is reduced to *triangular form*. Liu Hui then explains that the two remaining numbers in the left column (36 and 99) correspond to a divisor and a dividend and that, consequently, a simple division may be used to determine the value of one of the unknowns ($z = 99/36$). He then determines the other two unknowns by successive substitutions.

In essence, this first *fangcheng* technique is visibly nothing other than Gauss's method, even though Gauss's theories have little to do with those of the authors of the *Jiuzhang suanshu*, for the famous mathematician of Brunswick studied the question of the resolution of linear systems[65] in relation to the theory of the movement of heavenly bodies[66] and using the least-squares method. It is nevertheless true that, broadly speaking, the two approaches clearly amount to the same thing from the algorithmic point of view.

Of course, the above example is minimal, but it is not difficult to convince oneself that it has a paradigmatic value. The overall idea involves eliminating unknowns by means of additions or subtractions between columns of numbers, not randomly, but as Liu Hui explains,[67] by trying to reduce to a minimum the number of computations needed to find the solution ("generally, the more [computationally] economic a method is, the better it is" (*fan wei shu zhi yi yue sheng wei shan*)).[68]

Indeed, in his commentary on *JZSS* problem 8-18, the famous mathematician illustrates this idea by comparing the one with the other two different *fancheng* techniques relative to the same problem *JZSS* 8-18. In both cases, he counts precisely the number of operations needed *suan*. Unfortunately, it is difficult to determine the precise meaning of this *suan* (literally, "a counting-rod"). Does it refer to the number of additions, subtractions or other arithmetical operations needed to find the solution? Anyway, the fact that Liu Hui had in mind some sort of minimisation is beyond doubt.

In fact, the idea of reducing the volume of the computations to a minimum is so natural that we may presume that during the course of history numerous mathematicians have used Gauss's method without being aware of the fact.

In particular, this was the case for the Early French algebraist Jean Borrel whose *Logistica, quae et Arithmetica vulgo dicitur in libros quinque digesta* (Lyons, 1559) contains a problem solved by Gauss's method: page 190 of this work we find the following system in three unknowns A, B, C[69]:

$$1A, \tfrac{1}{3}B, \tfrac{1}{3}C, [14 \qquad E_1$$
$$1B, \tfrac{1}{4}A, \tfrac{1}{4}C, [\, 8 \qquad E_2$$
$$1C, \tfrac{1}{5}A, \tfrac{1}{5}B, [\, 8 \qquad E_3$$

where [and , stand for = and +, respectively. Borrel simplifies his system in a way very similar to that of Liu Hui (the main difference between the two approaches is due to the fact that the latter uses counting-rods and that intermediary computations disappear as the computations proceed, whereas Borrel keeps a record of his computations in writing). He subtracts equations from each other and obtains $150\,C\ [\ 750$, hence $C = 5$; then he finds the values of B and A by two successive substitutions (since his system has been "naturally" reduced to the triangular form).

---

[65] Cf. Goldstine (1), *1977*, p. 212 ff.
[66] Cf. Gauss (2), *1809*.
[67] QB, I, pp. 237 and 240.
[68] *JZSS* 4-11 (QB, I, p. 149)
[69] See also Kloyda (1), *1938*.

However, it goes without saying that the technique described above could not suffice in all cases, for at least two reasons:

(a) The division at the end of the calculations may be impossible to execute; this happens when the divisor is zero. In more mathematical terms, not all linear systems have a non-zero determinant.

(b) It is not certain that all the subtractions which are carried out during the direct reduction are always possible. For example, in the case of *JZSS* problem 8-3, which leads to the following diagram:

Left	Centre	Right
1		2
	3	1
4	1	
1	1	1

It is clear that the general directives of the *fangcheng* method cannot be applied. For how could the column on the right be subtracted from that on the left (or vice versa), given that it was impossible to subtract 2 from 1, 1 from "nothing" or 2 from 1?

Ancient and even medieval Chinese mathematics was a long way off being able to overcome the difficulty raised in (a) since this would probably have required the use of determinants in one form or another. However, this in turn would have required the mastery of an algebraic arsenal based not only numerical computations but also on literal calculations.

On the other hand, the second difficulty was negotiated more successfully. Surprisingly, for their time, the authors of the *Jiuzhang suanshu* managed to overcome this problem by introducing two new sorts of numbers: positive and negative numbers.

With the addition of the "rules of the positive and the negative" (*zheng fu shu*), the initial method could be applied with less restrictions, but there was still room for improvement for, the stage of iterated subtractions may be impossible to realise in practice. Suppose, for example, that we had to subtract 27 from 125 172 until the result was zero; if we followed the procedure to the letter, we would have to carry out 4636 subtractions! This is why Liu Hui proposed a more realistic way of eliminating the unknowns. Instead of carrying out a multiplication followed by a series of subtractions, he preferred to re-establish an operational symmetry by carrying out two multiplications followed by a single subtraction between columns. His new technique, which corresponds to the "addition method" of school algebra is self evident in the sequence of diagrams of *JZSS* problem 8-7 on which it is modelled[70]:

2	5
5	2
8	10

10	10
25	4
40	20

	10
21	4
20	20

	210
84	84
80	420

	210
84	
80	340

	1
1	
20	34
21	21

But, possibly because of the unexpected obstacles he met, Liu Hui also sought to broaden the range of the *fangcheng* techniques. For example, in relation to *JZSS* problem 8-18, he described what he called the "new *fangcheng* prescriptions" *fangcheng xin shu*, the originality of which consisted of the elimination of some of the right-hand sides so as to reduce the problem to an arithmetical question of proportional sharing.[71]

A generalisation of the *fangcheng* techniques in another direction involves linear systems with more unknowns than equations, for example, *JZSS* 8-13 which has 6 unknowns and 5 equations:

Suppose that five families share a well. 2 of A's ropes are short of the well's depth by 1 of B's ropes; 3 of B's ropes are short of the well's depth by 1 of C's ropes; 4 of C's ropes are short of the well's depth by 1 of D's ropes; 5 of D's ropes are short of the well's depth by 1 of E's ropes; 6 of E's ropes are short of the well's depth by 1 of A's ropes. What is the depth of the well and the length of each rope?[72]

Answer:

Depth of well:	7 *zhang*	2 *chi*	1 *cun*
Length of A's rope:	2 *zhang*	6 *chi*	5 *cun*
Length of B's rope:	1 *zhang*	9 *chi*	1 *cun*
Length of C's rope:	1 *zhang*	4 *chi*	8 *cun*
Length of D's rope:	1 *zhang*	2 *chi*	9 *cun*
Length of E's rope:		7 *chi*	6 *cun*

Here, with obvious notation:

$$2A + B = W$$
$$3B + C = W$$
$$4C + D = W$$
$$5D + E = W$$
$$6E + A = W$$

and the general solution of this problem is

$$W = 721k, \quad A = 265k, \quad B = 191k, \quad C = 148k, \quad D = 129k, \quad E = 76k.$$

---

[70] "Suppose that 5 oxen and 2 sheep are worth 10 taels (*liang*) and that 2 oxen and 5 sheep are worth 8 taels. What is the value of one ox and one sheep, respectively? Answer: 1 ox: (1+13/21) taels; 1 sheep: 20/21 taels." (Cf. QB, I, p. 228).

[71] Cf. Bai Shangshu (3'), *1983*, p. 287.

[72] QB, I, p. 232.

In that case, the solution proceeds as before but Liu Hui explains that the numbers of the answer are to be understood as *lü* ("models"). That means that if each number is multiplied by the same coefficient, another solution is obtained.

Still another generalisation of the *fangcheng* prescriptions occurred when the usual techniques of elimination were applied to pairs of polynomials rather than to linear expressions in the later *tianyuan* algebra. None the less, after the *Jiuzhang suanshu*, Chinese mathematicians never modified the *fangcheng* techniques themselves in any radical way. In particular, they never thought of the notion of determinant.

For other details, readers may refer to Schrimpf, (1), *1963*, p. 288 ff.; Libbrecht (2), *1973*, p. 152; Martzloff (2), *1981*, pp. 161–234.

13th-Century Chinese Algebra: the *Tianyuan Shu*

For the contemporary mathematician, the term "algebra" refers to the study of certain mathematical structures (groups, rings, fields, etc.), but for followers of Boole, Schröder or Couturat, it designate the mathematical analysis of logic, which is a completely different thing. For others, this word evokes the solution of certain equations, involving a succession of factorisations, simplifications, square roots, identities and discriminants; in other words, school algebra.

From a historical and strictly etymological point of view, "algebra" comes from the Arabic word "al-jabr" which is itself the beginning of "al-jabr wa al-muqābala," an expression found in the title of al-Khwārizmī's classical treatise on algebra and in those of many other subsequent Arabic works. Briefly, it consists of a written technique which involves adding the same term to the two sides of an equation or eliminating equal terms appearing on both sides of an equation.[73] For the mathematicians of the Italian Renaissance, algebra was a technique which enabled them to solve problems using special symbols which were manipulated in calculations as though they were known. This was "l'arte della cosa" ("the art of the thing," where "thing" denoted the symbol for the unknown). Finally, for certain 13th-century Chinese mathematicians, Li Zhi, Zhu Shijie and their emulators, algebra was the *tianyuan* technique.

The term *tianyuan*, which, strictly speaking, denotes the unknown, is not an invention *ex nihilo*, but an ancient expression charged with many calendrical, geomantic and even game-related connotations in which the notion of position relative to a temporal or spatial origin plays an essential role. Thus, among other things, the *tianyuan* represents not only "the primordial breath of the fundamental heavens" which the ancient sovereign used as the basis for fixing the origin of the calendar or, by extension, the sovereign himself;[74] it is also

---

[73]The etymology of the expression "al-jabr wa al-muqābala" is much more uncertain and complex than currently believed. Cf. Berggren (1), *1986*, p. 102; Saliba (1), *1972*; article "Algebra" in vol. 1 of the new edition of the *Encyclopedia of Islam*, Leiden: E.J. Brill, *1986*.

[74]*Shiji*, j. 26, p. 1256.

the trunk symbol associated with someone's birthday;[75] on a *weiqi* board (*weiqi* is the Chinese name for the game of go – a term borrowed from the Japanese language), it designates the central point of the grid on which the game is played.[76] But in mathematics, it is the generic appellation for various techniques which all attribute a special value to a certain reference place (or origin) on the counting surface. Sometimes, this reference place is reserved for sheer numbers (constants), sometimes for the coefficient of the first power of the unknown. But in all cases, the precise meaning attributed to a given number depends on its relative position with respect to an origin. For example, a number situated three ranks above the origin would designate the coefficient of the cube of the unknown.

Unlike the other kinds of algebra for which there is historical evidence, the *tianyuan* algebra does not use specific symbols such as $x, y, a, +, =$ or). The original texts dealing with this technique contain firstly verbal problems, drawings, geometrical figures, numerical tables and secondly diagrams of rod-numbers associated with computational instructions. By virtue of this second characteristic, Chinese algebra may be said to be rhetorical. But these written traces are always very laconic and tell only a small fraction of the whole story. Fundamentally, the *tianyuan* algebra is not a written technique but a set of procedures based on an instrument: counting-rods. This is why Chinese algebra is sometimes said to be an instrumental algebra.[77] The picture of a Chinese algebraist at work is the image of a person manipulating spillikins on the floor. A priori one would think that this person was playing go or some related game, one would not readily associate the scene with algebra.

Since techniques based on instruments are generally not concerned with the recording of the details of the computations, it is not easy to understand how the *tianyuan* algebra developed. We only know of it as a finished product which had already attained a high level of maturity and which is now accessible to us in a lapidary which seems typical of its strong links with a mathematical tradition more dependent on oral transmission that on complete recorded reasoning.

In fact, the set of *tianyuan* procedures which has been preserved seems to have been invented in Northern China towards the 11th century. Nowadays, we have access to these procedures through four main works:

(i) the *Ceyuan haijing* (1248) by Li Zhi;

(ii) the *Yigu yanduan* (1259) by Li Zhi;

(iii) the *Suanxue qimeng* (1299) by Zhu Shijie;

(iv) the *Siyuan yujian* (1303) by Zhu Shijie.

These works use a common terminological core, quote common sources and describe analogous mathematical techniques with a similar textual structure.

---

[75]Morgan (1), *1980*, p. 243.

[76]Zhao Zhiyun and Xu Wanyun (1'), *1989*, p. 95.

[77]Qian Bacong (13'), *1932*, p. 115.

Thus, to a first approximation the *tianyuan shu* is equivalent to this common basis.

In all cases, what first strikes one is the fact that the *tianyuan shu* is only found in the context of particular problems. It is often difficult to detect from the statements that these problems are about algebra, since other modes of solution might also be appropriate. The questions may be very varied, but they never concern the search for a means of solving a particular type of equation. This is not because the solutions do not contain such techniques, quite the contrary. The point to be stressed here is the fact that these techniques are used in an impromptu manner, as though their existence were self-evident.

As far as the solutions are concerned, these always involve a type of procedure analogous to that commonly used in school algebra, where the aim is to solve verbal problems. The ritual begins with the choice of the unknown(s) and continues with the expression of the problem in terms of rod-equations. This ceremony may be followed by a series of algebraic calculations intended to simplify the equation(s) of the problem in hand or to eliminate unknowns. In the second stage, the final equation is solved numerically using a particular algorithm.

The unknowns chosen depend on the problem; they may relate to the length of the hypotenuse of a town in the shape of a right-angled triangle, the diameter of a circular pond, a number of days, a rate of interest, etc.

Less mundanely, the unknowns may not be those which would apparently make for the greatest simplicity.

Let us consider the fairly typical example of problem III-5-1 of the *Siyuan yujian* which is formulated as follows:

Now we have the area of a [right-angled] triangle: 30 *bu*. It is only said that BASE-LEG-SUM: 17 *bu*. How much [is] BASE-HYP-SUM? Answer: 18 *bu*.[78]

If the two sides of the right-angle (x, y) are chosen as unknowns, this problem reduces to solving the system $xy = 60$ and $x + y = 17$. Thus, it is a matter of finding two numbers given their sum and product, which reduces to the solution of an equation of degree two. Zhu Shijie must have been aware of such a simple solution; however, he prefers to choose the two unknowns $z + x$ and x (where z denotes the hypotenuse of the right-angled triangle), which leads to the following equation[79]:

$$-t^4 + 34t^3 - 71t^2 - 3706t - 3600 = 0$$

where $t = z + x$ whose roots are -8, -1, 18 and 25; only $t = 18$ is retained. (The original derivation of this equation has not been recorded but, according to Luo Shilin's explanations,[80] it is the result of the elimination of the unknown x between

$$x^2 - 34x + 2tx + 289 - t^2 = 0 \qquad (1)$$

---

[78] *Siyuan yujian xicao*, op. cit., II, p. 845. Hoe (1), *1976*, appendix 2.
[79] Hoe (1), *1976*, appendix 2, p. 134.
[80] Ibid, pp. 846–848.

and

$$x^2 + 17x - 60 = 0 \qquad (2)$$

((1) comes from Pythagoras' theorem and (2) from $xy = 60$)).

In some other problems of the same chapter, the data remains identical but the unknowns are $t = x + y + z$, $z + (y - x)$, $(x + y) - z$ and so on.

This may be seen as a deliberate complication. By proceeding in this way, Zhu Shijie astutely makes use of the necessarily limited repertoire of problems known at that period. He uses judicious complications to obtain new algebraic situations in a relatively inexpensive way.

As one might expect, after the choice of the unknown(s), the expression of the problem in terms of equations uses a whole arsenal of different relationships. These relationships may be deduced from the particular statement of each problem, or they may also come from a stock of ready-made mathematical formulae. Thus, certain rules (Pythagoras's theorem, similarity of right-angled triangles, etc.) may be used to discover new ones. Using this material, the mathematician then attempts to construct two equal but different expressions in parallel, which he eventually systematically subtracts from one another to obtain what we would now call "the equations of the problem." This is the method of "equal accumulations" *ruji* (*ru* means equal and, according to the context, *ji* means area, volume, product or even "sum" (of a series)). We might use the word "accumulation" to render these various meanings with a single term).

After this preparative work, the initial verbal problem is sidelined and all that remains is the mathematical manipulation and the orchestration of skilful mathematical ballets of counting-rods; this may involve eliminating the unknowns, reducing the equations to a canonical form, etc.

These rod puppets which now chatter together are alive in their own right. They rise, fall, and turn to the left or the right, translating series of polynomial additions, subtractions, multiplications or divisions. When the red camp (positive) meet the black camp (negative) they kill one another, while respecting the sign rules. At the end of this small drama which does not depart from the laws of mathematical etiquette for a moment, the key character in the problem appears: the final resolutory equation.

In non-Chinese algebraic manuals, this character is treated with the utmost care and respect; it is studied in minute detail, in the search for the secrets of the algebraic formulae which unveil the results. This initially involves a study of equations of degree two, since these are the most docile. Then more fearsome beasts, equations of degree three are considered using the teratological formulae of Cardan. Equations of degree four (Ferrari's formulae) are even more difficult! Beyond this comes the long and desperate search for the philosophers's stone, which is the wall of impossibility of solution by radicals.

But in medieval China the degree of the equations is of little importance. The Chinese pass from one problem to the next, from degree two to degree four and from degree four to the degree six without drawing attention to the degree. Chinese equations are not exactly equations, but algebraic forms or schemes for extracting roots, which, as we saw in the section on the extraction of roots, consist of sequences of numbers to be operated on, as though one were extracting square, cube or nth roots. These Chinese equations are actually arithmetical operations. Thus, logically, they have only one result, in other words, a single root. Since in the 13th century these operations had long been part of Chinese wisdom, neither Li Zhi nor Zhu Shijie bothered to explain the calculations themselves. So unconcerned were they about solving the equations that they constructed their problems backwards so as only to obtain numbers which "fall out exactly" as roots.

Thus, in this respect, Chinese algebra may be viewed as a province of arithmetic. One consequence of this important fact is the existence of organic links between the *tianyuan shu* and the arithmetic of the Ten Computational Canons. These links are clearly apparent both in the notation and in the operational procedures.

As we know, Chinese techniques for extracting roots have always tended to respect the implicit algebraic hierarchy of the different types of number which appear during the course of the calculations. For example, the extraction of the square root involves schemes of type:

Quotient	200
Dividend	15225
Fixed divisor	$4000x$
Unit rod	$100x^2$

in which the coefficient of the unknown is below the dividend (constant term) and the coefficient of the square of the unknown is below the former.

In exactly the same way, Li Zhi and Zhu Shijie both use analogous schemes with columns.

Li Zhi sometimes uses descending hierarchies and sometimes ascending ones. In the first case, he denotes the numerical coefficients of the powers of the unknown in the order of increasing powers from top to bottom, while in the second case, he does the opposite. Sometimes, he also generalises the procedure by ordering the coefficients of the inverses of the powers of the unknowns in the same way. For his part, Zhu Shujie goes much further by using the two dimensions of the plane, which enable him to represent expressions in up to four unknowns (Fig.14.5).

In the arithmetic of these extractions of roots, each number had a particular name (the dividend *shi*, the divisor *fa*, etc.) but the *tianyuan shu* takes a more sober approach and settles on only one point of reference, which may indicate either the position of the unknown or that of the constant term. For this, it uses the terms *tai* and *yuan* which denote the constant term and the coefficient of the unknown, respectively. These are probably concepts borrowed from neo-Confucean philosophy and Taoism. *Tai* is an abbreviation for *taiji* which denotes a mass of energy which includes an organisational principle. *Yuan* is an abbreviation for *yuan yi* (primordial unit) a synonym for which is *tai yi* (the great unit).[81] This is the universal *dao* which envelops everything and gives birth to all beings.

Thus, the unknown is as omnipresent and as invisible as the *dao*. However, it is the unknown, which (like the central void which drives the cart spoken of by Laozi the father of Taoism)[82] which stimulates Chinese algebra. In this algebra, it is as physically absent as it is in our modern notation according to which the polynomial $a_0 + a_1x + a_2x^2 + \ldots + a_nx^n$ is written as $(a_0, a_1, \ldots, a_n)$. One can but admire the descriptive power of this medieval Chinese notion in which the *position* of a number by itself indicates the power of the unknown with which the number is associated.

In the same way that it borrows its notation from operational schemes for extracting roots, so the *tianyuan shu* is also inspired by other algorithms of the Nine Chapters. As in the ancient method of *fangcheng* prescriptions, in *tianyuan* algebra the algebraic expressions consisting of lines of numbers are called columns (*hang*). As in the venerable classic, calculations on fractions still exist but relate to more complex algorithmic objects.

For example, in problem I-6-13 of the *Siyuan yujian*, the solution of which depends on the equation[83]

$$576x^4 - 2\,640x^3 + 1729x^2 + 3\,960x - 1\,695\,252 = 0 \ .$$

Zhu Shijie first finds $x_1 = 8$ for the integral part of the root. The equivalent of the change of variable $x = 8 + x_2$ then leads him to the equation

$$576x_2^4 + 15\,792x_2^3 + 159\,553x_2^2 + 704\,392x_2 - 545\,300 = 0 \ ,$$

which he again transforms using the equivalent of $x_2 = \frac{y}{576}$ (method of fractions, *zhifen*), to obtain

$$y^4 + 15\,792y^3 + 91\,902\,528y^2 + 233\,700\,360\,192y - 104\,208\,452\,812\,800 = 0 \ .$$

This equation has an integral root $y = 384$. Finally $x = 8 + \frac{384}{576} = 8\frac{2}{3}$. In the same spirit, Zhu Shijie thinks of eliminating the unknown x from the two equations

$$
\begin{aligned}
P(y)x - R(y) &= 0 \\
Q(y)x - S(y) &= 0
\end{aligned}
$$

by forming the cross-product $P(y)S(y) - Q(y)R(y)$ as though he were calculating $ps - qr$ in order to subtract the two fractions p/q and r/s from one another. He called this method *huyin tongfen xiang xiao*, which means "reciprocal elimination [by] equalisation of the parts [fractions] which are mutually hidden."

In an even more elaborate way, using analogous principles, he handles even more complex cases involving more than two unknowns. For this, he uses the

---

[81] Cf. *DKW*, 1-1340: 27 and 3-5834: 10.

[82] *Daodejing*, ch. 11.

[83] Cf. *Siyuan yujian xicao*, I, p. 261 ff.

tifen method (dismembering of the parts) which involves dismembering (*ti*) the polynomials by taking the unknown to be eliminated as a factor (example: see later, page 270).

It can be shown that the other analogous techniques of the *Siyuan yujian* (for example), do not depend upon a single idea which is applicable in all cases.[84] In fact, they are *ad hoc* techniques which depend upon the specific characteristics of the particular case in question. If a simple addition is sufficient, the author does not seek further complication under the pretext of including his example in a predetermined general mould, instead, he lets himself be guided by the degree of simplicity of the range of solutions to which he could have recourse.

Because of this, given that Chinese authors do not always state their solutions explicitly, it is difficult to determine the respective roles of general rules and computational cleverness in any given problem.

There is another sort of difficulty which we would stress. When the ancient texts are updated (i.e. modernised), it is easy to make astounding comparisons. Such comparisons may be very enlightening or they may lead to errors, above all as a result of the associations of anachronistic ideas which they irresistibly suggest to readers. However, it would not be very satisfying to deny oneself completely any possibility of explaining the ancient texts using modern formulations, provided, of course, that readers are forewarned of the fact that convenient provisional short cuts are being used. Take, for example, the case of the equalisation of equations to zero. In Chinese texts, no "operation-equation" is ever set equal to zero. This is very logical: how can an arithmetical operation be equalised to 0? However, in the examples considered above, it is as though the equations were effectively equalised to zero. The Chinese notation scrupulously respects both the signs and the hierarchy of the coefficients.

What is surprising is not the fact that Chinese texts do not have the equivalent of our "$= 0$," but that they denote the equations according to uniform schemes of the type

$$f(x) \ [= 0] \ (\text{resp. } f(x,y) \text{ or } f(x,y,z) \text{ or } f(x,y,z,t) \ [= 0]).$$

In the absolute, how can one fail to recognise the superiority of this mode of Chinese notation over other medieval practices? For the simple equation of degree two, al-Khwārizmī and his emulators already distinguish five different cases[85] (squares equal to roots: $ax^2 = bx$; squares equal to numbers: $ax^2 = n$; roots and squares equal to numbers: $ax^2 + bx = n$; squares and numbers equal to roots: $ax^2 + n = bx$; squares equal to roots and to numbers: $ax^2 = bx + n$). With such a fragmented system it is impossible to say anything about equations of degree two without distinguishing a number of cases and subcases.

Historical phenomena turn out to be more singular than they appear when one reduces them to the common denominator of modern formulations. In particular, Chinese "equations" cannot be easily considered in isolation, out of context. In the *tianyuan shu* there is no formal distinction between the algebraic

---

[84]Hoe (2), *1977*, p. 172: "there is no unique method for obtaining the final equation."
[85]Rashed (1), *1984*, p. 38; Karpinski (1), *1915*.

expression $f(x)$ and the equation $f(x) = 0$. One can only specify the necessary details in context. Readers may refer to the example of a problem due to Zhu Shijie, which is discussed below and which clearly illustrates this phenomenon. Moreover, if it is desired to compare Chinese algebra with other algebras of the Middle Ages or the Renaissance period, one must also take the notation into account. Chinese algebra contains neither abbreviations nor algebraic symbols. It could be more interesting to consider the profound reasons for this absence of symbols. Of course, since we are talking about medieval texts, it is more normal not to find algebraic symbols than the converse. Thus, it seems that if there is anything to be explained, it is the presence rather than absence of symbols. Even so, one may wonder whether this absence of symbols in the Chinese case could stem from certain peculiarities of classical Chinese. In such a language, it is actually possible to express certain algebraic expressions in written characters using less signs than one would in elementary algebra! In other words, Chinese is sometimes so concise that to use algebraic symbols would, paradoxically, complicate everything.

For example, the two expressions *xian he he* 弦和和 and *xian jiao he* 弦較和 which literally read "hypotenuse–sum–sum" and "hypotenuse–difference–sum" mean, "the sum of the hypotenuse and the sum (of the two other sides of a right-angled triangle)" and "the sum of the hypotenuse and the difference (between the other two sides)," respectively, would be written as $z+(y+x)$ and $z+(y-x)$ (obvious notation), respectively. Here, classical Chinese requires only three written characters in each case, although seven characters are required in elementary algebra (three letters, two operational symbols, two brackets). This notation coincides with the so-called Polish notation invented by J. Lukasiewicz[86] (but expressions composed of more than three terms are not attested in classical Chinese). Yet non-sinologist readers will certainly object that the symbols *he* 和 and *jiao* 較 (sum and difference, respectively) are clearly more complex than $+$, $-$, x, y, z and (and thus that the comparison is invalid. Why is 和 counted as a single unit when it seems to consist of two units? This is incontestable. However, from a semantic point of view, the two characters in question each represent an indecomposable unit. Thus, the fact that they are not written in one block and consist of many strokes does not substantially alter the validity of the previous remark.

An Example of an Algebraic Problem: Problem 4 of the Introductory Chapter of the *Siyuan Yujian*

今有股乘五較與弦冪加句乘弦等。只云句除五和
與股冪減句弦較同。問黃方帶句股弦共幾何。

jin you gu cheng wu jiao yu xian mi jia gou cheng xian deng. zhi yun gou chu wu he yu gu mi jian gou xian jiao tong. wen huang fang dai gou gu xian gong jihe.

---

[86]Lukasiewicz (1), *1951*, p. 93.

We now have: LEG times FIVE-DIF equals HYP-SQUARED plus BASE times HYP. It is only said that FIVE-SUM divided by BASE is the same as LEG-SQUARED minus BASE-HYP-DIF. How much does YELLOW-SIDE together with BASE, LEG and HYP make? Answer: 14 bu.[87]

This problem concerns a right-angled triangle and the lengths of its sides called BASE, LEG and HYP, respectively. The YELLOW-SIDE denotes the diameter of the inscribed circle of such a triangle (based on a classical figure of the *Jiuzhang suanshu*).[88] In what follows, for clarity, we shall set: BASE $= x$, LEG $= y$, HYP $= z$, YELLOW-SIDE $= d$.

With this notation, the quantity referred to in the statement as BASE-HYP-DIF is just $(z - x)$. The five sums (resp. five differences) denote:

$$x + y, \quad x + z, \quad y + z, \quad x + y + z, \quad y + (z - x)$$

and

$$y - x, \quad z - x, \quad z - y, \quad (x + y) - z, \quad (x + z) - y$$

respectively. But when the statement talks of the FIVE-SUM (resp. FIVE-DIF), this should actually be taken to mean the sum of the five sums (resp. the sum of the five differences). Thus, FIVE-SUM is equal to $(x+y)+(x+z)+(y+z)+(x+y+z)+(y+z-x) = 2x+4y+4z$. Similarly, FIVE-DIF is equal to $2z$. Under these conditions, the statement asserts that (i) $y(2z) = z^2 + xz$ and that (ii) $2x + 4y + 4z = x(y^2 - z + x)$. If, to these two equations we add the relationships (iii) $x^2+y^2 = z^2$ (Pythagoras's theorem) and (iv) $d = x+y-z$ (classical property from the *Jiuzhang suanshu*),[89] we obtain a polynomial system in 4 unknowns. The aim of the author of the problem is to eliminate the unknowns x, y and z so as to obtain a final resolutory equation in $u = d + x + y + z = 2x + 2y$.

To enable the reader to obtain a precise idea of how the equations are derived and of the construction of the final equation from the point of view of the Chinese algebraist we also note that:

(a) Zhu Shijie uses four unknowns which he calls primordial units (*yuan yi*). He qualifies them as celestial (*tian*), terrestrial (*di*), human (*ren*) and material (*wu*).[90]

(b) The coefficients of the unknowns of algebraic expressions are written down in an indefinite two-dimensional table (Fig. 14.5). Moreover the character *tai* (which is just the *taiji* – the "Great Ultimate" of neo-Confuceans) is used to mark the position of the constant term and lastly, the products xu and yz occupy an irregular position. But otherwise the table is a perfectly regular matrix.

---

[87] *Siyuan yujian xicao*, op. cit., p. 60; *SY*, p. 285.

[88] Hoe (2), *1977*, p. 37.

[89] *JZSS*, j. 9, problem no. 16.

[90] The first three cases are reminiscent of the "three powers" (*san cai* – Heaven, Earth and Man) of the *Yijing*.

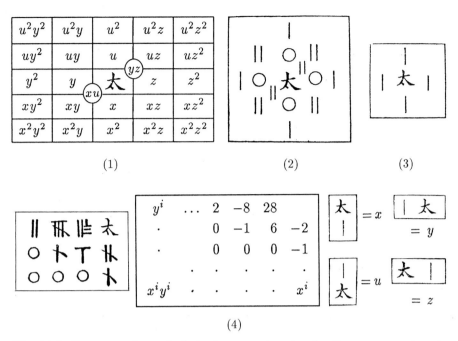

(1)   (2)   (3)

(4)

Fig. 14.5. Representations of algebraic expressions in the *Siyuan yujian* (1303). (1) Diagram showing the positions of the coefficients of the algebraic expressions (in the case of 4 unknowns). (2) Expansion of $(x + y + z + u)^2 = x^2 + y^2 + z^2 + u^2 + 2xy + 2xz + 2xu + 2yz + 2yu + 2zu$. (3) Representation of $x + y + z + u$. (4) Representation of the expression $2y^3 - 8y^2 - xy^2 + 28y + 6xy - x^2 - 2x$.

With this information, readers are now in a position to follow the various stages of the solution, step by step:[91]

"Detailed solution (*xicao*):

Take primordial celestial unit 太 $\genfrac{}{}{0pt}{}{}{1}$ $_{[x]}$ as BASE, primordial terrestrial unit

1 太 $_{[y]}$ as LEG, primordial human unit 太 1 $_{[z]}$ as HYP, primordial mate-

rial unit $\genfrac{}{}{0pt}{}{1}{太}$ $_{[u]}$ as value of the root. LEG 1 太 $_{[y]}$ times FIVE-DIF 太 2 $_{[2z]}$

gives 太 $_{[2yz]}$ deposit at LEFT. HYP-SQUARED 太 0 1 $_{[z^2]}$ plus BASE

times HYP 太 1 $_{[xz]}$ gives $\genfrac{}{}{0pt}{}{太\ 0\ 1}{1}$ $_{[z^2 + xz]}$; subtract that from LEFT to

---

[91]This is the translation of the "detailed solution" (*xicao*) due to Shen Qinpei (ca. 1829). Cf. *SY*, p. 287. Zhu Shijie's solution is too concise and would require too long explanations – however, see Hoe (2), *1977*, p. 219 ff.

obtain $\boxed{\begin{array}{ccc} & & \\ 太^{-2} & 0 & 1 \\ & 1 & \end{array}}_{[z^2 + xz - 2yz]}$, divide by the human primordiality [i.e. sim-

plify by z] to obtain the configuration "now" (*jinshi*) $\boxed{\begin{array}{ccc} -2 & 太 & 1 \\ & 1 & \end{array}}_{[x - 2y + z]}$ (1)

[this is the first equation of the problem $x - 2y + z = 0$].

LEG-SQUARED $\boxed{\begin{array}{ccc} 1 & 0 & 太 \end{array}}_{[y^2]}$, minus BASE-HYP-DIF $\boxed{\begin{array}{cc} 太 & 1 \\ -1 & \end{array}}_{[z - x]}$ gives

$\boxed{\begin{array}{cccc} 1 & 0 & 太 & -1 \\ & 1 & & \end{array}}_{[y^2 - (z - x)]}$ times BASE makes $\boxed{\begin{array}{cccc} & 太 & & \\ 1 & 0 & 0 & -1 \\ & 1 & & \end{array}}_{[xy^2 - xz + x^2]}$

subtracted from FIVE-SUM gives the configuration "it is said that" (*yun shi*)

$\boxed{\begin{array}{cccc} 4 & 太 & 4 & \\ -1 & 0 & 2 & 1 \\ & -1 & & \end{array}}$ (2)   [2nd equation of the problem].

$[2x + 4y + 4z - xy^2 + xz - x^2[= 0]]$

Add BASE-SQUARED and LEG-SQUARED then subtract HYP-SQUARED

to obtain the configuration "of the 3 primordial units" $\boxed{\begin{array}{ccccc} 1 & 0 & 太 & 0 & -1 \\ & & 0 & & \\ & & 1 & & \end{array}}$ (3)

$[x^2 + y^2 - z^2[= 0]]$

[Pythagoras's theorem, 3rd equation of the problem].

Add YELLOW-SIDE, BASE, LEG, HYP; obtain $\boxed{\begin{array}{cc} 2 & 太 \\ 2 & \end{array}}_{[2x + 2y]}$ [the

author uses the property $d = x + y - z$ and calculates $x + y + z + u = 2x + 2y$].

Subtract the root $\boxed{\begin{array}{c} 1 \\ 太 \end{array}}_{[u]}$; obtain the configuration of the material primordiality

(*wu yuan zhi shi*) $\boxed{\begin{array}{cc} -1 & \\ 2 & 太 \\ 2 & \end{array}}$ (4)   [4th equation of the problem].

$[2x + 2y - u[= 0]]$

Double the configuration "now"; and subtract the configuration of the material primordiality, obtain the upper position (*shang wei*)

$$\begin{array}{ccc} 1 & & \\ -6 & 太 & 2 \end{array}$$ (5)

$$[2z + u - 6y[= 0]]$$

[this involves eliminating x from the equations (1) and (4): $(2) \times (1) - (4) = (5)$].

Dismember (*ti*) the formula "it is said thatänd that of the "three primordialitiesthen carry out the elimination; obtain

$$\begin{array}{ccccc} 1 & 4 & 太 & 4 & -1 \\ -1 & 0 & 2 & 1 & \end{array}$$ (6)

$$[2x - xy^2 + xz + 4y + 4z + y^2 - z^2[= 0]]$$

[on the method of dismembering the arrays of numbers, see Hoe, op. cit., p. 236 ff.; see also *SY*, p. 183 ff. and the explanations below. In the present case, this reduces to adding (2) to (3) to eliminate x^2].

Then carry out the elimination from this configuration and that of the material primordiality. Obtain the central position (*zhong wei*)

$$\begin{array}{cccccc} 1 & 0 & -2 & -1 & & \\ -2 & -2 & -4 & 太^2 & -8 & 2 \end{array}$$ (7)

$$[2z^2 + 2yz - uz - 8z + uy^2 - 2y^3 - 2y^2 - 4y - 2u[= 0]]$$

[the author eliminates x from (4) and (6). For this, he calculates the cross-product $(2 - y^2 + z)(2y - u) - 2(4y + 4z + y^2 - z^2) \, [= 0]]$.

Dismember the configuration of the three primordialities and that of the material primordiality in order to carry out the mutual elimination. Obtain the lower position (*xia wei*)

$$\begin{array}{ccccc} & 1 & & & \\ -4 & 0 & & & \\ 8 & 0 & 太 & 0 & -4 \end{array}$$ (8)

$$[-4z^2 + 8y^2 - 4uy + u^2[= 0]]$$

[this involves elimination of x from (3) and (4). x^2 is eliminated by multiplying (4) by x and (3) by 2, then x can be eliminated from the resulting new equation and (4)].

Dismember the upper position and the central position in order to carry out the elimination. Obtain

$$\begin{array}{cccc} & -4 & & \\ -4 & 40 & -8 & \\ 8 & -88 & 112 & 太 \end{array}$$ (9)

$$[8y^3 - 4uy^2 - 88y^2 + 40uy + 112y - 4u^2 - 8u[= 0]]$$

[at this stage, z is eliminated from (5) and (7). These two equations may be written as

$$A_2(u, y)z^2 + A_1(u, y)z + A_0(u, y) \ [= 0] \tag{10}$$

$$B_2(u, y)z^2 + B_1(u, y)z + B_0(u, y) \ [= 0] \tag{11}$$

where $A_2(u, y) = 2$, $A_1(u, y) = 2y - u - 8$, $A_0(u, y) = uy^2 - 2y^3 - 2y^2 - 4y - 2u$, $B_2(u, y) = 0$, $B_1(u, y) = 2$, $B_0(u, y) = u - 6y$. According to Shen Qinpei (cf. SY, p. 183 ff.), Zhu Shijie used the term "dismember" to refer to the breaking of the arrays into pieces by separating the terms containing z from the others. This is equivalent to taking z as a factor, as follows:

$$(A_2(u, y)z + A_1(u, y))z + A_0(u, y) \ [= 0] \tag{10'}$$

$$(B_2(u, y)z + B_1(u, y))z + B_0(u, y) \ [= 0] \tag{11'}$$

in order to carry out a cross-multiplication, leading to a new equation with no term in z^2:

$$(A_2 B_0 - B_2 A_0)z + (A_1 B_0 - A_0 B_1) \ [= 0] \tag{12}$$

It now becomes possible to eliminate z from (10) and (11). Whence (9), above].

Simplify by 4: matter takes the place of sky; obtain the configuration "before" (qian shi)

$$
\begin{array}{cccc}
2 & -22 & 28 & 太 \\
- & 1 & 10 & -2 \\
 & & & -1
\end{array} \qquad (12')
$$

$$[2y^3 - uy^2 - 22y^2 + 10uy + 28y - u^2 - 2u[= 0]]$$

[(12') only contains the unknowns u and y and the following elimination will also result in another configuration in u and y only. Shen Qinpei writes that matter takes the place of sky, i.e. that, the coefficients of terms relative to u should be displayed as if u were x. A priori, this seems useless, but for Zhu Shijie, it is probably in order to bring the numbers into the canonical position for the final root extraction (in extractions of polynomial roots, the coefficients are usually set below the constant term (x column) rather than above it (u column)).] Nothing prevents us from proceeding in this way, at least when the x have been completely eliminated, because their place is then empty.

Dismember the upper and lower positions [arrays (5) and (8)]. Carry out the elimination; obtain

$$
\begin{array}{cc}
 & -32 \\
112 & 太
\end{array} \qquad (13).
$$

$$[112y - 32u[= 0]]$$

Simplify by 16; matter takes the place of sky; obtain the configuration "after" (hou shi)

$$
\begin{array}{cc}
-7 & 太 \\
 & 2
\end{array} \qquad (14)
$$

$$[-7y + 2u[= 0]]$$

Considering the configuration "after" as a "left" configuration, carry out the elimination from the configurations "left" and "before"; obtain the configuration

"right"
$$\begin{array}{cc} 0 & 294 \\ 8 & 3 \\ & -4 \end{array}$$
(15).

$$[8uy - 4u^2 + 3u + 294[= 0]]$$

[Here, the author partially eliminates y from (12′) and (14).] The product of internal columns gives
$$\begin{array}{c} 0 \\ 0 \\ 16 \end{array}$$
; that of the external columns gives
$$\begin{array}{c} -2058 \\ -21 \\ 28 \end{array}$$
.

The internal and external columns cancel each other out whence the configuration for extracting the square root (*kai fang shi*)
$$\begin{array}{c} 686 \\ -7 \\ 4 \end{array}$$
. Extract the square root; obtain 14 *bu* (paces), which corresponds to the question."

[Here, the author has eliminated y from (14) and (15). The expressions "left, right, external and internal" which are scattered throughout the text are self-explanatory when one knows that the numerical arrays were actually arranged as follows:

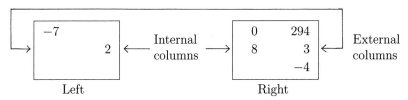

To carry out the elimination, Zhu Shijie calculates $(8u)(2u) = 16u^2$ and $-7(294 + 3u - 4u^2) = -2058 - 21u + 28u^2$, with the final result (16)

$$4u^2 - 7u - 686 \ (= 0)$$

(because $(2u)(8u) - (-7)(294 + 3u - 4u^2) = 16u^2 + 2058 + 21u - 28u^2 = -12u^2 + 21u + 2058 = 3(-4u^2 + 7u + 686))$.]

15. Geometry

A priori, it is difficult to conceive of geometry without definitions, axioms, postulates, theorems and proofs. A geometry which amounted to land surveying is often not perceived as geometry at all. As the famous British mathematician G. H. Hardy puts it: "The Greeks were the first mathematicians who are still 'real' to us today. Oriental mathematics may be an interesting curiosity, but Greek mathematics is the real thing."[1]

According to an opinion which has held sway for a long time Chinese geometry is insubstantial, and amounts only to land surveying, the indirect measurement of distances and Pythagoras' theorem. However, it has recently been shown that the commentaries of Liu Hui and his emulators on the *Jiuzhang suanshu* contain highly elaborate reasoning and perfectly convincing proofs, recorded in writing, even if these are not of the hypothetico-deductive type. Just as in the Greek case it is enlightening to study the pre-Greek foundations of geometry, one might also ask about the ancient bases of Chinese geometry.

In ancient China, two schools of thought could have played a role in this respect, namely the sophist school (disciples of Hui Shi (ca. 380–300 BC) or of Gongsun Long (ca. 320–250 BC)) and the Mohist school (disciples of Mozi, third–fourth centuries BC).

In the case of the sophists, the connection with later mathematical developments is not particularly obvious and remains to be precisely assessed. From all the evidence, the reasoning of Liu Hui and his successors is not based either closely or remotely on specious use of language; however it has long been noted that the Mohists defined certain geometrical objects.[2]

Let us consider these definitions, as they can be extracted from the Mohist Canon[3] as cited by A. C. Graham, save that we shall not retain this author's English translations, since these translations introduce numerous elements which do not exist at all in classical Chinese (for example, the verb "to be"). We prefer to adopt a "telegraphic-style translation" similar to that proposed by J. Hoe[4] which has the advantage of delivering English sentences no less concise than their Chinese counterparts with, in addition, English words assembled in a way which, most often, allows them to remain invariable whatever their function

---

[1] Hardy(1), *1940*, pp. 80–81.
[2] *SCC*, III, p. 91.
[3] Cf. Graham (3), *1978*, pp. 301–316.
[4] Hoe (2), *1977*.

within the sentence, exactly as in the Chinese case. But since ordinary words are not sufficient, we also use a punctuation mark – the colon – as a rough equivalent to the empty word *ye*:

(i) *ping tong gao ye* 平，同高也。 Level: same height.
tong zhang yi zheng xiang jin ye 同長，以正相盡也。 Same length: laid straight exhaust each other.
jian yu kuang zhi tong zhang ye zheng 楗與狂之同長也正。 The same lengths of the door-bar[5] and its door-frame:[6] straight lines.

(ii) *zhong tong chang ye* 中，同長也。 Centre: same length.
zi shi wang xiang ruo ye 自是望，相若也。 Outward from this: alike.

(iii) *hou you suo da ye* 厚，有所大也。 Bulky [voluminous]: having something bigger than.

(iv) *wei (duan?) wu suo da* 惟端無所大。 Alone [the starting point]: nothing bigger than.

(v) *ri zhong zheng nan ye* 日中，正 南也。 The Sun [at the] centre [at noon]: full South.

(vi) *zhi can ye* 直，參也。 Straight: three.[7]

(vii) *yuan yi zhong tong chang ye* 圜，一中同長也。 Round: same lengths from a centre.

(viii) *fang kuang yu si za ye* 方，匡隅四雜也。 Square: in a circuit [?] with four corners of a square container.

(ix) *bei wei er ye* 倍，爲二也。 To double: to make two [of them].

(x) *duan ti zhi wu hou er zui qian zhe ye* 端，體之無厚而最前者也。 Starting point [or extremity]: unit without thickness and furthest forward.

Clearly, some of these definitions can be compared with Euclidean definitions. For example, in (iv), we see that, like the Euclidean point the Mohist point has no size, in (x), we also see that this same point is an extremity (Euclid wrote that the extremities of a line are points). Is it possible that in both cases, this notion of extremity refers to the starting point, to the mark which the artisan makes as a reference? Certainly since the Mohist movement was rooted in the trades and crafts of the towns, [. . .].[8]

However, there are differences. Firstly, the Mohist definitions do not relate to a well defined science of geometry, but also refer to optics, mechanics and

---

[5]*jian*: door-bar. Cf. *Daodejing*, section 27: "a well-closed door does not need a door-bar."
[6]Graham, op. cit., pp. 304–305.
[7]This refers to the definition of a line by the visual alignment of three points.
[8]Graham op. cit., p. 6.

economics.[9] Secondly, as far as the definitions are concerned, certain objects of the physical world (the door-bar, the Sun) have "civic rights", which is poles apart from the Euclidean spirit; they also use verbs such as "to double", which Euclid never does. Finally, Mohist definitions have never been integrated into any deductive system whatsoever; they are not cited by any Chinese mathematical work.

In Chinese mathematics, terms which have special definitions in Greek mathematics essentially correspond to unreserved words borrowed from the everyday language.

For Chinese mathematicians, followers of the traditions of the Nine Chapters, a segment is either "the distance from [the man to] the bottom of the tree"[10], (*ren qu mu*) "the depth of the water"[11] (*shui shen*) or some other circumlocution, unless it is "the side of the square"[12] (*fang mian*); a point is the "bottom of a tree"[13] (*ben*), and so on. Moreover, for the Chinese, geometric notions such as parallelism or perpendicularity depend on the relative situation of real-world objects with respect to one another, which they express by bringing in horizontal/vertical frames (*zong heng*) such as the grids of geographical maps or the cardinal points. Hero of Alexandria (1st century AD) also does this: "The plane geometry consists of directions and points of reference and curves and angles [. . .]. And there are four directions: East, West, North and South."[14]

In addition, for geometers influenced by the Greek tradition and for those of the Chinese world, figures have a role to play, although, on the whole, this role is not the same in the two cases.

In the Greek tradition, all the elements of figures are most often named using letters of the alphabet. Using this rule, which is as regular as it is simple, even the most complicated figures may be refered to abstractly, without the need to refer to concrete characteristics, as in the Chinese case.

In the Chinese tradition, as we shall see later, the figures essentially refer not to idealities but to material objects which, when manipulated in an appropriate manner, effectively or in imagination, may be used to make certain mathematical properties visible.

Thus, we have two diametrically opposed situations. In the one case, figures tend to become useless and sometimes even harmful accessories. Distrusting deceptive appearances, Euclid often complicated his figures to the extent that they became difficult to decipher at first sight.[15] There is nothing like this in the Chinese case, and hardly any non-transparent figures. Instead, for numerous Chinese mathematicians, a geometrical property tends to be reread, or rather reviewed in each individual case and not cited in a reference to a treatise in

---

[9]Graham, op. cit., p. 301. "The corresponding series of propositions concern optics, mechanics and economics, but the definitions are of terms in geometry."

[10]*JZSS* 9-22 (QB, I, p. 257).

[11]*JZSS* 9-6 (QB, I p. 244).

[12]*JZSS* 4-16 (QB, I, p. 150).

[13]*JZSS* 9-7 (QB, I, p. 244).

[14]Cited from Bruins (1), *1964*, p. 2.

[15]Cf. Martzloff (6), *1984*, p. 47.

which it was established once and for all. In his treatise on planar trigonometry entitled *Ping sanjiao juyao*,[16] Mei Wending uses the equality of the complements of rectangles around the diagonal of a rectangle (special case of proposition I-43 of Euclid's *Elements*). Instead of proving this property once and for all, he rejustifies it each time. He takes the same approach in his *Jihe tongjie* (General explanation of [Clavius's version of Euclid's] geometry). His contemporaries (Fang Zhongtong, Du Zhigeng, etc.) also tend to do the same.

For all that, it is impossible to stick to the traditional judgement according to which the whole of Chinese geometry is simply a conglomerate of empirical procedures and practices.

As previously mentioned, certain texts by Liu Hui and other mathematicians contain reasoning which, while it is not Euclidean, is no less well constructed and completely exact. Moreover, although they are not numerous, these arguments appear all the more salient because they are without peers in other non-Euclidean mathematical traditions. But they also enable us to understand that Chinese mathematics is in part based on a small number of heuristic operational methods of a geometrical type. Above all, these methods are used to explain a vast set of Chinese results, including both geometrical, and algebraic or arithmetical results, in a unified way.

These methods are not all original. For example, those which are based on the similarity of triangles or which use infinitely small quantities are also found elsewhere outside China.

In fact, the most striking thing is the concrete appearance of the Chinese approach or rather the fact that abstract results are accessed via "concrete" means. Chinese proofs tend to be based on visual or manual illustration of certain relationships rather than on purely discursive logic.

In Chinese mathematics, one common approach involves the use of figures which are cleverly designed to make certain properties immediately visible, without (or almost without) discourse, where these properties would otherwise remain hidden.

Another equally common approach involves the manipulation of material objects, like the pieces of a jigsaw or a building kit, which take the place of planar or three-dimensional figures. This is based on the intuitive observation that if the pieces of a jigsaw are arranged in two different ways, without overlaps, the area or volume of the whole remains unchanged. As Liu Hui wrote *xing gui er shu jun* "the figures [take] strange [forms] but the numbers [which measure the areas or the volumes] remain equal."[17] If actual games (such as the game of tangrams) were involved, the pieces would be given at the beginning. But the same is not necessarily true in mathematics; often, the "mathemanipulator" begins with a given object *en bloc* and forms judicious cuts, sections and fragmentations reminiscent of how Hilbert envisaged the construction of elementary geometry.[18] But there is an enormous difference between the Hilbertian and the Chinese

---

[16]j. 3, p. 3a–b and j. 4, p. 2b–3a in *Meishi congshu jiyao*, 1874, j. 21.

[17]*JZSS* 9-5 (QB, I, p. 243).

[18]Hilbert (1), *1989*; Boltianskii (1), *1978*.

systems: with action rather than contemplation as its objective, the latter is based on a heuristic principle rather than an axiomatic theory. The sum total of Liu Hui's theory is practically contained in succinct formulae comprising only four written characters such as the following: *yi ying bu xu* ("to piece together (or fill in) the empty using the full"[19]; in other words, to show that a certain figure which is said to be "empty" because it is not yet physically constructed has the same area or volume as a figure, said to be "full", which is to be "unstitched" in order to piece together the other).

After this rapid survey, we shall now give a number of examples to show how these procedures are applied in various situations.

Planimetry

To find the area of a planar figure bounded by straight line segments, Liu Hui carries out a dissection in order to reconstitute a rectangle (*JZSS*, j. 1). We note in passing that a similar technique can already be found in the *Āpastamba-Śulba-sūtra* (second century BC).[20] Euclid himself also uses it.[21]

Such figures do not give rise to very difficult problems, but the same is not true of those which are bounded by curves. Of these, Chinese mathematicians only recognized those which are bounded by a circle or by the arcs of a circle (disk, ring, segment of a circle).

As far as the general history of mathematics is concerned, the question of the calculation of the area of the disc is of primary importance, because it is linked with the number π. Within Chinese mathematics, this same question has also attracted the interest of a large number of mathematicians.

The *Jiuzhang suanshu* proposes the following four formulae for calculating the area of the disc:

- $S = \frac{c}{2} \cdot \frac{d}{2}$ $\hfill (i)$

- $S = \frac{cd}{4}$ $\hfill (ii)$

- $S = \frac{3}{4}d^2$ $\hfill (iii)$

- $S = \frac{1}{12}c^2$ $\hfill (iv)$

where S is the area of the disc, d is the diameter and c is the length of the circumference.

The last two formulae are inaccurate (they "use" $\pi = 3$). However, the first two formulae are perfectly correct.

To prove (i), the commentator Liu Hui (end of third century AD) uses the approach which involves passing to the limit after having approximated the area of the disk using a sequence of regular inscribed polygons (the first

---

[19] *JZSS* 1-26, 1-28, 5-1, 9-3 (QB, I, pp. 102, 160 and 240, resp.).
[20] Cf. van der Waerden (3), *1983*, p. 26.
[21] Cf. Euclid's *Elements*, I-35.

polygon being a hexagon and each successive polygon having twice as many sides as its immediate predecessor). This technique is similar to that also used by Archimedes, both involve the approximation of the circle by regular polygons. The idea dates back to the fifth-century Athenian sophist and rival of Socrates, Antiphon.[22] However, apart from this common point the works of the Syracusean and those of the famous commentator from the Kingdom of Wei are very different, for many reasons and in particular because the one uses double reduction to the absurd and the other does not.[23] This difference alone leads us to conclude that epistemologically, Greek and Chinese mathematics belong to different universes which are far apart.

Indeed, Liu Hui's reasoning comprises a sequence of affirmations arising from the study of a figure:

按爲圖，以六觚之一面乘半徑，因而三之，得十二觚之冪。若又割之，
次以十二觚之一面乘半徑，因而六之，則得二十四觚之冪。割之彌細，
所失彌少。割之又割，以至於不可割，則與圓合體，而無所失矣。

an wei tu, yi liu-gu zhi yi mian cheng banjing, yin er san zhi, de shi'er-gu zhi mi. ruo you ge zhi, ci yi shi'er-gu zhi yi mian cheng banjing, yin er liu zhi, ze de ershisi-gu zhi mi. ge zhi xi mi, suo shi mi shao. ge zhi you ge, yi zhi yu bu ke ge, ze yu yuan he ti, er wu suo shi yi.

Commentary: Making a figure, taking the side of the 6-gon *liugu*,[24] multiplying it by half the diameter and tripling that we obtain the area of the 12-gon. If we then divide [the circle] again, taking the side of the 12-gon, multiplying it by half the diameter and multiplying that by six, we obtain the area of the 24-gon. The finer the division becomes, the more the loss decreases. By dividing again and again until it becomes impossible to divide (further) we obtain coincidence with the substance of the disk and there is no longer any loss.[25]

The figure to which Liu Hui alludes is irremediably lost. Yet, as we shall see in a moment, a convincing reconstruction is possible. What is of interest here, however, is not so much the reconstruction of a figure but the way the proof is set out.

Firstly, two particular results are merely asserted without more ado:

$$S_{12} = 3c_6 \left(\frac{d}{2}\right) \quad \text{and} \quad S_{24} = 6c_{12} \left(\frac{d}{2}\right)$$

(where S_i and c_i denote the the area and the side of the regular polygon inscribed in the circle, respectively).

---

[22]Michel (1), *1950*, p. 214 ff. Heath (1), *1921*, I, p. 222.

[23]Cf. Dijksterhuis (1), *1938*.

[24]The *gu* is an angular sacrificial vase; whence the idea of using its name to denote polygons. Note that the prefix *liu* (six) indicates the number of sides of the polygon.

[25]*JZSS* 1-23 (QB, I, p. 103).

Secondly, a general result, or rather a sort of conclusion relating to what happens when the number of sides of the polygons is increased, is once more enunciated in an assertive mood.

Considered in isolation, Liu Hui's first two assertions may appear more or less unrelated to the last one but when *JZSS* problem 1-23 together with Liu Hui's comments are taken into account, what is meant becomes clear for, on the one hand, *JZSS* problem 1-23 asserts that the area of the disc is

$$S = \frac{c}{2} \cdot \frac{d}{2} \, , \qquad (v)$$

while, on the other hand, the remaining part of Liu Hui's comment revolves around successive computations of S_6, S_{12}, S_{24}, S_{48}, S_{96} which all respect the formula

$$S_{2n} = \frac{n}{2} \cdot c_n \cdot \frac{d}{2} \qquad (vi)$$

(n = number of sides of the inscribed polygon).

Given that the result sought is (v), the underlying reasoning can be reconstituted as follows: reasoning analogically, as n increases indefinitely, the product S_{2n} tends to S and nc_n tends to c (where S_n and c_n denote the area and the side of the regular polygon with n sides, respectively). Liu Hui says nothing more about the formula (vi), but it is easy to prove. If we consider Fig. 15.1, we see that the area K of the "kite" $OABC$ with sides AC and BC both equal to c_{2n} and diagonals $OC = (d/2)$ and $AB = c_n$ may be expressed in two different ways (Fig. 15.1):

$$K = \frac{1}{2}CO \cdot AB = \frac{1}{2} \cdot \frac{d}{2} \cdot c_n$$

and

$$K = \frac{1}{n}S_{2n} \, .$$

(v) being so established or rather rendered plausible, it would seem that no more computations are needed. However, Liu Hui still gives the detail of specific computations of areas of inscribed polygons. But these also lead to other results going beyond the scope of *JZSS* problem 1-23.

Beginning with a circle of diameter 2 *chi* (feet), he iteratively calculates the apothems a_n and then the lengths c_{2n} of the sides of the polygons with 2×12, 2×24 and 2×48 sides. For this he applies (without justification) the fact that the side of the regular inscribed hexagon is equal to the radius (1 in the present case) and uses the following formulae which follow directly from Pythagoras's theorem applied to the triangles AOH and ACH (Fig. 15.1):

$$a_n = \sqrt{r^2 - \left(\frac{c_n}{2}\right)^2} \qquad c_{2n} = \sqrt{\left(\frac{c_n}{2}\right)^2 + (r - a_n)^2} \qquad \left(r = \frac{d}{2}\right).$$

On reaching c_{48} he applies the formula $S_{96} = 24c_{48}r$ which was suggested in the first part of the commentary and thus deduces an approximate value for S_{96}.

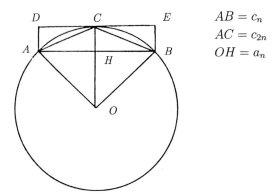

$AB = c_n$

$AC = c_{2n}$

$OH = a_n$

Fig. 15.1. In this figure reconstituted from Liu Hui's commentary on *JZSS* problem 1-23, alphabetical letters have been added. However, when Liu Hui refers to this figure, he only uses the usual Chinese terminology relative to the right-angled triangle. For him, OB, AC and AH, e.g. are thus the hypotenuse (of the triangle OHB), the "small" hypotenuse (of the small triangle AHC) and the base, respectively.

He then proceeds to calculate S_{192} from c_{96} in the same way and presents his results in the form

$S_{96} = 313\ 584/625$ [square] cun^{26} and $S_{192} = 314\ 64/625$ [square] cun.

Then, observing that the area of the disc can be entirely covered by 96 polygonal figures identical to $OADEB$, he notes that

$$S_{192} < S_{\text{disc}} < S_{96} + 2(S_{192} - S_{96});$$

or, numerically,

$$314\ \frac{64}{625} < S_{\text{disc}} < 314\ \frac{169}{625}\ .$$

For practical purposes, he abandons fractions and concludes that the ratio of the area of the disc to its circumscribed square is equal to $\frac{314}{400}$ or $\frac{157}{200}$. He also explains that the corresponding models (*xiang yu zhi lü*) of the diameter and the circumference are equal to 50 and 157, respectively (i.e. the diameter and the circumference vary proportionally in the ratio 50 to 157). Liu Hui did not think in terms of a specific number (π) but of a pair of numbers (one relating to the diameter, the other to the circumference). The other ancient Chinese mathematicians thought in the same way. None of them ever mentions a specific number, which would be π. All describe computational rules (sometimes exact, sometimes approximate) which are always adapted to the solution of particular problems and have no mandatory link with the idea of the constant π. But, the ratio between the diameter and the circumference is omnipresent in their works as in those of Liu Hui.

[26] $1\ cun = 1/10\ chi$. The text of the commentary on the *JZSS* says only *cun* and not "square *cun*". This may be because the calculator is interested in the formal rather than the concrete value of the numbers (while this dimension is incorrect, the calculations are correct).

For example, for Li Chunfeng "when the model of the diameter is 7, the model of the circumference is 22;"[27] for an anonymous commentator on the *JZSS* the pair in question is (1250, 3927);[28] while for Zu Chongzhi:

When the diameter of the circle, 1 *zhang*, is considered as made of *yi* [units][29] the excess value of the circumference of the disc is 3 *zhang* 1 *chi* 4 *cun* 1 *fen* 5 *li* 9 *hao* 2 *miao* 7 *hu* and its default value is 3 *zhang* 1 *chi* 4 *cun* 1 *fen* 5 *li* 9 *hao* 2 *miao* 6 *hu*; the exact value lies between these two excess and default values. [Hence] the "close models" *mi lü*, 113 for the diameter and 355 for the circumference, and the "rough models" *yue lü*, 7 for the diameter and 22 for the circumference, [respectively].[30]

There then arises the question of the origin of these various pairs. Did Zu Chongzhi obtain his legendary (113, 355), which corresponds to an expansion of π that is correct to six decimal places and which was not improved before the 15th century by al-Kāshī,[31] by extending Liu Hui's method of polygons to its ultimate bounds well beyond the polygon with 96 sides? This is highly likely (even though Zu Chongzhi's actual computations have not reached us) and would have required great patience and computational energy but no new theory. However, there are other possibilities. As A. P. Yushkevich judiciously remarked,[32] in the 16th century, the Dutchman Valentin Otho obtained the value $\frac{355}{113}$ by taking Ptolemy's value[33] for $\pi = \frac{377}{120}$ and Archimedes's value $\pi = \frac{22}{7}$ and subtracting the numerator (resp. denominator) of the latter from the numerator (resp. denominator) of the former to obtain a new numerator (resp. denominator), as follows:

$$\frac{377 - 22}{120 - 7} = \frac{355}{113} \ .$$

The same phenomenon could also have occurred in the case of China. The calendarist He Chengtian[34] used a method called "harmonisation of the divisor of the days"[35] *diaorifa* to obtain new fractional values from old ones by adding their numerators and denumerators in the manner of V. Otho. For example,

---

[27]QB, I, p. 107.

[28]QB, I, p. 106. Up to a coefficient of proportionality, these values correspond to those of Āryabhaṭa (20000, 62832). See Clark (1), *1930*, p. 28.

[29]Here, Zu Chongzhi uses a system of numeration according to which *yi* is equal to *wan wan* or a "myriad of myriad." Consequently, he implicitly expresses the length of the diameter of the circle in a unit such that 1 *zhang* = 10^8 times this unit. This unit is such that 1 *wei* = 10^8 *zhang*. But since the smallest unit he effectively uses is the *hu*, the computations stop at the *hu*, not at the *wei* (1 *hu* = 10 *wei*). The other units he considers are wholly decimal (1 *chi* = 10 *cun*, 1 *cun* = 10 *fen*, etc.).

[30]*Suishu*, j. 16, p. 388; Yan Dunjie (2′), *1936*.

[31]Berggren (1), *1986*, p. 7.

[32]Yushkevich (1), *1964*, p. 59.

[33]Of course, Ptolemy expresses numbers in the sexagesimal system. His value $\pi = 3$; $8'\,30''$ is equal to 377/120. Cf. Ptolemy's text in Toomer (1), *1984*, p. 302 (English translation).

[34]*CRZ*, j. 7; QB, *Hist*, pp. 87–88.

[35]The term "divisor" (*fa*) denotes the constant in the denominator of a fraction $\frac{a}{b}$ giving the surplus number of days needed to make a year when the number of days of a year is expressed as $365 + \frac{a}{b}$ days.

beginning with the values $\frac{22}{7}$ and $\frac{157}{50}$, attributed in China to Liu Hui and Li Chunfeng, we obtain the new value $\frac{179}{57}$ ($157+22 = 179, 50+7 = 57$). Reiterating the process, we finally obtain $\frac{355}{113}$ ($157 + (9 \times 22) = 355, 50 + (9 \times 7) = 113$). Of course, other reconstitutions (based on continued fractions, for example) are also possible. Anyway, in the case of reconstitutions based on Liu Hui's techniques it should be noted that there are special and interpretative pitfalls due to the special nature of Liu Hui's algorithm which is based on computations made with a fixed number of digits with truncation of certain intermediate results.[36]

Be that as it may, in China, as elsewhere, the calculation of π was to mark time following a number of spectacular results. The list of those who attempted to square the circle and their values of π (which exhibit more-or-less aberrant fluctuations) is never ending.[37]

In fact, the value $\pi = 3$ of the ancient *Jiuzhang suanshu* was still used in China up to the 19th century, not because it was thought to be correct, but because it was found to be sufficient for practical calculations.[38]

Certain mathematicians such as Minggatu[39] (?–1763), Xu Youren[40] (1800–1860) and Xiang Mingda[41] (1789–1850) used analytical methods of European origin. The advances in the accuracy of the values π obtained in this way were largely due to the sporting exploits of bold calculators such as Zeng Jihong (1848–1881), the youngest son of Zeng Guofan (1811–1872) who obtained π and $1/\pi$ correctly to 100 decimal places using decompositions in Arctan,[42] as Machin had done in 1706.[43]

Stereometry

Liu Hui uses four types of elementary component (qi) to calculate the volume of solids, namely: the cube *lifang*, the *qiandu*, the *yangma* and the *bienuan*.[44]

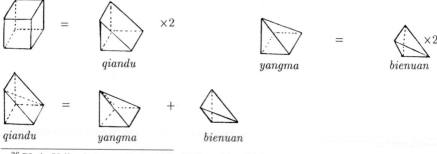

qiandu   yangma   bienuan

qiandu   yangma   bienuan

---

[36] Cf. A. Volkov's interesting article (Volkov (3), *1994*).
[37] Cf. Mikami (14), *1913*, p. 135 ff.; Hirayama (2′), *1980*.
[38] *CRZ*, j. 7, p. 85; QB, *Hist.*, pp. 87–88.
[39] *CRZ*, j. 48.
[40] *CRZ3B*, j. 4.
[41] Ibid., j. 3.
[42] QB, *Hist.*, pp. 337–338.
[43] Cf. Beckmann (1), *1971*, chapter 10.
[44] Wagner (1), *1975* and (4), *1979*.

As the figure shows diagonal cutting *xie jie* of a cube gives two *qiandu* while diagonal cutting of the *qiandu* gives a *yangma* plus a *bienuan*. Finally, a *yangma* decomposes in turn into two *bienuan*.

(i) 1 cube = 2 *qiandu*

(ii) 1 *qiandu* = 1 *yangma* + 1 *bienuan*

(iii) 1 *yangma* = 2 *bienuan*

It is highly likely that these terms take their names from real objects, although we do not know exactly which. The written characters *qian* and *du* suggest that the *qiandu* was a moat-wall.[45] As far as the *yangma* is concerned, this is said to be a technical term borrowed from architecture.[46] The etymology of the term *bienuan*, does not seem at all clear.

Whatever the exact etymologies, there is no doubt that here these terms have no more to do with architecture than terms like "pyramid" used in a mathematical context. However, it is certain that for Liu Hui, these objects were *qi* (components, fragments, pieces). These *qi* may have been used as pieces in a game before anyone thought of using them in mathematics. This is not an absurd supposition, for Liu Hui states that they were of distinctive red and black colours. Such a concrete characteristic fits well with the idea that they were pieces of a game or puzzle.

If there was actually such a game, we do not know its rules; however, the situation as far as mathematics is concerned is clear: the volume of a solid may be calculated using one or other of the following two heuristic methods:

(a) reconstruction of the given solid by combining basic components;

(b) decomposition of the given solid into non-arbitrary fragments belonging to one of the above four categories.

In the simplest of cases, this was sufficient, but most often other methods had to be associated with these heuristics.

For example, to calculate the volume of the *fang ting*[47] (square pavilion), Liu Hui begins by decomposing it into nine elementary components: 1 cube (in fact, a rectangular parallelepiped; in Liu Hui's commentary, the term, *lifang* whose general meaning is "cube" denotes more generally the rectangular parallelepiped), 4 *qiandu* and 4 *yangma* (Fig. 15.2). But, since the sum of these various volumes does not give him the desired result in an evident way, he continues his argument with an attempt to evaluate three times the desired volume.

---

[45] Wagner (4), *1979*, p. 166. In his commentary on *JZSS* problem 5-14 (QB, I, p. 166), Lui Hui writes that he was unaware of the origin of the term *qiandu* (generally speaking, *qian* denotes a moat outside a rampart and *du* a low wall).

[46] The term *yangma* is mentioned in j.5 of the *Yingzao fashi*, a famous treatise of architecture from the Song dynasty.

[47] *JZSS* 5-10 (QB, I, p. 164). The text is translated and explained in Wagner (4), *1979*.

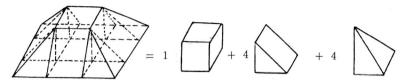

Fig. 15.2. Dissection of the square pavilion

Assembling $3 \times 9 = 27$ elementary components, he obtains three rectangular parallelepipeds of known dimensions (according to Wagner[48] he pieces together 1 cube and 4 *qiandu* for one of these – note that 2 *qiandu* placed together head-to-tail form a rectangular parallelepiped).[49] Thus, if a, b, h and V denote the sides of the "upper and lower squares," the height and the volume of the pavilion, respectively, he obtains:

$$3V = abh + a^2h + b^2h \,.$$

(Liu Hui implicitly assumes that the volume of a rectangular parallelepiped equals the product of its three dimensions). The calculation of the volume of the square pavilion depends on algebraic operations implemented geometrically, but there are cases in which that would not be sufficient. For example, the calculation of the volume of the pyramid requires other considerations.

This is why, when calculating the volume of the *yangma* (which is a special case of the pyramid), Liu Hui is not content with algebraic manipulations, but also uses the idea of passage to the limit.

Given a *yangma* obtained by dissecting not a cube but an arbitrary rectangular parallelepiped, (i) and (ii) above remain valid as equalities relating volumes, although it is not certain that (iii) (volume of the *yangma* = 2 × volume of the *bienuan*) still holds. Nevertheless, if it could be proved, by combining (i) and (ii) we would have

(iv) volume of the *yangma* = $\frac{1}{3}$ volume of the rectangular parallelepiped.

Consequently, to prove (iv), which is the desired result, Liu Hui concentrates all his effort on (iii) which he manages to prove in a convincing way.[50] For this, he dissects one of the *yangma* (Y) and one of the *bienuan* (B) arising from the dissection of the same rectangular parallelepiped as shown in Fig. 15.3.

It is a matter of showing that $Y = 2B$. Clearly, the components A, g and h arising from the dissection of Y at half its height have a total volume twice that of the components a and b together (which derive from B). Thus, to prove

[48]Wagner op. cit., p. 169.

[49]Liu Hui's terse explanations might be interpreted in many ways. For more details and different reconstructions cf. Schrimpf (1), *1963*, p. 275 ff.; Wagner, op. cit.; Bai Shangshu (3'), *1983*, p. 142 ff.; Guo Shuchun (4'), *1984*; Li Jimin (2'), *1990*, p. 305; Guo Shuchun (11'), *1992*, p. 207 ff.

[50]Cf. *JZSS* 5-15, (QB, I, p. 166); Wagner, ibid.

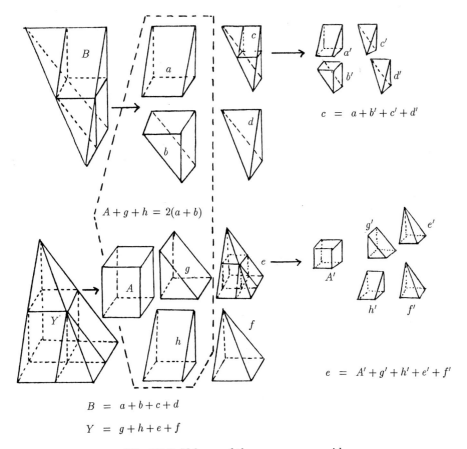

$$A + g + h = 2(a + b)$$

$$c = a + b' + c' + d'$$

$$e = A' + g' + h' + e' + f'$$

$$B = a + b + c + d$$

$$Y = g + h + e + f$$

Fig. 15.3. Volume of the *yangma* pyramid.

that $Y = 2B$, it is sufficient to show that the residual volumes e and f (resp. c and d) still satisfy the same volume ratio (one to two). Since these volumes are similar to the initial volumes, Liu Hui reiterates the same dissections as at the beginning, but this time with smaller components. Thus, he obtains more and more portions of the initial *yangma* which are themselves in the desired proportion. The remaining parts decrease constantly and, after passing to the limit, he concludes that $Y = 2B$. Of course, the reasoning is not rigorous, but it provides the desired result.

Until now, we have only mentioned polyhedral solids, but the Ten Computational Canons (*SJSS*) also discusses the cone, the truncated cone, the cylinder and the sphere (in the form of piles of grain, silos, caves, pillars and balls).

We shall not linger over the calculation of all the corresponding volumes, since the commentaries which explain them are not all very enlightening as far as mathematics is concerned. However, in China the calculation of the volume of the sphere has a long and rich history, which we shall now discuss.

Chapter 4 of the *Jiuzhang suanshu* contains two problems concerning the calculation of the diameter d of a sphere with a known volume V. In both cases, the text indicates that the solution is obtained by "extracting the spherical root" (*kai liyuan*), by calculating d using the formula

$$d = \sqrt[3]{\frac{16}{9} V}\,.$$

Much has been written on the question of the origin of this formula.

For an anonymous commentator on the *Jiuzhang suanshu*, it was the result of an empirical calculation: "a copper cube of diameter one inch (*cun*) weighs 16 ounces (*liang*), while a copper ball of the same diameter weighs 9 ounces: this is the origin of the models 16 and 9."[51]

For some historians of mathematics,[52] the logical explanation of this is likely to be associated with the history of π. By comparing the above formula with the exact formula $V = \frac{\pi d^3}{6}$, they conclude that the authors of the *Jiuzhang suanshu* used $\pi = 3\frac{3}{8}$. However, this explanation appears highly unlikely because, as D. B. Wagner notes,[53] as a general rule the *Jiuzhang suanshu* uses $\pi = 3$.

For Liu Hui however, this formula is the result of an elegant, but flawed argument.[54] Taking for granted the equivalent of "$\pi = 3$," the disc occupies $\frac{3}{4}$ of the area of the square which circumscribes it. When one passes, respectively from the disc to the right cylinder with height the diameter of the disk, and from the square to the cube with the same side, the ratio of the volumes is again $\frac{3}{4}$:

$$\text{volume of the cylinder inscribed in a cube} = \tfrac{3}{4} \text{ volume of the cube}\,. \qquad (i)$$

Assuming that the author of the formula in the *Jiuzhang suanshu* wrongly believed that the sphere inscribed in the cylinder also had volume $\frac{3}{4}$ that of the cylinder:

$$V_{\text{sphere}} = \frac{3}{4} V_{\text{cylinder}}\,. \qquad (ii)$$

Liu Hui then explains that by combining (i) and (ii) he obtains:

$$V_{\text{sphere}} = \frac{9}{16} V_{\text{cube}} = \frac{9}{16} d^3\,. \qquad (iii)$$

This last formula is equivalent to the formula to be explained.

But the mathematician of the Kingdom of Wei did not stop there. Continuing his exposition, he then explains that (ii) would be perfectly exact were it applied not to the sphere and the cylinder but to the sphere and another

---

[51] *JZSS* 4-24 (QB, I, p. 156, column 7).
[52] Yushkevich (1), *1964*, p. 61.
[53] Wagner (3), *1978* (the best article ever written on the volume of the sphere in China).
[54] QB, I, p. 155; Schrimpf (1), *1963*, p. 284.

object with volume less than that of the cylinder, which object he called *mouhe fanggai*[55] (or, literally, double (*mou*) assembly (*he*) of (two) vaults (or lids?) (*gai*) (with a) square (*fang*) (base)) (Fig. 15.4):

$$V_{\text{sphere}} = \frac{3}{4} V_{\text{vault}}. \qquad (iv)$$

However, he went no further.

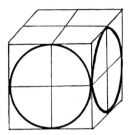 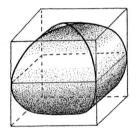

Fig. 15.4. The double vault (*mouhe fanggai*).

Two centuries later, Zu Xuan (son of Zu Chongzhi, the discoverer of the ratio $\frac{355}{113}$, mentioned above) finally managed to obtain the desired result by showing that:

$$V_{\text{vault}} = \frac{2}{3} V_{\text{cube}} = \frac{2}{3} d^3 \qquad (v)$$

and consequently that:

$$V_{\text{sphere}} = \frac{3}{4} \times \frac{2}{3} V_{\text{cube}} = \frac{1}{2} V_{\text{cube}} = \frac{1}{2} d^3. \qquad (vi)$$

Zu Xuan could have stopped there, however, he continued his argument by explaining that using the "precise models"[56] (*mi lü*) 22 and 7 (which for us corresponds to $\pi = 22/7$), the volume of the sphere was $V = \frac{11}{21} d^3$. To obtain this result he replaces the ratio $\frac{3}{4}$ (used by Liu Hui at the begin of his argument) by the more precise ratio $\frac{1}{4} \times \frac{22}{7}$.

Because of this, Zu Xuan can be said to have perceived the link between squaring the circle and cubing the sphere, in the sense that he was able to echo the progress achieved in solving the first problem for the second problem.

Up to this point, we have only presented the skeleton of the arguments. It now remains to explain how the Chinese authors justified the intermediate results (*iv*) and (*v*).

---

[55]According to the recent edition of the *Cihai* dictionary published in Shanghai in 1979, volume 1, page 1093, the written character *mou* can mean "double." We have no idea whether the *mouhe fanggai* is the name of a real object or whether it is a mathematical fiction.

[56]Note that the expression *mi lü* ("precise model") usually designates 35/113 and not 22/7 (the latter is more often called *yue lü* "rough model").

Since we are talking about the sphere, the dissection procedure would not have been sufficient on its own. Indeed, in both cases, the explanations of Liu Hui and Zu Xuan are implicitly based on a generalisation of Cavalieri's principle[57] to the case of volumes. In this case, the principle stipulates that[58] "if two solids which are both contained between two parallel planes are such that the sections cut from each at all levels have the same surface area then the volumes of these solids are equal."

But for the two Chinese commentators, if the ratio between the surface areas is the same at all levels, the volumes of the corresponding solids are also in the same ratio. That is precisely why, when proving (*iv*), Liu Hui states that

合蓋者，方率也。丸居其中，即圓率也。

hegaizhe, fang lü ye. wan ju qi zhong, ji yuan lü ye.

"The double-vault has the "model" *lü* of the square and the ball which is inside it has the model of the round." (if the sphere and the vault can be shown to be sectionable into a square and a circle always having the same proportion between them, their volumes will also have the same proportion).

But let us consider the geometry of the situation in more details.

Firstly, we note that, for us, the double vault is defined by the intersection of two orthogonal cylinders inscribed in the same cube. But, for Liu Hui, to whom the concepts of cylinder, orthogonality and intersection were unavailable, the double vault is an object the properties of which may be visualised.

Given a cube, we choose a face of the cube and draw the inscribed circle on this face, as Liu Hui's commentary advises. We then take this circle as a base and detach the cylinder with axis passing through centre of the present face and that of opposite face from the cube. We now apply the same operation, this time beginning with a face of the cube adjacent to the one we have just considered. Thus, at the centre of the cube, we obtain an object which is just the double vault and which is determined by the intersection of two orthogonal cylinders. It turns out that the sphere inscribed in the cube is also inscribed in this double vault. If we now take a section of the resulting object along a plane parallel to the axes of the two cylinders, we obtain a circle and the square circumscribing this circle. In other words, the plane in question cuts the double vault in a square and cuts the sphere in the circle which is inscribed in this square; moreover, this is true regardless of the level at which the section is taken.[59]

From there on, Liu Hui's intuition was to observe that because the areas of the sections of the double vault and the sphere are always in the same ratio ($\frac{3}{4}$

---

[57]Bonaventura Cavalieri (ca. 1598–1647): pupil of Galileo, professor of mathematics at the university of Bologna from 1629. See the article in the *DSB*.

[58]Cavalieri (1), *1635*, book 7, p. 485. See the translation and commentary on Cavalieri's principle in (e.g.) Struik (1), *1969*, p. 209 ff.

[59]Readers desiring a proof of this result should consult (e.g.): Edwards (1), *1982*, p. 73.

according to the approximation of the *Jiuzhang suanshu*), the volumes of these two objects must also be in the same proportion. As previously stated, Liu Hui describes this result by saying that the respective models of the vault and the sphere are 3 and 4; in other words that in terms of volume, 3 vaults may be proportionally exchanged for 4 spheres (the ratio of the volumes is $\frac{3}{4}$ – this is formula (*iv*)).

However, the connection between the double-vault and the sphere was more difficult to establish and had to wait for Zu Xuan, two centuries after Liu Hui.

This time, Zu Xuan takes as his starting point not the whole cube, but an eighth of it which he obtains by cutting the initial cube into 8 identical smaller cubes (Fig. 15.4). This approach may be explained by evident reasons of symmetry. To find the result corresponding to the large cube, it would be sufficient to multiply the findings for the small cube by 8.

By dissecting one of the small cubes by 2 cylindrical cuts (Fig. 15.5) he obtains 4 pieces: one double-vault piece which he calls the "internal piece" (*nei qi*) and 3 "external pieces" (*wai qi*): Reassembling these four pieces so as to re-form the cube, and taking a horizontal section (*hengduan*) of the latter, he obtains firstly a square (this is the section of the internal piece) and secondly a gnomon consisting of 2 equal rectangles and a small square (these are the sections of the 3 external pieces).

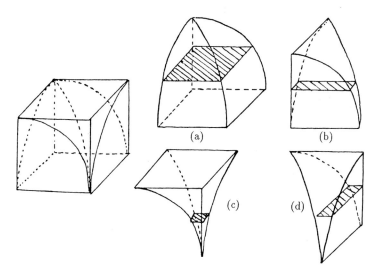

Fig. 15.5. The internal piece (a) and the external pieces (b, c, d).

He then determines the area of this gnomon, which is given by $S = c'^2 - b^2$ ($c' = $ side of the small cube = half the side of the large cube, or $c/2$; $b = $ side of the small square). Applying Pythagoras's theorem to the triangle marked in bold in Fig. 15.6, he obtains $b^2 = c'^2 - h^2$ (h = height at which the section is taken), whence:

$$S = h^2 .\qquad\qquad (vii)$$

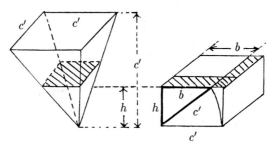

Fig. 15.6

Remarking then that the particular *yangma* (pyramid with a square base) for which both the length of a side of the square base and the height are equal to c' may be sectioned by planes parallel to its base into squares which always have area h^2, Zu Xuan concludes that the total volume of the 3 external pieces is the same as that of the pyramid. In fact, when they are put together, the 3 sections of these pieces form a gnomon which, as we have just seen, also has area h^2.

It is at precisely this point in the argument that Zu Xuan states his version of Cavalieri's principle[60]:

夫疊棊成立積，緣冪勢旣同，則積不容異 。

fu die qi cheng liji, yuan mi shi ji tong, ze ji bu rong yi.

The stacked blocks form the volumes. The *shi* of the areas being identical, the volumes cannot differ from one another.

Philological analysis of this text could take us a long way. Does *shi*, which in general means "appearance, manner, state, situation, circumstances" and even "potential" really refer to the fact that the areas are located "at the same level," in other words, at the same height? Or is it a specialised technical term? Only a complete study of all the commentaries on the *Jiuzhang suanshu* could possibly enable us to answer this question. What of this *qi*? We would like to be able to translate this by "surface", with the idea that it would be an element of area of infinitesimal thickness, an indivisible which, as in Cavalieri's principle would produce a volume by stacking (just as the thin pages of a book produce a volume when bound together). But *qi* means "block" or "piece." Could it be that the text of Zu Xuan which has come down to us includes an error and that *qi*

---

[60]QB, I, p. 158.

should really read *mi* (surface)?[61] In any case, despite these ambiguities which appear difficult to eliminate, the general meaning of this passage remains clear. It is a version of Cavalieri's principle for the case of volumes.

Thus, by virtue of this principle, Zu Xuan has shown that the total volume of the 3 external pieces is $V_{\text{ext}} = \frac{1}{3}c'^2$ (he knew the formula for the pyramid *yangma* due to Liu Hui).

Then, subtracting V_{ext} from the volume of the small cube c'^2, he obtains the volume of the internal piece:

$$V_{\text{int}} = \frac{2}{3}\, c'^3 \ .$$

But, since the volume of this internal piece (or ungula) is one eighth of the volume of the double vault, he concludes that

$$V_{\text{double vault}} = 8\left(\frac{2}{3}\right)c'^3 = 8\left(\frac{2}{3}\right)\left(\frac{c}{2}\right)^3 = \frac{2}{3}c^3 \ .$$

Whence, he finally obtains (*vi*), since the side of the cube has the same value as the diameter of the sphere inscribed in the cube.

In summary, the explanations of Liu Hui and Zu Xuan are based on two ideas: (a) recourse to a particular solid, the double vault, (b) Cavalieri's principle.

As in many other cases, all that occurs "miraculously" in the texts. The question of the origin of these Chinese ideas arises once again.

Of course, we may always assume that these magnificent ideas are the fruit of the reflections of Liu Hui and Zu Xuan.

However, as in the case of the problem of the area of the disc, it is again impossible in this case to fail to note that analogous ideas also appear in Greek mathematics.

More than half a millennium before Zu Xuan, Archimedes had already determined the volume of the double vault. At the beginning of the treatise known as the *Method Concerning Mechanical Theorems* he writes that: "if in a cube a cylinder be inscribed which has its bases in opposite squares and the surface of which touches the four other faces, and if in the same cube another cylinder be inscribed which has its bases in other squares and the surface of which touches the four other faces, the solid bounded by the surfaces of the cylinders, which is enclosed by the two cylinders, is two-thirds of the whole cube ... ".[62]

It would be interesting to compare Archimedes's proof of this result with that due to Zu Xuan. However, only a single incomplete manuscript of the text of *The Method* remains, namely that which was first announced in 1899 by the Greek palaeographer Papadopulos Cerameus in his report about the manuscripts of the Greek patriarchate of Jerusalem.[63] Unfortunately, the text

---

[61]Cf. Wagner (3), *1978*, p. 61.

[62]Cited from Dijksterhuis (1), *1938*, p. 314.

[63]From Reinach (1), *1907*, p. 913; see also Dijksterhuis, op. cit., p. 44.

of the proof in which we are interested is to be found in the missing part of this extraordinary palimpsest.

But Archimedes was not the only ancient author to have considered this question. In his *Metrica*, Hero of Alexandria (first century AD), quoting Archimedes, explains that: "The same Archimedes shows in the same book [i.e. in *The Method*] that if in a cube two cylinders penetrate, having the bases tangent to the edges of the cube, the common segment of the cylinders shall be two thirds of the cube."[64]

But while Hero does not give a proof of this proposition he does explain that the result in question is "useful for vaults being considered in this way, which often appear in springs and bathing houses, whenever the entrance or the windows are at all four sides, whereas it is not convenient to have the places covered with beams."[65]

Curiously, the Chinese term *mouhe fanggai* may be interpreted as containing the idea of a vault. Thus, one may wonder whether Liu Hui borrowed the term from the vocabulary of architecture. Unfortunately, our knowledge of the Chinese architecture of his period (end of the third century AD) is too scanty to enable us to throw light on this question.

Despite the fact that we are unable to compare the Archimedean and Chinese proofs directly, we may wonder about the nature of the lost proof due to the Syracusean. A priori the proof may be based on the principles of mechanics (theory of the lever, centre of gravity) together with some application of indivisibles, either for plane figures or for solids. The proofs of the existing part of *The Method* are based on principles of this type. As explained by E. J. Dijksterhuis: "[The Method] is further based on the view that the area of a plane figure is to be looked upon as the sum of the lengths of the line segments drawn therein in a given direction and of which the figure is imagined to be made up; this view will be extended to space in the following propositions in the sense that a solid, too, is conceived to be made up of all the intersections determined therein by a plane [. . .] and that subsequently also the volume of the solid is looked upon as the sum of the areas of those intersections."[66]

In the same spirit, many other historians, among them Th. Reinach,[67] C. B. Boyer,[68] P. H. Michel[69] and W. R. Knorr[70] have already noted Archimedes' use of indivisibles. But Archimedes is not the only Greek mathematician known for his recourse to indivisibles and Cavalieri's principle for the case of solids.[71] Still, there is no doubt that the style of the Chinese proof differs much from that of Greek ones, but the fact that all of them use a common ingredient –

---

[64] Bruins (1), *1964*, p. 302.
[65] Ibid.
[66] Dijksterhuis (1), *1987*, p. 318.
[67] Reinach, op. cit.
[68] Boyer (1), *1949*, p. 48 ff.
[69] Michel (1), *1950*, p. 231.
[70] Knorr (3), *1986*, p. 266.
[71] Cf. Tropfke (2), *1924*.

Cavalieri's principle – is obvious. Thus, the possibility of a Greek influence on Chinese mathematics cannot be ruled out.

After Liu Hui and Zu Xuan, the problem of the double vault arose again during the Italian Renaissance. At the end of one of his works entitled *De corporibus regularibus*, the painter and mathematician Pietro dei Franceschi[72] (1420?–1492) gives a correct proof of the formula for calculating the volume of this vault.[73] For this he quotes Archimedes's treatise *On the Sphere and Cylinder* rather than *The Method*, which was unknown at that time.

In China itself, after Zu Xuan until the end of 19th century no mathematicians studied either the vault problem or Cavalieri's principle.

Finally, in Japan, from the end of the 17th century many mathematicians were interested in determining volumes defined by the intersection of two solids (in other words, the very procedure which had been used to construct the double vault). But these Japanese problems were based on original methods which seem to owe nothing to either Chinese or Western mathematics.[74]

The Right-Angled Triangle

In general, when discussing right-angled triangles, we find ourselves involved with Pythagoras' theorem, resolutions of right-angled triangles (that is, determination of certain unknown elements pertaining to such a triangle from other elements which are assumed to be known) and lastly, number theory (determination of the sides or other elements of right-angled triangles in terms of integers).

While these themes constantly recur in the history of antique and medieval mathematics[75] they are even more ubiquitous in Chinese mathematics at all stages of its development from the origins to the beginning of the 20th century.

In Zhao Shuang's commentary on the *Zhoubi suanjing*[76] we find a list of prescriptions concerning 15 resolutions of right-angled triangles.[77] Most of these are found again (in various guises) in *juan* 9 of the *Jiuzhang suanshu*. All of them may be classified into nine distinct types as shown in the following table[78]

---

[72] Also known as Piero della Francesca.

[73] Original text in *Memorie della R. Academia del Lincei, Scienze Morali, 1915*, ser. 5, vol. 14, pp. 41–43. See also Coolidge (1), *1963*, p. 41.

[74] Kobayashi and Tanaka (1'), *1983*.

[75] Tropfke (1), *1922*, p. 56 ff; Tropfke (3), *1980*, p. 406 ff.; Busard (1), *1968*; Gericke (2), *1990*.

[76] QB, I, p. 18.

[77] Analysis in Wu Wenjun (1'), *1982*, p. 76 ff.

[78] Following an idea due to van der Waerden. Cf. van der Waerden (2), *1980*; Lam Lay-Yong and Shen Kangshen, (1), *1984*, p. 90.

Type	Given		Sought		JZSS no.
1	a, b		c		1, 5
2	b, c		a		2, 3, 4
3	or	$a, c - a$ $a, c - b\,^{78}$	or	b, c a, c	6, 7, 8, 9, 10
4	$c, b - a$		a, b		11
5	$c - a, c - b$		a, b, c		12
6	or	$a, b + c$ $b, a + c\,^{78}$		b a	13
7	$a, a + c = (\frac{7}{3})b$		b, c		14
8	a, b		e		15
9	a, b		d		16

(In this table a, b and c denote the lengths of the "small side" and the "large side" of the right-angle and that of the hypotenuse of a right-angled triangle, respectively. d denotes the diameter of the circle inscribed in the right-angled triangle with sides a, b and c; while e represents the length of the side of the inscribed square one of the right angles of which coincides with the angle of the given triangle and the opposing vertex of which lies on the hypotenuse).

Under the influence of the tradition of the Nine Chapters, most Chinese mathematicians enriched the initial collection with new problems such as the following in the table page 295. (The most complete collection of such problems is that contained in the *Shuli jingyun* (1723), "xia bian," j. 12, p. 609–678 in the edition of the text published in 1936 by Shangwu Yinshuguan in the collection *Guoxue jiben congshu*).

With the invention of the *tianyuan* algebra during the Song and Yuan periods, new problems much more complex than those studied previously were invented. The subject even became so important that in the *Siyuan yujian* 101 problems (out of 284) concerned right-angled triangles. Even more strikingly, all the problems of the *Ceyuan haijing* were restricted to the study of a particular right-angled triangle. Later, from the 17th century onwards, trigonometry was introduced into China as a consequence of Jesuit activities

---

[78] A priori, if we know how to solve the problem corresponding to a, $c - a$, there is no need to carry out new calculations for the problem with data a, $c - b$, since the two sides of the right angle play the same role. But this is not the case for Chinese authors, for whom, the different sides of a right angle have different names and are not interchangeable.

However, it is easy to pass from the solution corresponding to a, $c - a$ to that corresponding to a, $c - b$ by replacing a by b everywhere it occurs and b by a in the expression for the solution in the first case (this is an example of "parallelism").

Type	Given		Sought	Origin
10	or	$ab, c - a$ $ab, c - b$	a, b, c	$JGSJ$ no. 15, 16
11	or	$ac, c - b$ $bc, c - a$	a, b, c	$JGSJ$ no. 17, 18
12	$S(= ab/2),\ c$		a, b, c	$MSCSJY$ j. 17
13	$S, a + b$		a, b	ibid.
14	$S, c - (b - a)$		c	ibid.
15	$S, c + b - a$		a, b, c	ibid.

and questions of resolutions of triangles (right-angled or not) were developed further and a new notion never before considered in Chinese mathematics – that of angles – became prominent.

These various techniques are not equally efficient and it would be natural to imagine that, once available, algebra or trigonometry would have supplanted former Chinese methods. This is not at all the case, however, and in China all available techniques were always considered on an equal footing until the beginning of the 20th century. Significantly, a mathematical encyclopedia published in 1889 by the famous Tongwen guan College – the *Zhongxi suanxue dacheng* (Compendium of mathematics, Chinese and Western) – still contains two chapters on ancient Chinese techniques for the resolution of right-angled triangles *gougu*.[80] The same work also contains independent developments on algebra and trigonometry.

While algebraic or trigonometric techniques tend to render all solutions uniform and reduce any problem, however complex, to an equation, the difficulty is moved and centers on the question of solving equations. But Chinese *gougu* techniques are fundamentally based on searches for *ad hoc* geometrical representations whose function is the visualisation of algebraic relationships between the data and the unknowns. Within such a context, each particular problem constitutes a separate new case each time and inventive effort is constantly required. Suppose we have a problem which reduces to an equation of degree greater than three, failure seems probable since the difficulty of knowing the degree of complexity of the problem in question in advance together with the physical impossibility of representing figures of dimension greater than three seems to be a crippling restriction. It would seem that this sort of limitation could allow us to guess in advance what was within the reach of *gougu* techniques and what was not. It apppears, however, that this is not wholly the case and that developments in various directions were still possible. As is well known,

---

[80] *Zhongxi suanxue dacheng*, j. 29 and 30.

certain of these took place in the direction of algebra. But other innovations also touched upon number theory: problems (or rather exercices) tended to focus the difficulty on the main theme considered rather than on parasitic aspects such as those born of unwanted numerical complications. The sides, areas and other elements of right-angled triangles found in Chinese works were often the simplest possible and expressed in terms of integers, with or without fractions. For this reason, formulae for the computing of integer triangles are known in ancient China. Indeed, when astutely reinterpreted, the solution of *JZSS* problem 9-14 is essentially based on the following result. The lengths a, b and c of the sides of a right-angled triangle are proportional to $(m^2 - (m^2 + n^2)/2), mn, (m^2 + n^2)/2$ where m and n are given integers.[81] For example, taking $m = 7$ and $n = 6$ and coefficient of proportionality 2, we find the triplet (13, 84, 85) which occurs in problem no. 17 of the *Jigu suanjing* in disguised form: it uses a particular right-angled triangle with sides of length $14 \frac{3}{10}$, $92 \frac{2}{5}$ and $93 \frac{1}{2}$, respectively. These three fractional numbers actually come from the triplet of integers (13, 84, 85) and are found by multiplying 13, 84 and 85 by a common coefficient, namely $\frac{11}{10}$. It is quite possible that other methods for determining certain elements of right-angled triangles in terms of integers are concealed here and there in the corpus of Chinese mathematical texts. However, except very belatedly, in the last part of the 19th century, the subject never developed into an independent branch of knowledge.

Pythagoras' Theorem

In his commentary on the *Jiuzhang suanshu*,[82] Liu Hui explains how to prove that the area of the square on the hypotenuse is equal to the sum of the areas of the squares on the "base" *gou* and the "leg" *gu*:

句自乘爲朱方，股自乘爲青方，令出入相補，各從
其類，因就其餘不移動也。合成弦方之幂〔…〕

gou zicheng wei zhu fang, gu zicheng wei qing mi, ling churu xiangbu, ge cong qi lei, yin jiu qi yu bu yidong ye. hecheng xian fang zhi mi [...].

BASE-SQUARED makes the red square, LEG-SQUARED makes the azure square.[83] Let the OUT-IN MUTUAL PATCHING [technique] [be] applied according to the categories to which [these pieces] belong[84] by taking advantage of the fact that what remains does not move and form the surface of the hypotenuse.

---

[81]Cf. Lam Lay-Yong and Shen Kangshen (1), *1984*, p. 102. In his commentary, Liu Hui proves this result using a diagram.

[82]*JZSS* 9-3 (QB, I, p. 241).

[83]azure: *qing* (blue/green).

[84]Categories: *lei*. The context does not enable us to understand what Liu Hui means by categories. For Liu Hui, two figures may have been said to be of the same category if they were superposable, or if they were of the same area or the same colour, etc.

Although this text may be interpreted in a way other than that indicated by the present translation and although the geometrical figure needed to understand this text is now lost, it is clear that Liu Hui's proof involves the physical reconstruction of the square on the hypotenuse by covering this with pieces of the two squares on the two sides of the right-angle.

There are many ways of proceeding in conformance with this idea; the following figure shows one such approach[85]: This version of Pythagoras' theorem may be used to solve problems 1 and 2 above (table p. 294). But its main importance lies elsewhere: it is one of the keys to the vault of *gougu* techniques.

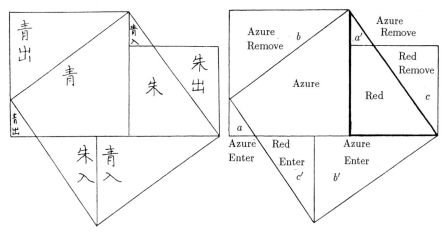

Fig. 15.7. Pythagoras' theorem. This figure shows the original right-angled triangle (in bold) together with the two squares constructed on the sides of the right angle; these squares are called azure and red on the Chinese figure since they correspond to jigsaw pieces of these colours. By construction, initially, the square on the hypotenuse is already partially covered by these azure and red squares. To show that these two square surfaces cover the square on the hypotenuse completely and exactly, as stipulated by Pythagoras' theorem, it is sufficient to move the pieces of the jigsaw as indicated on the figure: a, b and c should be removed and cover a', b' and c', respectively.

The Gnomon or Carpenter's Square (*ju*)

When associated with algebraic calculations, a knowledge of Pythagoras' theorem makes it easy to discover new relationships. From $c^2 = a^2 + b^2$, it is child's play to pass to $a^2 = c^2 - b^2$ (difference of two squares) and then to $a^2 = (c - b)(c + b)$ and

$$c = \frac{a^2}{c - b} - b\,,$$

etc. (incidentally, the last equation provides the key to solving problem 3 of

---

[85]From: Gu Guanguang. *Jiushu cungu* (The Nine Chapters, guardians of the tradition), 1892, , ch. 9, p. 4a, in the edition published by Jiangsu Shuju.

the above table, p. 294). From the point of view of ancient mathematics, the situation is very different.

Our difference of two squares corresponds to Euclid's gnomon or the carpenter's square.[86]

The same is true for Liu Hui.

Beginning with the squares on the hypotenuse (c^2) and on the leg (b^2), arranged as shown below, Liu Hui states that the area of the red gnomon is $c^2 - b^2$ (this is also a^2, according to Pythagoras). Transforming this gnomon into a rectangle of width $c - b$ and length $c + b$, he then obtains the property $a^2 = (c - b)(c + b)$.[87]

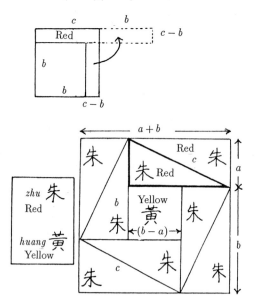

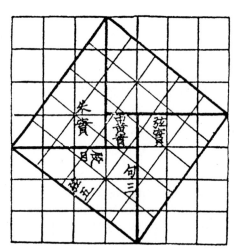

Fig. 15.8. The red gnomon.   **Fig. 15.9.** The hypotenuse figure. Source: QB, I, p. 17

The Hypotenuse Figure (*xian tu*)

The previous figures involve movements of pieces. But there also exist figures which may be read directly. The so-called "hypotenuse figure," which is one of the most ancient Chinese figures to have reached us is undoubtedly the most famous of these.

By noting that the area c^2 of the square on the hypotenuse (which is inscribed obliquely in the large square) is equal to the sum of the five areas of which it is composed, our Chinese authors obtain the formula

$$c^2 = (b - a)^2 + 4S, \qquad S = \frac{ab}{2}. \tag{i}$$

---

[86]Cf. van der Waerden (1), *1961*, p. 121.
[87]*JZSS* 9-5 (see Bai Shangshu (3′), *1983*, p. 311).

Applying the same procedure to the large square $(a + b)^2$, they also find that:

$$(b + a)^2 = (b - a)^2 + 4ab \ . \tag{ii}$$

The formula (ii) is interesting because it may be used to solve the following problem of degree two: find two numbers for which the sum (or the difference) and the product are known.[88]

A Figure Due to Yang Hui

The hypotenuse figure gives rise to additive decompositions of areas. But it is sometimes necessary to subtract areas from one another or even to divide them into a certain number of portions. Yang Hui does exactly that in his commentary on problem 9-11 of the *JZSS*.

(This problem asks the following question: "Suppose that the height of a door is 6 *chi* 8 *cun* greater than its width and that the opposing corners are 1 *zhang* apart. Determine the height and the width of the door)."[89]

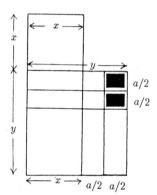

Fig. 15.10. A figure due to Yang Hui. From Yang Hui, *Xiangjie Jiuzhang suanfa*, p. 58 (Shanghai: Shangwu Yinshuguan), *1936.*

This problem may be transcribed algebraically as follows (x = width, y = height, z = hypotenuse)

$$
\begin{aligned}
x^2 + y^2 &= z^2 \\
y - x &= a \quad (= 6 \ chi \ 8 \ cun) \\
z &= d \quad (= 1 \ zhang)
\end{aligned}
$$

The algebraic calculations leading to the solution are scarcely difficult, but they are not immediate.

However, Yang Hui's approach is very different since, rather than establishing equations, he "establishes" a figure. Once this is in place, he uses it to obtain the desired result correctly, without calculations.

---

[88]Cf. Lam Lay-Yong (6), *1977*, p. 117.
[89]1 *zhang* = 10 *chi* and 1 *chi* = 10 *cun*.

Let us consider his figure (Fig. 15.10). By construction, it consists of the two squares constructed on the sides of the right-angle of the door. By Pythagoras's theorem, its total surface area is $x^2 + y^2 = z^2 = d^2$. If the two small shaded squares are removed from this area, it follows that half of what is left is equal firstly to $\frac{1}{2}\left[d^2 - 2\left(\frac{a}{2}\right)^2\right]$ and secondly to $\left(x + \frac{a}{2}\right)^2$ (or $\left(y - \frac{a}{2}\right)^2$). Whence the resolutory formulae:

$$x = \sqrt{\frac{1}{2}\left[d^2 - 2\left(\frac{a}{2}\right)^2\right]} - \frac{a}{2}$$

$$y = \sqrt{\frac{1}{2}\left[d^2 - 2\left(\frac{a}{2}\right)^2\right]} + \frac{a}{2}.$$

Inscribed Figures

The previous examples involve algebraic manipulations of areas by addition, subtraction or division. There are also cases of multiplication.

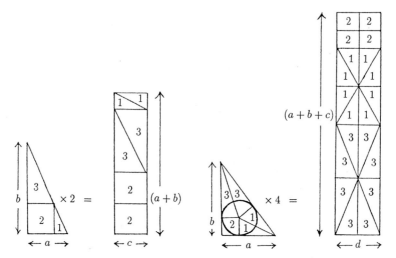

Fig. 15.11. Square and circle inscribed in a right-angled triangle. *Key:* 1 = red, 2 = yellow, 3 = azure, c = length of the side of the inscribed square, d = length of the diameter of the inscribed circle. From Li Huang, *Jiuzhang suanshu xicao tushuo* (Detailed solutions (of the problems) of the *Jiuzhang suanshu* with figures and explanations), Chengdu, 1896, ch. 9, p. 6b, 7a, 7b.

Problems 9-15 and 9-16 of the *Jiuzhang suanshu* involve the determination of the length of the side of a square inscribed in a right-angled triangle (resp. the length of the diameter of an inscribed circle). As Liu Hui indicates, the method of solution in both cases involves a rectangle the area of which may be described in two different ways. For this, he rearranges the pieces resulting from a dissection of 2 (resp. 4) identical examples of the initial triangle, as shown in Fig. 15.11. Thus, he creates a rectangle the area of which is equal to ab (resp. $2ab$) on the one hand and $c(a+b)$ (resp. $d(a+b+c)$) on the other, since the rectangle in question has width c (the side of the square) (resp. d (the diameter of the inscribed circle)) and length $a+b$ (resp. $(a+b+c)$). Whence:

$$c = \frac{ab}{a+b}, \qquad d = \frac{2ab}{a+b+c} \; .$$

In the case of the inscribed square, Liu Hui also obtains the same result based on a purely arithmetical argument[90]:

令句爲中方率，以並句股爲率，據股十二步而今有之，則中方又可知。

ling gou wei zhong fang lü, yi bing gou gu wei lü, ju gu shi'er bu er jin you zhi, ze zhong fang you ke zhi.

Let BASE [a] [be] taken for "model" *lü* of the inscribed square and BASE-LEG-SUM [$a+b$] for model. Let the rule "we now have" be applied with a LEG of 12 *bu*. Whence the length of the inscribed square.

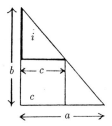

Fig. 15.12

In other words, if the length $a+b$ corresponds to the length b [$=12$] to what does the length a correspond? The answer follows from the rule of three (called the rule "we now have" *jin you* in the *Jiuzhang suanshu*), and the length $ab/(a+b)$ corresponds to the length a. In fact, since the triangle denoted by (i) in Fig. 15.12 is similar to the initial triangle and since the sum of the sides of the right-angle of (i) is equal to b, we have $b/(a+b) = c/a$. Whence we obtain c.

A Problem Due to Wang Xiaotong

Procedures based on arithmetic or the "geometrical algebra" of dissections can scarcely ever be applied to problems other than those of degree one, two or three (above this, one might think of representing areas by lengths, volumes by areas, etc., but that would undoubtedly be difficult to implement).[91] But even

---

[90] *JZSS* 9-15 (Bai Shangshu, op. cit., p. 330).

[91] However, in the 19th century, Li Shanlan does use such a technique in other problems (cf. Wang Yusheng (1'), *1983*).

in the case of mere volumes, what was to be done was often far from obvious. Thus, it is not surprising that mathematicians developed other procedures. For example, Wang Xiaotong explains how to solve a problem of degree three using computational tricks.

Problem 15 of the *Jigu suanjing*, asks the following question:

If BASE times LEG is seven hundred and six and one fiftieth and if HYP is thirty six and nine tenths more than BASE, what are the values of the three things[92]?"[93]

(In self-evident notation:

$$\begin{aligned} xy &= 706\,\tfrac{1}{50} &(= P) \\ z - x &= 36\,\tfrac{9}{10} &(= Q) \\ x^2 + y^2 &= z^2 &). \end{aligned}$$

Wang Xiaotong explains that

Area [*mi*] BASE times LEG: BASE-SQUARED times LEG-SQUARED. Thus, divided by double BASE-HYP-DIF $[2(z - x)]$ [this] gives one BASE and a half DIF, times BASE-SQUARED.

That is:

$$(xy)^2 = x^2 y^2, \qquad \frac{x^2 y^2}{2(z - x)} = \left(x + \frac{z - x}{2}\right) x^2 \, . \qquad (i)$$

Having obtained this result, he then lists the coefficients of the equation of degree three in x which follows from (i).

This reasoning is not completely explicit, but it is clearly based on two clever ideas:

(a) Transformation of the known quantity $x^2 y^2 / 2(z - x)$ into an expression containing only one unknown (the base x).

(b) Use of the identity $y^2 / (z - x) = z + x$ (gnomon) in the clever form $y^2 / (z - x) = 2x + (z - x)$ (divided by 2 and multiplied by x^2, this last formula gives (i)).

Series Summation

In his *Chengchu tongbian suanbao* (Treasury of calculations relative to the set of all aspects of multiplication and division) (1274), Yang Hui calculates the number of objects contained in a "pile with four corners" *siyuduo* using an algorithm equivalent to[94]:

$$S = \tfrac{1}{3} n(n + 1)(n + \tfrac{1}{2}) \, .$$

---

[92]The "three things": i.e. the base, the leg and the hypotenuse.
[93]QB, II, p. 524.
[94]Lam Lay-Yong (6), *1977*, p. 18.

The square pile in question actually consists of small cubes arranged in square slabs which are piled on top of one another and consist, respectively, of n^2, $(n-1)^2$, ..., 3^2, 2^2 and 1^2 small cubes (Fig. 15.3). Thus,

$$S = 1^2 + 2^2 + 3^2 + \ldots + n^2.$$

Yang Hui does not explain the *raison d'être* of his formula, but four centuries later a certain Du Zhigeng (fl. end of 17th century)[95] noted that three examples of the "pile with four corners" may be pieced together to obtain a rectangular parallelepiped surmounted by half a slab of cubes. But the half slab is essentially equivalent to a whole slab of half the height fitting exactly onto the upper face of the parallelepiped. Thus, we have a rectangular parallelepiped with dimensions n, $n+1$ and $n+\frac{1}{2}$, with volume equal to three times the desired sum.

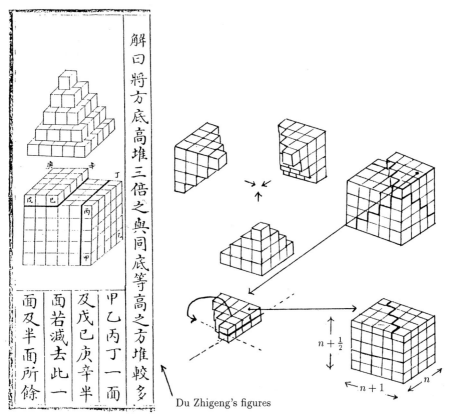

Du Zhigeng's figures

Fig. 15.13. The sum of the squares of the first n numbers (reconstitution).

---

[95]Du Zhigeng, *Shuxue yao* [The key to mathematics], j. 4, p. 41b.

Much more recently, historians of mathematics and educationalists such as Lurje, Xu Chunfang and others have proposed, either out of curiosity, or by way of a reconstitution of the Chinese formula or of an ancient Babylonian formula, a concrete proof analogous to that of Du Zhigeng.[96] This time, thanks to their diagrams, the manipulations may be understood on their own, without words.

Could the idea of justifying the formula for the sum of the squares of the first n numbers using Du Zhigeng's method be said to be original? The Chinese procedure is reminiscent of the figurative techniques used to calculate certain sums of figurate numbers. These techniques were common in European arithmetical treatises of the Middle Ages and the Renaissance period. In Du Zhigeng's day, China had had access to ideas of European origin for approximately a century. However, as far as we know, there is no evidence for the idea of calculating sums of series using three-dimensional figurate numbers analogous to those of Du Zhigeng in Europe itself before the 19th century.[97] The mathematician and educationalist Edouard Lucas (1842–1891) seems to have been the first to have used such procedures (see his *Récréations Mathématiques* in which he forms what he terms "piles of cobblestones" to calculate the sums of series of triangular or pyramidal numbers).[98]

A Recurrence Relation

In his *Siyuan Yujian* (1303), Zhu Zhijie takes advantage of the regularities made evident by the distributions of numbers in his "Pascal's triangle" to calculate the sums of the first n terms of certain series. As J. Hoe observed,[99] the famous algebraist uses a very unusual terminology to denote certain pairs of descending diagonals of his "Pascal's triangle." For example, while he refers to the sequence of numbers 1, 3, 6, 10, 15, ..., $n(n+1)/2$ as a *sanjiao duo* (triangular pile),[100] he at the same time refers to the adjacent sequence 1, 4, 10, 20, ... as *sanjiao luoyi duo* (triangular pile one rank lower). However, if we look at his "Pascal's triangle" (with Arabic entries rather than rod-numbers, for ease of reading, see Fig. 15.14), we see that if we move one rank down from 1 we obtain 1, if we move one rank down from $1 + 3$ we obtain 4, if we move one rank down from $1+3+6$ we obtain 10, etc. This property holds in general. In other words, one is led to think that Zhu Shijie had noticed that the sum of successive terms of the first n numbers of a descending diagonal is equal to the number located exactly one rank below the last number of the diagonal in question (for example, in Fig. 15.14: $1 + 2 + 3 + 4 = 10$; $1 + 5 + 15 = 21$, etc.).

---

[96] Yushkevich (1), *1964*, p. 82, note 1.

[97] Needless to say, in Europe, at the time in question, no mathematicians had any scientific reasons for seeking to calculate the sums of series in such a way, since much more appropriate techniques for that were available.

[98] Lucas (1), *1894*, IV, p. 60 ff.

[99] Cf. Hoe (2), *1977*, p. 306.

[100] Cf. Dickson (1), *1920*, II, p. 1.

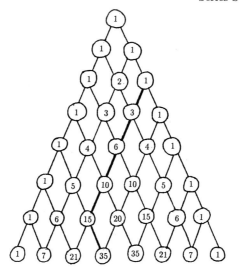

Fig. 15.14

Since we also know that he was aware of the general expression for binomial coefficients, it might be said that, intuitively, he had access to the following summation formula (which translates the visual recurrence shown in the figure for a number of special cases into a general formula)[101]:

$$\sum_{j=k}^{n} \binom{j}{k} = \binom{k}{k} + \binom{k+1}{k} + \ldots + \binom{n}{k} = \binom{n+1}{k+1} .$$

[101]Cf. Comtet (1), *1970*, I, p. 10.

16. Indeterminate Problems

Two categories of indeterminate problem are commonly found in Chinese mathematics:

(i) Problems which translate into systems of linear equations of degree one of type:

$$ax + by + cz = d \quad (= 100)$$
$$x + y + z = d \quad (= 100)$$

(ii) Problems which reduce to systems of simultaneous congruences:

$$x \equiv r_1 \pmod{m_1} \equiv r_2 \pmod{m_2} \equiv r_3 \pmod{m_3} \ldots$$

A problem of the first category is usually referred to as a "hundred fowls problem" even if its statement does not relate to farmyard birds, since the oldest problem of this type is that found in the *Zhang Qiujian suanjing* (Zhang Qiujian's Computational Canon, end of the fifth century AD) which involves cockerels, hens and chickens.[1]

As far as problems of the second category are concerned, these are usually referred to in modern books on number theory as "remainder problems" (since they involve the following question: find a number which when divided by m_1, m_2, m_3, ..., has remainders r_1, r_2, r_3, ..., respectively). Moreover, the solution of these problems is said to depend upon the "Chinese remainder theorem." This expression alludes to an algorithm which first appears in the *Shushu jiuzhang* (Computational Techniques in Nine Chapters, 1247) by Qin Jiushao. Less commonly, some works also refer to "Sunzi's problem," thereby alluding to a remainder problem in the *Sunzi suanjing* (Sunzi's Computational Canon, fourth or fifth century AD).

---

[1] The name "hundred fowls problem" is due to the Belgian historian Louis van Hée. See van Hée (1), *1913*.

The Hundred Fowls Problem

The *Zhang Qiujian suanjing* contains the following problem:

> If cockerels cost 5 *qian* [a *qian* is a copper coin] each, hens cost 3 *qian* each and 3 chickens cost 1 *qian* and if 100 fowls are bought for 100 *qian*, how many cockerels, hens and chickens are there?
>
> Answer: 4 cockerels costing (a total of) 20 *qian*, 18 hens costing 54 *qian* and 78 chickens costing 26 *qian*.
>
> Alternative answer: 8 cockerels, 11 hens and 81 chickens costing 40, 33 and 27 *qian*, respectively.
>
> Alternative answer: 12 cockerels, 4 hens and 84 chickens costing 60, 12 and 28 *qian*, respectively.[2]

From the point of view of elementary algebra, this problem is very easy to solve. It suffices to set:

$$5x + 3y + \frac{1}{3}z \;=\; 100 \qquad\qquad (a)$$

$$x + y + z \;=\; 100 \qquad\qquad (b)$$

Then, writing $z = 100 - x - y$ and substituting this expression for z in (a), we obtain $7x + 4y = 100$; whence $y = 25 - (7/4)x$. Since y must be an integer, it follows that x is a multiple of 4. Thus, $x = 4t$, $y = 25 - 4t$ and $z = 75 + 3t$ $(t = 0, 1, 2, 3)$. Incidentally, we note that Zhang Qiujian gives all these solutions except that corresponding to $t = 0$ (number of cockerels equal to zero).

However, from the point of view of the ancient Chinese techniques, the solution was difficult to justify. For a long time, Chinese authors looked in vain for rational approaches capable of explaining the answer. For example, the commentator Zhen Luan (fl. 570 AD) explains that one should "divide 100 by 9 to obtain the number of hens, then subtract the remainder 1 from the divisor 9 to obtain the number of cockerels" (i.e. 11 hens and 8 cockerels). As U. Libbrecht notes "this method does not make sense at all."[3] In fact, the first correct justification of the hundred fowls problem by a Chinese author was given in 1861.[4]

Consequently, this problem is not of great interest as far as the history of Chinese reasoning processes is concerned except, possibly, to emphasise the limitations of these processes. None the less, the remark is not very significant since similar limitations are also observed, in general, in medieval manuals of the same level as the *Zhang Qiujian suanjing*. Moreover, rather than an elaborate argument, we probably have here a mere mnemotechnic device. However, this

---

[2] QB, II, p. 402 ff.
[3] Libbrecht (2), *1973*, p. 279.
[4] Ibid. p. 278.

sort of problem does merit discussion in a very different context, namely that of the comparative history of mathematics.

As many historians have noted, in the Middle Ages, numerous problems analogous to that of Zhang Qiujian may be found in Indian, Arabic and European mathematics.[5]

The Hundred Fowls Problem

- Alcuin (ca. 735–804).[6] A certain master of a household has 100 people in his service to whom he proposes to distribute 100 bushels of corn: 3 bushels per man, 2 bushels per woman and 1/2 bushel per child. Can anyone say how many men, women and children there were?[7]

- Śrīdharacārya (ca. 850–950).[8] Pigeons cost 3 *rupas* for 5, cranes are 5 *rupas* for 7, swans are 7 *rupas* for 9 and peacocks are 9 *rupas* for 3. For the prince's pleasure, 100 birds costing a total of 100 *rupas* are to be bought.[9]

- Abū Kāmil (ca. 900).[10] You are given 100 drachma with which to buy 100 birds: ducks, hens and sparrows. A duck costs 5 drachma, 20 sparrows cost one drachma and a hen costs one drachma [...].[11]

Of course, these occur in various guises, and may, for example, involve men, woman and children or pigeons, swans and peacocks, etc. instead of cockerels, hens and chickens. However, the two totals each amount to one hundred. It is natural to think that this is explained by a diffusion phenomenon. Nothing is more plausible than to suppose that such an amusing problem stimulated the

---

[5]Tropfke (3), *1980*, p. 615.

[6]See Migne (1), *1851*, tome 101, vol. 3, column 1154. The same text contains 6 other similar problems, 3 of which use the number 100 twice in succession in their statements (no. 5, p. 1145 (boars, sows, piglets); no. 38, p. 1156 (horses, cattle, sheep); no. 39, p. 1156 (camels, donkeys, sheep)). See also Folkerts (1), *1977*.

[7]Alcuin gives a single answer (11, 15, 74). In fact, there are 7 different solutions involving strictly positive integers (since this is a problem of type $x+y+z = 100$, $3x+2y+(z/2) = 100$, we have $x = 3t - 100$, $y = 200 - 5t$ and $z = 2t$ for $t = 34, 35, \ldots, 40$).

[8]Cf. Shukla (1), *1959*, p. 50.

[9]In self-evident notation, this problem translates to $5x + 7y + 9z + 3u = 100$, $3x + 5y + 7z + 9u = 100$. There are 16 solutions involving strictly positive integers.

[10]Cf. Suter (1), *1910*, p. 102.

[11]This problem corresponds to $x + y + z = 100$, $5x + y/20 + z = 100$. Abū Kāmil solves it correctly by a technique of algebra in words in which he calls the unknown number of ducks the "thing."

imagination of many people, and thus spread far from its place of origin by word of mouth. But, as in many instances already noted in this book, so too in this case, it is difficult to pin down the direction of diffusion and its chronology.

The Remainder Problem

Exactly as in the case of the hundred fowls problem the way in which Chinese authors tackle the remainder problem does not tell us anything about Chinese arithmetical reasoning (although in the Middle Ages, the absence of proofs for these problems is a general rule, both in China and elsewhere). However, while in the first case (hundred fowls problem) the texts only contain numerical answers without logically consistent resolutory rules, this is not so in the second case (remainder problem), where the solutions are described by perfectly correct rules, albeit without justifications. Thus, we shall consider the most representative of these rules, namely that due to Sunzi and that due to Qin Jiushao.

Sunzi's Rule

The *Sunzi suanjing* contains the following problem:

Suppose we have an unknown number of objects. If they are counted in threes, 2 are left, if they are counted in fives, 3 are left and if they are counted in sevens, 2 are left. How many objects are there?

Answer: 23.

Rule: If they are counted in threes, 2 are left: set 140. If they are counted in fives, 3 are left: set 63. If they are counted in sevens, 2 are left: set 30. Take the sum of these [three numbers] to obtain 233. Subtract 210 from this total; this gives the answer.

In general: For each remaining unit when counting in threes, set 70. For each remaining unit when counting in fives, set 21. For each remaining unit when counting in sevens, set 15. If [the sum obtained in this way] is 106 or more, subtract 105 to obtain the answer.[12]

According to these directives, the desired number is obtained by calculating:

$$70 \times 2 + 21 \times 3 + 15 \times 2 - 210 = 23 \,. \tag{1}$$

23 is actually a solution of this problem. Moreover, the second part of the rule also indicates how to solve the same problem in the slightly

---

[12] *Sunzi suanjing*, 3-26 (QB, II, p. 318).

more general case of arbitrary remainders r_1, r_2 and r_3. In this case, we must calculate:

$$x = 70r_1 + 21r_2 + 15r_3 - 105n\,, \tag{2}$$

where r_1, r_2 and r_3 denote the remainders when counting in threes, fives and sevens, respectively (i.e. on division by 3, 5, 7, respectively).

Since the general form of this problem is:

$$x \equiv r_1 \pmod{m_1}$$

$$x \equiv r_2 \pmod{m_2} \tag{3}$$

$$x \equiv r_3 \pmod{m_3}$$

readers will observe that the above calculations are the same as those to be executed according to the Chinese remainder theorem which is found in most modern books on elementary number theory (see the statement of the theorem on page 312).

Surprisingly, the next known examples of remainder problems in a Chinese text occur eight centuries after the supposed date of the *Sunzi suanjing*. The *Xugu zhaiqi suanfa* (1275) by the arithmetician Yang Hui contains five versions of this problem.[13] In each case, the rules used are comparable in level with those of the *Sunzi suanjing*.[14]

Outside the Chinese world, there is evidence for the remainder problem in the Islamic countries, India and Europe.

For example, the historian of Arabic mathematics A.S. Saidan showed that Sunzi's problem is to be found in the works of an 11th-century Arabic author, Ibn Tāhir, with the same moduli, but with arbitrary remainders.[15] However, unlike Sunzi, this author attempts to justify the procedure using a technique which involves reducing a number modulo a given number for checking the correctness of an arithmetical operation.

The mathematician and physicist Ibn al-Haytham (965–1040) proposes solving the system

$$x \equiv 1 \pmod{m_i}, \quad x \equiv 0 \pmod{p}$$

with p prime and $1 < m_i \le p - 1$.

---

[13] Translation in Lam Lay-Yong (6), *1977*, pp. 151–153.
[14] Libbrecht (2), *1973*, pp. 283–285.
[15] Saidan (1), *1978*, p. 477 ff.

The Chinese Remainder Theorem

(a) Suppose that the positive integers $m_1, m_2, \ldots, m_k$ are pairwise coprime; in other words $\mathrm{hcf}(m_i, m_j)=1$ for all i, j where $i \neq j$, where hcf denotes the highest common factor. Then the set of congruences

$$x \equiv r_i \pmod{m_i} \text{ for } i = 1, 2, \ldots k \qquad (*)$$

has a unique common solution modulo M where $M = m_1 m_2 \ldots m_k$.

Set
$$M = m_1 M_1 = m_2 M_2 = \ldots = m_k M_k$$

Since M_i and m_i are coprime we can find integers $\mu_1, \mu_2, \ldots, \mu_k$ such that $M_1 \mu_1 \equiv 1 \pmod{m_1}, \ldots, M_k \mu_k \equiv 1 \pmod{m_k}$ and the general solution of $(*)$ is

$$x \equiv M_1 \mu_1 r_1 + M_2 \mu_2 r_2 + \ldots + M_k \mu_k r_k \pmod{M} \qquad (**)$$

(b) More generally, a necessary and sufficient condition for the system of simultaneous congruences $(*)$ to be soluble is that

$$r_i - r_j \equiv 0 \pmod{\mathrm{hcf}(m_i, m_j)}$$

for all pairs i, j with $i \neq j$.

(c) If condition (b) is satisfied, it is always possible to replace the original set $(*)$ of congruences by another equivalent set of simultaneous linear congruences

$$x \equiv r_i \pmod{m_i'}$$

the moduli of which are pairwise coprime. Then, for all i, $1 \leq i \leq k$, m_i' divides m_i and $\mathrm{lcm}(m_1', \ldots, m_k') = \mathrm{lcm}(m_1, \ldots, m_k)$, where lcm denotes the least common multiple.

　　A proof of (a) appears in almost every book on elementary number theory. (b) is dealt with, for example, in K. Malher, "On the Chinese Remainder Theorem," *Mathematische Nachrichten*, 1958, vol. 18, pp. 120–122 and in A. S. Fraenkel, "New Proof of the Generalized Chinese Remainder Theorem," *Proc. Am. Math. Soc.*, 1963, vol. 14, no. 4, pp. 790–791. A constructive proof of (c) akin to Qin Jiushao's original algorithm (i.e. relying only on the computation of hcfs and not on prime number decompositions) is contained in R. J. Stieltjes, *1918, Oeuvres Complètes*, Groningen: P. Noordhoff, vol. II, p. 280 ff. and p. 295 ff.

This is a special case of the Chinese remainder theorem. Stating that the system admits an infinity of solutions. Ibn al-Haytham proposes two methods of resolution. The first of these is

$$x = (p - 1)! + 1$$

and the fact that it gives a solution of the above system is a consequence of Wilson's theorem, a theorem in fact established by Ibn al-Haytham[16] (Wilson's theorem asserts that

$$(p - 1)! \equiv -1 \pmod{p}$$

for any prime p).

Finally, we note that Leonardo Fibonacci's *Liber Abaci* (1202) also contains Sunzi's problem, formulated with the same moduli and the same resolutory rule.[17]

We shall not continue the long enumeration of all the sources from the Middle Ages to the modern period which mention this problem; instead, we refer readers to the very detailed studies by Libbrecht (2), *1973*, p. 214 ff. and Tropfke (3), *1980*, p. 636.

Qin Jiushao's General *dayan* Rule *dayan zongshu shu*

The expression *dayan zongshu shu* (General *dayan* rule) is the name of the general rule for solving systems of simultaneous congruences which is to be found immediately after the statement of and answer to the first problem of chapter 1 of the *Shushu jiuzhang*.

Exactly as in the previous case, this rule is intended for solving problems similar to that of Sunzi (system of simultaneous congruences) but with a totally different degree of complexity and sophistication to that of Sunzi's problem. As, U. Libbrecht wrote:

> The only progress [with respect to Sunzi's rule] [...] was the *ta-yen* [*dayan*] rule of Ch'in-Chiu shao. We should not underestimate this revolutionary advance, because from this single remainder problem, we come at once to the general procedure for solving the remainder problem, even more advanced (from the algorithmical point of view) than Gauss's method, and there is not the slightest indication of gradual evolution.[18]

The nine artificial problems described by Qin Jiushao involve divination, the calendar, finances, military logistics, architecture and excavation work. These problems are factually very intricate.

The *dayan* rule is itself complex. This complexity relates partly to the length of the rule and partly to the nesting of the sub-rules which compose it.[19]

---

[16] Rashed (1), *1984*, pp. 227 ff.

[17] Libbrecht, op. cit., p. 236 ff.

[18] Libbrecht (1), *1972*, p. 183; Tropfke (3), *1980*.

[19] Partial translation of the rule in Libbrecht (2), *1973* pp. 328 ff.

This rule consists of 873 written characters[20] including 701 standard-size characters and 172 small-size characters. The former constitutes the main text of the rule itself and the latter represents notes on the main text, which, for example, cross-reference particular problems of the text, provide a numerical example, or give details of a particular instance of application.

The main text of the *dayan* rule essentially comprises the following stages:

(a) Conversion of the moduli into integers.

(b) Reduction of the moduli into pairwise coprime moduli.

(c) Solution of congruences of the type $ax \equiv 1 \pmod{m}$ (*dayan* rule for searching for unity).

(d) Completion of the calculations as above (congruence $**$) (Table p. 312).

From a modern point of view, there is no apparent need for the conversion of moduli into integers, but Qin Jiushao and his contemporaries did not exactly operate on integers like those the theory of numbers has accustomed us to but on numbers with units attached, which frequently involved fractions. Qin Jiushao distinguishes four types of number:

- *yuanshu* "primordial numbers" (integers).

- *shoushu* "gathered numbers," i.e. "those for which the tail (*wei*) contains *fen* and *li*"(since the *fen* and *li* are decimal units which are submultiples of a main unit, this refers to decimal numbers). The idea behind the term *shoushu* is probably that when assembled the one with the others, the various decimals form a homogeneous number rather than a disparate collection, as in the case of disparate numbers composed of integers (relative to a certain unit) and fractions.

- *tongshu* "communicating numbers" (or fractions), so-called because they are able to "communicate" *tong* with one another by addition (allusion to the operation of addition of fractions in the *Jiuzhang suanshu*).

- *fushu* "complex numbers," that is, those for which the tail contains ten, one hundred, one thousand (multiples of 10^n). These numbers had to be considered because, when converting the initial data to integers, the numbers had to be expressed in terms of the smallest unit of measurement possible, thus zeros had to be added to certain numbers.

The second stage of the *dayan* rule is much more intricate and its terse formulation has given rise to many, often contradictory, tentative interpretations.[21] These interpretations may be essentially divided into two families depending on whether the explanation is based on the decomposition of the moduli into prime factors or on the calculation of the hcf of the moduli.

---

[20]Counted in the *Yijiatang* edition (1842).

[21]Cf. Wu Wenjun (3'), *1987*. See also: Mei Rongzhao (8'), *1987*; Wang Yusheng (2'), *1987*; Wang Yixun (1'), *1987*.

As various historians[22] have recently remarked, explanations based on decomposition into prime factors raise a fundamental problem since neither Qin Jiushao nor his Chinese predecessors knew of the notion of prime numbers.

Conversely, explanations of the second type appear much more convincing since the notion of the hcf is present throughout the *Shushu jiuzhang* and was already known to the authors of the *Jiuzhang suanshu* of the Han. However, that does not mean to say that Qin Jiushao uniformly used a single technique to solve all his problems.

Without going into the detail of the particular procedures of each of the problems, we shall now consider the example of the third problem of chapter 1 of the *Shushu jiuzhang* which has the advantage of having a fully explained working with particular numbers.[23]

The moduli of this problem are 54, 57, 75 and 72. The process begins with the search for the hcf of these four numbers

$$\text{hcf}\,(54, 57, 75, 72) = 3$$

which is calculated using the method of repeated subtraction. The result found is then used to "reduce" *yue* all these moduli (by dividing by the hcf), except the smallest:

$$(54, 57, 75, 72) \rightarrow (54, 19, 25, 24)\,.$$

The reduction is followed by the search for the hcf of the new numbers, by successive consideration of all pairs with first element 24, then 25, and finally 19. If $\text{hcf}(x, y) = d \neq 1$ this pair is replaced by $(x, y/d)$. After all the pairs have been reviewed another pass is made during which the reduction becomes slightly more complicated: if d is not equal to 1, the pair (x, y) is replaced by $(x/d, yd)$. In the present case, this technique corresponds to the following calculations[24]:

$\text{hcf}\,(24, 25) = 1$
$\text{hcf}\,(24, 19) = 1$
$\text{hcf}\,(24, 54) = 6$ whence $(24, 54) \rightarrow (24, 9)$ and $(54, 19, 25, 24)$
$$\rightarrow (9, 19, 25, 24)$$

$\text{hcf}\,(25, 19) = 1$
$\text{hcf}\,(25, 9) \ = 1$
$\text{hcf}\,(19, 9) \ = 1$
$\text{hcf}\,(24, 9) \ = 3$ whence $(24, 9) \rightarrow (8, 27)$ and $(9, 19, 25, 24)$
$$\rightarrow (27, 19, 25, 8)$$

The final values of these reduced moduli are $(27, 19, 25, 8)$.

---

[22]Wu Wenjun, op. cit.
[23]*Shushu jiuzhang*, j. 1, pp. 18a–b, *Yijiatang* edition of the text.
[24]Wu Wenjun, op. cit, p. 233.

Subject to a number of readjustments, this process corresponds to the algorithm for reducing the moduli described by Li Jimin in his systematisation of Qin Jiushao's procedure.[25] This algorithm may be described in the programming language C as follows (where the yuan[i] are the initial moduli):

```
void modred (yuan, number)
int yuan[], number;
{
int i, j, d;
for (i=1; i<number; i ++)
{
for (j=i+1; j<=number; j++ )
{
yuan[j] = yuan[j]/(hcf(yuan[i],yuan[j]));
while (hcf(yuan[i],yuan[j]) != 1 )
{
d = hcf(yuan[i],yuan[j]);
yuan[i] = yuan[i]/d;
yuan[j] = yuan[j]*d;
}
}
}
}
```

This program fragment may be used to reduce the same moduli as those of the above example. However, the reduction is not unique and depends on the particular order in which the moduli are considered; for example, the initial moduli $(72, 105, 77, 399)$ lead to the following reductions, all of which are correct: $(72, 35, 11, 19)$, $(72, 5, 11, 133)$, $(72, 5, 77, 19)$.

Once the moduli are reduced, Qin Jiushao uses the procedure *dayan qiuyi shu* (*dayan* procedure for searching for unity) This procedure may be used to solve congruences of the type

$$ax \equiv 1 \pmod{m}$$

where a and m are given with $\mathrm{hcf}(a, m) = 1$ and x is unknown, which arise during the explicit solution of the Chinese remainder theorem after the stage of the reduction of the moduli.

In Qin Jiushao's terminology, the coefficient a of the unknown and the modulus m of the congruence are called the surplus (*qi*) and the "mother of the expansion" (*yanmu*), respectively. The term surplus is used because a is in fact a remainder modulo m; prior to the calculations, a was chosen to be as small as possible. The *yan* in *yanmu* reminds us that we are still in the framework of the *dayan* rule. Finally, use of the term *mu* (lit. the mother, but according to traditional Chinese terminology, the denominator) alludes to the

---

[25] Cf. Li Jimin, in Wu Wenjun, (op. cit.), p. 232.

fact that the calculations have something to do with calculations on fractions. Indeed, as we shall see, the resolutory procedure proposed by Qin Jiushao corresponds point by point to what for us would be a technique for expanding m/a as a continued fraction, involving both the calculation of the hcf of m and a (Euclidean algorithm, or more correctly, the procedure of alternating subtractions (antiphairesis) used in the *Jiuzhang suanshu* of the Han) and the explicit determination of a positive solution of the indeterminate equation $ax - my = 1$, equivalent to the congruence $ax \equiv 1 \pmod{m}$.

In practice, the instructions of the *dayan qiu yi shu* rule begin by placing the numbers a and m of the congruence to be solved above one another, as in the following table.[26]

Place the surplus *qi* at TOP-RIGHT and the DE-TERMINED MOTHER *dingmu* [i.e. the modulus] at BOTTOM-RIGHT. Set one PRIMORDIAL CELESTIAL UNIT *tianyuan yi* at TOP-LEFT and a void *kong* [i.e. a zero] at BOTTOM-LEFT.

1	a
0	m

Then divide m by a

$$m = aq_1 + r_1$$

and calculate

$$P_1 = q_1 . 1 = q_1 .$$

The results are positioned as shown in the table.

1	a
P_1	r_2
	q_1

Then divide a by r_2

$$a = r_2 q_2 + r_3$$

and calculate

$$P_2 = q_2 P_1 + 1 .$$

The results are positioned as shown in the table.

	q_2
P_2	r_3
P_1	r_2

The alternating divisions are continued as in the Euclidean algorithm until the process stops when the number in the place of the surplus becomes equal to 1; whence the name of the rule (search for unity). Note that the final position of q depends upon the parity of n.

	(q_n)
P_{n-1}	$r_n = 1$
P_{n-2}	r_{n-1}
	(q_n)

By interpreting the terse text of this algorithm using the numerical examples described explicitly in the original text of the *Shushu jiuzhang*, we may reconstitute the successive stages as follows[27]:

---

[26]The following explanations are essentially those of the text of the *Shushu jiuzhang*, j. 1, pp. 5a–b.

[27]Here, we adopt the notation of Vinogradov (1), *1954*, p. 11 and p. 62.

$$P_0 = 1 \quad tianyuan\ yi$$

$$m = aq_1 + r_2 \qquad P_1 = q_1$$

$$a = r_2q_2 + r_3 \qquad P_2 = q_2P_1 + P_0$$

$$r_2 = r_3q_3 + r_4 \qquad P_3 = q_3P_2 + P_1$$

$$\cdots \qquad \cdots$$

$$r_i = r_{i+1}q_{i+1} + r_{i+2} \qquad P_{i+1} = q_{i+1}P_{i-1} + P_{i-2}$$

$$\cdots \qquad \cdots$$

$$r_{n-2} = r_{n-1}q_{n-1} + r_n \qquad P_{n-1} = q_{n-1}P_{n-2} + P_{n-3}$$

with $r_n = 1$.

Here, the P_i are the numerators of the convergents in the expansion of m/a as a continued fraction; P_{n-1} is the penultimate convergent of this expansion.

This P_{n-1} plays a crucial role here, since elementary number theory tells us that the desired solution is precisely[28]:

$$x \equiv (-1)^{n-1}P_{n-1} \pmod{m}.$$

Yet this result is meaningless as far as Qin Jiushao is concerned, since the mathematics of his period did not permit negative numbers as solutions of equations.

According to Qian Baocong's interpretation[29] taken up again by Li Jimin, Yuan Xiangdong and Li Wenlin,[30] Qin Jiushao's solution may be expressed as follows:

- If n is odd, $x = P_{n-1}$
- If n is even, $x = (r_{n-1} - 1)P_{n-1} + P_{n-2}$.

The solution x thus obtained is the smallest positive solution of the congruence $ax \equiv 1 \pmod{m}$. The procedure is particularly well adapted to the effective calculations, since the P_i are always less than the modulus m. It is also useful to compare this technique with that based on Euler's theorem:

$$x \equiv a^{\phi(m)-1} \pmod{m}.$$

Unlike the above technique, this purely theoretical solution is of little interest as far as the effective calculation of the solutions is concerned, since it requires a knowledge of the decomposition of m as a product of prime factors.

By way of illustration, the table below shows the sequential numerical values which occur when solving the congruence

$$377873x \equiv 1 \pmod{499067},$$

which Qin Jiushao treats in full.[31]

---

[28] Ibid., p. 62.
[29] QB, *Hist.*, p. 207.
[30] Wu Wenjun, op. cit., pp. 140 and 166.
[31] *SSJZ*, j. 3, p. 16a–20a.

s	0	1	2	3	4	5	6	7	8	9	10
q_s		1	3	8	2	12	3	1	1	6	11
r_s			121194	14291	6866	559	158	85	73	12	1
P_s	1	1	4	33	70	873	2689	3562	6251	41068	499067
Q_s	0	1	3	25	53	661	2036	2697	4733	31095	377873

In the table, the P_s, q_s and r_s are defined as above; moreover, $Q_0 = 0$ and $Q_s = q_s Q_{s-1} + Q_{s-2}$, the $\frac{P_s}{Q_s}$ are the convergents of the expansion of $\frac{377873}{499067}$ in continued fractions. Here $r_{10} = 1$ and the solution is obtained by taking $x = (r_9 - 1)P_9 + P_8 = 11 \cdot 41068 + 6251 = 457999$.

The final stage of the calculations needs no special commentary.

We shall now consider the problem II-1 of the *Shushu jiuzhang* and its full numerical solution:

Suppose we have three first class peasants (*shang nong san ren*). The yields of their paddy-fields, when making use of the full *dou* [a dry measure for grain], are all equal. All of them go to different places to sell it. After selling his rice on the official market of his own prefecture A (*jia*) is left with 3 *dou* and 2 *sheng*.[32] After selling his rice to the villagers of Anji[33] B (*yi*) is left with 7 *dou*. After selling his rice to a middleman from Pingjiang[34] C (*bing*) is left with 3 *dou*. How much rice did each peasant have initially and how much did each one sell?

Answer: total amount of rice: 738 *dan*[35] to be divided among 3 men or 246 *dan* each; amount of rice sold by A: 296 *dan*; by B: 223 *dan*; by C: 182 *dan*.

[...] Note: the *hu* of the Crafts Institute[36] is worth 83 *sheng*, that of Anji is worth 110 *sheng* and that of Pingjiang is worth 135 *sheng*.

(Note: 7 *dou* = 70 *sheng* and 3*dou* = 3 *sheng*)

Thus, we have

$$x \equiv 3\ dou\ 2\ sheng \qquad (\text{mod } 83\ sheng)$$
$$x \equiv 7\ dou \qquad (\text{mod } 110\ sheng)$$
$$x \equiv 3\ dou \qquad (\text{mod } 135\ sheng)$$

where x denotes the amount of rice possessed by each peasant. Qin Jiushao's solution is the following:

---

[32] 1 *dou* = 10 *sheng* (measures of dry capacity).

[33] Name of a prefecture in Zhejiang province.

[34] Name of a prefecture in Hunan province.

[35] Unit of capacity (1 *dan* = 10 *dou*).

[36] The Crafts Institute (*wensi yuan*) is a eunuch-staffed workshop for production of jewelry, fine brocades, etc., for use by the Emperor and his wives." (Cf. Hucker (1), *1985*, no 7724, p. 568).

[Solution] with the *dayan* method: set the "models" *lü* of the local administrations [i.e. set 83, 110 and 135] as primordial numbers [*yuanshu* – stage (i) p. 321]. Find their common "equal" [hcf]. Since these numbers cannot all be simplified at the same time they are simplified separately. Find the "equals" two by two in chain. Simplify the odds, but not the evens.[37] If in certain cases, number of a same category remain, find again their "equals" and multiply them back by them. Obtain the determined mothers [*dingmu*[38] – stage (ii) p. 321]. The product of these mothers gives the mother of the expansion [stage (iii) p. 321]. The mutual products give the "numbers of the expansion" *yanshu* [stage (iv) p. 321]. Suppress that which "fills" the "determined mothers"[39] [stage (v) p. 321]. Find the units [stage (vi); solution of congruences]. Deduce the "multiplicative models" *cheng lü* [the solutions of these congruences]. Multiply these by the "numbers of the expansion" to give the "useful numbers" *yongshu* [stage (vii)]. Multiply the rice remainders [stage (viii)] by the corresponding useful numbers. Add the results [stage (ix)]. Suppress that which fills the mother of the expansion *yanmu* [i.e. calculate the sum modulo M]. This gives the parts of rice [all equal] times the number of men, whence the total amount of rice.

In more familiar terms, Qin Jiushao first notes that the hcf of 83, 110 and 135 taken together is 1. He then calculates the hcfs of 83 and 110, 110 and 135 and 83 and 135. To do this, he carries out the alternating subtractions until he obtains two equal numbers. For example, for 135 and 110: $135 - 110 = 25$; $110 - 25 = 85$; $85 - 25 = 60$; $60 - 25 = 35$; $35 - 25 = 10$; $25 - 10 = 15$; $15 - 10 = 5$; $10 - 5 = 5$. The results of the last two subtractions are equal, consequently the hcf of 135 and 110 is equal to this number 5.

Dividing 135 by 5, Qin Jiushao replaces the pair (110, 135) by (110, 27). He then obtains a triplet of mutually prime moduli: 83, 110 and 27.

Next, he carries out the calculations summarised by the table page 321. These correspond to the modern algorithm of the Chinese remainder theorem except that Qin Jiushao gives only one solution, rather than infinitely many. This is not an indeterminate problem as far as he is concerned (unless one interprets the fact that Qin Jiushao qualifies his solution by *lü* to mean that there are potentially infinitely many solutions).[40]

This type of highly sophisticated mathematics which "falls from the heavens" may leave one perplexed.

Did Qiu Jiushao have innate knowledge (he was a genius) as G. Sarton believed?[41] Conversely, did he only borrow his rule from others (for example,

---

[37]This expression is difficult to interpret. In this context, "even" seems to be the opposite of "odd" not in the sense of multiples and non-multiples of 2, but in that of the singular as opposed to the multiple. Cf. Li Jimin's remark in Wu Wenjun (3'), *1987*, note 1, p. 222.

[38]In general, "mother" means "denominator" but there is a possibility of a shift in meaning from "denominator" to "divisor." Here, the "determined mothers" are the moduli of the congruences which have been made pairwise coprime.

[39]"Suppress" means "reduce modulo M_i."

[40]See above, p. 258.

[41]G. Sarton writes that Qin Jiushao was "one of the greatest mathematicians of his race, of his time and indeed of all times." (quoted by Ho Peng-Yoke in the notice Ch'in Chiu-shao in the *DSB*).

		Stages in the Solution of the System of Simultaneous Congruences (According to Qin Jiushao (see above, p. 319 ff.))			

$$x \equiv 32 \quad (\mathrm{mod}\,83)$$
$$x \equiv 70 \quad (\mathrm{mod}\,110)$$
$$x \equiv 30 \quad (\mathrm{mod}\,135)$$

	Terminology		Numerical values			Remarks
(i)	*yuanshu* Primordial numbers y_i	元 數	83	110	135	Moduli (in the current sense)
(ii)	*dingmu* Determined mothers m_i	定 母	83	110	27	Coprime moduli m_i
(iii)	*yanmu* Mother of the expansion M	衍 母	246510			M = product of the m_i
(iv)	*yanshu* Numbers of the expansion M_i	衍 數	2970	2241	9130	$M_i = M/m_i$
(v)	*qishu* Surplus N_i	奇 數	65	41	4	$M_i = \mathrm{mult}.m_i + N_i$ (N_i = remainder after division of M_i by m_i)
(vi)	*cheng lü* Multiplicative numbers	乘 率	23	51	7	The μ_i are the smallest positive solutions of the congruences ($N_i\mu_i \equiv 1 \bmod m_i$)
(vii)	*yongshu* Useful numbers F_i	用 數	68310	114291	63910	$F_i = M_i\mu_i$
(viii)	*yu* Remainders r_i	餘	32	70	30	Remainders of the initial congruences
(ix)	*zong* Totals T_i	總	2185920	8000370	1917300	$T_i = F_i r_i$
(x)	*zongshu* Sum S	總 數	12103590			Sum of T_i
	Reduce 12 103 590 modulo 246 510 [43] (12103590 = 49 × 246510 + 24600) whence the solution of the system of congruences given in the text: $x = 24600$					

---

[43]This number is the product of the reduced moduli (27 × 83 × 110)

Indian mathematicians), as the German historian H. Hankel has suggested?[43] Or, did he simply construct his examples backwards, beginning with the answers and ending with the questions?

The first supposition would settle the difficult question of the origin of the *dayan* rule without further ado. Everything would be explained by a miracle due to the intervention of an extraordinary demiurge capable of mastering on his own a complicated mathematical phenomenon which would only be understood several centuries later.

When one considers the way in which Qin Jiushao's contemporaries and even his successors, including Fibonacci (1202), Regiomontanus (ca. 1460), Beveridge (1669), Van Schooten (1657), Euler (1743), Gauss (1801) and many others[44] solved the remainder problem one can but marvel at the consistency and generality of the *dayan* algorithm. After the Greek miracle, here we have the Chinese miracle.

It is certainly true that, historically, exceptional individuals have existed. However, as it stands, this explanation is still very unsatisfactory. In the same way that historians of science have begun to perceive the genesis of Greek mathematics and the emergence of the concept of proof, a rational explanation of how the *dayan* rule came into existence is now required.

On this matter, current research stresses the probable importance of questions of calendrical astronomy. One central theme of the Chinese calendar is the determination of the origin of certain natural and artificial periodic time cycles such as the year, the month and the artificial 60-day cycle. Let x denote the number of days separating the beginning of the cycles from an instant for which r_1 days of the yearly cycle (Y), r_2 days of the lunar cycle (L) and r_3 days of the sexagesimal cycle (S) have elapsed since these cycles last began. Then we have

$$x \equiv r_1 \pmod{Y}$$
$$x \equiv r_2 \pmod{L}$$
$$x \equiv r_3 \pmod{S}$$

It is reasonable guess to conjecture that information on the question is buried somewhere in the immense corpus of Chinese treatises on calendrical astronomy. The probability of an interpretation of such sources, however, in a way which would help us to understand the logical genesis of the *dayan* rule seems very weak for these are merely composed of compilations of prescriptive rules.

---

[43]Libbrecht (2), *1973*, p. 359.

[44]In 1202, Fibonacci (Leonardo of Pisa), who is usually thought of as the greatest mathematician of the Middle Ages proposed a solution of the remainder problem similar to that due to Sunzi, in other words, incomparably less powerful than that due to Qin Jiushao. See his *Liber Abaci*, vol. 1, p. 304 (Rome, 1857, ed. Boncompagni): "suppose we ask someone to think of a number, divide it by 3, 5 and 7 and tell us the remainder after each division. We retain 70 times the number of units remaining after division by 3, 21 times the number of units remaining after division by 5 and 15 times the number of units remaining after division by 7 and successively subtract 105 from the number found while this remains greater than 105. The result will be the number first thought of." (See also Libbrecht, op. cit., p. 231 ff.).

Thus, the hypothesis of a foreign influence seems more likely since the problem of remainders is well documented in Indian mathematics (problems relating to astronomical cycles).[45] In fact, Qin Jiushao, who does not claim to be the inventor of the *dayan* rule, explains that he learnt it from the calendarists working in the Bureau of Astronomy at Hangzhou. He also says that the latter used the rule without understanding it.

However, according to Libbrecht,[46] the Indian rules for solving simultaneous congruences are not based on the same overall principle as that of the *dayan* rule. Thus, this supposition must be rejected. The Indian rules of the *kuṭṭaka* (pulveriser) are based on both the Euclidean algorithm (but applied not in the same way as in the *Shushu jiuzhang*) and algebraic substitutions. Here is a brief explanation of the principle:

Given

$$a_1 x_1 + r_1 = a_2 x_2 + r_2 = \ldots = a_n x_n + r_n$$

the technique begins with a search for the smallest value of x_1 satisfying the indeterminate equation

$$a_1 x_1 + r_1 = a_2 x_2 + r_2$$

based on the Euclidean algorithm. Thus, we have $x_1 = \alpha_1 + a_2 t$. This is then followed by the substitution $a_1(\alpha_1 + a_2 t) + r_1 = a_3 x_3 + r_3$, leading to another indeterminate equation to which the same process is applied, and so on, step by step.

Finally, the hypothesis of a discovery of the solution of the remainder problem by deriving artificially the problem from its answer appears very unlikely, since in individual cases of the remainder problem there is no simple relationship between the path from the data to the solution and that from the solution to the data. This is not a banal algebraic problem in which simply replacing multiplication by addition and addition by subtraction (and conversely) would enable us to follow the path in both directions. The question of the origin of Qin Jiushao's *dayan* rule is thus left open.

---

[45] Cf. Datta and Singh (1), *1935*, I, p. 134 ff.
[46] Libbrecht, op. cit., p. 362 ff.

17. Approximation Formulae

Geometrical Formulae

Area of an Arbitrary Quadrilateral

The problems I-14 of the *Wucao suanjing*[1] and I-6 of the *Xiahou Yang suanjing*[2] both describe, without commentary, a procedure for calculating the area of a field with four unequal sides (*si bu deng tian*), which involves taking the product of the half sums of the opposite sides (which we denote by a, b, c and d, respectively):

$$S = \frac{a + b}{2} \cdot \frac{c + d}{2} \ .$$

This coarse formula always gives an overestimate,[3] except, of course, if the quadrilateral in question is a rectangle (in which case it is of no further interest). Moreover, given that a specific quadrilateral cannot by determined from the lengths of its sides alone, since, of course, it is deformable, the formula is logically flawed. However, it at least has the advantage of simplicity relative to other imaginable methods of calculation, which doubtless explains its prolonged success among surveyors.

It already occurs in certain Babylonian mathematical texts,[4] in Egyptian mathematics[5] and, later, in the writings of Roman surveyors,[6] in Alcuin's *Propositiones ad acuendos juvenes*,[7] and in Gerbert's geometry.[8]

In addition to the above references, we also note that Yang Hui, an arithmetician of the late 13th century, quotes this formula in order to criticise its inaccuracy and proposes replacing it by a decomposition of the quadrilateral in question into two triangles.[9]

---

[1] QB, II, p. 414.

[2] QB, II, p. 569.

[3] For an elementary proof of this property see (for example), Bruins and Rutten (1), *1961*, p. 5.

[4] See the Babylonian text *YBC* 4675 (in Bruins and Rutten, op. cit.).

[5] Gericke (1), *1984*, p. 35.

[6] Ibid., p. 35; Mortet (1), *1896*. On the Roman surveyors see also Dilke (1), *1971*.

[7] Migne (1), *1851*, vol. 101, Tome 3, column 1151, problem no. 23 (*Propositio de campo quadrangulo*). See also Gericke (2), *1990*, p. 65.

[8] Bubnow (1), *1899*, p. 353.

[9] Cf. Lam Lay-Yong (6), *1977*, p. 111.

Area of a Segment of a Circle

In the *Jiuzhang suanshu*, the area of a segment of a circle (*hu tian*) (field in the shape of a bow), determined from its chord b (*xian*) and its arrow h (*shi*) (Fig. 17.1), is calculated using the prescription:

Take chord times arrow then square the arrow, add [the results], of two [parts] the one.[10]

Or:
$$S = \frac{bh + h^2}{2}.$$

Liu Hui's corresponding commentary explains this formula by means of a geometrical figure which involves a circle and its circumscribed square, its inscribed regular dodecagon as well as certain coloured areas whose exact nature seems difficult to ascertain since all the figures of the *JZSS* are lost. A possible reconstruction of the underlying reasoning assumes that the area of the half disc may be approximated by default by that of half the regular inscribed dodecagon, which Liu Hui knows to be equal to exactly 3/4 of the area of the square which circumscribes it.[11] Thus:

$$\text{Area of the semi-dodecagon} = \frac{1}{2}\frac{3d^2}{4} = \frac{1}{2}\left[d\left(\frac{d}{2}\right) + \left(\frac{d}{2}\right)^2\right]$$

hence the desired formula follows by replacing the diameter d by the base of the segment b, and the radius $d/2$ by the arrow h.[12]

One may wonder about the degree of accuracy of this formula.

Given that the exact formula for the area of the segment of the circle is

$$S' = 2\left(\frac{b^2 + 4h^2}{8h}\right)^2 \arctan\frac{2h}{b} - \frac{b(b^2 + 4h^2)}{16h} + \frac{bh}{2},$$

it can be shown that the relative error $\delta = (S' - S)/S'$ arising from the use of the Chinese formula is

$$\delta = \frac{(1 + e^2)^2 \arctan e - e(1 + e^2 + e^3)}{(1 + e^2)^2 \arctan e - e(1 - e^2)}$$

where $e = \frac{2h}{b}$.

But, e is related to the angle α subtended at the centre by the chord b (Fig. 17.1) by the equation $\tan(\alpha/4) = e$. Thus, the error δ depends only on this angle.

---

[10] JZSS 1-36 (QB, I, p. 109). A partial translation of Liu Hui's commentary is given in Schrimpf (1), *1963*, p. 252. See also Bai Shangshu (3'), *1983*, p. 56 ff.

[11] As we have already seen (see above, p. 278), Liu Hui knows that the area of the inscribed regular dodecagon = 3 × side of the regular inscribed hexagon × radius. Since he also knows that the side of the hexagon is equal to the radius, he deduces that the area of the dodecagon is $3/4\,d^2$.

[12] To simplify the description, we describe the arguments in algebraic form here, but in fact, Liu Hui's arguments are geometrical.

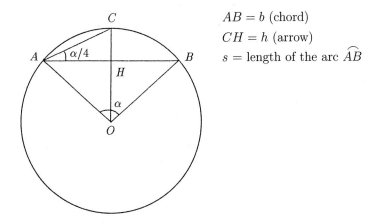

AB = b (chord)

CH = h (arrow)

s = length of the arc $\overset{\frown}{AB}$

Fig. 17.1. Figure relating to the area of a segment of a circle and to the arc-length of a circle.

If we study the variation of δ as a function of e (or, equivalently, as a function of α), we see that for $0 \leq \alpha \leq 180°$, $0.045 \leq \delta < 0.25$. This means that the relative error in the value of the area of segments for which the arrow is shorter than or equal to the radius[13] is never less than 4.5% and may approach 25%. We also note that the error becomes less as the segment approximates to the half disc.

However, the Chinese authors acted as though their formula e were generally applicable. The examples of applications given in the *Jiuzhang suanshu*[14] and the *Zhang Qiujian suanjing*[15] give poor results since in both cases, the corresponding errors are of the order of 14%.[16]

It is true that Liu Hui did realise that something was wrong, because he explains how to refine the procedure by approximating the area of the segment of the circle using a sequence of inscribed polygons.[17] But, since his calculations do not lead to anything simple, his successors continued to apply the inaccurate formula for a long time. However, in 1303, Zhu Shijie in his *Siyuan yujian* uses the following improved version[18]:

$$S = \frac{1}{2}\frac{bh + h^2}{2} + \frac{\pi - 3}{8}b^2$$

with two values of π, 157/50 and 22/7.

---

[13]Here, we only consider segments of a circle within a half-disc.

[14]*JZSS* 1-36 (QB, I, p. 109).

[15]*ZQJSJ* 2-22 (QB, II, p. 371).

[16]In the problems of the *JZSS* and the *ZQJSJ* referred to, the lengths of the chord and the arrow are, respectively: $b = 78$ *bu* 1/2 (68 *bu* 3/5) and $h = 13$ *bu* 7/9 (12 *bu* 2/3). This corresponds to relative errors of 13.9% and 13.4%, respectively.

[17]On this, see for example, Bai Shangshu, op. cit.; Xu Chunfang (1'), *1952*, p. 38.

[18]*Siyuan yujian*, III- 3-11 and 12; Hoe (2), *1977*, pp. 295–296.

As in the case of the formula for the area of a quadrilateral, the formula for the area of the segment of the circle is often attested to outside ancient China. In particular, it may be found:

- in the writings of the Roman surveyor Columella;[19]

- in Hero of Alexandria's *Metrica*.[20]

Length of a Circular Arc

In chapter 18 of his collection of notes, the *Mengqi bitan* (1086),[21] Shen Gua describes a new procedure for calculating the length s of a circular arc as a function of the chord b subtending the arc, the arrow h and the diameter of the circle d:

$$s = b + \frac{2h^2}{d}.$$

The famous polymath does not say anything about the logical origin of this formula which he claims to have invented. However, it is easy to imagine a plausible reconstitution (Fig. 17.1).

Firstly, by comparing the sector $OABC$ with an isosceles triangle with base the arc $\overset{\frown}{ACB}$, we see that the area of the sector is approximately $(1/2)s(d/2)$. On the other hand, writing the area of the same sector as equal to the triangle OAB plus the area of the segment of the circle ACB, calculated using the approximation formula described in the last section, we find that

$$\text{Area of sector} = \frac{1}{2}b\left(\frac{d}{2} - h\right) + \frac{1}{2}(bh + h^2).$$

Finally, reconciling these two relations, we find the desired formula.

Although it is based on approximations, this formula is, however, better than the previous one. In this case, considering only half the circumference, the relative error is never greater than 5.5%.

Calculation of the Arrow of an Arc

The formula of the previous section together with Pythagoras's theorem constitute all the ingredients needed to derive one of the most astonishing formulae of all Chinese mathematics, namely the formula for calculating the arrow ($HC = x$) of an arc ($\overset{\frown}{AB} = 2a$) as a function of this arc and the diameter ($d = 2r$) of the circle containing it (Fig. 17.2).

---

[19] Gericke (2), *1990*, p. 42. See also Mortet. op. cit., p. 22.
[20] *Metrica*, 1-32 (cf. Bruins (1), *1964*, p. 266).
[21] Hu Daojing (1'), *1987*, II, p. 585.

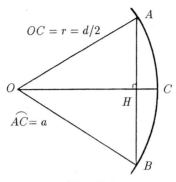

Fig. 17.2

According to Shen Gua's formula:

$$\overset{\frown}{AC} = \frac{\overset{\frown}{AB}}{2} = \frac{s}{2} = \frac{b}{2} + \frac{x^2}{d}. \tag{i}$$

According to Pythagoras's theorem:

$$AH^2 = OA^2 - OH^2 = \left(\frac{b}{2}\right)^2 = r^2 - (r-x)^2 = dx - x^2. \tag{ii}$$

Eliminating the half chord $(b/2)$ from (i) and (ii), we obtain:

$$x^4 + (d^2 - 2ad)x^2 - d^3x + a^2d^2 = 0. \tag{iii}$$

This is one of the major formulae of Guo Shoujing's (1231–1316) so-called Chinese spherical trigonometry.[22] The term "spherical" is quite justified by the particular context in which this formula first appears: spherical astronomy.[23] However, on the one hand, this sort of trigonometry is neither based on angles and exact formulae but rather on arcs, chords and sagitta computed by means of approximate formulae and, on the other hand, it never involves triangles with three circular sides but only triangles with at most one circular side as well as plane triangles.

Guo Shuojing[24] became famous as a hydrographer specialising in canal drainage, a clockmaker and a calendarist astronomer author of the calendrical comput *Shoushi li* (lit. Season-Granting Calendar). This comput provide us with access to the astronomical/calendrical science of the Chinese of the Mongol period, whence its importance.

Unfortunately, the original writings of Guo Shoujing have all been lost and we can now only access the works of the famous calendarist via late documents. Amongst these, the oldest and most reliable source is the calendrical part of the

---

[22]Cf. Gauchet (1), *1917*.

[23] *Yuanshi*, j. 54, p. 1200 ff. and *Mingshi*, j. 32, p. 551 ff.

[24]On Guo Shoujing, see: Li Di (1'), *1976*.

Yuanshi (Annals of the Yuan) (the compilation of these Annals was completed ca. 1370). However, this source is difficult to exploit because, as L. Gauchet wrote[25]: "The Annals of the Yuan describe the use of Kuo [Guo]'s Cheou-che-li [*Shoushi li*] at length and include the corresponding tables but are almost silent on the procedures of the inventor. Thus, it is natural to seek to turn to other, possibly less ancient, sources which are less economical with their explanations. But, since most of the known sources date from the 17th century, a new difficulty arises in this case, since at that period Chinese astronomers had access to European knowledge of mathematical astronomy. Thus, it is possible that some of them, such as Mei Wending (1633–1721), rewrote the history of Chinese astronomy by introducing foreign elements. Did the master of Xuancheng not reinterpret Guo Shoujing by drawing geometrical three-dimensional figures in the European manner? Did he not prove formulae of spherical trigonometry using projections in the manner of Clavius in his *Astrolabium*?[26]

To obtain a clear view of all this, one has to study not only all the Chinese authors who touched on the question, but also Japanese authors, including Seki Takakazu (?–1708) and Takebe Katahiro (1664–1739) in particular.[27]

Given that the present book does not aim to study Chinese astronomy, we shall not enter these historical quicksands and shall adhere as far as possible to the mathematical aspect of Guo Shoujing's formula. Since the calendrical chapters of the History of the Ming Dynasty (*Mingshi*) which deals with this subject describe the calculations very clearly without introducing mathematical techniques of European origin at this point, we shall use these.[28]

Let us now return to formula (*iii*) where we left off. The section of the *Mingshi* which relates to it is called *geyuan qiushi shu*, "Method for determining the arrow [born] of the division of the circle." The method in question is intended to calculate the root of the polynomial (*iii*) using an arithmetical procedure analogous to that of Horner's method, thus obtaining the figures of the root one by one up the desired accuracy.[29]

The calculations are carried out for a circumference called "the celestial periphery" *zhou tian* consisting of 365.25 degrees, that is, as many degrees as there are days in the sideral year (thus, each Chinese degree is equivalent to $360/360.25$ sexagesimal degrees; to avoid confusion, we shall denote the former by $^{\circ\circ}$). Fractions of a degree are denoted in a centesimal system (thus, for example, the expression $63.4567^{\circ\circ}$ corresponds to that which the original texts denote by 63 *du* 45 *fen* 67 *miao* or, literally, 63 degrees, 45 minutes 67 seconds). The "celestial periphery" (i.e. the trigonometrical circle) is also measured in degrees, like arcs. Since $\pi = 3$, this diameter is equal to $d = 365.25/3^{\circ\circ} = 121.75^{\circ\circ}$. Finally, the trigonometric quadrant is equal to $365.25/4 = 91.31^{\circ\circ}$.

---

[25]Gauchet (1), op. cit., p. 152.

[26]Cf. *MSCSJY*, j. 29 onwards.

[27]On these authors, see: *Meijizen*, vol. 2, p. 133 ff. and 266 ff., resp. See also Nakayama (1), *1969*.

[28]*Mingshi*, j. 32, p. 551 ff.

[29]Ibid.

Let us, for example, take an arc $a = 63^{\circ\circ}$. The successive coefficients of the polynomial (iii) are: $\mathrm{coeff}(x^4) = 1$, $\mathrm{coeff}(x^3) = 0$, $\mathrm{coeff}(x^2) = d(d - 2a) = -517.4375$, $\mathrm{coeff}(x) = -d^3 = -1804707.859$, constant term $= a^2 d^2 = 58832735.06$. One of the roots of this polynomial is 32.9409, which is the value of the arrow corresponding to an arc of $63^{\circ\circ}$. Similarly, for $a = 1^{\circ\circ}$, $20^{\circ\circ}$, $34^{\circ\circ}$ and $90^{\circ\circ}$, we find values of 0.0082, 3.3472, 9.8520 and 59.5625, respectively for the arrow.

A program based on Bairstow's method (a method used to determine all the roots of a polynomial) may be used to check that for $0 < a < 91.31^{\circ\circ}$ the polynomial (iii) always has two real positive roots and two complex roots. The smaller of the two positive roots is the only one considered by the Chinese texts (the other positive root is greater than the arc and thus could not represent a value of the arrow of this arc).

These calculations of the arrow of an arc are used in conjunction with other mathematical formulae, in order, in particular, to solve questions useful in positional astronomy. We shall consider only one of the most fundamental of these problems, which is used to convert ecliptic coordinates into equatorial coordinates.

On Fig. 17.3,[30] A represents the vernal equinox, D the Summer solstice, $\overset{\frown}{DE} = \varepsilon$ the obliquity of the ecliptic, AOD the plane of the ecliptic with the Sun at B, AOE the plane of the equator, $\overset{\frown}{BD} = a$ the co-arc of longitude of the Sun, $\overset{\frown}{CB} = y$ the declination of the Sun and $\overset{\frown}{CE} = z$ the co-arc of right ascension of the Sun. The problem is to determine y and z given a.

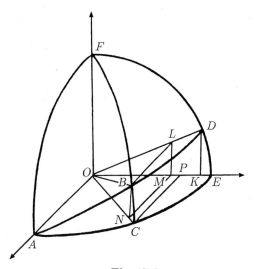

Fig. 17.3

---

[30]This figure is a reconstitution. Cf. QB, *Hist.*, p. 211.

From a "modern" point of view, y and z are determined using the following two formulae:

$$\sin y \;\; = \;\; \sin \varepsilon \cos a$$

$$\tan z \;\; = \;\; \frac{\tan a}{\cos \varepsilon} \; .$$

The Chinese techniques of Guo Shoujing involve a series of exact or approximate transformations between the arc a, its arrow f and other segments analogous to our sine and cosine. We may reconstitute the logic of this last approach as follows:

First the following elements are calculated in terms of $\varepsilon(=\overset{\frown}{DE})$:

$$KE \;=\; f(\varepsilon) \qquad\qquad \text{arrow of arc } \overset{\frown}{DE}= \varepsilon$$
$$OK \;=\; r - f(\varepsilon) \qquad\qquad \text{'}\cos \varepsilon\text{'}$$
$$DK \;=\; \sqrt{f(\varepsilon)(d - f(\varepsilon))} \qquad \text{'}\sin \varepsilon\text{' .}$$

(We note that, for some unknown reason, the texts use two different values of ε simultaneously, namely $\varepsilon = 24^{\circ\circ}$ and $\varepsilon' = 23.9^{\circ\circ}$; this can be seen in formulae (v) and (vi) below).

We then calculate the arrow $f(a)$ corresponding to the given arc, as the smallest positive square root of the polynomial of degree four (iii), then:

$$LB = MN = \sqrt{f(a)(d - f(a))} \; . \qquad\qquad (iv)$$

Since the right-angled triangles OLM and ODK are similar:

$$OM = \frac{OL \cdot OK}{OD} = \frac{OL \cdot OK}{r} = \frac{(r - f(a))(r - f(\varepsilon))}{r}, \qquad (v)$$

$$BN = LM = \frac{OL \cdot DK}{OD} = \frac{(r - f(a))\sqrt{f(\varepsilon')(d - f(\varepsilon'))}}{r}, \qquad (vi)$$

then:

$$ON = \sqrt{MN^2 + OM^2}, \qquad\qquad (vii)$$

$$NC = r - ON, \qquad\qquad (viii)$$

and finally

$$y = \overset{\frown}{CB} = BN + \frac{NC^2}{d} \qquad \text{(Shen Gua's formula) .} \qquad (ix)$$

The procedure for determining z as a function of the arc a may be reconstituted as follows:

By virtue of the similarity of the right-angled triangles OPC and OMN:

$$CP = \frac{MN \cdot OC}{ON} = \frac{MN}{ON}r, \qquad (x)$$

$$OP = OC\frac{OM}{ON} = r\frac{OM}{ON},$$

thus:

$$PE = OE - OP = r - r\frac{OM}{ON}. \qquad (xi)$$

Since, MN, ON and OM are known from the previous formulae, the new equalities (x) and (xi) may be used to calculate CP and PE. Thus, by virtue of Shen Gua's formula

$$z = \overset{\frown}{CE} = CP + \frac{PE^2}{d}. \qquad (xii)$$

These calculations are actually based on two approximations (calculation of $f(a)$ and Shen Gua's formula) and on three exact procedures (similarity of right-angled triangles, Pythagoras's theorem and property (iv)).

To be specific, in the original texts, the various quantities which occur in the above calculations are tabulated once and for all in a series of tables in which the arc a, which is used as an argument, is called *chidao jidu* (accumulated degrees of the "red road," i.e. of the equator) and varies in degree steps from $0^{\circ\circ}$ to $91^{\circ\circ}$, a final entry being reserved for $a = 91.31^{\circ\circ}$ in order to complete the quadrant. All the calculations are based on these tables, but non-integer values are calculated by linear interpolation (the texts provide for the possibility of a more exact calculation directly based on the formulae used to construct the tables).

More precisely, in the *Mingshi*, the three tables corresponding to the above calculations are the following (here, T_{ij} denotes the jth entry of the ith table, which collects together the calculation of several values on the same page)[31]:

$$
\begin{aligned}
T_{11}(a) &= f(a) \\
T_{12}(a) &= f(a+1) - f(a) \\
T_{13}(a) &= \sqrt{f(a)(d - f(a))} \\
T_{14}(a) &= \frac{(r - f(a))(r - f(\varepsilon))}{r} \\
T_{15}(a) &= \sqrt{T_{13}(a)^2 + T_{14}(a)^2} \\
T_{21}(a) &= r\frac{T_{13}(a)}{T_{15}(a)}
\end{aligned}
$$

[31] *Mingshi*, j. 32, pp. 553–574.

$$T_{22}(a) = r - r\frac{T_{14}(a)}{T_{15}(a)}$$

$$T_{23}(a) = T_{21}(a) + \frac{T_{22}(a)^2}{d} = z$$

$$T_{31}(a) = r - T_{15}(a)$$

$$T_{32}(a) = \frac{(r - f(a))T_{13}(\varepsilon')}{r}$$

$$T_{33}(a) = \frac{T_{31}(a)^2}{d} + T_{32}(a) = y.$$

Accuracy of Guo Shoujing's Calculations. The calculation of the coefficients of the polynomial (iii), above, requires the use of an approximation to π since the expressions for the coefficients involve both the diameter and the circumference of the "trigonometrical circle." The approximation for π which Guo Shoujing used to calculate his tables from (iii) is just $\pi = 3$, since he computes the diameter of the circle by dividing its circumference by 3. The choice of such a coarse approximation for π is a priori quite surprising, especially when one considers that much better values of π were available in Guo Shoujing's day (for example, Zu Chongzhi's value $\pi = 355/113$ (fifth century)).

However, against all expectations, Guo Shoujing's results turn out to be far better than one could have imagined. To convince oneself of this, one has only, for example, to draw a graph of the variations of the arrow $f(a)$ as a function of a and to compare it with the exact graph derived using rigorous trigonometrical methods to see that the two curves are remarkably close together.

But there is better to come. Recently, the Japanese historian T. Sugimoto has suggested that Guo Shoujing's tables provide globally better approximations when they are calculated using the value $\pi = 3$ rather than more exact values of π as a base![32]

For this comparison, Sugimoto comes down in all cases to a circle of unit radius divided into 360° sexagesimal degrees. Let us consider this author's comparison for the case of the calculation of the declination of the Sun (y). The curves of Fig. 17.4 represent the differences between the exact and the approximated values of y, determined using $\pi = 3$, $\pi = 3.0708$ and $\pi = 3.1416$, respectively, in the calculation of the coefficients of the constant term and of the coefficient of x^2 of (iii) – Guo Shoujing's polynomial. The graph shows that the curve corresponding to $\pi = 3$ appears more regular and well-balanced than the two others. But, above all, the evaluation of the areas bounded by each curve and the x axis reveals that the minimal area is obtained with the curve built using $\pi = 3$. In other words, the best approximation to y is obtained when π is taken equal to 3. Of course, these observations do not prove than $\pi = 3$ is mathematically 'the' value of π which really gives the best results with respect to Guo Shoujing's approximate computations but at any rate, the best possible value is certainly very close to $\pi = 3$.

---

[32] Cf. Sugimoto (1'), *1987* and (2'), *1987*.

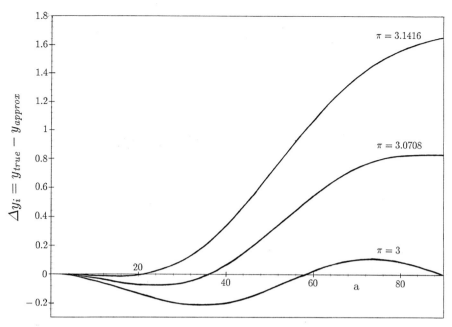

Fig. 17.4. Error curves $\Delta y_i = y_{true} - y_{approx}$ for the computation of the declination y of the Sun using Guo Shoujing's approximation formulae with three values of π.

How Guo Shoujing could have derived the best possible approximation remains to be established. This is a difficult question. The famous astronomer may have had access to trigonometrical tables of islamic origin, which he could have used for comparison. But, if this were the case, he would have had to have known that these tables were better than his; that being so, it is unclear why he should not have preferred them. More likely, any comparison that was made would have been between theoretical calculated values and observed values.

Another Approximation Formula

The monograph of the *Songshi* on calenderics and mathematical astronomy contains several theoretical formulae for the computation of the length of the shadow of the sun cast by a gnomon 8 *chi* long, each day of the year, at its meridian passage, from one winter solstice to the next, and for a place whose latitude is $\varphi = 34°48'45''$.[33]

Contrary to what might be expected, these formulae are not based on exact trigonometrical computations but on algebraical approximations. In the original sources,[34] everything is stated 'rhetorically' but the text is such that it allows a straightforward reformulation using modern symbolism. In the following, t

---

[33] According to Chen Meidong (3'), *1989*, p. 22.
[34] *Songshi j.* 76 and 79, pp. 1763 ff. and 1863 ff., resp.

denotes the number of days separating a given day of the year from the last winter solstice and

$$s_1(t) = 12.83 - \frac{20000t^2}{100(100617 + 100t + \frac{10000}{725}t^2)}$$

$$= 12.83 - \frac{200t^2}{100617 + 100t + \frac{400}{29}t^2}$$

$$s_2(t) = 1.56 + \frac{10000t^2}{100(\frac{900}{4}t + 198075)}$$

$$= 1.56 + \frac{4t^2}{7923 + 9t}$$

$$s_3(t) = 1.56 + \frac{10000t^2}{100((\frac{900}{4}t + 198075) + \frac{100(182.62-t)(t-91.31)}{77})}$$

$$= 1.56 + \frac{7700t^2}{13584271.78 + 44718t - 100t^2} \, .$$

With this notation, the length $l(t)$ of the shadow cast by the sun can be computed as follows:

$$l(t) = \begin{cases} s_1(t), & \text{if } 0 \le t < 62.2; \\ s_3(182.62 - t), & \text{if } 62.2 \le t < 91.31; \\ s_2(182.62 - t), & \text{if } 91.31 \le t < 182.62; \\ s_2(t - 182.62), & \text{if } 182.62 \le t < 273.93; \\ s_3(t - 182.62), & \text{if } 273.93 \le t < 303.04; \\ s_1(365.24 - t), & \text{if } 303.04 \le t < 365.25 \, . \end{cases}$$

(The various intervals of time defined here are determined by theoretical dates of astronomical events, such as equinoxes and solstices, computed by Chinese astronomers during the Chongning era (1102-1106)).

According to Chen Meidong's study (op. cit.) the lengths of shadows evaluated using the above approximations, are almost as much precise as those computed backwards using modern solar theory.

Interpolation Formulae

Linear Interpolation (Double-False-Position)

To explain the double-false-position rules *ying bu zu shu* (lit. "prescriptions of Too Much and Not Enough")[35], chapter 7 of the *Jiuzhang suanshu* describes

---

[35] Cf. above, p. 134.

a series of cases in which the two false assumptions (*liang she* or *jialing* — *liang* and *she* (resp. *jialing*) mean "double" and "supposition" (resp. "false supposition")) lead to a certain value which is too small, too large or equal with respect to a known value when computations made in accordance with the statement of a given problem are performed.

The chapter begins by describing four problems, of which the prototype is the following:

Suppose that goods are bought collectively.[36] If each man [resp. each family][37] were to pay a sum of money x_1, the "Too Much" paid would be y_1. If each man paid x_2 the "Not Enough" paid would be y_2. The question is how many men are there and what is the price of the goods?

It then gives the following resolutory rule:

Set the models (*lü*) of the sums paid [x_1 and x_2], the resulting Too Much and Not Enough [y_1 and y_2] being respectively placed below these. Cross multiply [the Too Much and the Not Enough and the models of the sums paid], then add the results obtained, which gives the dividend; add the Too Much and the Not Enough, which gives the divisor. If the dividend is equal to the divisor, then "one;" if there are fractions, make them communicate.[38]

Before explaining this prescription, we note that the expression "if the dividend is equal to the divisor then 'one' " *shi ru fa er yi* means that a division is to be carried out. This seems to be the beginning of a division table which would continue as follows: "if the dividend is so many times the divisor then the quotient is equal to that many."[39] "To make fractions communicate" means to reduce them to the same denominator.

With these explanations in mind, the above prescriptions correspond to

$$
\begin{array}{c c}
x_1 \; x_2 & \\
& \longrightarrow \\
&
\end{array}
\qquad
\begin{array}{c}
x_1 \; x_2 \\
y_1 \; y_2
\end{array}
\; \longrightarrow \;
\begin{array}{c l}
x_1 y_2 + x_2 y_1 & \text{Dividend} \\
y_1 + y_2 & \text{Divisor}
\end{array}
\; \longrightarrow
$$

Perform the division $\longrightarrow$ in case of fractions (at some stage of the computation) reduce them to the same denominator

The prescriptions continue as follows:

Set the models of the sums paid [x_1 and x_2] use the small to decrease the large and take the remainder [$x_1 - x_2$] to simplify the dividend and the divisor; the dividend gives the price of the goods and the divisor gives the number of men.

Denoting the sum to be paid by each man by x_0, the price of the goods by A and the number of men by B, this rule states that:

---

[36] *JZSS* problem 7-1 is about "things." The problem 7-2 involves hens, 7-3 involves jade-like stones and 7-4 cattle.

[37] The expression "each family" (*jia*) only occurs in *JZSS* problem 7-4.

[38] *JZSS* 7-4 (QB, I, p. 206).

[39] From QB, *Hist.*, p. 37.

$$\left(x_0 = \frac{x_1 y_2 + x_2 y_1}{y_1 + y_2} \right) \quad A = \frac{x_1 y_2 + x_2 y_1}{x_1 - x_2} \quad B = \frac{y_1 + y_2}{x_1 - x_2}.$$

After that, the text successively examines cases in which the false hypotheses are in the same direction (either both "too much" or both "not enough"); neither does it forget to mention the case in which the answer might be "just enough." At each point it indicates the various formulae which are suitable and which the pupil should memorise one by one.

In the first eight problems the statements indicate the values of the false hypotheses (or false suppositions). Of course, this is a pedagogic trick, intended to familiarise the pupil with the various possible situations which he may meet. But, from problem no. 9, the statements do not indicate which false suppositions would be appropriate, and the pupil is left to stand on his own two feet and to imagine what he will.

Within the large variety of the problems, it appears that the part of this rule which is used to calculate x_0 is that which is the most often applied.

One may wonder if this technique is limited to what we would call linear problems, when one considers problems such as the following (*JZSS* 7-11).

Suppose that a reed grew 3 *chi* [feet] on the first day and that a bulrush grew one *chi* on the first day. If the growth of the reed decreases by half on each successive day, while that of the bulrush doubles, after how long will the two [plants] have the same length?

The author of the *JZSS* applies his double false position rule by assuming firstly that the answer is 2 days, then he starts again, assuming it to equal 3 days. Thus, he deduces a Not Enough of 15 and a Too Much of $17\frac{1}{2}$ and obtains an answer of $2\frac{6}{13}$ days. According to certain authors[40] this answer is a little too low, because after x days, the length of the reed is $3 + \frac{3}{2} + \frac{3}{4} + \frac{3}{8} + \ldots = 3(1 + \frac{1}{2} + \frac{1}{2}^2 + \ldots + (\frac{1}{2})^{x-1}) = 6(1 - \frac{1}{2^x})$ and that of the bulrush is $2^x - 1$. Consequently, $6(1 - \frac{1}{2^x}) = 2^x - 1$, whence $x = \log_2 6 = 2.59$.

However, this modernised interpretation of the problem surreptitiously introduces notions which are absent from Han mathematics and moreover the statement of *JZSS* problem 7-11 says nothing about the growth of plants for a non-integer number of days.[41] It seems thus preferable to interpret the texts in the sense of linear variations for non-integer values of the variable inasmuch as linearity is a concept which permeates the totality of the *JZSS*. If this last interpretation is correct, we note that the double false position rule is applied here in the case of variations linear by intervals. Consequently, the correct interval of applicability of the rule has first to be determined: with false suppositions other than 2 or 3 days, the result would be erroneous. Perhaps, such a complication in fact means that the above problem is a disguised problem about interpolation between the successive values taken from a numerical table.

---

[40] Wu Wenjun, (1′), *1982*, p. 267; Bai Shangshu (3′), *1983*, p. 242.
[41] I am indebted to Michel Guillemot (University, Toulouse) for this remark.

Quadratic and Cubic Interpolation

Analysis of the computational techniques on which the Chinese ephemerides from the sixth century AD are based often reveals the presence of quadratic or cubic interpolation formulae.

In all cases, these are ready-made formulae, the interpolatory nature of which becomes evident when one examines the structure of the numerical tables and studies the verbal descriptions accompanying these.

Given that the terminology used often depends closely on the nature of the intended application, it is particularly difficult to describe the mathematical aspect of these formulae without going into long discussions, since, the meanings of many astronomical terms must first be defined. Thus, in what follows, we shall only summarise the known facts.[42]

- (i) In his calendar *Huangji*, the astronomer Liu Zhuo (544–610)[43] used interpolation formulae of degree two, with constant step length.[44]

- (ii) The Tantric monk and astronomer Yi Xing (683–727) used a quadratic interpolation formula applicable to non-equidistant points; he also used a formula of degree three.[45]

- (iii) The calendarist Guo Shoujing (1231–1316) used a cubic interpolation formula and brought the corresponding finite differences to the fore.[46]

- (iv) The mathematicians Qin Jiushao[47] and Zhu Shijie[48] included exercises involving such formulae in their manuals.

Do these formulae reveal Chinese originality? This is by no means certain, since, according to O. Neugebauer, they date back to early antiquity and can already be found in cuneiform astronomical tablets containing numerical tables relating to the movement of the planets.[49] Such formulae are also known to appear simultaneously in China and in India,[50] precisely at the time when Indian and Chinese astronomers were working together at the court of the Tang.[51] Not to mention the fact that, as Kennedy notes, quadratic interpolation techniques were also common currency in the Islamic countries in the Middle Ages.[52]

---

[42]See Li Yan (52'), *1957*; Ang Tian-se (3), *1976*; Martzloff (2), *1981*, pp. 235–245.

[43]*CRZ*, j. 12, p. 148 ff.

[44]Li Yan, op. cit., p. 23 ff.

[45]Ang Tian-Se (5), *1979*; Cullen (1), *1982*.

[46]Li Yan, op. cit., p. 62.

[47]Ibid., p. 55.

[48]Hoe (2), *1977*, p. 306; *SY*, p. 193.

[49]Interpolatory schemes of degree 2 or 3 enable us to interpret the structure of these tables. Cf. Neugebauer (4), *1975*, p. 413 (ephemerides of Jupiter), p. 419, (ephemerides of Mercury). See also Neugebauer (2), *1953*, vol. 3, pp. 299, 313, 328, 355.

[50]See Gupta (2), *1969*.

[51]Cf. Cullen, op. cit.

[52]Kennedy (1), *1964*.

Problem II-10-5 of the *Siyuan yujian*

(Cf. Hoe (1), *1976*)

Soldiers recruited in cubes. First day: side of cube −3 *chi* [feet]. Following days: 1 *chi* more per day. At present side [of cube]: fifteen. Each soldier receives two hundred and fifty *guan* [a string of 1000 cash] per day. What is the number of soldiers recruited and what is the total amount paid out? Answer: twenty three thousand four hundred soldiers, twenty three million four hundred and sixty two thousand *guan*.

In this problem, the number of soldiers recruited on day x is equal to $f(x) = (2 + x)^3$ $(1 \leq x \leq 15)$, and we have to calculate

$$F(x) = f(1) + f(2) + \ldots + f(15).$$

In his solution, Zhu Shijie uses an "upper difference" (*shang cha*), a "second difference" (*er cha*), a "third difference" (*san cha*) and a lower difference (*xia cha*). According to the Chinese text, it can be shown that these differences correspond exactly to the successive differences D_1, D_2, D_3 and D_4 (= 27, 37, 34 and 6, resp.) of the function F, which may be tabulated as follows:

Days	Numbers of soldiers recruited	First differences	Second differences	Third differences	Fourth differences
0	0				
1	27	$3^3 = 27$	37		
2	91	$4^3 = 64$	61	24	6
3	216	$5^3 = 125$	91	30	6
4	432	$6^3 = 216$	127	36	
		$7^3 = 343$			

After calculating these differences, Zhu Shijie explains that the number of soldiers recruited is found by multiplying the differences by what he calls "products" (*ji*). (In fact, his products correspond to binomial coefficients stemming from sums of series). In our notation, his solution amounts to calculating:

$$F(n) = nD_1 + \frac{n(n-1)}{2!}D_2 + \frac{n(n-1)(n-2)}{3!}D_3 + \frac{n(n-1)(n-2)(n-3)}{4!}D_4$$

and

$$S(n) = 250\left(\frac{n(n+1)}{2!}D_1 + \frac{(n-1)n(n+1)}{3!}D_2 + \frac{(n-2)(n-1)n(n+1)}{4!}D_3 + \frac{(n-3)(n-2)(n-1)n(n+1)}{5!}D_4\right).$$

Thus, $F(15) = 23\,400$ and $S(15) = 23\,462\,000$.

18. Li Shanlan's Summation Formulae

In Louis Comtet's *Advanced Combinatorics, The Art of Finite and Infinite Expansions*,[1] in the part of volume I devoted to exercises (p. 173), the author gives the following formula

$$\sum_{0 \leq j \leq k} \binom{k}{j}^2 \binom{n+2k-j}{2k} = \binom{n+k}{k}^2 \tag{18.1}$$

the proof of which is not particularly self-evident, which he attributes to a certain Li Jen-Shu.

In his bibliography, Comtet refers to a paper by a Czech mathematician, Josef Kaucký from Bratislava, containing a proof of (i).[2] Most interestingly, Kaucký also provides a bibliography and a German summary of the history of (18.1). From this, the astonishing history of Li Jen-Shu's formula can be reconstituted. In 1937, as a result of the deterioration of the political situation in his country following the rise of Nazism, the Hungarian mathematician György Szekeres took refuge in Shanghai. In this great Chinese metropolis he met Zhang Yong (1911–1939), a young mathematician who had studied in Europe and was keenly interested in the history of mathematics. The latter told him about the existence of the formula in question, which he had tried to prove, but without success. The Hungarian refugee was intrigued and wrote to Paul Turan who wrote back with a rather sophisticated (since based on Legendre polynomials) proof of the difficult formula. Much later, in 1954, Turan published his proof in a Hungarian journal in an article which aroused the curiosity of other mathematicians including Kaucký.[3] But Zhang Yong also had his own proof of the above formula published in the Chinese journal *Kexue* (Science)[4] and from this we learn that Li Jen-Shu is in fact Li Shanlan ("Li" is the surname of Li Shanlan and "Jen-Shu" [Renshu] his alias *bie hao* while Shanlan corresponds to his school name *xiangming*. "Li Shanlan" is the name under which this mathematician is most often quoted in Chinese publications (in Western publications, the reader will often find him cited under a bewildering variety of fanciful spellings such as Li Zsen-Su or Shoo Le-Jen, but in the sequel we shall uniformly call him Li Shanlan to avoid confusion)).

---

[1]Comtet (1), *1974*.
[2]Kaucký (1), *1963*.
[3]Turan (1), *1954*.
[4]Zhang Yong (1'), *1939*.

As Zhang Yong explains, the source of Li Shanlan's formula is the third chapter of the *Duoji bilei* (Heaps [i.e. finite series] summed (*ji*) using analogies (*bilei*)) which constitutes the fourth part of Li Shanlan's collected works, the *Zeguxizhai suanxue* (1867). Most interestingly, the *Duoji bilei* also includes other noteworthy results: recurrence formulae for Stirling numbers of the first type[5] and for Eulerian numbers,[6] Worpitzki's formula[7] among other things.

The whole appears all the more remarkable since Chinese traditional mathematics is usually considered worthy of historical reflections but not of mathematical research. Thus, the fact that 20th century mathematicians not concerned with the history of mathematics could have taken an interest in Li Shanlan's mathematics incites us once again to raise the question of the nature of Li Shanlan's mathematics which is certainly non-trivial.

The *Duoji bilei* begins with a preface in which Li Shanlan expounds the background to his research and his motives (Fig. 18.1):

The piling up of heaps *duoji* [i.e. the summation of series] constitutes a branch of [the part of Chinese traditional mathematics] called *shaoguang* (Decrease [in length] to the benefit of the width).[8] The [works of] the Great Astrologer *taishi* Guo Shoujing [(1231–1316)] concerning the inequalities of the solar and lunar motions,[9] Master Wang Lai's[10] iterated sums, Master Dong Fangli's[11] cyclotomical computations[12] and lastly the summations of series which appear in the algebra and the differential calculus of the Westerners[13] constitute the major part of this [fourth chapter]. The usefulness of [the mathematical techniques] contained herein is indeed multifarious. None the less, mathematical books seldom deal directly with these. Master Zhu Shijie from the Yuan dynasty is the only one who has made use of prescriptions relating the piling up of heaps in [the chapters of his *Siyuan yujian* entitled] "Bundles of water oat piled up into various heaps", "Recruiting soldiers according to fixed patterns" and "Heaps of fruits."[14] But his intention was only to expound the *tianyuan yi* [algebra] and for that reason he presents [the piling up of heaps] neither precisely nor methodically. As for Master Wang [Lai] and Dong [Youcheng] [mentioned above] they are methodical but the scope of their works is very limited since the former deals only with triangular heaps[15] while the latter does not go beyond quadrangular heaps.[16] None of their studies is comprehensive. [However], the present study [i.e. the *Duoji bilei*] is composed of: numerical tables *biao*; drawings *tu* and methods *fa*.

---

[5]Comtet, op. cit., II, p. 243.

[6]Ibid., p. 84.

[7]Worpitzki (1), *1883*.

[8]i.e. the fourth chapter of mathematics in Nine Chapters which was initially devoted to root extractions but later augmented with questions relating to the summation of series.

[9]Allusion to interpolation formulae which are the basis of Guo Shoujing's mathematical astronomy.

[10]Horng (4'), *1993*.

[11]i.e. Dong Youcheng (1791–1823). Cf. QB, *Hist.*, p. 306; *CRZ*, j. 51, p. 690.

[12]In fact, expansion into infinite series of trigonometrical functions.

[13]Allusion to Chinese translations of Augustus De Morgan's *Elements of Algebra* and Elias Loomis's *Elements of Analytical Geometry and of the Differential and Integral Calculus* (See Appendix I, below).

[14]Hoe (2), *1977*, pp. 250 and 252.

[15]i.e. the finite series whose general term is $n(n+1)/2$ (or triangular numbers).

[16]i.e. the finite series whose general term is $n(n+1)(n+2)/3$!

Fig. 18.1. Chinese text of Li Shanlan's preface to the *Duoji bilei*.

The whole is divided into sections and has new subdivisions; everything is expounded precisely and meticulously in order to enable those who study mathematics to grasp the meaning of the prescriptions relating to the piling up of heaps. Beginning with Li Shanlan, a new territory[17] beyond that of the Nine Chapters [has been conquered].!"[18]

From this preface we deduce that the mathematics of the *Duoji bilei* could have been inspired by that of Augustus De Morgan and Elias Loomis since Li Shanlan's reference to "the algebra and differential calculus of the Westerners" could hardly refer to any other works (Li Shanlan was the translator of works on these subjects by these authors).

As noted by Li Yan[19] summation techniques are developed to some extent in De Morgan and Loomis's works. In particular, Loomis's treatise contains

---

[17]More literally, "a new pennon."

[18]Integral translation, *Duoji bilei*, j. 1, pp. 1a–b.

[19]*ZSSLC-T*, III, p. 286.

Taylor's formula. Moreover, since Li Shanlan was a translator, he evidently had direct or indirect access to material of Western origin. However, judging from the actual content of the *Duoji bilei* and its redactional style, it would seem that Li Shanlan never cared much about Western sources.

One might also imagine Japanese influences. Traditional Japanese and Chinese mathematics have much in common, from the point of view of both their content and their form. In Japan, mathematics was often written in classical Chinese, so that, in theory, Japanese work was directly accessible to the Chinese. Moreover, some Japanese authors prior to Li Shanlan, such as Wada Yasushi (1787–1840), solved questions similar to those dealt with by Li Shanlan such as the following result[20]:

$$1^p + 2^p + \ldots + n^p = c_1 \binom{n+p}{p+1} + c_2 \binom{n+p-1}{p-1} + \ldots + c_p \binom{n+1}{p+1} \quad (18.2)$$

where the $\binom{n}{p}$ are the usual binomial coefficients.

A priori, it is possible that Li Shanlan may have heard of this Japanese mathematician. As far as we know, however, Chinese mathematicians became aware of ancient Japanese mathematics at a very late stage (towards 1890).

Zhu Shijie's mathematics, however, is omnipresent in the *Duoji bilei* and consists essentially of: the *tianyuan* algebra (but limited to polynomials in one unknown), "Pascal's triangle", interpolation formulae and results on finite summation formulae whose general terms are[21]:

$$\mu_n = \frac{n(n+1)(n+2)\ldots(n+s-1)}{s!}$$

(stated in the case of particular values of s) or

$$\mu_n = n^2$$

$$\mu_n = n\left(1 + 3 + 6 + 10 + \ldots + \frac{n(n+1)}{2!}\right)$$

$$\mu_n = \begin{cases} \frac{(3n)^2+3}{12} & \text{if } n \text{ is odd} \\ \frac{(3n)^2}{12} & \text{if } n \text{ is even} \end{cases}$$

$$\mu_n = (a+n-1)(b+n-1)$$

$$\mu_n = n[9 + 3(n-1)]$$

$$\mu_n = [3 + (n-1)]^2$$

---

[20] *Meijizen*, V, p. 82.
[21] From Hoe (1), *1976*, Appendix 2, pp. 95 ff. and p. 76.

Yet, Li Shanlan departs from Zhu Shijie's mathematical style, for the text of the *Duoji bilei* is not structured by artificial problems but by prescriptions (or algorithms) devoted to the summation of finite series.

More precisely, as announced by Li Shanlan in his preface, the *Duoji bilei* consists of four *juan* which are uniformly subdivided into a number of sequences made up of the following elements:

(a) double-entry diagrams *biao* based on the version of Pascal's triangle found at the beginning of the *Siyuan yujian*, each diagram being accompanied by details of how it was constructed, that is, in most cases, by a statement of the recurrence relation which allows one to fill it in from initial values.

(b) Drawings (*tu*) representing stacks or piles of small disks, marbles or cubic dice showing the successive terms of series which the author wishes to sum (Fig. 18.2).

(c) Rules (*fa*) formulated in written characters.

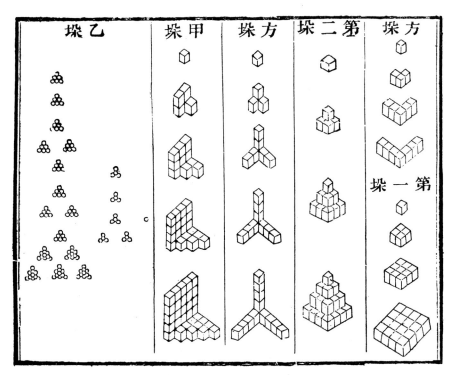

Fig. 18.2. Sequence of figurate numbers which Li Shanlan presents without comment. From the *Duoji bilei* (1867), j. 1, p. 6b and 11a and j. 3, p. 21b, resp.

The whole work includes 15 different diagrams (numerical "triangles"). These are constructed in three different ways:

(i) Generalised Pascal's triangles (total of eight). These all satisfy the Pascal recurrence

$$\binom{n}{p} = \binom{n-1}{p-1} + \binom{n-1}{p} \qquad n, p \geq 1 \qquad (18.3)$$

but are not all constructed from the same initial values, in the sense that their left sides may contain only 1's (like the usual version of Pascal's triangle) or other sequences of numbers such as 1, 2, 2, ...; 1, 10, 19, 20, 20, 20, ...; etc.

(ii) Triangles satisfying a more complex recurrence. There are two such: one contains the Stirling numbers of the first type; the other contains the Eulerian numbers.[22]

(iii) Triangles defined without using a recurrence relation. There are five such: the first contains the powers n^p (beginning of *juan* 2); the second contains the squares of the binomial coefficients (beginning of *juan* 3); the last three contain the products of the binomial coefficients by $(n - p + 1)^m$, $(m = 1, 2, 3)$ (*juan* 4).

The diagrams are not easy to analyse since they are almost wholly without comment. Most of them appear to be intended to illustrate the self-evident recurrence formula:

$$\binom{k}{k} + \binom{k+1}{k} + \ldots + \binom{n}{k} = \binom{n+1}{k+1} \qquad (18.4)$$

already cited above (p. 305) with reference to Zhu Shijie's mathematics, where the $\binom{n}{p}$ denote not the usual binomial coefficients but generalised binomial coefficients which occur naturally when the usual Pascal's triangle is replaced by one or more general triangles built using the same fundamental recurrence (18.3) but with new initial values. This formula, which is particularly useful for calculating sums, is probably one of the keys to the vault of the *Duoji bilei*.

Finally, the rules (or algorithms) are used to resolve two types of problem which are the inverse of one another, and which occupy an equal proportion of the remaining text.

(a) Summation of the first n terms of a given series (here, the term 'series' refers to one of the diagonals of one of the above diagrams). This is the most innovative part of the book.

---

[22] Euler, *Institutiones Calculi Differentialis*, in *Leonhardi Euleri Opera Omnia*, 10th vol., Leipzig and Berlin, 1913, ch. 7, p. 373 (Euler introduces these numbers for very different reasons from Li Shanlan (differential equations)).

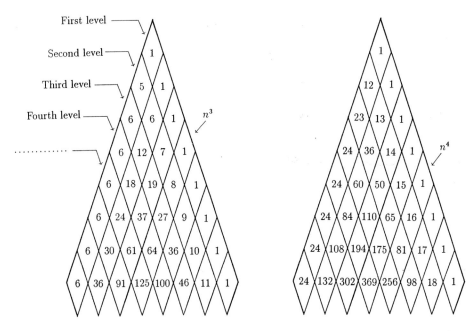

Fig. 18.3. Li Shanlan's generalised versions of Pascal's triangle where a diagonal is filled up with the successive cubes (resp. fourth powers) of natural numbers.

(b) Conversely, determination of the number of terms of a given series knowing the sum of its first n terms. This is the "trivial" part of the book, which is based almost solely on self-evident algebraic calculations (for us). Li Shanlan's solutions of these problems in some way illustrate his respect for tradition; on the one hand he uses the medieval algebra *tianyuan*, while on the other hand, he returns to a *modus operandi* typical of the *Siyuan yujian*.

There are approximately 120 formulae for each type of problem. Regardless of the question in hand, Li Shanlan never does more than state particular numerical results. However, this approach does lead to general formulations, since the accumulation of a sufficient number of well-chosen particular cases does highlight the emergence of certain paradigms. In fact, he passes from the particular to the general not by sequences of logical deductions, but by a series of consecutive numerical formulae which themselves are merely stated without more ado, and never feels it necessary to justify anything discursively. Everything is based on the observation of structured patterns and not on causal sequences of successively deduced arguments. Still, the *Duoji bilei* is not a chaos of results: series are regrouped in a particular order and analogies between them are suggested by this very fact (let us not forget that Li Shanlan highlights this aspect in the title of his book). More specifically, Li Shanlan observes that everything which applies to the ordinary version of Pascal's triangle also applies to more general triangles, subject to slight modifications which are apparent from various analogies.

For example, in the second chapter of the *Duoji bilei*, Li Shanlan set out to sum

$$1^p + 2^p + 3^p + \ldots + n^p.$$

The chapter begins with a generalised Pascal's triangle where the usual $\binom{n}{p}$ are replaced by the powers of natural numbers. But since the above summation recurrence (*iv*) does not apply in such a case, he replaces the power triangle with a sequence of other triangles built from the fundamental Pascalian recurrence (*iii*) but with one of their diagonals filled with the n^p (n variable and p constant. Cf. Fig. 18.3).[23]

From this, he states successively that[24]:

$$x^2 = \binom{x+1}{2} + \binom{x}{2} \tag{18.5}$$

$$x^3 = \binom{x+2}{3} + 4\binom{x+1}{3} + \binom{x}{3} \tag{18.6}$$

$$x^4 = \binom{x+4}{3} + 11\binom{x+2}{4} + 11\binom{x+1}{4} + \binom{x}{4} \tag{18.7}$$

$$x^5 = \binom{x+4}{5} + 26\binom{x+3}{5} + 66\binom{x+2}{5} + 26\binom{x+1}{5} + \binom{x}{5} \tag{18.8}$$

and beyond these particular decompositions of powers, he obtains summation formulae for squares, cubes and other powers by using (18.4) above.

For example, from (18.6) he notes that:

$$\sum_{i=1}^{n} x^3 = \binom{n+3}{4} + 4\binom{n+2}{4} + \binom{n+1}{4}. \tag{18.9}$$

Then he observes that for the following powers "one can continue in the same way" (*yi xia ke lei tui*) using the diagram of Eulerian numbers which he gives together with its general recurrence rule.[25]

---

[23] *Duoji bilei*, j. 2, pp. 8a and 15a, resp. (the triangles containing the n^2 ($n = 1, 2, 3, \ldots$) appear in chapter 1, p. 6a.).

[24] Li Shanlan does not explain how these formulae come about. For a plausible reconstruction of his approach, see Luo Jianjin (1′), *1982*.

[25] $A(n+1, k) = (n-k+2)A(n, k-1) + kA(n, k)$ ($k \geq 2$). *Duoji bilei*, ch. 2, p. 3b.

In Li Shanlan's own wording, (18.9) is formulated exactly as follows:

Prescriptions for the summation of derived heaps born of [the table] of multiplicative powers.

Here, the "derived heaps" *zhi duo* are those which appear in the diagonals of triangles containing the successive integers raised to a given power such as in Fig. 18.3; the expression "multiplicative power" merely refers to powers. In particular, the "bimultiplicative powers" *ercheng fang* denote the cubes of integers which are so called because their computation involves two multiplications.

Each heap has one SIDE *fang*, four EDGES *lian*, and one CORNER *yu*.

We have already met the terms SIDE, EDGE and CORNER in connection with root extractions but here these represent algebraical expressions in one variable having 1, 4 and 1 as coefficients, namely $\binom{n+3}{4}$, $\binom{n+2}{4}$ and $\binom{n+1}{4}$, respectively.

The SIDE has LEVEL   as [its] HEIGHT;
The EDGE has LEVEL minus one as [its] HEIGHT]
The CORNER has LEVEL minus two as [its] HEIGHT.

Here, the HEIGHT *gao* is a variable equal to the number of terms summed (denoted by n in the sequel) and the LEVEL *ceng* is another variable defined without words (Fig. 18.3).

The SIDE, EDGE and CORNER of the FIRST HEAP are dealt with using the [summation] prescription relating to the TRIMULTIPLICATIVE TRIANGULAR HEAP of the TRIANGLE.

The TRIANGLE refers here to Pascal's ordinary triangle.[26] The TRIMULTIPLICATIVE TRIANGULAR HEAP is that composed of the numbers 1, 4, 10, 20, 35, ... so called because its summation formula involves three multiplications which correspond to the computation of

$$\frac{n(n+1)(n+2)(n+3)}{4!},\qquad (18.10)$$

where n is the number of terms added together (HEIGHT). Thus, (18.10) enables us to compute $\sum_{i=1}^{n} x^3$ once noted that "The SIDE has LEVEL as [its] HEIGHT".

$$\text{SIDE} = \frac{n(n+1)(n+2)(n+3)}{4!} = \binom{n+3}{4}.$$

The EDGE has LEVEL minus one as [its] HEIGHT;

that is, EDGE is always computed by means of (18.10) but with n replaced by $(n-1)$.

---

[26] *Duoji bilei*, j. 1, p. 1b.

$$4 \text{ EDGES} = 4\frac{(n-1)(n)(n+1)(n+2)}{4!} = 4\binom{n+2}{4}.$$

The CORNER has LEVEL minus two as [its] HEIGHT

$$\text{CORNER} = \frac{(n-2)(n-1)n(n+1)}{4!} = \binom{n+1}{4}$$

and the sum 1 SIDE + 4 EDGES + 1 CORNER gives formula (18.9) above.

Similarly, in chapter 3, Li Shanlan constructs generalised versions of Pascal's triangle, for which one of the "diagonals," chosen in an appropriate way, contains the squares of certain binomial coefficients (Fig. 18.4). He then notes that the general terms of the simplest of the triangles considered are equal to:

$$Z_n^p(3) = \binom{n-3}{p} + 9\binom{n-2}{p} + 9\binom{n-1}{p} + \binom{n}{p}$$

$$Z_n^p(4) = \binom{n-4}{p} + 16\binom{n-3}{p} + 36\binom{n-2}{p} + 16\binom{n-1}{p} + \binom{n}{p}$$

Lastly, Li Shanlan explains that the generalisation to the following cases is self-evident, since the coefficients appearing in the formulae are just the squares of the binomial coefficients: $1^2, 3^2, 3^2, 1^2$; $1, 4^2, 6^2, 4^2, 1^2$.

Hence Li Shanlan's formula is obtained after generalisations based on observations of particular cases of summation formulae.

The same observation applies also in the case of the last chapter of the *Duoji bilei* which contains the following results[27]:

$$\sum_{i=p}^{i=p+n-1} (i-p+1)\binom{i}{p} = \binom{n+p+1}{p+2} + p\binom{n+p}{p+2}$$

$$\sum_{i=p}^{i=p+n-1} (i-p+1)^2\binom{i}{p} = \binom{n+p+2}{p+3} + (3p+1)\binom{n+p+1}{p+3} + p^2\binom{n+p}{p+3}$$

$$\sum_{i=p}^{i=p+n-1} (i-p+1)^3\binom{i}{p} = \binom{n+p+3}{p+4} + (7p+4)\binom{n+p+2}{p+4}$$

$$+ (6p^2 + 4p + 1)\binom{n+p+1}{p+4} + p^3\binom{n+p}{p+4}.$$

---

[27] *Duoji bilei*, j. 4, pp. 3b, 10b, 17b.

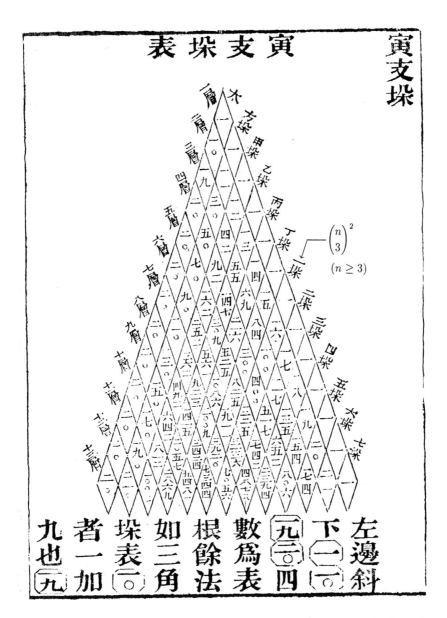

Fig. 18.4. One of Li Shanlan's generalised versions of Pascal's triangle *Duoji bilei*, j. 3, recto of page 7.

19. Infinite Series

The first known occurrence of infinite series expansions in Chinese works is in 1759, when the mathematician Mei Juecheng[1] (1681–1763) published a revised version of the works of Mei Wending, his illustrious grandfather,[2] at the end of which he inserted two chapters (*juan*) of his own.[3] In the first of these, entitled *Chishui yizhen* (pearls recovered from the Red River[4] bequeathed to posterity), he reflects on various subjects (units of length in the classics, a formula of spherical trigonometry,[5] comparison between the algebra of the *jiegenfang* and the Chinese medieval algebra of the *tianyuan*, geometrical construction of the golden ratio, etc.). He also shows how to calculate:

(a) the length of the circumference given its diameter

(b) the sine of an arc (*zhengxian*)

(c) the versed sine of an arc (*zhengshi*)

using infinite series which he does not prove and refers to as "formulae from the Western scholar Du Demei." Du Demei is the Chinese name of the French Jesuit missionary Pierre Jartoux[6] (1669–1720), who arrived in China in 1701, became famous as an engineer, mathematician and geographer and died in Tartary.

In Mei Juecheng's original formulation (a) is expressed as follows (Fig. 19.1):

Given DIAMETER, 20 *yi* [*yi* = 10^8] find [the length] of the circumference. The more numerous the number of digits of the diameter are, the more precise the tail-number [i.e.

---

[1] Cf. Yan Dunjie (20′), *1959*.

[2] This refers to *Meishi congshu jiyao* (1759). On Mei Wending, cf. Hummel, p. 570; Li Yan (5′), *1925*; Hashimoto (1′), *1970* and (3′), *1973*; Martzloff (2), *1981*; Li Di and Guo Shirong (1′), *1988*.

[3] *Meishi congshu jiyao*, j. 61 and 62.

[4] Allusion to chapter 12 of the *Zhuangzi*: "The Yellow Emperor was walking to the North of the Red River. He climbed mount Kun Lun and as he was about to return to the South, he realised that he had lost his 'fine pearls.' " The pearls recovered from the Red River to which Mei Juecheng alludes in his title are the fine mathematical techniques (such as algebra, trigonometry, series, etc.) which, according to him, were once known in China and then lost, to be found much later in European hands through the contacts between China and the European missionaries (myth of the Chinese origin of mathematics and, more, generally, of all the sciences).

[5] $\sin^2(A/2) = (\sin(p - b)\sin(p - c))/(\sin b \sin c)$.

[6] Cf. Dehergne (1), *1973*, pp. 131–132.

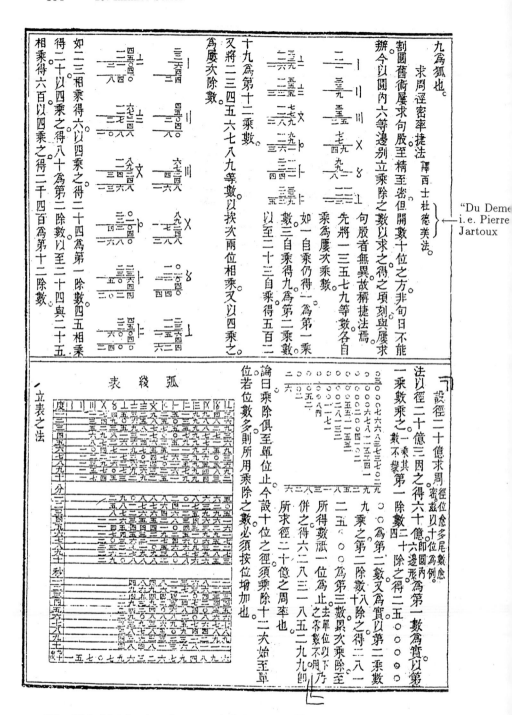

Fig. 19.1. Mei Juecheng's text on infinite series. The manually inserted corner marks delimit the part of the text whose translation appears on pp. 353–355.

the decimals] are. Here we take ten digits as an example.[7] Method: DIAMETER multiplied by 3. Result: 60 *yi* [That is, the length of the periphery of the regular hexagon inscribed in the circle] as FIRST NUMBER taken as DIVIDEND. Multiply it by the first multiplier [i.e. '1'] ['1' multiplied by this number: nothing changes.]. Divide it by the first divisor [24]. Result: 25000000 as SECOND NUMBER again taken as DIVIDEND. Multiply it by the second multiplier [...] Multiplications and divisions stop when units are obtained [...].[8]

This computation corresponds to

$$\text{circumference} = 3d \left(1 + \frac{1^2}{3!4} + \frac{1^2 \cdot 3^2}{5!4^2} + \frac{1^2 \cdot 3^2 \cdot 5^2}{7!4^3} + \dots \right). \tag{19.1}$$

(d = diameter) and those for the sine and the versed sine to[9]:

$$\alpha - \frac{\alpha^3}{3r^2} + \frac{\alpha^5}{5!r^4} - \frac{\alpha^7}{7!r^6} + \dots \tag{19.2}$$

and

$$\frac{\alpha^2}{2!r} - \frac{\alpha^4}{4!r^3} + \frac{\alpha^6}{6!r^5} + \frac{\alpha^8}{8!r^8} + \dots \tag{19.3}$$

(α = arc, r = radius), respectively.

According to Pfister[10] Jartoux is the author of two manuscripts describing the above three formulae, the *Zhoujing mi lü* (The precise ratio between the circumference and the diameter) and the *Qiu zhengxian zhengshi jiefa* (Quick method of determination of the sine and versed sine). Although not a single mention of these manuscripts in historical sources from the Kangxi period has ever been found, almost nobody has ever doubted their existence and this has given rise to endless speculation: is the content of Jartoux's manuscripts limited to the above three series as their title strongly suggests? Or, did Jartoux also cite other new series such as those for the determination of the arc given the sine or the versed sine, which were later studied at length in China by mathematicians such as Minggatu? Again, did Jartoux merely state ready-made formulae as Mei Juecheng does or does he give proofs too?[11] And of what kind? As the case may be, certain research of later Chinese mathematicians might be considered original or not.

---

[7]Here we respect the Chinese traditional practice according to which footnotes are inserted in the main text itself and not at the bottom of the page.

[8]*Meishi congshu jiyao*, j. 61, p. 6b.

[9]Mei Juecheng, op. cit., p. 7a, explains that the computations are with an arc α = 21°19′51″ and a radius 10^8. With this data he first computes $\alpha = \frac{10^8 \pi a}{180} \simeq 37229325$ and, using (*ii*) finds $\sin \alpha = 36375254$.

[10]Pfister (1), *1934*, notice on Jartoux. See also Bernard-Maître (7), *1960*, pp. 373–374, notices no. 609 and 610 (but note that Bernard-Maître merely repeats Pfister and that Pfister does not give any precise information on his sources).

[11]According to Chen Jixin (cited below), it would seem that it was not the case.

In fact, even if Jartoux's manuscripts have never existed, it is certain that from 1700, Chinese mathematicians who entered the service of Kangxi and other Emperors could have had access to publications on infinite series, especially thanks to mathematical books of European origin preserved in the Beitang library in Peking (the Beitang [Pé-T'ang] library is the former library of the Jesuits in Peking which was active during the two centuries from the arrival of Matteo Ricci in China in 1583 until the suppression of the Society of Jesus decided by Clement XIV in 1773 and promulgated in China two years later, in 1775). Its present Catalogue – the *Catalogue de la Bibliothèque du Pé-T'ang* (*CBPT* in the sequel)[12] – which lists what has survived from the initial library until now, contains no less than 4101 books in various European languages). Although to my knowledge this has never previously been noticed by historians of Chinese mathematics, the *CBPT* lists many sources relating to infinite series such as the following:

(a) Various mathematical publications from the French Académie des Sciences published from 1666 to 1755.[13]

(b) A treatise on calculus by the French academician Louis Carré (1663–1711), *Méthode pour la mesure des surfaces, la dimension des solides, leurs centres de pesanteur, de percussion et d'oscillation par l'Application du Calcul intégral.* Paris: Chez Jean Boudot, Librairie de l'Académie Royale des Sciences, 1700.[14]

(c) A treatise on fluxions by James Hodgson (1672–1755), *The doctrine of fluxions, founded on Sir Isaac Newton's method, Published by Himself in his tract upon the quadrature of curves.* London, 1736.[15]

(d) Various treatises by Leonhard Euler (1707–1783), including the influential *Introductio in analysin infinitorum.* Lausanne, 1748.[16]

(e) The Marquis de l'Hospital's famous *Analyse des infiniments petits*, in a late edition published in Avignon in 1768.[17]

(f) Treatises on logarithms including those by Adriaen Vlack [Vlacq] and Henry Briggs.[18]

Nevertheless, the linguistic obstacle was formidable; not a single Chinese mathematician from that period is known to have mastered Latin or any other European language. Even so, the decoding of certain isolated mathematical techniques found in European books and expressed through the medium of mathematical notation independent of any particular language was perhaps not

---

[12]See Mission Catholique des Lazaristes à Pékin (1), *1949*.

[13]*CBPT*, notices nos. 9 to 21.

[14]Ibid., notice no. 161.

[15]Ibid., notice no. 4079.

[16]Ibid., notice no. 1563.

[17]Ibid., notice no. 440. (first edition: Paris, 1696).

[18]Ibid., notices nos. 3061 to 3063.

beyond the reach of courageous mathematicians. As the French Jesuit Jean-François Foucquet reports in one of his manuscripts (1716), certain Chinese were so eager to access foreign knowledge that they sometimes threatened to resort to force against Jesuits in order to gain possession of European mathematical books: once, a certain Prince from the Court of Kangxi forced Jesuits to give them a volume of Napier's treatise on logarithms; in Peking, on another occasion, while searching for European books, Chinese rummaged the rooms of the Jesuits.[19] More commonly yet, Chinese asked Jesuits to explain them the meaning of certain mathematical techniques from European books which is probably what Mei Juecheng did to Jartoux. In a limited number of cases, Chinese translations of recent European mathematical treatises were undertaken. As already noted,[20] the treatise on logarithms of the *Shuli jingyun* was adapted from a book by Henry Briggs and later became an important source of information on computational techniques such as the binomial theorem.

Under these conditions, Chinese research into infinite series from the 18th and 19th centuries was most probably influenced by European developments, whence the presence of certain European techniques such as the addition, subtraction, multiplication and division of series, the reversion of series and the binomial theorem in the works of Minggatu, Dong Youcheng (1791–1823), Xiang Mingda (1789–1850) and Dai Xu (1805–1860). On the whole, the achievements of these is impressive and covers infinite series for trigonometric functions direct and inverse, logarithms and the rectification of curves such as the ellipse.

For all that, nothing like calculus ever emerged from this Chinese research and many features characteristic of previous Chinese mathematics (especially analogical reasoning) continued to predominate. The works of Minggatu are fairly typical in this respect.

The Mongol mathematician Minggatu (?–1764) expended considerable energy attempting to prove the validity of the above series (*i*), (*ii*) and (*iii*); it is said that he worked on the question for thirty years.[21]

Contrary to what one might think, Minggatu's research is not based on hypothetical 'pre-existing' Mongol mathematics.[22] This mathematician acquired his knowledge from a number of mathematical and astronomical works of European origin which were adapted into Chinese during the first half of the 18th century on the initiative of the Jesuit missionaries. Many traces of this may be found in his mathematical work, including his use of the Euclidean notion of 'continuous proportion' *lian bili* and of the so-called *jiegenfang*, an algebra of European origin (recognisable, amongst other things, from his typical use of the '=' sign and the way in which algebraic operations are presented as though they were arithmetic operations). Moreover, it is known that Minggatu was taught these methods by the Emperor Kangxi in person. It is also known that he was involved in the revision of the *Lixiang kaocheng* (Compendium of

---

[19]Cf. Foucquet (1), *1716*, p. 81.

[20]Cf. above, p. 165.

[21]*CRZ*, j. 48.

[22]On Mongol mathematics, cf. Shagdarsüren (1), *1982*.

observational and computational astronomy) (1742) and that in 1756 he took part in the topographical survey of the new territories of the West (the present-day Xinjiang) which were newly conquered by Qianlong. Finally, in 1759, shortly before his death, he became President of the Bureau of Astronomy. On his death, Minggatu left an incomplete manuscript. This precious manuscript fell into the hands of his best pupil, Chen Jixin:

> In his youth, Master Ming'antu [Minggatu], Director (*jianzheng*) of the Bureau of Astronomy (*qintianjian*) learnt mathematics under the aegis of the Emperor Kangxi. He then devoted his whole life to this discipline. When he fell gravely ill, he confided the manuscript [of his work] to his youngest son Jingzhen and asked me, Jixin, to continue it until it was complete. He said to me: "this work entitled *Geyuan milü jiefa* (Fast method for obtaining the precise ratio of the division of the circle) contains three procedures for finding, resp.: the length of the circumference as a function of the diameter and the chord and the arrow as a function of the arc. All that comes from writings by P. Jartoux. Indeed, nothing like it exists in ancient or modern mathematics. I would have liked my colleagues to have been able to use them, unfortunately, Jartoux's text contains recipes without justifications. Therefore, I did not wish to divulge them, for fear of providing those who might find them with the 'golden needle' without the secret to the way of using it.[23] I have built up arguments (*jie*) over many years without succeeding in completing the task I gave myself. Thus, I would like my work to be continued." I, Chen Jixin, continued this research after the death of my master [...]. When I encountered difficulties I talked about them with Jingzhen and Zhang Liangting,[24] a student [of my master]. In 1774, I finally managed to complete the work [...].[25]

However, the text remained in the manuscript state for a very long time before it was first xylographed in 1839.[26] However, Minggatu's text was studied seriously well before that date. In 1808 and in 1819, the mathematician Zhu Hong showed a manuscript copy of the text to Wang Lai (1768–1813) and to Dong Youcheng[27] (1791–1823). This explains the fact that writings on the subject appeared well before 1839.[28]

When working on infinite series expansions, Chinese mathematicians all used *ad hoc* methods without ever having recourse to the techniques of infinitesimal calculus which were then common currency in Europe. They attacked this type of question with perseverance and tenacity, using tools such as elementary geometrical considerations (similarity of triangles, metrical relationships, manipulations using proportions, algebraic calculations using polynomials with numerical coefficients (Fig. 19.2), trigonometry).

---

[23]Classical expression due to the poet Yuan Haowen (1190–1257) meaning "to teach someone a technique without breathing a word about the principles on which the technique is based". Cf. *DKW* 11-40152: 621, p. 12009.

[24]i. e. Zhang Gong (he was an intendant in the ministry of finance). Cf. *CRZ*, j. 48, p. 628.

[25]Chen Jixin's preface to *Geyuan milü jiefa* in *Guanwoshengshi huigao* (1839).

[26]By Luo Shilin in the collection *Guanwoshengshi huigao*.

[27]*CRZ*, j. 48, notice on Ming'antu [Minggatu].

[28]Cf. H. Kawahara (3′), *1989*, p. 245.

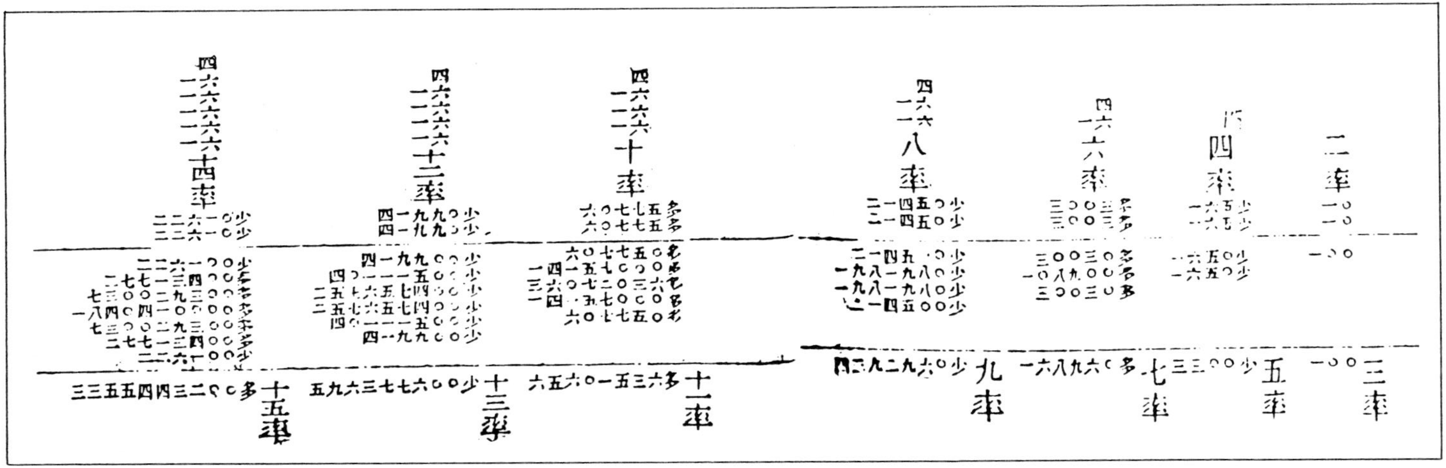

Fig. 19.2. Minggatu's computation of

$$\left(10\,x - \frac{165\,x^3}{4} + \frac{3003\,x^5}{4^3} - \frac{21450\,x^7}{4^5} + \frac{60775\,x^9}{4^7} - \frac{41990\,x^{11}}{4^9} - \frac{22610\,x^{13}}{4^{11}}\right)^2$$

$$= 100\,x^2 - \frac{3300\,x^4}{4} + \frac{168960\,x^6}{4^3} - \frac{4392960\,x^8}{4^5} + \frac{65601536\,x^{10}}{4^7} - \cdots$$

Source: *Geyuan milü jiefa*, 1839, j. 3, pp. 34b–35a. Detailed analysis: see Luo Jianjin (6'), *1990*, pp. 202 ff. Note that, in particular, *duo* 多 and *shao* 少 (lit. "much" and "short of", resp.) mean + and –. Note also the way powers of 4 are noted using columns of 4 and 16 superposed.

Later, around 1845, they also used the binomial expansion for non-integer powers. They relied heavily on reasoning by analogy, replacing numbers such as 0.3333 by 1/3 after audaciously taking limits in a purely intuitive manner, and never worried about questions of convergence.[29]

For example:

To calculate the length of the chord C of an arc (denoted here by $2a$), Minggatu and his pupils began by looking for a formula which would give C as a function of the chord C_n subtending the nth part of the arc in question. In so doing, they discovered that the parity of n plays an important role. For odd n, all the series obtained contained only a finite number of terms, while for even n, the series were infinite. For example[30]:

$$\text{For } n = 2 \qquad C = 2C_2 - \frac{C_2^3}{4r^2} - \frac{C_2^5}{4 \cdot 16 \cdot r^4} - \frac{C_2^7}{4 \cdot 16^2 \cdot r^6} - \cdots \qquad (i)$$

$$\text{For } n = 3 \qquad C = 3C_3 - \frac{C_3^3}{r^2} \qquad (ii)$$

$$\text{For } n = 5 \qquad C = 5C_5 - 5\frac{C_5^3}{r^2} + \frac{C_5^5}{r^4} \qquad (iii)$$

Applying (iii) to C_2 and C_{10}, they obtained (iv)

$$C_2 = 5C_{10} - 5\frac{C_{10}^3}{r^2} + \frac{C_{10}^5}{r^4} \qquad (iv)$$

and replacing C_2 by its value in (i):

$$C = 10C_{10} - 165\frac{C_{10}^3}{4 \cdot r^2} + 3003\frac{C_{10}^5}{4 \cdot 16 \cdot r^4} - \cdots$$

In other words, knowing C as a function of C_2 and C as a function of C_5, they were able to calculate C as a function of C_{10}. Using the same logic to combine the formulae for even indices with those for odd indices, they successively obtained C as a function of C_{100}, C_{1000} and even C_{10000}. Then, assuming that 10000 small chord portions C_{10000} are practically identical to the arc, they deduced that:

$$C = 2a - 0.166666665\frac{(2a)^3}{4r^2} + 0.03333333\frac{(2a)^5}{4 \cdot 16 \cdot r^4}$$

$$-0.003174603\frac{(2a)^7}{4 \cdot 16^2 \cdot r^6} + \cdots$$

---

[29] Cf. the descriptive presentation of Chinese works on infinite series in QB, *Hist*, p. 300 ff.; *ZSSLC-T*, III, p. 197 ff.; He Shaogeng (1') to (4'), *1982 to 1989*; Luo Jianjin (6'), *1990*; Kawahara (3'), *1989*.

[30] For this, they took a geometrical approach, generalising the procedure for trisecting arcs given in the *Shuli jingyun*, "xiabian" (i.e. second part), j. 16, p. 745.

Finally, replacing approximate values by exact fractions (for example, 0.16666666 by 1/6, 0.03333333 by 1/30), they found that:

$$C = 2a - \frac{(2a)^2}{4 \cdot 3! r^2} + \frac{(2a)^5}{4^2 \cdot 5! r^4} - \frac{(2a)^7}{4^3 \cdot 7! r^6} + \ldots$$

In the same vein, Li Shanlan's brother Li Xinmei explained the infinite series expansion of $1 - \sqrt{1 - x^2}$ by taking $x^2 = 10^{-8}$ and observing that[31]

$$
\begin{aligned}
1 - \sqrt{1 - 10^{-8}} &= 1 - \sqrt{0.99999999} \\
&= 0.000000005000000012500000006250000003906250027343750\ldots \\
&= 0.5 \times 10^{-8} + 0.125 \times 10^{-16} + 0.0625 \times 10^{-24} + 0.0390625 \\
&\quad \times 10^{-32} + 0.02734375 \times 10^{-40} + \ldots
\end{aligned}
$$

whence

$$1 - \sqrt{1 - x^2} = \frac{1}{2} x^2 + \frac{1}{4!!} x^4 + \frac{3!!}{6!!} x^6 + \frac{5!!}{8!!} x^8 + \frac{7!!}{10!!} x^{10} + \ldots$$

(here, if n is odd (respec., even), $n!!$ denotes the product of all the successive odd (resp., even) numbers from 1 (resp., 2) to n). Other Chinese mathematicians were not afraid of calculations using enormous numbers and, for example, Minggatu also uses numbers with up to 52 figures.[32] This feature presents a material difficulty as far as the historical study of these texts is concerned and it would undoubtedly be advantageous to use software capable of carrying out formal calculations and manipulating integers with unlimited accuracy in order to reconstitute the content in a true and accurate translation.

---

[31] From Mei Rongzhao (11′), *1990*, p. 355.
[32] From Luo Jianjin (5′), *1988*, p. 243.

20. Magic Squares and Puzzles

In China, the origin of magic squares[1] is lost in the mist of legends. Zou Yan (305–240 BC), the patron of magicians, is said to have manipulated the square of order three;[2] the same can apparently be said of Zhang Heng (78–139 AD), the patron of *yinyang*, who is better known as a calendarist and constructor of armillary spheres.[3]

In fact, the first indisputable reference to a magic square occurs in the *Da Dai Liji* (Record of Rites by Tai the Elder), which was compiled in the first century AD from older sources.[4] A short sentence of this work assigns nine numbers to each of the "Nine Halls" of the Ming Tang (royal house of the calendar),[5] not in an arbitrary way, but in the order 2, 9, 4; 7, 5, 3; 6, 1, 8 (so that the sums are all equal to 15). Later, in the sixth century, Zhen Luan, the commentator on the *Shushu jiyi*, explains in a vivid way that of the "Nine Halls"[6] *jiu gong*:

2 and 4 are the shoulders, 6 and 8 the feet, 3 the left and 7 the right, 9 the head, 1 the shoe and 5 the centre.[7]

This anthropomorphic description also corresponds to that of a square of order three once noted that numbers are laid out from right to left in accordance with the direction of Chinese writing.

Later still, in the 10th century, this same square is represented in the well-known pseudo archaic form (*Luoshu* (The Luo River Diagram, Fig. 20.1)) of small black or white circles which may or may not be interlinked.[8]

There is no evidence for larger magic squares in China until much later; they apparently occur first in Yang Hui's *Xugu zhaiqi suanfa* (1275).[9] In this

---

[1]A series of numbers arranged in a square is said to form a "magic square" if the partial sums of these for all lines, columns or diagonals are all identical. By extension, an arbitrary figure consisting of numbers is said to be "magic" if the sums of certain groups of numbers of this figure taken in certain directions or along certain paths are all identical.

[2]Cammann (3), *1962*, note 1, p. 14.

[3]Cammann (1), *1960*, p. 118. On Zhang Heng see also Ngo Van Xuyet (1), *1976*, p. 43.

[4]This is an apocryphal work. See the notice on page 107 of vol. 1 of Yamane (1′), *1983*.

[5]Granet (1), *1934*, p. 90 ff.

[6]On the "Nine Halls," see Morgan (2), *1981*.

[7]QB, II, p. 544.

[8]The *Luoshu* pattern is first mentioned in the *Yijing*, but the relationship between this pattern and the magic square of order 3 seems unclear.

[9]Complete English translation of this text in Lam Lay-Yong (6), *1977*, p. 145 ff.

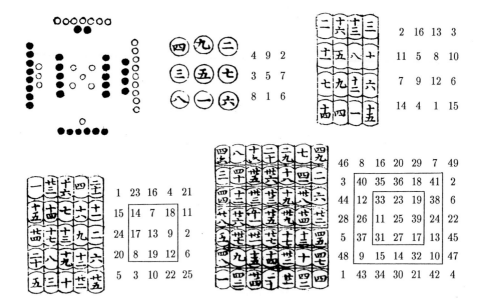

Fig. 20.1. The *Luoshu.* Yang Hui's magic squares. From Yang Hui's *Xugu zhaiqi suanfa, j.* 1 (p. 71 in Kodama (1'), *1966*)).

late book, it seems that the divinatory tradition is already dead and that the author only wanted to use arithmetical curiosities to whet the interest of his readers. In fact, Yang Hui never explains how magic squares may be put to esoteric uses and, moreover, the famous arithmetician gives them a generically neutral name *zonghengtu* (lit. transverse/longitudinal diagrams) which does not suggest that they are in some way magical. Indeed, except in the case of the 3 × 3 square, he does not represent his squares by configurations analogous to those of the *Luoshu* and reminiscent of the *yinyang* polarities, but prosaically uses simple figures written inside identical cells (Fig. 20.1).

Generally speaking, Yang Hui does not explain the construction of his squares, except for the squares of order 3 and 4.[10] Consequently, one may speculate at leisure on the rules he used to construct them. All this could involve long discussions of dubious historical interest, however, it is impossible not to note the curious fact that two of his squares of order five and seven, respectively, are bordered[11] (Fig. 20.1). There is already evidence for this particular type of structure in the works of certain mathematicians of the Islamic world (notably in the works of al-Būnī, who died in 1225) prior to Yang Hui. This could be a pure coincidence, even though the tradition of magic squares in the Islamic

---

[10]Ibid., p. 293.

[11]The term "concentric" is also used. This refers to squares of order n, constructed from a magic square of order $n - 2$ by surrounding it with cells containing suitable numbers. Thus, these are squares which remain magic when one or more outer edges are removed, where the edge includes all the peripheral cases. See Cazalas (1), *1934*, note 4, p. 8.

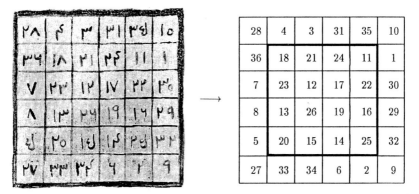

28	4	3	31	35	10
36	18	21	24	11	1
7	23	12	17	22	30
8	13	26	19	16	29
5	20	15	14	25	32
27	33	34	6	2	9

Fig. 20.2. Metallic plaque engraved with a magic square in Arabic figures (discovered in 1956 in the ruins of the palace of the Prince of Anxi, in the present-day suburb of Xi'an, dating back to the Mongol period). Cf. *SY, 1966*, pp. 263–264.

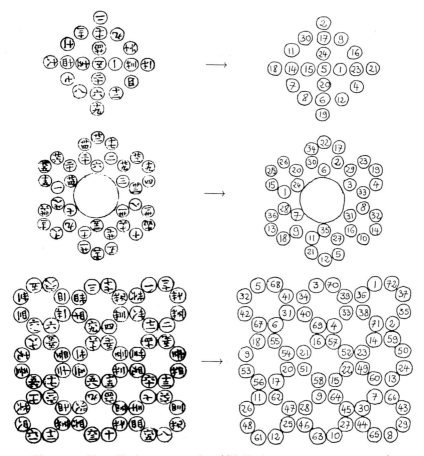

Fig. 20.3. Yang Hui's magic circles. (Cf. Kodama, op. cit., pp. 72–73).

countries dates back to 900 AD, while in China there is no evidence for anything other than the square of order three before the end of the 13th century. However, in 1956, during archaeological excavations in Xi'an, a bordered magic square of order six engraved in Arabic numbers on five metallic plaques was discovered (Fig. 20.2).[12]

In any case, in addition to squares, Yang Hui also describes other magic figures consisting of interlaced patterns of circles with constant sums[13] (Fig. 20.3). As far as we know, this type of figure is not found anywhere outside China during the medieval period.

After Yang Hui, many Chinese authors studied the same subject.

Puzzles

Puzzles and pastimes are never mentioned in traditional Chinese mathematical books and what can be gleaned about them is widely disseminated throughout the ocean of Chinese general literature. For this reason, the subject has never been thoroughly investigated.

Tait's Counter Puzzle

"Tait's counter puzzle"[14] (Tait, i.e. the British mathematician Peter Guthrie Tait (1831–1901). See the entry 'Tait' in the *DSB*) includes $2n$ counters, n white and n black continuously placed in a row. Any two adjacent counters may be moved jointly over others and occupy together some blank space: it is required to form a line of alternately black and white counters in a certain number of moves. According to Hayashi Tsuruichi (1873–1935), a famous Japanese historian of mathematics, this puzzle (or rather its reverse where the initial and final states are exchanged) occurs in a Japanese source dated 1743[15] but according to Liu Dun[16] it was solved correctly still earlier for $n = 3, 4, \ldots, 10$ in Chu Renhuo's collection of short and casual notes typical of *biji* literature, the *Jianhu ji* (literally, "Collection of the solid gourd"), cf. Fig. 20.4. The "solid gourd" alluded to here is intended to mean "something useless, a trifle": a solid gourd is not hollow and thus cannot serve

---

[12]Cf. *SY*, p. 264. On the history of magic squares in the Islamic world, see the papers by W. Ahrens in the journal *Der Islam*, *1917*, vol. 7, pp. 186–250, 1922, together with Sesiano (1), *1980* and (3), *1987*. On the history of magic squares in India, see Hayashi (Takao) (1′), *1986* and Roşu (1), *1989*. Finally, on the history of magic squares in Europe, see Folkerts (2), *1981*.

[13]Cf. Li Yan, *ZSSLC-T*, III, pp. 59–110.

[14]See P. G. Tait, *Collected Scientific Papers*, Cambridge, II, *1890*, p. 93.

[15]See Hayashi (Tsuruichi) (1′), *1937*, II, pp. 633–634 and 735–747.

[16]Liu Dun (4′), *1989* and (5′), *1993*, p. 353 ff.

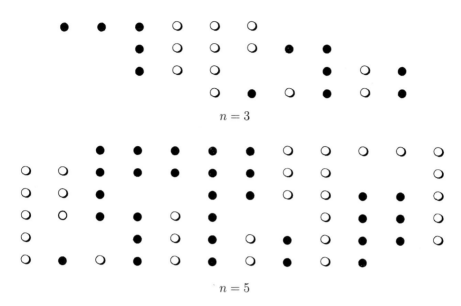

Fig. 20.4. Reconstitution of Chu Renhuo's solution of the so-called Tait's counter puzzle for $n = 3$ and $n = 5$ (cf. Liu Dun (4′), *1989*, p. 50).

any useful purpose as a vessel.[17] This image was quite adequate because *biji* literature but was often considered futile for Confucean scholars. Chu Renhuo (ca. 1630– ca. 1705) is best known for having revised the *Sui-Tang yanyi* (Romance of the Sui and Tang).

The *qiqiaotu* (Seven Subtle Shapes) or Tangram

The tangram is a planimetric puzzle composed of seven pieces (5 triangles, 1 square and 1 lozenge) born of a geometrical dissection of a square. Assembled in astute ways, these pieces elegantly generate an incredible number of various shapes given in advance (animals, plants, human beings, things from everyday life, Chinese characters, geometrical shapes, etc.) (Fig. 20.5).

As explained in Joost Elffers's excellently documented book *Tangrams* (Penguin, 1976), the oldest known printed source on tangrams dates back to the beginning of the 19th century. However, as recently shown by two historians of Chinese mathematics, Liu Dun and Guo Zhengyi,[18] a few puzzles, similar but not exactly identical to tangrams, are attested in *biji* literature from 1617. In

---

[17]See W.K. Liao (transl.), *The Complete Works of Han Fei Tzu*, II, London (Probsthain's Oriental Series vol. 26), *1959*, p. 40.

[18]See Liu Dun (4′), *1989*, p. 52 and Guo Zhengyi (1′), *1990*.

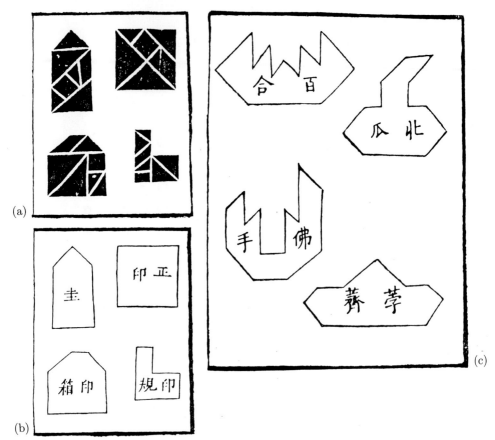

Fig. 20.5. (a) shows the 7 basic shapes of the puzzle assembled in the form of a square; (b) and (c) show various shapes realisable using the puzzle (with solutions). Source: Joost Elffers, *Tangrams*, Paris: Société Nouvelle des Editions du Chêne, pp. 13, 14. Original edition: Joost Elffers, Köln, 1973.

particular, a short verbal description of one of them appears in *juan 3* of Liu Xianting's (1648–1695) *Guangyang zaji*.[19] In Europe a similar puzzle is referred to by various authors from antiquity, among whom Archimedes.[20]

From the beginning of the 19th century, tangrams have known a wide diffusion all over the world. In 1862, a variant of tangram called *yizhitu* (the shapes increasing wisdom) was published in China. This new puzzle is still based on a dissection of a square but into 15 planar pieces of which 2 are a half disc (Fig. 20.6). But only tangrams have enjoyed a lasting success in Occidental countries.

---

[19]See Martzloff (2), *1981*, p. 294; Liu Dun, op. cit., p. 52.
[20]See Dijksterhuis (1), *1987*, pp. 408–412.

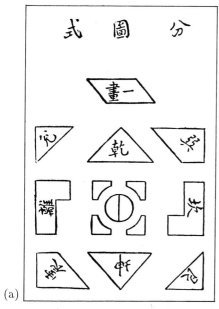

(a)

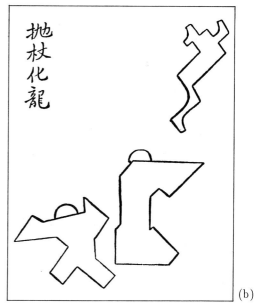

(b)

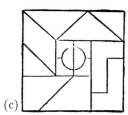

(c)

Fig. 20.6. A variant of tangrams called *yizhitu*. (a) shows the 15 basic pieces of the puzzle, two of which are half discs. Some pieces are marked with Chinese characters (the eight trigrams of the *Yijing*). (c) shows the pieces assembled to form a square. (b), (d) and (e) represent some of the multifarious shapes allowed by the puzzle (without solutions). Source: *Yizhitu, 1878.* (cited from the reprint of the text published by Shangwu Yinshuguan ca. 1930 (no place name, no date).

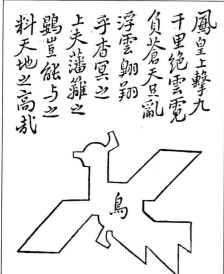

(d)

(e)

The Linked Rings

Among other puzzles attested in China, one of the best known is the *jiulianhuan* (nine linked rings) and its variants (Fig. 20.7). Incidentally, this puzzle is mentioned in passing in chapter 7 of *Hongloumeng* (the dream of the red chamber), a famous novel first published in its 120-chapter form in 1791.

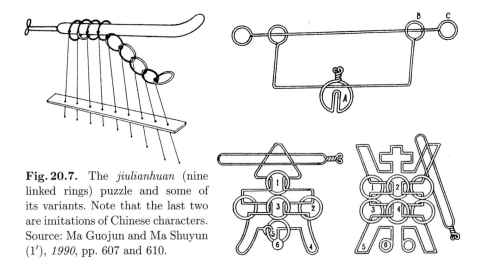

Fig. 20.7. The *jiulianhuan* (nine linked rings) puzzle and some of its variants. Note that the last two are imitations of Chinese characters. Source: Ma Guojun and Ma Shuyun (1′), *1990*, pp. 607 and 610.

In their original context, these Chinese rings, tangrams and other puzzles were looked upon much more as pleasant pastimes than as devices susceptible of mathematical analysis. However, mathematical treatments of various Chinese puzzles have been devised. In particular, the French mathematician Edouard Lucas has shown that the steps necessary to separate the rings from a set of linked rings can be related to binary numbers. The interested reader will find an adequate bibliography in Singmaster (1), *1988*.

Appendix I: Chinese Adaptations of European Mathematical Works (from the 17th to the Beginning of the 20th century)

This list, which is not exhaustive, contains 46 titles of works classified according to subject. It is essentially based on the works of Li Yan (13′), *1927* and (42′), *1937*; Xu Zongze (1′), *1958*; Henri Bernard-Maître (7), *1945–1960* and Adrian Arthur Bennett (1), *1967*, not forgetting Maurice COURANT's *Catalogue des livres chinois, coréens, japonais, etc. de la Bibliothèque Nationale* (Paris 1902–1912) and other fundamental sources on the subject such as Carlos SOMMERVOGEL, *1890, Bibliothèque de la Compagnie de Jésus*, Brussels–Paris; Louis PFISTER, *1932–1934, Notices biographiques et bibliographiques sur les Jésuites de l'ancienne Mission de Chine*, 2 vols, Shanghai or Joseph Dehergne (1), *1973*.

Although we did not judge it useful to list the complete content of these various secondary sources (for they are readily accessible but not always reliable), information which cannot be found herein has been included.

The sources of Chinese adaptations of European works have been given whenever known. In general, however, it is very difficult to determine the precise edition of a particular European work on which adaptations are based. For this reason, whenever possible, we provide indications about successive editions of European works.

No attempt has been made to list the publishers of Chinese books and the dates of their successive editions. However, the reader will readily find information on the subject in Ding Fubao and Zhou Yunqing (1′), *1957*.

In what follows: Bennett = Bennett (1), *1967*; BN = Bibliothèque Nationale (Paris); HB = Bernard-Maître (7), *1945–1960*; Xu Zongze = XU Zongze (1′), *1958*; *CBPT = Catalogue de la Bibliothèque du Pé-T'ang* (full reference in the bibliography of Western works under the entry "Mission Catholique des Lazaristes à Pékin (1)"). The numbers given after BN, HB and *CBPT* refer to the serial or classification numbers of the corresponding books; in particular, the BN number corresponds to the classification mark of the corresponding Chinese book as found in the the "Fonds Chinois" of the Bibliothèque Nationale.

Algebra

ca. 1696 • Original Chinese title unknown.

Treatise of algebra, by Antoine Thomas, in Manchu, intended for lessons to the Emperor Kangxi. It is not known whether this manuscript is still extant. Mentioned in Mme Yves de THOMAZ DE BOSSIERRE, *1977, Un Belge Mandarin à la Cour de Chine aux XVIIe et XVIIIe siècles, Antoine Thomas (1644–1709), Ngan To P'ing-Che.* Paris: Les Belles Lettres, p. 164.

HB, no. 558, gives the date 1690 and mistakenly identifies this manuscript with another one by Jean-François Foucquet (see the next notice).

1712 • *A'erribala xinfa* 阿爾日巴拉新法

New method of algebra
Biblioteca Apostolica Vaticana: *Borg. Cin. 319 (4).*
HB no. 558 (wrong identification of the author of the manuscript)
Manuscript due to the initiative of Jean-François Foucquet, S. J. (1665–1741) in the context of special lessons given by the missionaries to the Emperor Kangxi.

See John W. WITEK, S. J. *1982, Controversial Ideas in China and in Europe, A Biography of Jean-François Foucquet, S. J. (1665–1741)*, Rome, 1982, Institutum Historicum S. I., p. 453. See also Jami (1), *1986* and Foucquet's French manuscript on algebra, also preserved in the Bibliotheca Apostolica Vaticana, Borg. lat. 566, 78–91v.

ca. 1700 • *Jiegenfang suanfa jieyao* 借根方算法節要

Summary of the computational methods of the 'borrowing of roots and powers' (a sort of algebra).
A copy of this manuscript is preserved in the library of the Tōhoku University (Sendai, Japan).
See also Li Yan (42'), p. 220.

1859 • *Daishuxue* 代數學

Algebra
The term *daishu* coined by Alexander Wylie and Li Shanlan is translated from 'substitutionary arithmetic', a term invented by Isaac Newton (*dai* means 'to substitute, substitution' and *shu* 'arithmetical prescriptions'; *xue* means 'knowledge').
BN no. 4885
Translated by Alexander Wylie and Li Shanlan from Augustus DE MORGAN, *1835, The Elements of Algebra Preliminary to the Differential Calculus*, London: J. Taylor (other edition in 1837).
See Wang Ping (1), *1962.*
On De Morgan (1806–1871) see the notice in the *DSB*.
Japanese reprinted in 1872 (*Meijizen*, V, p. 429).

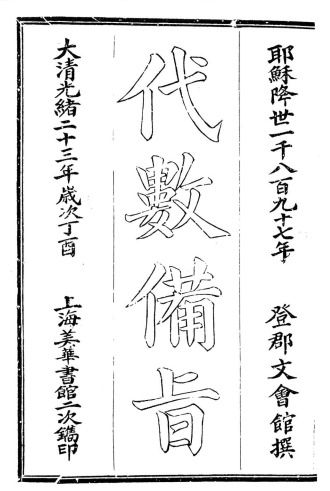

Fig. A.1. Title page of the second edition of the *Daishu beizhi* (1897).

Special feature: English preface by Alexander Wylie incorporated in the Chinese edition of the text.

1873 ● *Daishu shu* 代數術

Litt. 'Algebraic computational prescriptions.'
Translation by John Fryer and Hua Hengfang.
Source: article on *algebra* in the 8th edition of the *Encyclopaedia Britannica* by William Wallace (Bennett, p. 84).

1879 ● *Daishu nanti jiefa*, 16 j. 代數難題捷法

Solution of difficult algebraic problems.
Translation by John Fryer and Hua Hengfang.

Source: Thomas LUND, *1860, A Companion to Wood's Algebra, containing solutions of various questions and problems in algebra, and forming a key to the chief difficulties found in the collection of examples appended to Wood's Algebra.* Third edition, London: Green, Longman and Roberts (previous editions in 1840 and 1847).

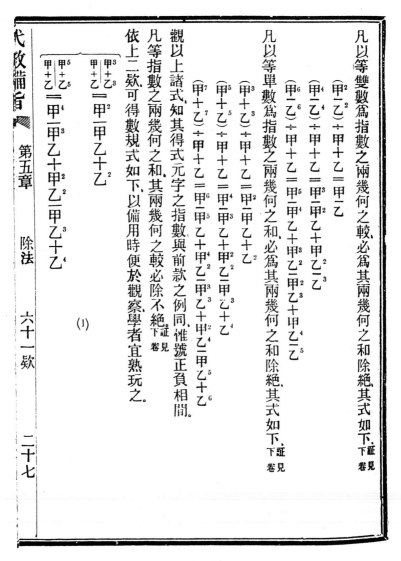

Fig. A.2. A page of algebra from the *Daishu beizhi* (op. cit.). Note the simultaneous use of European and Chinese algebraic symbols in the composition of formulae. The second and third columns from the right, respectively, mean: $\frac{a^2-b^2}{a+b} = a - b$ and $\frac{a^4-b^4}{a-b} = a^3 - a^2b + ab^2 - b^3$.

1887 • *Daishu xu zhi*, 13 j. 代數須知

What one needs to know about algebra (Bennett, p. 85).

1891 • *Daishu beizhi* 代數備旨

Complete Explanation of Algebra
Translation by Calvin W. Mateer, Zhou Liwen and Sheng Fuwei. See Li
Yan, op. cit., p. 288.
See Figs. A.1 and A.2.

1889 • *Suanshi jiefa* 算式解法

The solution of mathematical formulae.
Translated by John Fryer and Hua Hengfang (Bennett, p. 85).
Source: Edwin James HOUSTON an A. E. KENNELLY, *1898, Algebra made
easy: being a clear explanation of the mathematical formulae found in Prof.
Thompson's dynamo-electric machinery and polyphase electric currents.*
New York (other edition in 1900).

1900 • *Daishu shu bushi* 代數術補式

Complements of Algebra.
Source: Not identified but the author is probably the same Wallace as the
one mentioned above. Translated by John Fryer and Hua Hengfang.

Arithmetic

1614 • *Tongwen suanzhi (qian bian, tong bian)*, 2 j. + 8 j. 同文算指

Combined learning mathematical indicator ('preliminary part' and
'general part').
HB nos. 79 and 99; BN nos. 4861–63; Xu Zongze p. 265.
Translation by Matteo Ricci and Li Zhizao.
Sources: Clavius, *1585, Epitome Arithmeticae Practicae*, Rome (other
editions in 1592 and 1593). See Clavius's *Opera matematica* published
in 5 vols in 1611–1612 (The *Arithmeticae practicae* is in vol. 2). See also
the *CBPT*, nos. 1288 and 1296.
As already noted, besides Clavius, the *Tongwen suanzhi* is also based
on numerous arithmetics from the Yuan and Ming periods (cf. Takeda
(3′), *1954*). Note also that the *Tongwen suanzhi* was reprinted in 1936
(Shanghai: Shangwu Yinshuguan) in 3 volumes the first of which not only
reproduces the text of the *Tongwen suanzhi qianbian* but also another
Chinese mathematical text which has nothing to do with the *Tongwen
suanzhi*, the *Ding Ju suanfa* (Ding Ju's arithmetic, ca. *1355*). This has
given rise to an error in a modern historical work by Karl Menninger (see
Fig. A.3).

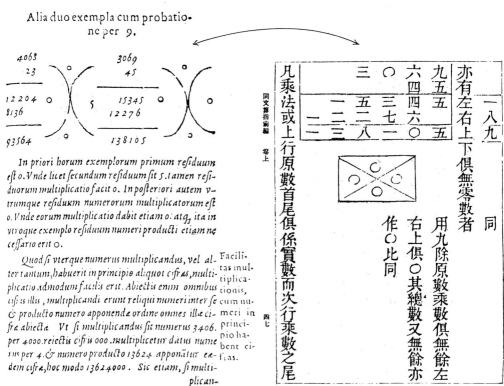

altero numero multiplicato. Sed hoc examen planius
intelligetur, cum Diuisio explicata fuerit.

Alia duo exempla cum probatio-
ne per 9.

In priori horum exemplorum primum residuum
est o. Vnde licet secundum residuum sit 5. tamen resi-
duorum multiplicatio facit o. In posteriori autem v-
trumque residuum numerorum multiplicatorum est
o. Vnde eorum multiplicatio dabit etiam o: atque ita in
vtroque exemplo residuum numeri producti etiam ne-
cessario erit o.

Quod si vterque numerus multiplicandus, vel alter tantum, habuerit in principio aliquot cifras, multiplicatio admodum facilis erit. Abiectis enim omnibus
cifris illis, multiplicandi erunt reliqui numeri inter se
& producto numero apponendæ ordine omnes illæ ci-
fræ abiectæ. Vt si multiplicandus sit numerus 3406.
per 4000. reiectis cifris 000. multiplicetur datus numerus per 4. & numero producto 13624. apponatur ea-
dem cifra, hoc modo 13624000 . Sic etiam, si multi-
plican-

Facili-
tas mul-
tiplica-
tionis,
cum nu-
meri in
princi-
pio ha-
bent ci-
fras.

Fig. A.3. The multiplication 3069×45 = 138105 computed on paper, by writing, with its proof by casting out nines, first presented as an example in Clavius's *Epitome Arithmeticae Practicae* (cited from the 1592 edition, p. 45) and translated into Chinese in the *Tongwen suanzhi qianbian* (1614) (cited from the Shangwu Yinshuguan (Commercial Press) reprint of the text, Shanghai, 1936, p. 47). In his celebrated *Number Words and Number Symbols, A Cultural History of Numbers*, fig. 270, p. 462 (cited from P. Broneer's English translation, New York, Dover, 1992), Karl Menninger erroneously attributes this example of multiplication to the *Ding Ju suanfa* (Ding Ju's arithmetic), a work which was published ca. 1355 and which has nothing to do with the *Tongwen suanzhi*. The reason for this error is most probably a consequence of the fact that in its 1936 edition, the text of the *Tongwen suanzhi qianbian* is edited together with that of the *Ding Ju suanfa* in the same volume.

ca. 1700   • *Suanfa zuanyao zonggang.* 算法纂要總綱

General survey of important methods of computation.
The content of this arithmetical manuscript is similar to that of the corresponding part of the *Shuli jingyun.*
According to *CRZ*, j. 40, p. 506, its unknown author may have been a member of Nian Xiyao's clan. In addition, the content of the *Suanfa zuanyao zonggang* is wholly the same as that of the Manchu manuscript no. 191 of the Bibliothèque Nationale (cf. Martzloff (7), *1984*). Thus, the *Suanfa zuanyao zonggang* is quite probably a Chinese translation of the Manchu manuscript no. 191 (or conversely). But the latter hypothesis seems more probable for the text of the Manchu manuscript no. 191 is clearly unfinished whereas its Chinese counterpart is hardly distinguishable from a printed text.
The notice no. 191 of the *Catalogue du Fonds Mandchou* (Paris, Bibliothèque Nationale, *1979*) asserts that the *Suanfa zuanyao zonggang* is a translation of Cheng Dawei's *Suanfa tongzong* (1592). According to what precedes, this identification is clearly incorrect. In particular, the way arithmetical operations are performed in the Manchu manuscript is close to that of the *Shuli jingyun,* and not at all close to that of the *Suanfa tongzong.* All this is probably the result of the activity of missionaries during the late Kangxi period. See Fig. A.4.

1853   • *Shuxue qimeng,* 2 j. 數學啓蒙

Compendium of arithmetic,
Alexander Wylie (Li Yan (42'), *1937*, p. 287).

1879   • *Shuxue li,* 9 j. + 1 j. 數學理

The principles of mathematics (litt. 'the science of numbers'. *shu =* number, *xue =* knowledge, science, *li =* principles).
Translation by John Fryer, (Bennett, p. 116).
Source: Augustus DE MORGAN, *1830, The Elements of Arithmetic,* London (other editions: 1832, 1835, 1846, 1848, 1851, 1853, 1854, 1857, 1858, 1869, 1876).

1886   • *Xin suan qimeng,* 1 j. 心算啓蒙

Introduction to mental arithmetic.
Published in Shanghai by H.V. NOYES (Bennett, p. 116; Li Yan, op. cit., p. 287).

1892   • *Bisuan shuxue,* 4 j. 筆算數學

The mathematics of computation with the brush (written computation).
Translated by Calvin W. Mateer (Li Yan, op. cit., p. 287).

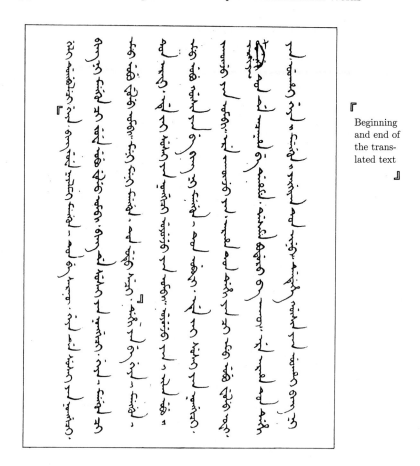

Beginning
and end of
the trans-
lated text

Fig. A.4. An arithmetical problem from the Manchu manuscript preserved at the Bibliothèque Nationale (Fonds Mandchou, no. 191). The text (which begins in the first column on the left, p. 453 of this manuscript) reads: "giya bing juwe niyalma i menggun i ton be sarkū. giya de susai *yan* nonggici, bing ni menggun ci juwe ubu fulu ombi, bing de susai *yan* nonggici, giya i menggun ci emu ubu fulu ombi. meni meni menggun i ton udu seci". Translation: two men, A and C, possess an unknown amount of *liang* [a monetary unit usually translated as 'ounce' or 'tael']. If A receives 50 *liang*, then he becomes 3 times as rich as C; if C receives 50 *liang*, then he becomes twice as rich as A. How many *liang* do A and C possess, respectively? [Note: what have been translated here by A and C corresponds to the cyclical characters *jia* and *bing* which are the first and the third 'celestial trunks' (*tian gan*), respectively].

This problem (in particular) is the same as the 13th of those reproduced in Chinese in the tenth section of the *Suanfa zuanyao zonggang* (cited from the copy of this manuscript preserved in the library of the Tōhoku University (Sendai, Japan)). See the Chinese text as follows:

又設如甲丙二人有銀不知其數。
若是與甲五十兩則比丙多二倍。
若與丙五十兩則比甲多一倍。
問甲丙二人各銀若干。

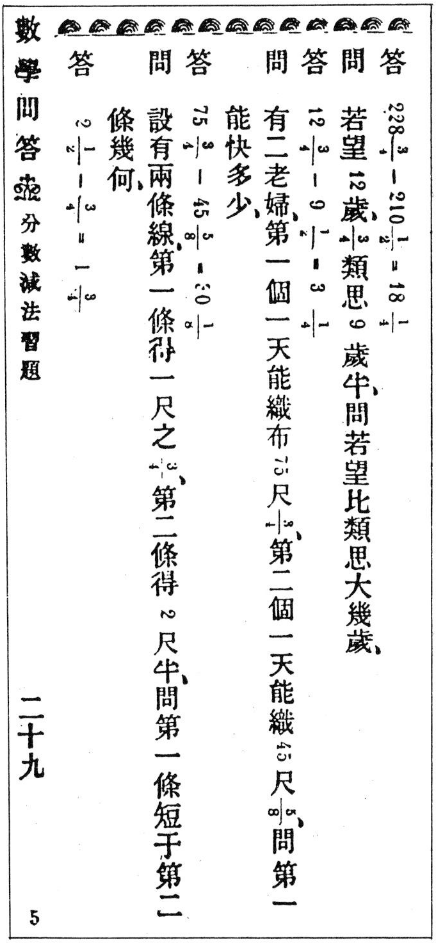

答 228 3/4 − 210 1/2 = 18 1/4
問 若望 12 歲 3/4、類思 9 歲半、問若望比類思大幾歲、
答 12 3/4 − 9 1/2 = 3 1/4
問 有二老婦、第一個一天能織布 75 尺 3/4、第二個一天能織 45 尺 5/8、問第一能快多少、
答 75 3/4 − 45 5/8 = 30 1/8
問 設有兩條線、第一條得一尺之 3/4、第二條得 2 尺半、問第一條短于第二條幾何、
答 2 1/2 − 3/4 = 1 3/4
數學問答 卷 分數減法習題
二十九
5

Fig. A.5. Page 29 of the *Shuxue wenda* (1912 reprint). This little primer was intended for the teaching of arithmetic in Jesuit elementary schools. Its use of Arabic numbers is noteworthy. Moreover, the book contains an interesting multilingual (English, German, French and Chinese) glossary of arithmetical terms (see Fig. A.6).

1901 • *Shuxue wenda* 數學文問答

Arithmetical problems with solutions.
P. F. Scherer S. J. (Li Yan, op. cit., p. 288).
See Figs. A.5 and A.6.

Differential and Integral Calculus

1859 • *Daiweiji shiji*, 18 j. 代微積拾級

Elements of algebra and differential and integral calculus.
Translation by Alexander Wylie and Li Shanlan.
Source: Elias Loomis, *1852*, *Elements of Analytical Geometry and of the Differential and Integral Calculus*, New York (other editions: 1853, 1856, 1857, 1858, 1859, 1860, 1864, 1867, 1868, 1869, 1879).

數學問答彙解 數學名目 華法英德合表 六十八

No.	華文	文德 (German)	文英 (English)	文法 (French)
1	數學	Arithmetik	Arithmetic	Arithmétique
2	數目	Eine Zahl	A numbre	Un nombre
3	整數	Eine Integralzahl	An integer	Un nombre entier
4	分數	Eine Bruch	A fraction	Une fraction
5	小數	Eine Decimalzahl	A decimal number	Un nombre décimal
6	號碼	Eine Ziffer	A figure	Un chiffre
7	號碼之值	Der Werth einer Ziffer	The value of a figure	La valeur d'un chiffre
8	號碼之絕對值	Der absolute Werth einer Ziffer	The absolute value of a figure	La valeur absolue d'un chiffre
9	號碼之相對值	„ relative „ „ „	The relative value of a figure	La valeur relative d'un chiffre
10	加法	Die Addition	Addition	L'addition
11	減法	Dio Subtraction	Subtraction	La soustraction
12	乘法	Die Multiplication	Multiplication	La multiplication
13	除法	Die Division	Division	La division
14	總數	Die Summe	The Sum	La somme ou le total
15	減法大數	Der Minuend	The minuend	Le nombre dont on soustrait
16	減法小數	Der Subtrahend	The subtrahend	Le nombre à soustrait
17	餘數或差數	Die Differenz	The remainder or the difference	Le reste ou la différence
18	乘法實數	Der Multiplicand	The multiplicand	Le multiplicande
19	乘法法數	Der Multiplicator	The multiplier	Le multiplicateur
20	積數	Das Product	The product	Le produit
21	除法實數	Der Dividend	The dividend	Le dividende
22	除法法數	Der Divisor	The divisor	Le diviseur
23	除法得數	Der Quotient	The quotient	Le quotient
24	覆驗	Die Probe machen	To make the proof	Faire la preuve
25	帶分數	Ein gemischter Bruch	A mixed number	Un nombre fractionnaire
26	假分數	Ein unaechter Bruch	An improper fraction	Une expression fractionnaire
27	真分數	Ein aechter Bruch	A proper fraction	Une fraction proprement dite
28	分子	Der Zaehler	The numerator	Le numérateur
29	分母	Der Nenner	The denominator	Le dénominateur
30	公分母	Der gemeine Nenner	The common denominator	Le dénominateur commun
31	約分	Den Bruch vereinfachen	To simplify a fraction	Simplifier une fraction
32	約分至最簡	Den Bruch zum kleinsten Ausdruck bringen	To reduce a fraction to ist lowest terms	Réduire une fraction à sa plus simple expression
33	最大公約數	Des groeste gemeine Divisor	The greatest common measure G. C. M,	Le plus grand commun diviseur
34	最小公倍數	Der kleinste gemeine Dividend	The least common multiple L. C. M.	Le plus petit commun multiple
35	原數	Die Primzahlen	A prime number	Un nombre premier

Fig. A.6. The multilingual glossary of the *Shuxue wenda* (1912 reprint, pp. 68 and 69). Note that here *shuxue* is given as equivalent to 'arithmetic', whereas in modern Chinese, *shuxue* generally means 'mathematics',

ALGEBRAIC GEOMETRY, WITH DIFFERENTIAL

AND INTEGRAL CALCULUS.

The present work, which is a translation of Loomis' ANALYTICAL GEOMETRY, AND DIFFERENTIAL AND INTEGRAL CALCULUS, is issued in pursuance of a project formed some time since, as the continuation of a course of mathematics, the first of which, a
1 Compendium of Arithmetic, was published by the undersigned in 1854. The next in order is a Treatise on Algebra, which should have preceded this, but in consequence of unavoidable delays in the publication, it will not be issued till some weeks later. A tolerable acquaintance with the last-named treatise, will put the student in a position to understand the work now presented to the public. Although this is the first time that the principles of Algebraic Geometry have been placed before the Chinese (so far as the translator is aware), in their own idiom, yet there is little doubt that this branch of the science will commend itself to native mathematicians, in consideration of its obvious utility; especially when we remember the readiness with which they adopted Euclid's Elements of Geometry, Computation by Logarithms, and other novelties of European introduction. A spirit of inquiry is abroad among the Chinese, and there is a class of students in the empire, by no means small in number, who receive with avidity instruction on scientific matters from the West. Mere superficial essays and popular digests are far from adequate to satisfy such applicants; and yet when anything beyond that is attempted, the want of a common medium of communication at once appears as an insuperable obstacle; and it is evident that how clearly soever we may be enabled to lay results before the native mind, yet until they understand something of the processes by which such results are obtained, thinkers of the above class can scarcely be supposed to appreciate the achievements of modern science, to repose absolute confidence in the results, or to rest satisfied till they are in a position to some extent to verify the statements which are laid before them. It is hoped that the present translation will in some measure supply what is now a desideratum ; and the translator, while taking this opportunity to testify to the exceeding care and accuracy displayed in the work of Pro-
2 fessor Loomis, considers it is but justice to the native scholar Le Shen-lan, who has assisted in the translation throughout, to state that whatever degree of perfection this version may have attained, is almost entirely due to his efforts and talents.

A list of technical terms used in the works above-named is subjoined.

Fig. A.7. Alexander Wylie's English preface inserted at the beginning of the *Daiweiji shiji*. *1* Compendium of Arithmetic = *Shuxue qimeng*, *2* Le Shen-lan = Li Shanlan.

On Loomis, see the notice in the *DSB*.
See also Mei Rongzhao (1'), *1960*.
Japanese reprint in 1872 (*Meijizen*, V, p. 430).
Special feature: English preface by Alexander Wylie incorporated in the Chinese edition of the text (see Fig. A.7).

1874 • *Weiji suyuan*, 6 j. 微積溯源

The fundamental principles of differential and integral calculus.
Translation by John Fryer and Hua Hengfang.
Source: article on 'fluxions' in the 8th edition of the *Encyclopaedia Britannica* (Bennett, p. 84).

1888 • *Weiji xuzhi*, 1 j. 微積須知

What one needs to know about differential and integral calculus.
Translation by John Fryer (Bennett, p. 85).
Source unknown.

Conics [1]

1888 • *Quxian xu zhi*, 1 j. 曲線須知

What one needs to know about curved lines.
Translation by John Fryer (Bennett, p. 85).
Source unknown.

1893 • *Yuanzhui quxian* 圓錐曲線

The curved lines of the cone.
Source: perhaps a book by E. Loomis.
See Li Yan, op. cit., p. 288.

Geometry

1607 • *Jihe yuanben*, 6 j. 幾何原本

Elements of [Euclidean] geometry.
Translation by Matteo Ricci and Xu Guangqi limited to the first six books of Euclid's *Elements*.
HB nos. 75 and 91; BN nos. 4861–63; Xu Zongze, p. 258; d'Elia (2), *1956*; Martzloff (1), *1980* and (13), *1993*.
Source: Clavius (2), *1574/1591*.
CBPT, nos. 1297, 1298.

---

[1]It should be noted that some information on conic sections had been included in Chinese adaptations of Western works well before 1888, as early as 1630. See Li Yan (49'), *1947*.

1607 • *Gougu yi*, 1 j. 句股義

Explanation of the BASE-LEG [i.e. of the right-angled triangle].
HB no. 73; BN no. 4866; Xu Zongze, p. 272.
Translation by Ricci and Xu Guangqi
(15 solutions of right-angled triangles explained in the Euclidean style).

1614 • *Yuan [Huan] rong jiao yi*, 1 j. 圜容較義

Litt. "the meaning of comparisons between [areas and volumes] of that
which is contained 'in the round' [i.e. 'in the circle or the sphere' or even
'in the heavens' – the Chinese allows all these various interpretations at
the same time]". The text is about isoperimetric figures.
HB nos. 81 and 98; BN no. 4864; Xu Zongze, p. 274.
Translation by Matteo Ricci and Li Zhizao.
Source: Clavius, *1585* [Third edition] *In Sphaeram Ioannis de Sacro Bosco
commentarius.*
CBPT, nos. 1307–1311.
Special feature: a large part of this text is cited in the *CRZ*, j. 44,
pp. 566–567 ('biography' (*zhuan*) of Matteo Ricci).
This work concerns the establishment *more geometrico* of the idea of the
divine perfection, which is apparent in the perfect rotundity of the celestial
sphere, the sphere having the greatest volume of all geometrical figures
with the same surface area. Contains 18 propositions. Includes the first
known Chinese transcription of Archimedes's name in a Chinese text.

1631 • *Jihe yaofa*, 1 j. 幾何要法

Important geometrical constructions.
Translation by Giulio Aleni (1582–1649) and Qu Shigu. HB. no. 184; BN
nos. 4869–4871.
This is a very elementary textbook explaining how to erect a
perpendicular, how to bisect an angle and other such topics using the
ruler and the compass.
Source unknown.

ca. 1721 • *Jihe yuanben*, 2 j., subdivided into 10 chapters.

Elements of geometry.
Despite its title, this book is completely different from the *Jihe yuanben*
mentioned above. It is partly based on Ignace Gaston PARDIES, S. J.,
*1683 Elemens de geometrie, où par une methode courte et aisée l'on peut
apprendre ce qu'il faut savoir d'Euclide, d'Archimède, d'Apollonius, et les
plus belles inventions des anciens et des nouveaux Geomètres.*
CBPT, no. 549.
As already noted (see above, p. 26), this work was reprinted and translated
many times into various European languages. See Ziggelaar (1), *1971*. The
Chinese text referred to here is that found at the beginning of the *Shuli
jingyun* (1723). Between 1690 and 1723, several manuscripts, Chinese and

Manchu were written and used to teach geometry to the Emperor Kangxi.
The *Jihe yuanben* included in the *Shuli jingyun* derives from them.
See Chen Yinke (1'), *1931*.

1635 • *Celiang quanyi*, 10 j. 測量全義

Complete explanation of measurements.
Source: C. CLAVIUS, *1604*, *Geometria Practica*.
See *CBPT*, no. 1300.
Contains a translation of Archimedes's treatise on the mensuration of the
circle (j. 5), a proof of Hero's formula (for the area of the triangle given
its sides) together with the definition of conics and regular polyhedra.
Spherical trigonometry.
See Bai Shangshu (4'), *1984*.

1859 • *Xu Jihe yuanben*, 17 j. 續幾何原本

A sequel to the *Elements* [of Euclid's geometry].
Continuation of the incomplete translation by Ricci and Xu Guangqi
(1607). Subsequently, the complete translation of the *Elements* was
published under the generic title *Jihe yuanben*.
Source: various English editions of the *Elements* (rather than Clavius's
work) (according to Wylie's preface).
Detailed analysis in Horng (1), *1991*, pp. 366–379. *See also p. 386*

1877 • *Suanshi jiyao*, 4 j. 算式集要

Essential mathematical formulae.
Translation by John Fryer and Jiang Heng.
Source: Charles Haynes HASWELL, *1858*, *Mensuration and Practical Geometry; containing tables of weights and measures, vulgar and decimal
fractions, mensuration of areas, lines, surfaces, solids ... To which is
appended a treatise on the carpenter's slide-rule and gauging*, New York:
Harper and brothers (other editions: 1863, 1866, 1875, 1884).

1871 • *Yungui yue zhi*, 3 j. 運規約指

Abridged guide to the use of the compass.
Translation by John Fryer and Xu Jianyin.
Source:
Richard BURCHETT, *1855*, *Practical geometry, a course of construc-
tion of plane geometrical figures for use of art-schools* (other editions:
1858 (second ed.), 1867 (9th ed.), 1874, 1877).
See Bennett, p. 84.

1884 • *Xing xue beizhi*, 10 j. 形學備旨

The complete meaning of the science of figures.
Source unknown (probably from a textbook by Elias Loomis).
See Li Yan, op. cit., p. 287. See Fig. A.8.

Fig. A.8. Title page of the sixth edition (1903) of the *Xingxue beizhi*. The first column of text, on the right, indicates that the first edition of this book was published in 1885.

Instruments

1607 • *Hungai tongxian tushuo*, 2 j. 渾蓋通憲圖說

    Illustrated explanation of the current model of the astrolabe.
    HB no. 74; BN no. 4899; Xu Zongze, p. 263.
    Source: possibly Clavius, *1593, Astrolabium.*
    CBPT, no. 1291.

1607 • *Celiang fayi*, 1 j. 測量法義

    Explanation of practical geometrical methods.
    HB nos. 76 and 108; BN no. 4865; Xu Zongze, p. 268.
    Source: Clavius, *1604, Geometria Practica.*
    CBPT, no. 1300.
    Contains 15 propositions about the use of the geometrical square.

1628 • *Chousuan*, 1 j. 籌算

Computation with [Napier's] rods.
HB no. 164; BN no. 4867.
Included in the *Chongzhen lishu*.
Translated by Giacomo Rho (1592–1639).
Source: John NAPIER, *1617, Rabdologiae, sev numerationis per virgulas.*
These rods are Napier's rods which are used to facilitate multiplication.
They are not based on logarithms but on the ordinary multiplication table.

1650 • *Biligui jie*, 1 j. 比例規解

Explanation of the proportional compass.
HB no. 180; BN no. 4868.
Included in the *Chongzhen lishu*. Translated by Giacomo Rho and Adam
Schall (1592–1666).
Source: GALILEO, *1606, La operazioni del compasso geometrico e militare,*
Padova, per P. Frambotto; Galileo, *1612, De proportionum instrumento a
se invento* [...] *tractatus* [...] *ex italica in latinam nunc primum translatus,*
Strasburg (This work is a Latin translation of the preceding one).
CBPT, no. 1655.
See Yan Dunjie (23'), *1964.*

Logarithms

1653 • *Bili duishu biao*, 1 j. 比例對數表

Litt. 'Tables of corresponding proportional numbers.'
Translation by Nikolaus Smogulecki (1610–1656) and Xue Fengzuo.
42 pages of logarithms to base ten with a 6-figure mantissa. See QB, *Hist.*,
p. 247.

1723 • *Shuli jingyun* 數理精蘊

Collected basic principles of mathematics (j. 5 of the part devoted to
tables).
Logarithms with a 10-figure mantissa.
Source: Henry BRIGGS, *1624, Arithmetica logarithmica.*
See Han Qi (1'), *1992.*

Important note: in a recent and outstanding article, the source of the
Chinese translation of the last nine books of Euclid's *Elements* has been
convincingly identified. Unexpectedly, it turns out that it was Henry
Billingsley's English translation of Euclid's *Elements* (1570). Cf. Xu Yibao,
2005. "The first Chinese translation of the last nine books of Euclid's
Elements and its sources", *Historia Mathematica*, Vol. 32, no. 1, pp. 4–32.

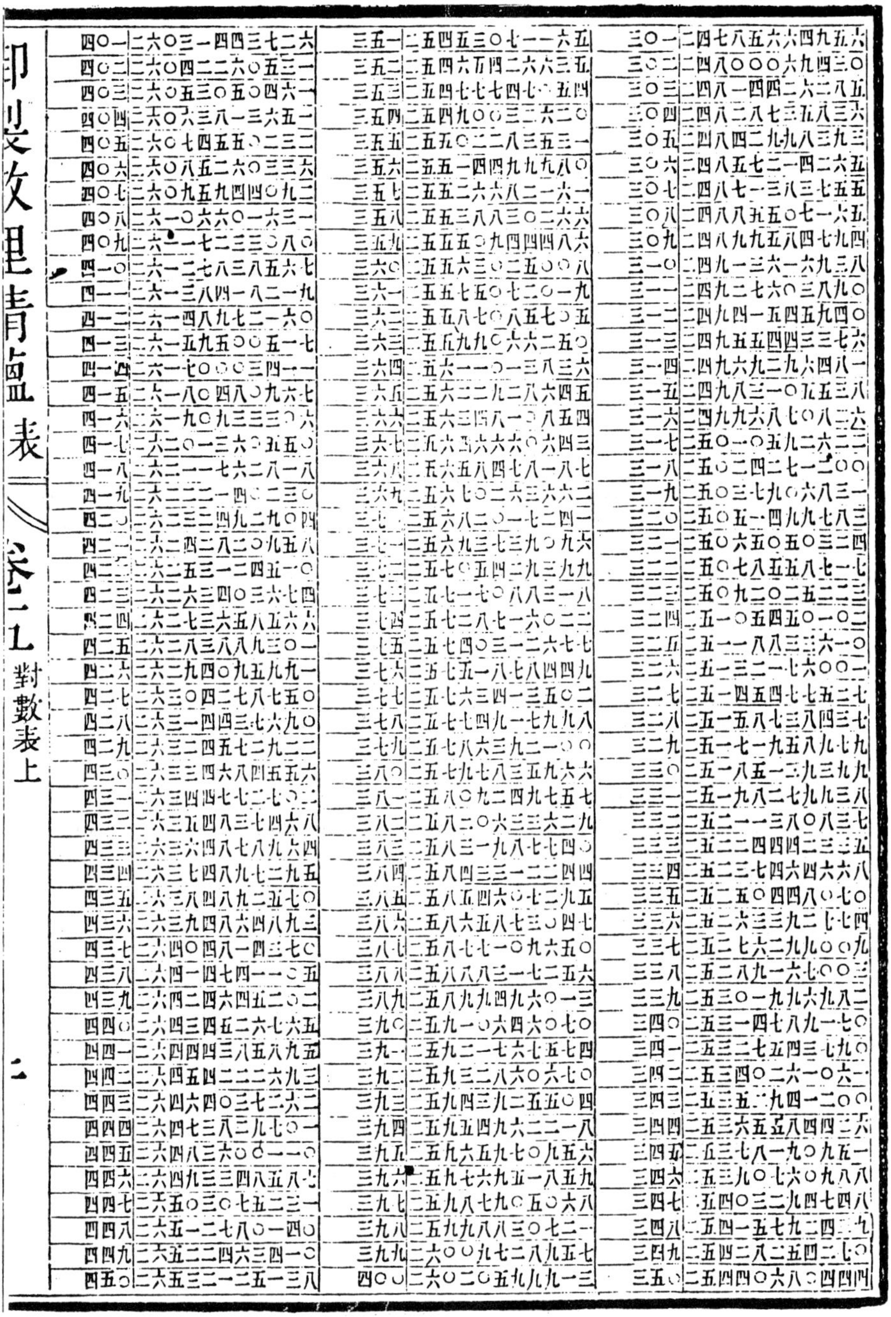

御製數理精蘊表 卷[illegible] 對數表上

三〇一	二四七八五六六四九五六	三五一	二五四五三〇七一一六五	四〇一	二六〇三一四四三七二六
三〇二	二四八〇〇〇六九四三〇	三五二	二五四六五四二六六三五	四〇二	二六〇四二二六〇五三一
三〇三	二四八一四四二六二八五	三五三	二五四七七七四七〇五四	四〇三	二六〇五三〇五〇四六一
三〇四	二四八二八七三五八三六	三五四	二五四九〇〇三二六二〇	四〇四	二六〇六三八一三六五一
三〇五	二四八四二九九八三九三	三五五	二五五〇二二八三五三一	四〇五	二六〇七四五五〇二三二
三〇六	二四八五七二一四二六五	三五六	二五五一四四九九九八〇	四〇六	二六〇八五二六〇三三六
三〇七	二四八七一三八三七五五	三五七	二五五二六六八二一六一	四〇七	二六〇九五九四四〇九二
三〇八	二四八八五五〇七一六五	三五八	二五五三八八三〇二六六	四〇八	二六一〇六六〇一六三一
三〇九	二四八九九五八四七九四	三五九	二五五五〇九四四四八六	四〇九	二六一一七二三三〇八〇
三一〇	二四九一三六一六九三八	三六〇	二五五六三〇二五〇〇八	四一〇	二六一二七八三八五六七
三一一	二四九二七六〇三八九〇	三六一	二五五七五〇七二〇一九	四一一	二六一三八四一八二一九
三一二	二四九四一五四五九四〇	三六二	二五五八七〇八五七〇五	四一二	二六一四八九七二一六〇
三一三	二四九五五四四三三七六	三六三	二五五九九〇六六二五〇	四一三	二六一五九五〇〇五一七
三一四	二四九六九二九六四八一	三六四	二五六一一〇一三八三六	四一四	二六一七〇〇〇三四一一
三一五	二四九八三一〇五五三八	三六五	二五六二二九二八六四五	四一五	二六一八〇四八〇九六七
三一六	二四九九六八七〇八三六	三六六	二五六三四八一〇八五四	四一六	二六一九〇九三三三〇六
三一七	二五〇一〇五九二六二二	三六七	二五六四六六六〇六四三	四一七	二六二〇一三六〇五五〇
三一八	二五〇二四二七一二〇〇	三六八	二五六五八四七八一八七	四一八	二六二一一七六二八一八
三一九	二五〇三七九〇六八三一	三六九	二五六七〇二六三六六二	四一九	二六二二二一四〇二三〇
三二〇	二五〇五一四九九七八三	三七〇	二五六八二〇一七二四一	四二〇	二六二三二四九二九〇四
三二一	二五〇六五〇五〇三二四	三七一	二五六九三七三九〇九六	四二一	二六二四二八二〇九五八
三二二	二五〇七八五五八七一七	三七二	二五七〇五四二九三九九	四二二	二六二五三一二四五一〇
三二三	二五〇九二〇二五二二三	三七三	二五七一七〇八八三一八	四二三	二六二六三四〇三六七四
三二四	二五一〇五四五〇一〇二	三七四	二五七二八七一六〇二二	四二四	二六二七三六五八五六六
三二五	二五一一八八三三六一〇	三七五	二五七四〇三一二六七七	四二五	二六二八三八八九三〇一
三二六	二五一三二一七六〇〇一	三七六	二五七五一八七八四四九	四二六	二六二九四〇九五九九一
三二七	二五一四五四七七五二七	三七七	二五七六三四一三五〇二	四二七	二六三〇四二七八七五〇
三二八	二五一五八七三八四三七	三七八	二五七七四九一七九九八	四二八	二六三一四四三七六九〇
三二九	二五一七一九五八九七九	三七九	二五七八六三九二一〇〇	四二九	二六三二四五七二九二二
三三〇	二五一八五一三九三九九	三八〇	二五七九七八三五九六六	四三〇	二六三三四六八四五五六
三三一	二五一九八二七九九三八	三八一	二五八〇九二四九七五七	四三一	二六三四四七七二七〇三
三三二	二五二一一三八〇八三七	三八二	二五八二〇六三三六二九	四三二	二六三五四八三七四六八
三三三	二五二二四四四二三三五	三八三	二五八三一九八七七四〇	四三三	二六三六四八七八九六四
三三四	二五二三七四六四六六八	三八四	二五八四三三一二二四四	四三四	二六三七四八九七二九五
三三五	二五二五〇四四八〇七〇	三八五	二五八五四六〇七二九五	四三五	二六三八四八九二五七〇
三三六	二五二六三三九二七七四	三八六	二五八六五八七三〇四七	四三六	二六三九四八六四八九三
三三七	二五二七六二九八九〇九	三八七	二五八七七一〇九六五〇	四三七	二六四〇四八一四三七〇
三三八	二五二八九一六七〇〇三	三八八	二五八八八三一七二五六	四三八	二六四一四七四一一〇五
三三九	二五三〇一九九六九八二	三八九	二五八九九四九六〇一三	四三九	二六四二四六四五二〇二
三四〇	二五三一四七八九一七〇	三九〇	二五九一〇六四六〇七〇	四四〇	二六四三四五二六七六五
三四一	二五三二七五四三七九〇	三九一	二五九二一七六七五七四	四四一	二六四四四三八五八九五
三四二	二五三四〇二六一〇六一	三九二	二五九三二八六〇六七〇	四四二	二六四五四二二二六九三
三四三	二五三五二九四一二〇〇	三九三	二五九四三九二五五〇四	四四三	二六四六四〇三七二六二
三四四	二五三六五五八四四二六	三九四	二五九五四九六二二一八	四四四	二六四七三八二九七〇一
三四五	二五三七八一九〇九五一	三九五	二五九六五九七〇九五六	四四五	二六四八三六〇〇一一〇
三四六	二五三九〇七六〇九八八	三九六	二五九七六九五一八五九	四四六	二六四九三三四八五八七
三四七	二五四〇三二九四七四八	三九七	二五九八七九〇五〇六八	四四七	二六五〇三〇七五二三一
三四八	二五四一五七九二四二九	三九八	二五九九八八三〇七二一	四四八	二六五一二七八〇一四〇
三四九	二五四二八二五四二七〇	三九九	二六〇〇九七二八九五七	四四九	二六五二二四六三四一〇
三五〇	二五四四〇六八〇四四四	四〇〇	二六〇二〇五九九九一三	四五〇	二六五三二一二五一三八

二

Fig. A.9. A page from the table of decimal logarithms reproduced in the *Shuli jingyun* (1723).

Mechanics

1865 • *Chongxue*, 12 j. 重學

> The science of weight.
> Translated by Joseph Edkins and Li Shanlan.
> Source: William WHEWELL, *1819, An Elementary treatise on mechanics,* Cambridge (other editions: 1824, 1828, 1852, 1836, 1841, 1847). On Whewell, see the notice in the *DSB.*

Perspective

1729 • *Shixue* 視學

> Litt. 'the science of vision'.
> HB no. 631; BN no. 5533 (lessons on perspective).
> Translation probably by Giuseppe Castiglione (1688–1766) and Nian Xiyao (?–1738).
> Source: Andrea POZZO, *1706, Perspectiva pictorum atque architectorum [...],* Rome.
> *CBPT,* no. 2511–2512.

Probability

1896 • *Jueyi shuxue*, 10 j. 決疑數學

> Litt. 'the science of numbers [i.e. mathematics] designed to determine what is uncertain.
> Translation by John Fryer and Hua Hengfang.
> Source: an article on probability in the 8th edition of the *Encyclopaedia Britannica* (Bennett, p. 85).
> See also an article by Yan Dunjie included in Mei Rongzhao (11'), *1990,* pp. 421–444.

Trigonometry

1631 • *Da ce*, 2 j. 大測

> The great measurement [i.e. the measurement of celestial distances and angles which concerns the heavens, something great comparatively to common surveying].
> HB no. 192; BN no. 4876.
> Translation by Johann Schreck (Terrentius) (1576–1630).
> Included in the *Chongzhen lishu.*
> Source: Bartholomaeus PITISCUS, *1612, Trigonometriae Sive, de Dimensione Triangulorum Libri quinque [...].*
> See also Bai Shangshu (1'), *1963* and (1), *1984.*

1635 • *Geyuan baxian biao* 割圓八線表

Tables of the eight lines of the division of the circle.
Translated by Johann Schreck and Adam Schall.
Included in the *Chongzhen lishu*.

1723 • Trigonometric tables included in the *Shuli jingyun*

1877 • *Sanjiao shuli* 三角數理

The principles of trigonometric calculations.
Translation by John Fryer.
Source: John HYMERS, *1847, A treatise on plane and spherical trigon-ometry and on trigonometrical tables and logarithms together with a selection of problems and their solutions*, third edition, London: Whittaker and Co. (other editions in 1837, 1841 and 1858).
See Bennett, p. 116.

1894 • *Baxian beizhi*, 4 j. 八線備旨

Complete explanation of trigonometry.
Translation by A. P. Parker.
Source unknown.

Appendix II: The Primary Sources

Currently, the most complete bibliographic catalogue of Chinese mathematics is that of DING Fubao[1] and ZHOU Yunqing (1'), *1957*. This is a compilation of earlier catalogues most of which were composed at the end of the 19th century. The following information is given for each work listed:

- list of earlier editions (and sometimes also of manuscripts).

- extracts from the prefaces of the work in question.

The catalogue also contains a single index of titles and proper names, classified jointly. It includes approximately a thousand titles, although it is difficult to give an exact number since the index contains repeated entries (titles are listed separately in their own right and under each author). The catalogue shows the following distribution of works according to period:

- First millennium AD: very approximately ten texts.

- 1248–1303: approximately fifteen texts.

- Ming dynasty (1368–1644): approximately 20 texts.

- Qing dynasty (1644–1911): The number of mathematical works increases very rapidly, especially after 1850; the total output of this period is in the order of one thousand titles.

(This rapid overview only takes into account texts which have been handed down to us in some form or another).

In addition, one might also consult ancient catalogues such as that of the *Siku quanshu zongmu tiyao* (Critical Notices of the General Catalogue of the Complete Library of the Four Branches of Books) (1782), the bibliographic monographs of the dynastic annals, the *congshu* catalogues (collections of various works), the catalogues of ancient and modern bibliophiles such as QIU Zhongman (1'), *1926*, and catalogues of manuscripts (for example, M. Soymié ed., *Catalogue des Manuscrits Chinois de Touen-Houang [Dunhuang] Fonds Pelliot de la Bibliothèque Nationale*, vol. III, Paris: Editions de la Fondations Singer-Polignac, *1983* (notices concerning mathematical manuscripts)).

---

[1] On Ding Fubao (1874–1952) cf. M. Boorman (ed.), 1970, *Biographical Dictionary of Republican China*, New York and London: Columbia Univ. Press, vol. III, pp. 269–270.

For particular periods, one might refer to specialised works such as those of
KODAMA Akihito (1'), *1966* and (2'), *1970*.

Catalogues of large libraries such as those of Pekingese libraries[2] , of
Shanghai, Taipei, Sendai (Tōhoku university) or Kyoto (Jinbun Kagaku
kenkyūsho (The Research Institute for Humanistic Studies)) are also very useful.
Other libraries such as the Seikado Bunko (Tokyo), the Bibliothèque Nationale
(Paris), the Bibliothèque de l'Institut des Hautes Etudes Chinoises (Paris) or
the Vatican Library and the Roman Archives of the Society of Jesus[3] are no
less fundamental. This list is clearly very incomplete.

---

[2]In particular, the very rich collection of ancient works of Chinese mathematics by Li
Yan (1892–1963) now belongs to the library of the Institute for Research into the History of
Science of the Academia Sinica (*Ziran kexue shi yanjiusuo*).

[3]On this, see, in particular, Martzloff (11), *1989*.

Index of Main Chinese Characters

Administrative Terms

bei sitiantai	北司天台
bing cao	兵曹
bishu shaojian	秘書少監
bishujian	秘書監
buzhengsi	布政司
cang cao	倉曹
dali pingshi	大理評事
guozijian	國子監
ji cao	集曹
jianzheng	監正
jiaoxi	教習
jin cao	金曹
jinshi	進士
juren	舉人
ming suan	明算
muliao	幕僚
qintianjian	欽天監
suanxue boshi	算學博士
suanxue guan	算學館
tian cao	田曹
tongzhi lang	通直郎
wei	尉
zhou	州
zhupu	主薄
zongli yamen	總理衙門
zuoban dianzhi	左班殿直

Calendars

Datong li	大統曆
Huihui li	回回曆
Jiyuan li	紀元曆
Shoushi li	授時曆
Wannian li	萬年曆
Yingtian li	應天曆

Geographical Terms

Daxing	大興
Huaining	懷寧
Jiangling	江陵
Lumuguo	魯穆國

Mathematical Terms

anma	暗碼
ayuduo	阿庾多
baodao	堢壔
bei	倍
ben	本
bian	辯
biancheng	徧乘
biao	表
bienuan	鱉臑
bilei	比類
bisuan	筆算
bu	步
cao	草
cewang	測望
cewang lei	測望類
chana	刹那
chen	塵
cheng	程
cheng zhi fa	乘之法
chidao jidu	赤道積度
chou	籌
chu	初
chu zhi fa	除之法
cong	從
cong fa	從法
da xie	大科
daicong kaifang fa	帶從開方法
daqian	大千
dayan	大衍
dayan qiuyi shu	大衍求一術
dayan zongshu shu	大衍總數術
de, shi	得失
deng	等
diaorifa	調日法
ding fa	定法
dou shu	都術
du	度

Other Terms

Titles of Books

A'erribala xinfa	阿爾熱巴拉新法
Baifutang suanxue congshu	白芙堂算學叢書
Bice huihan	秘册彙函
Biligui jie	比例規解
Bishujian zhi	秘書監志
Celiang fa yi	測量法義
Celiang jiyao	測量集要
Ceyuan haijing	測圓海鏡
Celiang quanyi	測量全義
Chengchu tongbian suanbao	乘除通變算寶
Chishui yizhen	赤水遺珍
Chongwen zongmu	崇文總目
Chongxue	重學
Chongzhen lishu	崇禎曆書
Chouren zhuan	疇人傳
Chouren zhuan san bian	疇人傳三編
Chousuan	籌算
Cihai	辭海
Da Dai liji	大戴禮記
Dace	大測
Dai kanwa jiten	大漢和辭典
Daishuxue	代數學
Daiweiji shiji,	代微積拾級
Daodejing	道德經
Duishu tanyuan	對數探源
Duixiang siyan zazi	對相四言雜字
Duoji bilei	垛積比類
Fangcheng lun	方程論
Fangyuan chanyou	方圓闡幽
Gao seng zhuan	高僧傳
Geyuan baxian zhuishu	割圓八線綴術
Geyuan milü jiefa	割圓密率捷法
Gougu geyuan ji	勾股割圜記
Gougu yi	勾股義
Guanwoshengshi huigao	觀我生室彙稿
Guangyang zaji	廣陽雜記
Gujin suanxue congshu	古今算學叢書
Gujin tushi jicheng	古今圖書集成
Guoxue jiben congshu	國學基本叢書
Haidao suanjing	海島算經
Hanshu	漢書
Hefang tongyi	河防通議
Hongloumeng	紅樓夢
Huangdi jiuzhang	黃帝九章
Huangdi jiuzhang suanfa	黃帝九章算法
Huangdi jiuzhang suanfa xicao	黃帝九章算法細草

Long Expressions

can suo tu si wei hu	蠶所吐絲爲忽
Chen Chunfeng deng qin an	臣淳風等謹按
cheng cheng zhi shu ye	乘成之數也
er suo lun zhe duo jin yu ye	而所論者多近語也
fan guang-cong xiang-cheng wei zhi mi	凡廣從相乘謂之冪
fan mu hu cheng zi wei zhi qi	凡母互乘子謂之齊
fan wei shu zhi yi yue sheng wei shan	凡爲術之意約省爲善
fazhe suo qiu zhi jia ye	法者所求之價也
gaoming neng suan shi	高明能算士
Li Madou suo yi, yin wenfa bu ming, xianhou	李瑪竇所譯,因文法不明 先後
nan jie, gu ling yi	難解,故另譯
li tong ben ti fan yan zhi	理同本題反言之
qi suanfa yong zhu	其算法用竹
qie lifenzhongmoxian dan you qiuzuo zhi fa	且理分中末線但有求作之法
er mo zhi suo yong	而莫知所用
qun mu xiang cheng wei zhi tong	群母相乘謂之同
shan qiao ji, you zhi li yingyang, tianwen lisuan	善巧機尤致力陰陽天文曆算
shu bu jin yan, yan bu jin yi	書不盡言,言不盡意
si nan er shi fei nan	似難而實非難
suan zhi gangji	算之綱紀
suanshi zhi zui zhe	算氏之最者
suanshu tianwen xiang wei biaoli	算術天文相爲表裏
xi wan wu zhi li	析萬物之理
xian zhi suo zhi chu yue dian	線之所誌處曰點
xiangming shuli	詳明數理
xing gui er shu jun	形詭而數均
yi shan suan ming shi	以善算命世
yi xia ke lei tui	以下可類推
yi ying bu xu	以盈補虛
yong suan er bu zhan	用算而布𣗝
zong heng lie yu ji'an	縱橫列於幾案

References

Bibliographical Orientations

Insofar as Chinese mathematics developed in a generally little known historical context, its intelligibility supposes the consideration of diverse sources of information; whence the variety of topics covered by the present bibliographies. But since it might not be easy to work one's way through what has now become a huge documentation, we shall provide some preliminary indications.

Western Sources

In seeking information on Chinese mathematics, ancient or modern, it should be noted that, even with no access to non-Western sources, the available documentation is already quite considerable.

General encyclopedias on the history and philosophy of mathematics, such as that recently edited in 2 volumes by Grattan-Guiness ((1),*1994*), provide various clues which are all the more interesting when they are not considered in isolation but in relation to the history of mathematics in general and seen from a large variety of angles. In the same way, the *Dictionary of Scientific Biography*, with its biographies of Liu Hui, Li Zhi, Qin Jiushao and others will also be of great service (Gillipsie (1), 1970–1980).

But other general dictionaries and catalogues are also of interest even though, *a priori*, these might be believed to be irrelevant to the study of Chinese mathematics. First and foremost, one must cite Hummel (1), *1943*, as well as Goodrich and Fang (1), *1976*. Although already ancient, the former contains much information not easily obtainable elsewhere on mathematicians from the Qing dynasty and the latter on those from the Ming. In the same spirit, the fundamental but rarely cited *Catalogue de la Bibliothèque du Pé-T'ang* published by the Mission Catholique des Lazaristes à Pékin in 1949 and reprinted in 1969 by Les Belles Lettres (Paris) and Dehergne (1), *1973*, are both indispensable tools for the study of the history of the Jesuit mission in China from the double point of view of the bibliography of European mathematical books imported into China during the 17th and 18th centuries by Jesuit missionaries and of the biographies of Jesuits. For a later period, i.e. that corresponding to the activity of Protestant missionaries in China,

Biggerstaff (1), *1961*, Fairbank (1), *1954*, Bennett (1), *1967* and Duus (1), *1966*, are indispensable.

As for journals regularly publishing various articles on the history of Chinese mathematics, their number is too large to pass them in review one by one. But among them, the two most important are probably *Historia Mathematica* (Toronto) and *Archive for History of Exact Sciences* (Berlin). Each has its distinctive style and, whereas the former tends to emphasize an historical approach, the latter often prefers papers based on an internal epistemology of mathematics. However, the reader interested in a particular subject should above all consult bibliographical journals devoted to the history of mathematics or to sinology such as the *Zentralblatt für Mathematik und ihre Grenzgebiete* (Berlin) or the *Revue Bibliographique de Sinologie* (Paris), not fogetting *Chinese Science* (Philadelphia and Los Angeles). But beyond these, one should also note that Chinese and Japanese journals sometimes publish papers in Western languages. In addition, important Chinese and Japanese journals generally give an English abstract of the articles they publish. That is the case, for example, of *Ziran kexue shi yanjiu* (Peking) and *Kagaku shi kenkyū* (Tokyo).

Chinese and Japanese Sources

The abundance and, even more, the rhythm of publication of Chinese (and, to a lesser extent, of Japanese) publications on the history of Chinese mathematics is such that the access to that immense literature has become a real problem.

Until recently, it was potentially impossible not only to have rapid access to but even to keep oneself reasonably well-informed of the existence of what has been published without great effort and waste of time. But now the situation has slightly changed and new bibliographical tools have become available. Among these the most comprehensive is certainly the *Zhongguo kexue shi tongxun* (Newsletter for the History of Chinese Science) published twice a year in Taiwan by the Institute of History of the Tsing Hua University (Hsinchu 30043, Taiwan, Republic of China). It should be noted that the scope of this publication is very large since it mentions practically everything published on the history of Chinese science everywhere in the world. Moreover, the various sciences are classified by subject so that what concerns mathematics is easily retrievable. Each year the whole of it amounts to approximately 400 pages; most articles and books are summarised or abstracted (sometimes in English) by the authors themselves.

In addition, since 1986, articles on the history of Chinese science (and hence of mathematics) published between 1900 and 1982 have their own catalogue (see Yan Dunjie (30'), *1986*). 1000 pages long, this very useful catalogue has several indices for the titles of papers, authors and journals. Last but not least, we also mention the important Japanese *Annual Bibliography of Oriental Studies* [Japanese title: Tōyōgaku bunken ruimoku, Kyoto: Kyoto University (The Research Institute for Humanistic Studies [Jinbun kagaku kenkyūsho]).

Books and Articles in Western Languages

Abbreviations (Books and Articles)

ActaAsiat	Acta Asiatica (Kyoto)
ActaOr	Acta Orientalia (Copenhagen)
AF	Altorientalische Forschungen (Berlin)
AHES	Archive for History of Exact Sciences (Berlin)
Altertum	Das Altertum (Berlin)
AMM	American (The) Mathematical Monthly (New York)
AMP	Archiv der Mathematik und Physik (Leipzig)
AnNYAS	Annals of the New York Academy of Sciences (New York)
AnSci	Annals of Sciences (London)
ArIntHS	Archives Internationales d'Histoire des Sciences (Paris)
Asia Major	Asia Major (Princeton)
AW	Antike Welt (Zurich)
BEFEO	Bulletin de l'Ecole Française d'Extrême-Orient (Paris)
BibMath	Bibliotheca Mathematica (Stockholm)
BJHS	The British Journal for the History of Science (London)
BT	Bantai xuebao (Kuala Lumpur)
C&M	Classica et Mediaevalia (Copenhagen)
CACM	Communications of the ACM (Association for Computing Machinery) (New York)
CBPT	*Catalogue de la Bibliothèque du Pé-T'ang*, Peking, 1949 (Cf. below, the entry Mission Catholique des Lazaristes à Pékin).
CCJ	Chong Ji Journal (Hong Kong)
Centaurus	Centaurus (Copenhagen)
ChinSci	Chinese Science (Philadelphia and Los Angeles)
CMSB	China Mission Studies Bulletin (1550–1800) (Cedar Rapids, Coe College, Iowa, USA)
DSB	C.C. Gillispie (ed.), *Dictionary of Scientific Biography*, New York, 1970–80.
EOEO	Extrême-Orient Extrême-Occident (Paris)
FEQ	The Far East Quarterly (Ann Arbor)
FundSci	Fundamenta Scientia (Strasbourg)
HistMath	Historia Mathematica (Toronto)
HJAS	Harvard Journal of Asiatic Studies (Cambridge, Mass.)
HistRel	History of Religions (Chicago)
HistSci	History of Science (London)
HS	Historia Scientiarum (Tokyo)
IJHS	Indian Journal of History of Science (Calcutta)
Isis	Isis (Washington)
JA	Journal Asiatique (Paris)
JAH	Journal of Asian History (Wiesbaden)
JS	Journal des Savants (Paris)
Janus	Janus (Leyde)
JAOS	Journal of the American Oriental Society (New Haven)
JDMV	Jahresberichte der Deutschen Mathematiker Vereinigung (Leipzig)
JHA	Journal for the History of Astronomy (Cambridge, England)

JOS	Journal of Oriental Studies (Hong Kong)
JRAM	Journal für die Reine und Angewandte Mathematik (Berlin)
JRAS/NCB	Journal of the North China Branch of the Royal Asiatic Society
JSHS	Japanese Studies in the History of Science (Tokyo)
KJ	Korea Journal (Seoul)
LishiXB	Lishi Xuebao (Taipei)
Maebashi	Maebashi Shiritsu Kōgyō Tanki Daigaku Kiyō (Maebashi)
MATFYZCAS	Matematico-Fyzikalny Časopis SAV (Bratislava)
Mathesis	Mathesis (Mons)
MatLap	Matematikai Lapok (Budapest)
MathMag	Mathematics Magazine (Washington)
MEFRIM	Mélanges de l'Ecole Française de Rome, Italie et Méditerranée (Rome)
Minerva	Minerva, A Review of Science, Learning and Policy (London)
MingStud	Ming Studies (Minneapolis)
MiscStor	Miscellanea Storica Ligure (Milan)
MN	Mathematische Nachrichten (Berlin)
MRDTB	Memoirs of the Research Department of the Toyo Bunko (Tokyo)
MS	Monumenta Serica (Los Angeles)
MT	Mathematics Teacher (Washington)
NAM	Nouvelles Annales de Mathématiques (Paris)
Needham	Li Guohao *et al.*, *Explorations in the History of Science and Technology in China*, Shanghai: Chinese Classics Publishing House, 1982 (a volume compiled in honour of Dr. Joseph Needham).
OE	Oriens Extremus (Wiesbaden)
OL	Orientalia Lovaniensa (Louvain)
Osiris	Osiris (Bruges)
PCW	Problems of the Contemporary Word (Moscow)
PFEH	Papers on Far Eastern History (Canberra)
PHSTJ	Philosophy and History of Science: A Taiwanese Journal (Taipei)
PP	Past and Present (Oxford)
PTRS	Philosopical Transactions of the Royal Society (London)
QSGM/B	Quellen und Studien zur Geschichte der Mathematik (Abteilung B, Astronomie und Physik) (Berlin)
RGSPA	Revue Générale des Sciences Pures et Appliquées (Paris)
RHS	Revue d'Histoire des Sciences (Paris)
RSR	Recherches de Science Religieuse (Paris)
SciCont	Science in Context (Jerusalem)
SGSK	Sūgaku shi kenkyū (Tokyo)
Sinologica	Sinologica (Basle)
Speculum	Speculum, A Journal of Medieval Studies (Cambridge, Mass.)
StudCop	Studia Copernicana (Warsaw)
SudhArch	Sudhoffs Archiv (Wiesbaden)
THG	Tōhōgaku (Kyoto)
TP	T'oung Pao (Leyde)
VistasAstro	Vistas in Astronomy (London and New York)
ZAS	Zentral-Asiatische Studien (Wiesbaden)
ZDMG	Zeitschrift der Deutschen Morgenländischen Gesellschaft (Stuttgart)

Books and Articles

ADAMO Marco (1), 1968. "La matematica nell'antica Cina." *Osiris*, vol. 15, pp. 175–195.

AGASSI Joseph (1), 1981. *Science and Society, Studies in the Sociology of Science.* Dordrecht: D. Reidel.

AHRENS Wilhelm (1), 1901. *Mathematische Unterhaltungen und Spiele.* Leipzig: B.G. Teubner.

AIGNER Martin (1), 1979. *Combinatorial Theory.* Berlin: Springer.

ANG Tian-Se (1), 1969. *A Study of the Mathematical Manual of Chang Ch'iu-chien.* Unpublished MA Thesis. Kuala Lumpur: University of Malaya.

ANG Tian-Se (2), 1972. "Chinese Interest in Indeterminate Analysis and Indeterminate Equations." *BT*, no. 5, pp. 105–112.

ANG Tian-Se (3), 1976. "The Use of Interpolation Techniques in Chinese Calendar." *OE*, vol. 23, no. 2, pp. 135–151.

ANG Tian-Se (4), 1978. "Chinese Interest in Right-angled Triangles." *HistMath*, vol. 5, no. 3, pp. 253-266.

ANG Tian-Se (5), 1979. *I-Hsing (683–727 AD): His Life and Scientific Work.* Unpublished PhD Dissertation, Kuala Lumpur: University of Malaya.

ANG Tian-Se and SWETZ F.J. (1), 1986. "A Chinese Mathematical Classic of the Third Century: The Sea Island Manual of Liu Hui." *HistMath*, vol. 13, pp. 99–117.

ARCHIBALD R. C. (1), 1915. *Euclid's Book on Divisions of Figures.* Cambridge: Cambridge University Press.

ARCHIMEDES (1). ca. 287–212/1970–1972. *Archimède, Texte établi et traduit par Charles Mugler*, 4 vols. Paris: Les Belles Lettres.

ARSAC Jacques (1), 1985. *Jeux et casse-tête à programmer.* Paris: Dunod.

BACHET Claude Gaspar (1), 1612/1959. *Problèmes plaisants et délectables qui se font par les nombres.* Paris: A. Blanchard (Reprint of the 5th ed.).

BAG A. K. (1), 1979. *Mathematics in Ancient and Medieval India.* Delhi: Chaukhambha Orientalia.

BAI Shangshu (1), 1984. "Présentation de la première trigonométrie Chinoise: le *Dace* [lit. 'The Great Surveying' (i.e. trigonometry, plane and spherical, applied to astronomy)]." *CMSB*, no. 6, pp. 43–50 (Transl. by J. C. Martzloff).

BAI Shangshu (2), 1988. "An Exploration of Liu Xin's value of π from Wang Mang's Measuring Vessel." *SGSK*, no. 116, pp. 24–31 (Cf. Bai Shangshu (2')).

BARNES Barry (1), 1974. *Scientific Knowledge and Sociological Theory.* London: Routledge and Kegan Paul.

BARNES Barry (2), 1982. *T. S. Kuhn and Social Science.* London: The Macmillan Press.

BASTID Marianne (1), 1971. *Aspects de la réforme de l'enseignement en Chine au début du XXe siècle d'après les écrits de Zhang Jian.* Paris, La Haye: Mouton.

BECKMANN P. (1), 1971. *A History of π.* Boulder: The Golem Press.

BEN-DAVID Joseph (1), 1971. *The Scientist's Role in Society, a Comparative Study.* Chicago and London: The University of Chicago Press.

BENNETT Adrian Arthur (1), 1967. *John Fryer, The Introduction of Western Science and Technology into Nineteenth Century China.* Cambridge (Mass.): Harvard University Press; Rev.: N. Sivin, *Isis*, vol. 61, no. 2, pp. 280–282.

BERGGREN J.L. (1), 1986. *Episodes in the Mathematics of Medieval Islam.* Berlin: Springer.

BERNARD[-MAÎTRE] Henri (1), 1935. "Les étapes de la cartographie scientifique pour la Chine et les pays voisins." *MS,* vol. 1, pp. 428-477.

BERNARD[-MAÎTRE] Henri (2), 1935. *L'apport Scientifique du Père Matthieu Ricci à la Chine.* Tientsin: Procure de la Mission de Sienshien.

BERNARD[-MAÎTRE] Henri, (3), 1937. "L'encyclopédie astronomique du Père Schall, *Chhung-Chên Li Shu* [*Chongzhen lishu*] (+1629) et *Hsi-Yang Hsin Fa Li Shu* [*Xiyang xinfa lishu*] (+1645): La Réforme du Calendrier Chinois sous l'Influence de Clavius, Galilée et Képler." *MS,* vol. 3, pp. 35–77 and 441–527.

BERNARD[-MAÎTRE] Henri (4), 1940. "Ferdinand Verbiest continuateur de l'oeuvre scientifique d'Adam Schall, quelques compléments à l'édition récente de sa correspondence." *MS,* vol. 5.

BERNARD[-MAÎTRE] Henri, (5), 1942. *Lettres et Mémoires d'Adam Schall SJ.* Tientsin.

BERNARD[-MAÎTRE] Henri (6), 1942. *Le voyage du Père de Fontaney au Siam et à la Chine 1685-1687.* Tientsin: Cathasia.

BERNARD[-MAÎTRE] Henri, (7), 1945–1960. "Les Adaptations Chinoises d'Ouvrages Européens: Bibliographie Chronologique. I. Depuis la Venue des Portugais à Canton jusqu'à la Mission Française de Pékin (1514–1688); II. Depuis la Mission Française de Pékin jusqu' à la Mort de l'Empereur K'ien Long (1689–1799)." *MS,* 1945, vol. 10, pp. 1–57 and 309–388; 1960, vol. 19, pp. 349–383.

BERNARD[-MAÎTRE] Henri, (8), 1950. *Pour la Compréhension de l'Indochine et de l'Occident.* Paris: Les Belles Lettres.

BERTRAND J. (1), 1869. "Les mathématiques en Chine." *JS,* pp. 317–464 (translation of Biernatzki (1)).

BERYOZKINA E. (1), 1981. "Studies in the History of Ancient Chinese Mathematics." *PCW,* 1981, no. 96, pp. 162–178.

BEURDELEY C. and M. (1), 1971. *Castiglione Peintre Jésuite à la Cour de Chine.* Fribourg: Office du Livre.

BIERNATZKI K.L. (1), 1856. "Die Arithmetik der Chinesen." *JRAM,* vol. 52, pp. 59–94.

BIGGERSTAFF Knight (1), 1961. *The Earliest Modern Government School in China.* Ithaca (New York): Cornell University Press.

BIELENSTEIN Hans (1), 1980. *The Bureaucracy of Han Times.* Cambridge: Cambridge University Press.

BLOOR David (1), 1976. *Knowledge and Social Imagery.* London: Routledge and Kegan Paul.

BODDE Derk (1), 1991. *Chinese Thought, Society and Science. The Intellectual and Social Background of Science and Technology in Pre-Modern China.* Honolulu: University of Hawaii Press.

BOLTIENSKII V.G. (1), 1978. *Hilbert's Third Problem.* New York: J. Wiley.

BOYER C.B. (1), 1949/1959. *The History of the Calculus and its Conceptual Development.* New York: Dover.

BOYER C.B. (2), 1968. *A History of Mathematics.* New York: J. Wiley.

BRAINERD B. and PENG F. (1), 1968. "A Syntactic Comparison of Chinese and Japanese Numerical Expressions." in H. BRANDT and CORSTIUS eds., *Grammars for Number Names.* Dordrecht: D. Reidel, pp. 53–81.

BRAUDEL Fernand (1), 1969. *Ecrits sur l'histoire.* Paris: Flammarion.

BRAUNMÜHL Anton von (1), 1900–1903. *Vorlesungen über Geschichte der Trigonometrie.* Leipzig: Teubner.

BRAY Francesca (1), 1991. "Some Problems concerning the Transfer of Scientific and Technical Knowledge." in Thomas H. C. LEE ed., *China and Europe, Images and Influences in Sixteenth to Eighteenth Centuries.* Hong Kong: The Chinese University Press, pp. 203–219.

BRENDAN Gillon S. (1), 1977. "Introduction, Translation and discussion of Chao Chün-Ch'ing's 'Notes to the Diagrams of Short Legs and Long Legs and of Squares and Circles'." *HistMath,* vol. 4, no. 3, pp. 253–293.

BRIGGS Henry (1), 1624. *Arithmetica Logarithmica, Sive Logarithmorum Chiliades Triginta, pro Numeris Naturali Serie Crescendibus ab Unitate ad 20 000 et ab 90 000 ad 100 000.* London.

BROCARD H. and MANSION P. (1), 1889. "La trigonométrie sphérique réduite à une seule formule." *Mathesis,* tome 9, pp. 161–164, 181–182, 265–267.

BROWN S. C. (1) ed., 1984. *Objectivity and Cultural Divergence.* Cambridge: Cambridge University Press.

BRUINS Evert M. transl. (1), 1964. *Codex Constantinopolitanus.* Leiden: E. J. Brill (contains an English translation with commentary on Hero's *Metrica* and other mathematical texts by the same author).

BRUINS Evert M., (2), 1984. "Requisite for the Interpretation of Ancient Mathematics." *Janus,* vol. 71, no. 1–4, pp. 107–134.

BRUINS Evert M. and RUTTEN M. (1), 1961. *Textes Mathématiques de Suse.* Paris: P. Geuthner.

BUBNOW N. (1) ed., 1899. *Gerberti, postea Silvestri II papae Opera Mathematica.* Berlin: Friedländer.

BUSARD Hubert L. L. (1), 1968. "L'algèbre au Moyen Age, le *Liber mensurationum* d'Abū Bekr." *JS,* April–June 1968, pp. 65–124.

CAJORI Florian (1), 1928–1929. *A History of Mathematical Notations.* 2 vols. La Salle (Illinois): Open Court Publishing Co.

CAMMANN Schuyler (1), 1960. "The Evolution of Magic Squares in China." *JAOS,* vol. 80, pp. 116-124.

CAMMANN Schuyler (2), 1961. "The Magic Square of Three in Old Chinese Philosophy and Religion." *HistRel,* vol. 1, pp. 37–80.

CAMMANN Schuyler (3), 1962. "Old Chinese Magic Squares." *Sinologica,* vol. 7, no. 1, pp. 14–53.

CANTOR Moritz Benedikt (1), 1875. *Die Römischen Agrimensoren und ihre Stellung in der Geschichte der Feldmesskunst.* Leipzig.

CANTOR Moritz Benedikt (2), 1880–1908. *Vorlesungen über Geschichte der Mathematik.* 4 vols. Leipzig.

CAUTY A. (1), 1987. *L'énoncé mathématique et les numérations parlées.* Thesis in mathematics (University of Nantes).

CAVALIERI B. (1), 1635. *Geometria Indivisibilibus.* Bologna.

CAZALAS E. (1), 1934. *Carrés magiques au degré n, Séries numérales de G. Tarry avec un aperçu historique et une bibliographie des figures magiques.* Paris: Hermann.

CHAN Wing-tsit (1), 1963/1969. *A Source Book in Chinese Philosophy.* Princeton: Princeton University Press.

CHAN Wing-tsit (2), 1969. "The Evolution of the Neo-confucean Concept *li* as Principle" in *Essays by Wing-tsit Chan* (Compiled by Charles H. K. CHEN), Hanover Amherst New York Hong Kong: Oriental Society.

CHANG Fu-Jui (1), 1962. *Les Fonctionnaires des Song, Index des Titres.* Paris: Mouton.

CHANG Jih-ming (1), 1980. *Les Musulmans sous la Chine des Tang (618–907).* Taipci.

CHEMLA Karine (1), 1990. "Du parallélisme entre énoncés mathématiques, analyse d'un formulaire en Chine au XIIIe siècle." *RHS*, vol. 43, no. 1, pp. 57–80.

CHEMLA Karine (2), 1992. "Méthodes infinitésimales en Chine et en Grèce anciennes." in H. SINACEUR and J. M. SALANSKIS eds., *Le labyrinthe du continu.* Berlin: Springer, pp. 31–46.

CHEMLA Karine (3), 1992. "Des nombres irrationnels en Chine entre le premier et le troisième siècle." *RHS*, Tome 45, no 1, pp. 135–140.

CHEN Cheng-Yih (1), 1987. *Science and Technology in Chinese Civilization.* Singapore: World Scientific.

CH'ÊN Yüan (1), 1966. *Western and Central Asians in China under the Mongols, their Transformation into Chinese.* Monumenta Serica Monographs, vol. 15.

CHENG-YIH Chen: see CHEN Cheng-Yih.

CHIH André (1), 1962. *L'Occident chrétien vu par les Chinois vers la fin du XIXe siècle.* Paris: Presses Universitaires de France.

CHINN William and LEWIS John (1), 1984. "Shiing-Shen Chern [Chen Xingshen]." in Donald J. ALBERS and G. L. ALEXANDERSON (eds.), *Mathematical People, Profiles and Interviews.* Basel: Birkhäuser.

CLAGETT Marshall (1), 1979. *Studies in Medieval Physics and Mathematics.* London: Variorum Reprints.

CLARK Walter Eugen transl. (1), 1930. *The Āryabhaṭīya of Āryabhaṭa, an Ancient Indian Work on Mathematics and Astronomy.* Chicago: The University of Chicago Press.

CLAVIUS Christoph (1), 1611–1612. *Opera Mathematica.* 5 vols. Cf. *CBPT*, notice no. 1288.

CLAVIUS Christoph (2), 1574/1591. *Euclidis Elementorum libri XV. Accessit liber XVI. de Solidorum Regularium cuiuslibet intra quodlibet comparatione. Omnes perspicuis Demonstrationibus, accuratisque Scholiis illustrati, ac multarum rerum accessione locupletati.* (First edn.: Rome, 1574). Cf. *CBPT*, notice no. 1297.

CLAVIUS Christoph (3), 1585. *Epitome Arithmeticae Practicae.* Rome. Cf. *CBPT*, notice no. 1296.

CLAVIUS Christoph (4), 1604. *Geometria Practica.* Rome. Cf. *CBPT*, notice no. 1300.

COHEN Floris H. (1), 1994. *The Scientific Revolution, A Historiographical Inquiry.* Chicago: The University of Chicago Press. (On Needham cf. ch. 5, pp. 378 ff. "The Nonemergence of Early Modern Science Outside Western Europe.")

COHEN Paul A. (1), 1987. *Between Tradition and Modernity: Wang T'ao and Reform in Late Ch'ing China.* Cambridge (Mass.): Harvard University Press.

COLEBROOKE Henry Thomas transl. (1), 1817. *Algebra with Arithmetic and Mensuration from the Sanscrit of Brahmegupta and Bhascara.* London: John Murray.

COLLANI Claudia von (1), 1985. *P. Joachim Bouvet S.J. Sein Leben und sein Werk.* Nettetal: Steyler Verlag.

COMTET Louis (1), 1970. *Advanced Combinatorics, The Art of Finite and Infinite Expansions.* Revised and enlarged edition. Dordrecht: D. Reidel Publishing Cie.

COOLIDGE J. L. (1). 1963. *The Mathematics of Great Amateurs.* New York: Dover.

COSENTINO Giuseppe (1), 1970. "Le mathematiche nella *Ratio Studiorum* della Compagnia di Gesù." *MiscStor*, new ser., no. 2, pp. 171–213.

CROSSLEY John N. and LUN Anthony W.-C. transl. (1), 1987. LI Yan and DU Shiran, *Chinese Mathematics, A Concise History.* Oxford: Oxford University Press.

CULLEN Christopher (1), 1982. "An Eighth Century Chinese Table of Tangents." *ChinSci*, no. 5, pp. 1–33.

DATTA B. and SINGH A. N. (1), 1935–1938. *History of Hindu Mathematics.* 2 vols. Lahore.

DAUBEN W. Joseph (1), 1993. "Mathematics: an Historian's Perspective." *PHSTJ*, vol. 2, no. 1, pp. 1–21.

DAUMAS M. (1). 1953. *Les Instruments Scientifiques aux XVIIe et XVIIIe siècles.* Paris: Presses Universitaires de France.

DEDRON P. and ITARD J. (1), 1959. *Mathématiques et Mathématiciens.* Paris: Magnard.

DEHERGNE Joseph (1), 1973. *Répertoire des Jésuites de Chine de 1552 à 1800.* Rome and Paris: Institutum Historicum S.I. and Letouzey et Ané.

D'ELIA Pasquale M. (1), 1942–1949. *Fonti Ricciane, Storia dell'Introduzione del Cristianismo in Cina.* 3 vols. Rome.

D'ELIA Pasquale M. (2), 1956. "Prezentazione della Prima Traduzione Cinese di Euclide." *MS*, vol. 15, pp. 161–202.

DELATTRE Pierre (1), 1949–1957. *Les Etablissements des Jésuites en France depuis quatre siècles.* 5 vols. Enghien (Belgium).

DEMIÉVILLE Paul (1), 1925. "Edition photolithique de la *Méthode d'Architecture de Li Ming-tchong des Song.*" *BEFEO*, tome 25, pp. 213–264 (this review of an edition of the *Yingzao fashi* is fundamental for any study of the history of Chinese architecture).

DEYDIER Christian (1), 1976. *Les jiaguwen, essai bibliographique et synthèse des études.* Paris: Ecole Française d'Extrême-Orient.

DHOMBRES Jean (1), 1978. "Aperçus sur un Développement Parallèle (des mathématiques) en Chine." In *Nombre, Mesure et Continu, Epistémologie et Histoire.* Paris: Nathan, pp. 285–304.

DICKSON L. E. (1), 1919–1923/1971. *History of the Theory of Numbers.* 3 vols. New York: Chelsea.

DICKSON L. E. (2), 1929/1957. *Introduction to the Theory of Numbers.* New York: Dover.

DIJKSTERHUIS E. J. (1), 1938/1987. *Archimedes.* Transl. by C. DICKSHOORN (1956), reprinted with a new bibliographic essay by Wilbur R. KNORR, 1987. Princeton: Princeton University Press.

DILKE O. A.W. (1), 1971. *The Roman Land Surveyors. An Introduction to the Agrimensores.* Newton-Abbot: David and Charles.

DOOLITTLE Justus (1), 1872. *Vocabulary and Handbook of the Chinese Language Romanized in the Mandarin Dialect.* 2 vols. Foochow: Rozario, Marcal and Co.

DREYER J. L. E. (1), 1906/1953. *A History of Astronomy from Thales to Kepler.* New York: Dover.

DU HALDE Jean-Baptiste (1), 1735. *Description géographique, historique, chronologique, politique et physique de l'Empire de la Chine et de la Tartarie Chinoise* [...]. 4 vols. Paris: P.-G. Le Mercier. Cf. *CBPT*, notice no. 257.

DURT Hubert (1), 1979. "Chū"[rods], in *Hōbōgirin*, Dictionnaire Encyclopédique du Bouddhisme d'après les Sources Chinoises et Japonaises. Cinquième fascicule. Paris Tokyo: Maisonneuve, pp. 431–456.

DUUS Peter (1), 1966. "Science and Salvation in China: The Life and Work of W.A.P. Martin (1827–1916)." In LIU Kwang-ching ed., *American Missionaries in China*. Harvard East Asian Monographs, pp. 11–41.

EDWARDS A.W. F. (1), 1987. *Pascal's Arithmetical Triangle*. London: Charles Griffin and Co.

EDWARDS C.H. (1), 1982. *The Historical Development of the Calculus*. Berlin: Springer.

ELIA Pasquale M.: see D'ELIA.

ELMAN B. A. (1), 1984. *From Philosophy to Philology, Intellectual and Social Aspects of Change in Late Imperial China*. Cambridge (Mass.) and London: Harvard University Press.

ETIEMBLE (1), 1988–1989. *L'Europe Chinoise*. 2 vols. I. De l'Empire Romain à Leibniz; II. De la sinophilie à la sinophobie. Paris: Gallimard.

EUCLID: See HEATH (3).

EVANS G.W. (1), 1917. "Cavalieri's Theorem in his Own Words." *AMM*, vol. 24, pp. 447–451.

FAIRBANK John K. (1), 1954/1970. *China's Response to the West, A Documentary Survey 1839–1923*. New York: Atheneum (Ninth printing).

FANG J. and TAKAYAMA K. P. (1), 1975. *Sociology of Mathematics – a prolegomenon*. Hauppauge: Paideia Press.

FELLMANN Rudolf (1), 1983. "Römische Rechentafeln aus Bronze." *AW*, vol. 14, no. 1, pp. 36–40.

FERGUSON John C. (1), 1941. "The Chinese Foot Measure." *MS*, vol. 6, pp. 357–382.

FEUCHTWANG Stephan D. R. (1), 1974/1982. *An Anthropological analysis of Chinese Geomancy*. Taipei: Nantian Shuju.

FOLKERTS Menso (1), 1977. *Die älteste mathematische Aufgabensammlung in lat. Sprache: Die Alkuin zugeschriebenen Propositiones ad acuendos iuvenes*. Vienna: Österr. Akademie der Wissenschaften, math.-naturwiss. Kl. – Denkschriften, vol. 116).

FOLKERTS Menso (2), 1981. "Zur Frühgeschichte der Magischen Quadrate in Westeuropa." *SudArch*, vol. 65, no. 4, pp. 313–338.

FOUCQUET Jean-François (1), 1716. *Relation exacte de ce qui s'est passé à Péking par raport à l'astronomie européane depuis le mois de juin 1711 jusqu'au commencement de novembre 1716*. Autograph manuscript preserved at the Roman Archives of the Society of Jesus (Archivum Romanum Societatis Jesu, *Jap. Sin. II, 154* (83p.)).

FOWLER David H. (1), 1992. "Dýnamis, mithartum and square." *HistMath*, vol. 19, no. 4, pp. 418–419.

FRISINGER H. (1), 1977. *The History of Meteorology to 1600*. New York: Science History Publications.

GATTY Janette (1), 1976. "Les recherches de Joachim Bouvet (1656–1730)." in *Actes Du Colloque International de Sinologie, La Mission Française de Pékin aux XVIIe et XVIIIe siècles, Chantilly, 20–22 September 1974*, Paris: Les Belles-Lettres, pp. 141-162.

GAUCHET L. (1), 1917. "Note sur la Trigonométrie Sphérique de Guo Shoujing." *TP*, vol. 18, pp. 151–174.

GAUSS Carl Friedrich (1), 1801/1986. *Disquisitiones Arithmeticae*. Transl. by A.A. CLARKE. Berlin: Springer-Verlag.

GAUSS Carl Friedrich (2), 1809/1857/1963. *Theoria Motus corporum coelestium in sectionibus conicis solem ambientum*. Hamburg. Engl. transl. by C.H. DAVIS, *Theory of the motion of the Heavenly Bodies moving about the Sun in Conic Sections*. Boston, 1857; reprinted New York, 1963.

GERBERT: see BUBNOW

GERICKE Helmuth (1). 1984. *Mathematik in Antike und Orient*. Berlin: Springer.

GERICKE Helmuth (2), 1990. *Mathematik im Abenland von den römischen Feldmessern bis zu Descartes*. Berlin: Springer-Verlag.

GERNET Jacques (1), 1972. "A propos des contacts entre la Chine et l'Europe aux XVIIe et XVIIIe siècles." *ActaAsiat* , no. 23, pp. 78–92.

GERNET Jacques (2), 1985. *a History of Chinese Civilization*. Cambridge: Cambridge University Press.

GERNET Jacques (3), 1980. "Christian and Chinese Visions of the World in the Seventeenth Century." *ChinSci*. no. 4, pp. 1–17.

GERNET Jacques (4), 1982. *China and the Christian Impact: a Conflict of Cultures*. Cambridge: Cambridge University Press.

GERNET Jacques (5), 1994. *L'intelligence de la Chine, Le Social et le mental*. Paris: Gallimard.

GILLINGS Richard J. (1), 1972/1982. *Mathematics in the Time of the Pharaohs*. New York: Dover.

GILLISPIE Charles Coulston (1), 1970–1980. *Dictionary of Scientific Biography*. 16 vols. New York: Ch. Scribner.

GOLDSTINE Herman H. (1), 1977. *History of Numerical Analysis from the 16th through the 19th Centuries*. Berlin: Springer-Verlag.

GONG Gerrit W. (1), 1984. *The Standard of 'Civilization' in International Society*. Oxford: Clarendon Press.

GOODRICH Luther Carrington and FANG Chao-ying eds. (1), 1976. *Dictionary of Ming Biography (1368-1644)*. 2 vols. New York: Columbia University Press.

GOODY Jack (1), 1977. *The Domestication of the Savage Mind*. Cambridge: Cambridge University Press.

GOULD S.H. (1), 1955. "The Method of Archimedes." *AMM*, vol. 62, no. 7, pp. 473–476.

GRAHAM A.C. (1), 1959. "Being in Western Philosophy compared with *shi/fei you/wu* in Chinese Philosophy." *Asia Major*, new ser., vol. 7, pp. 79–112.

GRAHAM A.C. transl. (2), 1960. *The Book of* Lieh-Tzu. London: John Murray.

GRAHAM A.C. (3), 1978. *Later Mohist Logic, Ethics and Science*. Hong Kong: The Chinese University Press.

GRAHAM A.C. (4), 1989. "Rationalism and Anti-Rationalism in Pre-Buddhist China." in BIDERMAN and SCHARFSTEIN eds., *Rationality in Question, On Eastern and Western Views of Rationality*. Leiden: E.J. Brill, pp. 141–164.

GRAHAM A.C. transl. (5), 1981. *Chuang Tzu, the Seven Inner Chapters and Other Writings from the Book* Chuang Tzu. London: G. Allen and Unwin.

GRAHAM A.C. (6), 1986. "Yin-yang and the Nature of Correlative Thinking." Unpublished paper for the Fourth International Congress on the History of Chinese Science, Sydney.

GRANET Marcel (1), 1934/1968. *La Pensée Chinoise*. Paris: Albin Michel.

GRANGER Gilles-Gaston (1), 1976. *La Théorie Aristotélicienne de la Science*. Paris: Aubier Montaigne.

GRANT Edward ed. (1), 1974. *A Source Book in Medieval Science*. Cambridge (Mass.): Harvard University Press.

GRATTAN-GUINNESS Ivor ed. (1), 1994. *Companion Encyclopedia of the History and Philosophy of the Mathematical Sciences*. 2 vols., London: Routledge.

GUITEL Geneviève (1), 1975. *Histoire comparée des numérations écrites*. Paris: Flammarion.

GUPTA R.C. (1), 1967. "Bhāskara I's Approximation to Sine." *IJHS*, vol. 2, no. 2, pp. 121–136.

GUPTA R.C. (2), 1969. "Second Order Interpolation in Indian Mathematics up to the Fifteenth Century." *IJHS*, vol. 4, nos. 1–2, pp. 86–98.

GUSDORF Georges (1), 1966/1977. *Les Sciences Humaines et la Pensée Occidentale, I – De l'Histoire des Sciences à l'Histoire de la Pensée*. Paris: Payot.

GUSDORF Georges (2), 1969. *Les Sciences Humaines et la Pensée Occidentale. III – La Révolution Galiléenne*, Tome II, Paris: Payot.

GUY R. Kent (1), 1987. *The Emperor's Four Treasuries, Scholars and the State in the Late Ch'ien-lung Era*. Cambridge Mass.: Harvard University Press.

HALL Rupert (1), 1962. "The Scholar and the Craftsman in the Scientific Revolution." in M. CLAGETT ed., *Critical Problems in the History of Science*. Madison: University of Wisconsin Press.

HARDY G.H. (1), 1940/1977. *A Mathematician's Apology*. Cambridge: Cambridge University Press.

HASHIMOTO Keizo (1), 1987. "Longomontanus's *Astronomia Danica* in China." *JHA*, vol. 18, pp. 95–110.

HASHIMOTO Keizo (2), 1988. *Hsu Kuang-Ch'i [Xu Guangqi] and Astronomical Reform – The Process of the Chinese Acceptance of Western Astronomy 1629–1635*. Osaka: Kansai University Press.

HAYASHI Takao (1), 1991. "A Note on Bhāskara I's Rational Approximation to Sine." *HS*, 42, pp. 45–48.

HAYASHI Takao, KUSUBA T., YANO Michio (1), 1990. "The Correction of the Mādhava Series for the Circumference of a Circle," *Centaurus*, 33, pp. 149–174.

HE Shi (1), 1990. "On the Li Shan-lan [Li Shanlan] Identity." in M. LOTHAIRE, *Mots, Mélanges Offerts à M.-P. Schützenberger*. Paris: Hermès, pp. 254–264.

HEATH Sir Thomas (1), 1921/1981. *A History of Greek Mathematics*. 2 vols. New York: Dover.

HEATH Sir Thomas (2), 1912. *The Works of Archimedes edited in Modern Notation with Introductory Chapters by T.L. Heath, with a Supplement*, The Method *of Archimedes, recently discovered by Heilberg*. New York: Dover.

HEATH Sir Thomas transl. (3), 1908/1956. *The Thirteen Books of Euclid's* Elements. *Translated with an Introduction and Commentary by Sir Thomas Heath*. 3 vols. New York: Dover.

HEIBERG J.L. and ZEUTHEN H.G. (1), 1906–1907. "Eine Neue Schrift des Archimedes." *BibMath*, vol. 3, pp. 321–363.

HELMS Mary W. (1), 1988. *Ulysses' Sail, An Ethnographic Odyssey of Power, Knowledge and Geographical Distance.* Princeton: Princeton University Press.

HENDERSON John B. (1), 1984. *The Development and Decline of Chinese Cosmology.* New York: Columbia University Press.

HENRICKS Robert G. transl. (1), 1983. *Philosophy and Argumentation in Third Century China, the Essays of Hsi K'ang* [Ji Kang]. Princeton: Princeton University Press.

HERMELINK Heinrich (1), 1958. "Die ältesten magischen Quadrate höherer Ordnung und ihre Bildungsweise." *SudhArch*, vol. 42, pp. 199–217.

HIGHTOWER James R. (1), 1959. "Some characteristics of parallel prose." in Søren EGEROD and Else GLAHN, eds., *Studia Serica Bernhard Karlgren Dedicata,* Copenhagen, pp. 60–91.

HILBERT David (1), 1899/1971. *Foundations of Geometry.* Transl. by L. UNGER from the German edition, 10th edn., revised and enlarged by P. BERNAYS, La Salle (Illinois): Open Court.

HO Peng-Yoke (1), 1965. "The Lost Problems of the Chang Ch'iu-chien Suan Ching, a Fifth-Century Chinese Mathematical Manual." *OE*, vol. 12, no. 1, pp. 37–53.

HO Peng-Yoke (2), 1966. *The Astronomical Chapters of the Chin Shu [Jinshu].* Paris: Mouton.

HO Peng-Yoke (3), 1969. "The Astronomical Bureau in Ming China." *JAH*, vol. 3, pp. 137–157.

HO Peng-Yoke (4), 1971. "Ch'in Chiu-shao" [Qin Jiushao]. in *DSB*, 1971, vol. 3, pp. 249–256.

HO Peng-Yoke (5), 1971. "Chu Shih-chieh" [Zhu Shijie]. in *DSB*, 1973, vol. 8, pp. 418–425.

HO Peng-Yoke (6), 1973. "Liu Hui." in *DSB*, vol. 8, pp. 418–425.

HO Peng-Yoke (7), 1973. "Li Chih." [Li Zhi] in *DSB*, vol. 8, pp. 313–320.

HO Peng-Yoke (8), 1976. "Yang Hui." in *DSB*, vol. 14, pp. 538–546.

HO Peng-Yoke (9), 1972. "The Earliest Chinese Magic Squares and Magic Squares in the Islamic Word." *BT*, no. 5, pp. 95–104.

HOE Jock (1), 1976. *L'algèbre Chinoise à la fin du XIIIe siècle à travers l'étude des systèmes d'équations-polynômes traités par Zhu Shijie dans son livre "Le Miroir de Jade des Quatre Inconnues,"* Siyuan yujian, *de 1303. Thèse pour le doctorat de spécialité en études extrême-orientales.* Université de Paris VII, 338 p. + appendix no. 1 [modern algebraical translation of the 284 problems of the original]: 154 p. + appendix no. 2 ["semi-symbolic" translation of the same 284 problems]: 194 p. + appendix no. 3 [Chinese original text]: 177 p.

HOE John [sic, in fact "Jock"], (2), 1977. *Les systèmes d'équations-polynômes dans le* Siyuan yujian *(1303).* Paris: Collège de France, Institut des Hautes Etudes Chinoises (Mémoires de l'Institut des Hautes Etudes Chinoises, vol. 6).

HOE Jock (3), 1981. "Zhu Shijie and his Jade Mirror of the four unknowns." in *History of Mathematics, Proceedings of the First Australian Conference, Monash Univ., Victoria,* pp. 1–24.

HOLZMAN D. (1), 1958. "Shen Kua [Shen Gua] and his *Meng-ch'i pi-tan [Mengqi bitan].*" *TP*, vol. 46, pp. 260–292.

HOMINAL François (1), 1980. *Terminologie mathématique en Chinois moderne.* Paris: Editions de l'Ecole des Hautes Etudes en Sciences Sociales.

HORIUCHI Annick (1), 1991. "The Development of Algebraic Methods of Problem Solving in Japan in the Late Seventeenth and the Early Eighteenth Centuries." *Proceedings of the International Congress of Mathematics, Kyoto/Japan, 1990*, vol. II, pp. 1639–1649.

HORIUCHI Annick (2), 1994. *Les mathématiques japonaises à l'époque d'Edo (1600-1868). Une étude des travaux de Seki Takakazu (?-1708) et de Takebe Katahiro (1664-1739)*. Paris: J. Vrin.

HORNER W.G. (1), 1819. "A new method of Solving Numerical Equations of all Orders by Continuous Approximation." *PTRS*, vol. 109, pp. 308–335.

HORNG Wann-Sheng [HONG Wansheng] (1), 1991. *Li Shanlan: The Impact of Western Mathematics in China During the Late Nineteenth Century*. Unpublished dissertation submitted to the Graduate Faculty in History for the degree of Doctor in Philosophy, The City University of New York.

HORNG Wann-sheng (2), 1993. "Chinese Mathematics at the Turn of the 19th Century: Jiao Xun, Wang Lai and Li Rui." in Cheng-hung LIN and Daiwie FU eds., *Philosophy and Conceptual History of Science in Taiwan*. Dordrecht: Kluwer Academic Publishers, pp. 167–208.

HØYRUP Jens (1), 1987. "The Formation of 'Islamic Mathematics' Sources and Conditions." *SciCont*, vol. 1, no. 2, pp. 281–329.

HØYRUP Jens (2), 1990. "Sub-scientific Mathematics: Observations on a Pre-modern Phenomenon." *HistSci*, vol. 28, part 1, no. 79, pp. 63–87.

HØYRUP Jens (3), 1990. "Sub-scientific Mathematics: Undercurrents and Missing Links in the Mathematical Technology of the Hellenistic and Roman World." Quoted from a preprint of an article written for *Aufstieg und Niedergang der römischen Welt*, vol. 2, no. 37, 3 (55 p.).

HØYRUP Jens (4), 1990. "Algebra and Naive Geometry. An Investigation of Some Basic Aspects of Old Babylonian Mathematical Thought (I and II)." *AF*, vol. 17, nos. 1 and 2, pp. 27–69 and 262–354.

HØYRUP Jens (5), 1994. "Babylonian Mathematics." in Ivor GRATTAN-GUINNESS ed., *Companion Encyclopedia of the History and Philosophy of the Mathematical Sciences*. 2 vols. London: Routledge, vol. 1, pp. 21–29.

HU Shih (1), 1992. *The Development of Logical Method in Ancient China*. Shanghai: The Oriental Book Co.

HUANG Shijian (1), 1986. "The Persian Language in China during the Yuan Dynasty." *PFEH*, vol. 34, pp. 83–95.

HUCKER Charles O. (1), 1985. *A Dictionary of Official Titles in Imperial China*. Stanford: Stanford University Press.

HUMMEL Arthur W. ed. (1), 1943/1970. *Eminent Chinese of the Ch'ing Period (1644-1912)*. 2 vols. Taipei: Ch'eng Wen Publishing Company.

IANNACCONE I. and TAMBURELLO A. eds. (1), 1990. *Dall'Europa alla Cina: contributi per una storia dell'Astronomia*. Napoli: Università degli Studi "Federico II." Istituto Universitario Orientale.

Institut National de la Communication Audiovisuelle (L') (1), 1983. *Mémoire Joseph Needham*. Paris.

Institute of the Natural Sciences, Chinese Academy of Sciences (The) (1), 1983. *Ancient China's Technology and Science*. Peking: Foreign Languages Press.

JAKI Stanley L. (1), 1978. *The Origin of Science and the Science of its Origin*. Edinburgh: Scottish Academic Press.

JAMI Catherine (1), 1986. *Jean-François Foucquet et la modernisation de la science en Chine, la* Nouvelle Méthode d'Algèbre. Paris: Université de Paris VII, Mémoire de Maîtrise préparé sous la direction de F. Martin.

JAMI Catherine (2), 1988. "Une histoire chinoise du nombre π." *AHES*, vol. 38, no. 1, pp. 39–50.

JAMI Catherine (3), 1988. "Western Influence and Chinese Tradition in an Eighteenth Century Chinese Mathematical Work." *HistMath*, 1988, vol. 15, no. 4, pp. 311–331.

JAMI Catherine (4), 1990. *Les* Méthodes rapides pour la trigonométrie et le rapport précis du cercle *(1774), Tradition Chinoise et Apport Occidental en Mathématiques.* Paris: Collège de France, Institut des Hautes Etudes Chinoises (Mémoires de l'Institut des Hautes Etudes Chinoises, vol. 32).

JIANG Xiaoyuan (1), 1988. "The solar Motion Theories of Babylon and Ancient China." *VistasAstro*, vol. 31, pp. 829–832.

JOSEPH George Gheverghese (1), 1994. "Tibetan Astronomy and Mathematics." in Ivor GRATTAN-GUINNESS ed., *Companion Encyclopedia of the History and Philosophy of Mathematics*, London: Routledge, vol. 1, pp. 131–136.

JUSCHKEWITSCH: See YUSHKEVICH

KALINOWSKI Marc (1), 1983. "Les instruments astro-calendériques des Han et la méthode liu-ren." *BEFEO*, Paris, tome 72, pp. 309–412.

KARPINSKI Louis C. (1), 1915. *Robert of Chester's Latin Translation of the Algebra of Al-Khowarizmi.* New York: Macmillian.

KAUCKÝ Josef (1), 1963. "O jednom Problému Z Dějin Čínské Matematiky" (On a problem from the history of Chinese mathematics). *MATFYZCAS*, vol. 13, no. 1 , pp. 32–40.

KENNEDY E.S. (1), 1964/1977. "The Chinese-Uighur Calendar as Described in the Islamic Sources." *Isis* (December 1964); reprinted in N. SIVIN ed., *Science and Technology in East Asia.* New York: Science History Publications, 1977, pp. 191–199.

KIM Yong-Sik (1), 1982. "Natural Knowledge in a Traditional Culture: Problems in the Study of the History of Chinese Science." *Minerva*, vol. 20, pp. 83–104.

KIM Yong-Woon (1), 1973. "Introduction to Korean Mathematical History." *KJ*, vol. 13, no. 7, pp. 16–23; no. 8, pp. 27–32; no. 9, pp. 35–39.

KITCHER Philip (1), 1983. *The Nature of Mathematical Knowledge.* New York Oxford: Oxford University Press.

KLEIN Jacob (1), 1934–36/1968. *Greek Mathematical Thought and the Origin of Algebra.* (transl. from the German by Eva BRANN). Cambridge (Mass.): The MIT Press.

KLINE Morris (1), 1972. *Mathematical Thought from Ancient to Modern Times.* New York: Oxford University Press.

KLOYDA Mary Thomas (1), 1935. *Linear and Quadratic Equations 1550–1660.* PhD Dissertation, Ann-Arbor: University of Michigan.

KNOBLOCH Eberhard (1), 1980. *Der Beginn der Determinantentheorie Leibnizens nachgelassens Studien zum Determinantenkalkül; im Zusammenhang mit dem gleichnamigen Abhandlungsband fast ausschließlich zum ersten Mal nach den Originalschriften.* Hildesheim: Gersteinberg Verlag.

KNOBLOCH Eberhard (2), 1988. "Sur la vie et l'oeuvre de Christophore Clavius (1538–1612)," *RHS*, vol. 41, nos. 3-4, pp. 331–356.

KNORR Wilbur Richard (1), 1978. "Archimedes and the *Elements*, Proposal for a Revised Chronological Ordering of the Archimedean Corpus." *AHES*, vol. 19, no. 3, pp. 211–290.

KNORR Wilbur Richard (2), 1985. "The Geometer and Archeoastronomers: On the Prehistoric Origins of Mathematics." *BJHS*, vol. 18 (part 2), no. 59, pp. 197–211.

KNORR Wilbur Richard (3), 1986/1993. *The Ancient Tradition of Geometric Problems.* New York: Dover.

KNUTH Donald Ervin (1), 1972. "Ancient Babylonian Algorithms." *CACM*, vol. 15, no. 7, pp. 671–677.

KNUTH Donald Ervin (2), 1981. *The Art of Computer Programming.* 3 vols. Reading (Mass.): Addison-Wesley.

KOGELSCHATZ Hermann (1), 1981. *Bibliographische Daten zum frühen mathematischen Schrifttum Chinas im Umfeld der 'Zehn mathematischen Klassiker' (1. Jh. v. Chr. bis 7. Jh. n. Chr.).* Munich: Veröffentlichungen des Forschungsinstituts des Deutschen Museums für die Geschichte der Naturwissenschaften und der Technik, Reihe B.

KOJIMA Takashi (1), 1954/1978. *The Japanese Abacus, its Use and Theory.* Tokyo: Charles E. Tuttle.

KOKOMOOR F.W. (1), 1928. "The Distinctive Features of Seventeenth Century Geometry." *Isis*, vol. 10, no. 34, pp. 367–415.

KOYRÉ Alexandre (1), 1957. *From the Closed World to the Infinite Universe.* Baltimore: Johns Hopkins Press.

KRAYER Albert (1), 1991. *Mathematik im Studienplan der Jesuiten.* Stuttgart: Franz Steiner Verlag.

KRENKEL Werner (1), 1969. "Das Rechnen mit römischen Ziffern." *Das Altertum,* vol. 15, no. 1, pp. 252–256.

LAM Lay-Yong (1), 1966. "On the Chinese Origin of the Galley Method of Arithmetical Division." *BJHS*, vol. 3, no. 9, pp. 66–69.

LAM Lay-Yong (2), 1969. "The Geometrical Basis of the Ancient Chinese Square Root Method." *Isis*, vol. 61, no. 1, pp. 96–102.

LAM Lay-Yong (3), 1969. "On the Existing Fragments of Yang Hui's *Hsiang Chieh Suan Fa [Xiangjie suanfa].*" *AHES*, vol. 6, no. 1, pp. 82–88.

LAM Lay-Yong (4), 1972. "The *Jih yung suan fa [Riyong suanfa]*: an Elementary Textbook of the Thirteenth Century." *Isis*, vol. 63, pp. 370–383.

LAM Lay-Yong (5), 1974. "Yang Hui's Commentary on the *Ying nu* Chapter of the *Chiu chang suan shu.*" *HistMath*, vol. 1, pp. 47–64.

LAM Lay-Yong (6), 1977. *A Critical Study of the Yang Hui Suan Fa, a Thirteenth-century Mathematical Treatise.* Singapore: Singapore University Press; Rev.: Jock Hoe, AnSci, 1982, vol. 39, pp. 491–504.

LAM Lay-Yong (7), 1979. "Chu Shih-chieh's *Suan hsüeh Ch'i-meng* [Introduction to Mathematical Studies]." *AHES*, vol. 21, no. 1, pp. 1–31.

LAM Lay-Yong (8), 1980. "The Chinese Connexion between the Pascal Triangle and the Solution of Numerical Equations of any Degree." *HistMath*, vol. 7, no. 4, pp. 407–424.

LAM Lay-Yong (9), 1982. "Chinese Polynomial Equations in the Thirteenth Century." in *Needham*, pp. 231–272.

LAM Lay-Yong (10), 1986. "The Conceptual Origin of our Numeral System and the Symbolic Form of Algebra." *AHES*, vol. 36, no. 3, pp. 183–195.

LAM Lay-Yong (11), 1987. "Linkages: Exploring the Similarities Between the Chinese Rod Numeral System and our Numeral System." *AHES*, vol. 37, no. 4, pp. 365–392.

LAM Lay-Yong (12), 1988. "A Chinese Genesis: Rewriting the History of Our Numeral System." *AHES*, vol. 38, no. 2, pp. 101–108.

LAM Lay-Yong and ANG Tian-Se (1), 1984. "Li Ye and His *Yi Gu Yan Duan* (Old Mathematics in Expanded Sections)." *AHES*, vol. 29, no. 3, pp. 237–266.

LAM Lay-Yong and ANG Tian-Se (2), 1986. "Circle Measurements in Ancient China." *HistMath*, vol. 13, pp. 325–340.

LAM Lay-Yong and ANG Tian-Se (3), 1986. "The Earliest Negative Numbers: How they Emerged from a Solution of Simultaneous Linear Equations." *ArIntHS*, vol. 37, pp. 222–262.

LAM Lay-Yong and ANG Tian-Se (4), 1992. *Fleeting Footsteps, Tracing the Conception of Arithmetic and Algebra in Ancient China.* Singapore: World Scientific.

LAM Lay-Yong and SHEN Kangshen (1), 1984. "Right-angled Triangles in Ancient China." *AHES*, vol. 30, no. 2, pp. 87–112.

LAM Lay-Yong and SHEN Kangshen (2), 1986. "Mathematical Problems on Surveying in Ancient China." *AHES*, vol. 36, no. 1, pp. 1–20.

LATOURETTE, K.S. (1), 1929. *A History of Christian Missions in China.* New York.

LAU Chung Him (1), 1958/1980. *The Principles and Practice of the Chinese Abacus.* Hong Kong: Published by LAU Chung Him and Co.

LECOMTE L. (1), 1701. Nouveaux mémoires sur l'Etat Présent de la Chine. Paris.

LEOÑARDO OF PISA (or FIBONACCI) (1), ca. 1170–1240. *Scritti di Leonardo Pisano.* Edited by B. Boncompagni, Rome, 1857–1862.

LESLIE D.D. (1), 1986. *Islam in Traditional China: a Short History to 1800.* Belconnen (Australia): Canberra College of Advanced Education.

LESLIE D.D. and GARDINER K.H.J. (1), 1982. "Chinese Knowledge of Western Asia during the Han." *TP*, vol. 68, pp. 254–308.

LÉVI-STRAUSS Claude (1), 1962/1985. *La Pensée Sauvage*, Paris: Plon.

LI Guohao, ZHANG Mengwen and CAO Tianqin eds. (1), 1982. *Explorations in the History of Science and Technology in China, Compiled in Honour of the Eightieth Birthday of Dr. Joseph Needham.* Shanghai: Shanghai Chinese Classics.

LI Yan (1), 1956. "The Interpolations Formulas of Early Chinese Mathematicians." in *Proc. VIIIth Int. Congress of the History of Science, Florence*, pp. 70–72.

LIBBRECHT Ulrich (1), 1972. "The Chinese *Ta-yen* Rule: a Comparative Study." *OL*, vol. 3, pp. 179–199.

LIBBRECHT Ulrich (2), 1973. *Chinese Mathematics in the Thirteenth Century: the Shu-shu chiu-chang of Ch'in Chiu-shao* [Qin Jiushao]. Cambridge (Mass.): The MIT Press: Rev.: K. Yabuuchi, *THG*, 1975, no. 50, pp. 124–128.

LIBBRECHT Ulrich (3), 1974. "Indeterminate analysis, Historical Relations Between China, Islam and Europe." in *Proc. XIVth Int. Congress of the History of Science, Tokyo, 1974*, pp. 311–314.

LIBBRECHT Ulrich (4), 1980. "Joseph Needham's Work in the Area of Chinese Mathematics." *PP*, no. 87, pp. 30–39.

LIBBRECHT Ulrich (5), 1982. "Mathematical Manuscripts from the Dunhuang Caves." in *Needham*, pp. 203–229.

LIBRI Guillaume (1), *1838–1841. Histoire des Sciences Mathématiques en Italie.* 4 vols. Paris.

422 References

LINDBERG David C, ed. (1), 1978. *Science in the Middle Ages.* Chicago and London: The University of Chicago Press.

LOEWE M. (1), 1961. "The Measurement of Grain During the Han Period." *TP*, vol. 49, pp. 64–95.

LOH Shiu-chang, HING Hing-Sum, KONG Luan *et al.* (1), 1976. *A Glossary of the Mathematical and Computing Sciences (Chinese–English).* Hong Kong: Machine Translation Project, The Chinese University of Hong Kong.

LORIA Gino (1), 1929/1950. *Storia delle Matematiche d'all'Alba della Civilta al Tramonto del Secolo XIX.* Turin (2nd edn., Milan, 1950).

LUCAS Edouard (1), 1883–1894/1960. *Récréations mathématiques.* 4 vols. Paris: Albert Blanchard.

ŁUKASIEWICZ Jan (1), 1951. *Aristotle's Syllogistic from the Standpoint of Modern Formal Logic.* Oxford: Oxford University Press.

MAHLER K. (1), 1958. "On the Chinese Remainder Theorem." *MN*, vol. 18, pp. 120–122.

MARROU H.I. (1), 1954/1975. *De la Connaissance Historique.* Paris: Editions du Seuil.

MARTIN W.A.P. (1), 1896. *A Cycle of Cathay or China, South and North, with Personal Reminiscences.* New York: Fleming H. Revell Co.

MARTZLOFF Jean-Claude (1), 1980. "La Compréhension Chinoise des Méthodes Démonstratives Euclidiennes au Cours du XVIIe Siècle et au Début du XVIIIe." in *Actes du IIe Colloque International de Sinologie: Les Rapports entre la Chine et l'Europe au Temps des Lumières, Chantilly, 16–18 Sept. 1977.* Paris: Les Belles Lettres, pp. 125–143.

MARTZLOFF Jean-Claude (2), 1981. *Recherches sur l'Oeuvre Mathématique de Mei Wending (1633–1721).* Paris: Collège de France, Institut de Hautes Etudes Chinoises (Mémoires de l'Institut des Hautes Etudes Chinoises, vol. 16), Stanislas Julien Prize 1982.

MARTZLOFF Jean-Claude (3), 1981. "La Géometrie Euclidienne selon Mei Wending." *HS*, no. 21, pp. 27–42.

MARTZLOFF Jean-Claude (4), 1983. "Matteo Ricci's Mathematical Works and their Influence." in *Int. Symp. on Chinese–Western Cultural Interchange in Commemoration of the 400th Anniversary of the Arrival of Matteo Ricci S.J. in China. Taipei, 11–16 Sept, 1983*, Taipei: Furen Daxue Chubanshe, pp. 889–895.

MARTZLOFF Jean-Claude (5), 1983. "Notice no. 3349." in Michel SOYMIÉ ed., *Catalogue des Manuscrits chinois de Touen-Houang, Fonds Pelliot Chinois de la Bibliothèque National,* vol. 3. Paris: Editions de la Fondation Singer-Polignac, pp. 283–284.

MARTZLOFF Jean-Claude (6), 1984. "Sciences et Techniques dans l'Oeuvre de Ricci." *RSR*, tome 72, no. 1, pp. 37–49.

MARTZLOFF Jean-Claude (7), 1984. "The Manchu Mathematical Manuscript *Bodoro arga i oyonggongge be araha uheri hešen i bithe* of the Bibliothèque Nationale, Paris: Preliminary Investigations." Unpublished Communication presented to the 3rd Int. Conf. on the History of Science in China, Peking, August, 1984.

MARTZLOFF Jean-Claude (8), 1985. "Sciences et Techniques en Chine (Mathématiques, Astronomie, Techniques)." in *Encyclopedia Universalis,* Paris, pp. 889–895.

MARTZLOFF Jean-Claude (9), 1985. "Aperçu sur l'Histoire des Mathématiques Chinoises telle qu'elle est pratiquée en République Populaire de Chine". *HS*, no. 28, pp. 1–30.

MARTZLOFF Jean-Claude (10), 1991. "Les contacts entre les astronomies et les mathématiques arabes et chinoise vues principalement à partir des sources chinoises – état actual des connaissances." in *Deuxième Colloque Maghrébin sur l'Histoire des Mathématiques Arabes, Tunis, 1–3 Dec. 1988*, Tunis, 1991, University of Tunis I, ISEFC and ATSCM, pp. 164–182.

MARTZLOFF Jean-Claude (11), 1989. "La science astronomique européenne au service de la diffusion du catholicisme en Chine. L'oeuvre astronomique de Jean-François Foucquet (1665–1741)." *MEFRIM*, tome 101–2, pp. 973–989.

MARTZLOFF Jean-Claude (12), 1990. "A Survey of Japanese Publications on the History of Japanese Traditional Mathematics (*Wasan*) from the Last 30 Years." *HistMath.*, vol. 17, no. 4, pp. 366–373.

MARTZLOFF Jean-Claude (13), 1993. "Eléments de réflexion sur les réactions chinoises à la géometrie Euclidienne à la fin du XVIIe siècle – le *Jihe lunyue* de Du Zhigeng vu principalement à partir de la préface de l'auteur et de deux notices bibliographiques rédigées par des lettrés illustres." *HistMath*, vol. 20, no. 2, pp. 160–179 and no. 3, pp. 460–463.

MARTZLOFF Jean-Claude (14), 1994. "Chinese Mathematics." in Ivor GRATTAN-GUINNESS ed., *Companion Encyclopaedia of the History and Philosophy of the Mathematical Sciences*. 2 vols. London: Routledge, vol. 1, pp. 93–103.

MARTZLOFF Jean-Claude (15), 1994. "Les Mathématiques Japonaises." in A. BERQUE ed., *Dictionnaire de la Civilisation Japonaise*, Paris: Hazan.

MARTZLOFF Jean-Claude (16), 1994. "Space and time in Chinese Texts of Astronomy and Mathematical Astronomy in the Seventeenth and Eighteenth Centuries." *ChinSci.*, no. 11, pp. 66-92.

MASPERO Henri (1), 1928. "Note sur la Logique de Mo-Tseu et de son école." *TP*, vol. 25, pp. 1–64.

MAY Kenneth O. (1), 1973, *Bibliography and Research Manual of the History of Mathematics*. Toronto: University of Toronto Press.

MAZARS Guy (1), 1974. "La notion de sinus dans les mathématiques indiennes." *FundSci*, no. 15, pp. 1–23.

MICHEL Paul-Henri (1), 1950. *De Pythagore à Euclide, Contribution à l'Histoire des Mathématiques Préeuclidiennes*. Paris: Les Belles Lettres.

MICHIWAKI Yoshimasa and KOBAYASHI Tatsuhiko (1), 1989. "Influence of *Li-suan Chuan-shu* [*Lisuan quanshu*] in Japanese Mathematics and Different Ways of Thinking Between Mei Wending and Japanese Mathematicians." in *Jōbu daigaku keiei jōhō gakubu kiyō* [Bulletin of the Department of Management and Information Science, Jōbu Univ.], no. 1, pp. 135–142.

MIGNE, Jacques Paul ed., (1), 1851. *Patrologiae Lat.* Paris.

MIKAMI Yoshio (1), 1909. "A Remark on the Chinese Mathematics in Cantor's *Geschichte der Mathematik.*" *AMP*, vol. 15, p. 68.

MIKAMI Yoshio (2), 1911. "The Influence of Abaci on Chinese and Japanese Mathematics." *JDMV*, vol. 20, pp. 380–393.

MIKAMI Yoshio (3), 1911. "Further Remarks on the Chinese Mathematics in Cantor's *Geschichte.*" *AMP*. vol. 18, p. 209.

MIKAMI Yoshio (4), 1913/1974. *The Development of Mathematics in China and Japan.* New York: Chelsea.

MISSION CATHOLIQUE DES LAZARISTES À PÉKIN (1), 1949/1969. *Catalogue de la Bibliothèque du Pé-T'ang.* Peking: Imprimerie des Lazaristes; Reprinted in 1969 (Paris: Les Belles Lettres).

MONTUCLA J.F. (1), 1798/1968. *Histoire des Mathématiques.* 4 vols. Paris: Blanchard.

MOON P. (1), 1971. *The Abacus, its History, its Design, its Possibilities in the Modern World.* New York: Gordon and Breach.

MORGAN Carole (1), 1980. *Le tableau du boeuf du printemps, étude d'une page de l'almanach chinois.* Paris: Collège de France, Institut des Hautes Etudes Chinoises (Mémoires de l'Institut des Hautes Etdues Chinoises, vol. 14).

MORGAN Carole (2), 1981. "Les 'Neuf Palais' dans les manuscrits de Touen Houang [Dunhuang]." in *Nouvelles Contributions aux Etudes de Touen-Houang, sous la direction de Michel Soymié*, Geneva: Droz, pp. 251–260.

MORROW Glenn R. transl. and comment. (1), 1970. *Proclus, a Commentary on the First Book of Euclid's Elements.* Princeton: Princeton University Press.

MORTET Victor (1), 1896. *Un Nouveau Texte des Traités D'Arpentage et de Géométrie d'Epaphroditus et de Vitruvius Rufus, Publié d'Après le MS. Latin de la Bibliothèque Royale de Munich par M. Victor Mortet avec une Introduction de M. Paul Tannery, Tiré des Notices et Extraits de la Bibliothèque Nationale et autres Bibliothèques.* Paris: Imprimerie Nationale.

MOULE G.E. (1), 1873. "The Obligations of China to Europe in the Matter of Physical Science acknowledged by Eminent Chinese; being Extracts from the Preface to Tsang Kwo-Fan [Zeng Guofan]'s Edition of Euclid with brief introductory observations." *JRAS/NCB*, vol. 7, pp. 147–164.

MOUNIN G. (1), 1970. *Introduction à la sémiologie.* Paris: Les Editions de Minuit.

MUELLER Ian (1), 1974. "Greek Mathematics and Greek Logic." in J. CORCORAN ed., *Ancient Logic and its Modern Interpretations, Proceedings of the Buffalo Symposium on Modernist Interpretations of Ancient Logic, 21 and 22 April 1972,* Dordrecht Boston: D. Reidel Publishing Company, pp. 35–70.

MUELLER Ian (2), 1981. *Philosophy of Mathematics and Deductive Structure in Euclid's Elements.* Cambridge (Mass.): The MIT Press.

MURRAY A. (1), 1978/1986. *Reason and Society in the Middle Ages.* Oxford: Clarendon Press.

NAGEL Ernest (1), 1979. "Some Reflexions on the Use of Language in the Natural Sciences." in E. NAGEL, *Teleology Revisted and Other Essays in the Philosophy and History of Science*, New York: Columbia University Press, pp. 49–63.

NAKAYAMA Shigeru (1), 1969. *A History of Japanese Astronomy: Chinese Background and Western Impact.* Harvard: Harvard University Press.

NAKAYAMA Shigeru (2), *1984.* *Academic and Scientific Traditions in China, Japan and the West.* Tokyo: University of Tokyo Press.

NAPIER John (1), 1617/1990. *Rabdologiae.* Transl. by William Frank RICHARDSON. Introduction by Robin E. RIDER. Charles Babbage Institute, Reprint Series for the History of Computing, 15, Cambridge (Mass.)/London: The MIT Press.

NAUX Charles (1), 1966. *Histoire des Logarithmes de Neper à Euler*, Tome I: *La découverte de logarithmes et le calcul des premières tables.* Paris: A. Blanchard.

NEEDHAM Joseph (1), 1958. "Chinese Astronomy and the Jesuit Mission: an Encounter of Cultures." London: The China Society, 20 p.

NEEDHAM Joseph (2), 1959. *Science and Civilisation in China*, vol. 3: *Mathematics and the Sciences of the Heavens and the Earth*. Cambridge: Cambridge University Press.

NEEDHAM Joseph (3), 1967. "The Roles of Europe and China in the Evolution of Oecumenical Science." *JAH*, vol. 1, no. 1, pp. 3–32.

NEUGEBAUER O. (1). 1935–37. *Mathematische Keilschrift-Texte*. 3 vols. Berlin: Springer, *Quellen und Studien zur Geschichte der Mathematik*.

NEUGEBAUER O. (2), 1953. *Astronomical Cuneiform Texts*. 3 vols. London: Lund Humphries.

NEUGEBAUER O. (3), 1957/1969. *The Exact Sciences in Antiquity*. New York: Dover.

NEUGEBAUER O. (4). 1975. *A History of Ancient Mathematical Astronomy*. 3 vols. Berlin: Springer.

NGO VAN XUYET (1), 1976. *Divination, Magie et Politique dans la Chine Ancienne*. Paris: Presses Universitaires de France.

NIENHAUSER William H. Jr. (1), 1986. *The Indiana Companion to Traditional Chinese Literature*. Bloomington: Indiana University Press.

PATON W.R. transl. (1), 1918/1979. *The Greek Anthology*. Cambridge (Mass.): Harvard University Press and London: William Heinemann.

PELLIOT Paul (1), 1948. "Le Hōja et le Sayyid Husain de l'histoire des Ming." *TP*, vol. 38, pp. 81–292.

PENG Rita Hsiao-fu (1), 1975. "The K'ang-hsi Emperor's absorption in Western Mathematics and its Extensive Applications of Scientific Knowledge." *LishiXB*, no. 3, pp. 1–74.

PETERSON Willard J. (1), 1973. "Western Natural Philosophy Published in Late Ming China." in *Proceedings of the American Philosophical Society*, vol. 117, no. 4, pp. 295–321.

PETERSON Willard J. (2), 1975. "Fang I-chih: Western Learning and the 'Investigation of Things'." in Theodore de Barry ed., *The Unfolding of Neo-Confucianism*. New York: Columbia University Press, pp. 369–411.

PETERSON Willard J. (3), 1986, "Calendar Reform Prior to the Arrival of Missionaries at the Ming Court." *MingStud*, no. 21.

PFISTER Louis (1), 1934. *Notices biographiques et bibliographiques sur les Jésuites de l'ancienne Mission de Chine 1552-1773*. 2 vols. Shanghai.

PORTER J. (1), 1980. "Bureaucracy and Science in Early Modern China: the Imperial Astronomical Bureau in the Ch'ing Period." *JOS*, vol. 18, nos. 1 and 2, pp. 61–76.

PORTER J. (2), 1982. "The Scientific Community in Early Modern China." *Isis*, vol. 73, no. 269; pp. 529–544.

PULLMAN J.M. (1), 1968. *The History of the Abacus*. New York: Praeger.

RASHED Roshdi (1), 1984. *Entre Arithmétique et Algèbre, Recherches sur l'Histoire des Mathématiques Arabes*. Paris: Les Belles Lettres.

RAWSKI Evelyn (1), 1979. *Education and Popular Literacy in Ch'ing China*. Ann Arbor: University of Michigan.

REIFLER Erwin (1), 1965. "The Philological and Mathematical Problems of Wang Mang's Standard Grain Measures." in *Qingzhu Li Ji Xiangsheng qishi sui lunwen ji* [Collected Papers in honour of Dr. Li Ji's seventieth birthday], Taipei, pp. 387–402.

REINACH Th. (1), 1907, "Un Traité de Géométrie Inédit d'Archimède." *RGSPA*, no. 22, pp. 911–928 and no. 23, pp. 954–961.

RICHÉ Pierre (1), 1979, *Les écoles et l'enseignement dans l'Occident chrétien de la fin du Ve siècle*. Paris: Aubier Montaigne.

RITTER Jim (1), 1989. "Chacun sa vérité." In Michel SERRES ed., *Eléments d'Histoire des Sciences*, Paris: Bordas, pp. 39–61.

RITTER Jim (2), 1992, "Metrology and the Prehistory of Fractions." In P. BENOIT et al., *Histoire de Fractions, Fractions d'Histoire*. Basel: Birkhäuser.

ROME A. ed. (1), 1936. *Théon d'Alexandrie, Commentaire sur les livres 1 et 2 de l'Almageste*, Tome 2. Rome: Città del Vaticano.

ROSSABI Morris (1), 1981. "The Muslims in the Early Yüan Dynasty." in J. LANGLOIS ed., *China Under Mongol Rule*, Princeton: Princeton University Press, pp. 257–295.

ROSSABI Morris ed. (2), 1983. *China among Equals, The Middle Kingdom and its Neighbors, 10th-14th Centuries*. Berkeley: University of California Press.

ROȘU Arion (1), 1989, "Les carrés magiques et l'histoire des idées en Asie." *ZDMG*, vol. 139, no. 1, pp. 120–158.

des ROTOURS Robert (1), 1932, *Le Traité des Examens traduit de la Nouvelle Histoire des Tang (chap. XLIV, XLV)*. Paris: Ernest Leroux.

des ROTOURS Robert (2), 1975. "Le *Tang lieou tien* [Tang liu dian] Décrit-il Exactement les Institutions en Usage sous la Dynastie des Tang?" *JA*, tome 263, fasc. 1–2, pp. 183–201.

RUFFINI Paolo (1), 1804. *Sopra la determinazione delle radici nelle equazioni numeriche di qualunque grado*. Modena.

RYBNIKOV K.A. (1), 1956. "On the Role of algorythms [sic] in the History of Mathematical Analysis." in *Actes du VIIIe Congrès International d'histoire des sciences*, pp. 142–145.

SAIDAN A.S. transl. and comment (1), 1978. *The Arithmetic of Al-Uqlīdisī — The Story of Hindu–Arabic Arithmetic as told in* Kitāb al-Fuṣūl fī al-Ḥisāb al-Hindī *by Abū al-Ḥasan Aḥmad ibn Ibrāhīm al-Uqlīdisī, written in Damascus in the year 341 [AD 952/3]*. Dordrecht/Boston: D. Reidel Publishing Co.

SALAFF Stephen (1), 1972. A Biography of Hua Lo-keng [Hua Luogeng], *Isis*, vol. 63, no. 217, pp. 143-183.

SALIBA George A . (1), 1972. "The meaning of al-jabr wa'l muqābalah." *Centaurus*, vol. 17, pp. 189–204.

SANFORD Vera (1), 1927. *The History and Significance of Certain Standard Problems in Algebra*. New York: Bureau of Publications, Teachers College, Columbia University.

SCHRIMPF Robert (1), 1963. *La Collection Mathématique "Souan King Che Chou," Contribution à l'Histoire des Mathématiques chinoises des origines au VIIe siècle de notre ère*. Unpublished Doctoral Dissertation, Rennes.

SCHUH Dieter (1), 1972. *Untersuchungen zur Geschichte der Tibetischen Kalenderrechnung*. Wiesbaden: Franz Steiner.

SEMEDO Alvaro [Alvarez] (1), 1645. *Histoire Universelle du grand royaume de la Chine*. Paris: S. et G. Cramoisy.

SESIANO Jacques (1), 1980, "Herstellungsverfahren magischer Quadrate aus islamischer Zeit (I)." *SudhArch*, vol. 64, no. 22, pp. 187–196.

SESIANO Jacques (2), 1985, "The Appearance of Negative Solutions in Mediaeval Mathematics." *AHES*, vol. 32, *1985*, pp. 105–150.

SESIANO Jacques (3), 1980, "Herstellungsverfahren magischer Quadrate aus islamischer Zeit (II)." *SudhArch*, vol. 71, no. 1, pp. 78–89.

SESIANO Jacques (4), 1987. "Survivance médiévale en Hispanie d'un problème né en Mésopotamie." *Centaurus*, vol. 30, pp. 18–61.

SHAGDARSÜREN Cevelijn (1), 1989. "Die mathematische Tradition der Mongolen." in W. HEISSIG and C.C. MÜLLER eds., *Die Mongolen*, Innsbruck: Pinguin-Verlag, pp. 266–267.

SHELBY (1), Lon R., 1972. "The Geometrical Knowledge of Mediaeval Master Masons." *Speculum*, vol. 47, pp. 395–421.

SHEN Kangshen (1), 1988. "Historical Development of the Chinese Remainder theorem." *AHES*, vol. 38, no. 4, pp. 285–305.

SHIMODAIRA Kazuo (1), 1977. "Recreative Problems in the *Jingōki*." *JSHS*, no. 16, pp. 95–103.

SHIMODAIRA Kazuo (2), 1981, "On Idai of *Jingōki*, *HS*, no. 21, pp. 87–101.

SHIMODAIRA Kazuo (3), 1982, "Approximate Formulae in the Early Edo Period." *Maebashi*, vol. 17, pp. 1–12.

SHUKLA Kripa Shankar Transl. and Comment (1), 1959. *The Patiganita of Sridharacarya with an Ancient Sanskrit Commentary.* Lucknow: Lucknow University (Department of Mathematics and Astronomy).

SHUKLA Kripa Shankar (2), 1976. *Āryabhaṭīya of Āryabhaṭa with the commentary of Bhāskara I and Someśvara.* New Delhi: Indian National Science Academy.

SINGMASTER David (1), 1989. *Sources in Recreational Mathematics, An Annotated Bibliography.* London: Department of Computing and Mathematics. South Bank Polytechnic, London.

SIU Man Keung (1), 1981. "Pyramid Pile and Sum of Squares." *HistMath*, vol. 8, no. 1, pp. 61–66.

SIVIN Nathan (1), 1973, "Copernicus in China." *StudCop.* vol. 6, pp. 63–122.

SIVIN Nathan (2) ed., 1977, *Science and Technology in East Asia.* New York: Science History Publications (Selections from *Isis* with an introduction by N. Sivin).

SIVIN Nathan (3), 1982. "Why the Scientific Revolution Did not Take Place in China — Or didn't It?" *ChinSci*, vol. 5, pp. 45–66.

SIVIN Nathan (4), 1989, "On the Limits of Empirical Knowledge in Chinese and Western Science." in BIDERMAN and SCHARFSTEIN ed., *Rationality in Question, On Eastern and Western Views of Rationality*, Leiden: E.J. Brill, pp. 165–189.

SMITH David Eugene (1), 1925/1958. *History of Mathematics.* 2 vols., New York: Dover.

SMITH David Eugene and MIKAMI Yoshio (1), 1914. *A History of Japanese Mathematics.* Chicago: Open-Court.

SOLOMON B.S. (1), 1954. "One is No Number in China and the West." *HJAS*, vol. 17, nos. 1 and 2, pp. 253–260.

SOOTHILL William Edward and HODOUS Lewis (1), 1976. *A Dictionary of Chinese Buddhist Terms with Sanskrit and English Equivalents and a Sanskrit–Pali Index.* Taipei: Ch'eng wen Publishing Co. (reprint).

SPENCE Jonathan D. (1), 1974/1977. *Emperor of China, Self-portrait of K'ang-hsi.* New York: Penguin books.

ŚRĪDHARĀCARYA See SHUKLA (1).

SRINIVASIENGAR C.N. (1), 1957. *The History of Ancient Indian Mathematics.* Calcutta: The World Press Private LTD.

STENDHAL (1), 1832/1973, *Vie de Henry Brulard.* Paris: Gallimard.

STRUIK D.J. ed. (1), 1969. *A Source Book in Mathematics, 1200–1800*. Cambridge (Mass.): Harvard University Press.

SUBBARAYAPPA B.V. and SARMA K.V. eds., (1), *1985. Indian Astronomy, a Source-Book (based primarily on Sanskrit Texts)*. Bombay: Nehru Centre.

SUGIMOTO Masayoshi and SWAIN, David L. (1), 1978. *Science and Culture in Traditional; Japan, A.D. 600–1854.* Cambridge (Mass.) and London: The MIT Press.

SUNG Z.D. transl. (1), 1935/1975. *The text of* Yi Jing *(and its Appendixes)*. Taipei: Wenhua Shuju.

SUTER H. (1), 1910–1911. "Das Büch der Seltenheiten der Rechenkunst von Abū Kāmil el Miṣrī." *BibMath*, 3rd ser., no. 11, pp. 100–120.

SWETZ Frank J. (1), 1972. "The Amazing *Chiu Chang Suan Shu*." *MT*, vol. 65, pp. 425–430.

SWETZ Frank J. (2), 1974. "The Introduction of Mathematics in Higher Education in China, 1865–1867." *HistMath*, vol. 1, pp. 167–179.

SWETZ Frank J. (3), 1974. *Mathematics Education in China, its Growth and Development*. Cambridge (Mass.): The MIT Press.

SWETZ Frank J. (4), 1977, "The 'Piling up of Squares' in Ancient China." *MT*, vol. 70, pp. 72–79.

SWETZ Frank J. (5), *1978*. "Mysticism and Magic in the Number Squares of Old China." *MT*, vol. 71, pp. 50–56.

SWETZ Frank J. (6), 1979, "The Evolution of Mathematics in Ancient China." *MathMag*, vol. 52, pp. 10–19.

SWETZ Frank J. and ANG Tian-Se (1), 1984. "A Brief Chronological Guide to the History of Chinese Mathematics. *HistMath*, vol. 11, no. 1, pp. 39 56.

SWETZ Frank J. and KAO T.I. (1), 1977. *Was Pythagoras Chinese? An Examination of Right Triangle Theory in Ancient China*. University Park and London: The Pennsylvania State University Press.

SZABÓ Arpad (1), 1974, "How to Explore the History of Ancient Mathematics." in J. HINTIKKA and U. REMES, *The Method of Analysis, its Geometrical Origin and its General Significance*. Dordrecht: D. Reidel.

TAAM Cheuk-Woon (1), 1935/1977. *The Development of Chinese Libraries under the Ch'ing Dynasty, 1644–1911*. San Francisco: Chinese Materials Center.

TACCHI VENTURI Pietro (1), 1911-1913. *Opere storiche del P. Matteo Ricci [...]*. 2 vols. Macerata.

TAISBAK C.M. (1), 1965. "Roman Numerals and the Abacus. *C&M*, vol. 26, pp. 147–160.

TASAKA Kōdō (1), 1957. "An Aspect of Islam Culture Introduced into China." *MRDTB*, no. 16, pp. 75–160.

TEBOUL Michel (1), 1983. *Les Premières Théories Planétaires Chinoises*. Paris: Collège de France, Institut des Hautes Etudes Chinoises (Mémoires de l'Institut des Hautes Etudes Chinoises, vol. 21).

TENG Ssu-yü transl. (1), 1968. *Family Instructions for the Yan Clan* Yen-shih chia-hsün [Yanshi jiaxun] *by Yen Chih-t'ui* [Yan Zhitui]. Leiden: E.J. Brill (monographies du *T'oung Pao*, vol. 4).

TERQUEM Olry (1), 1862. "'Arithmétique et Algèbre des Chinois' par M.K.L. Biernatzki, Docteur à Berlin." *NAM*, no. 1, p. 35 and no. 2, p. 529.

THEON of Alexandria (1), ca. 370 AD/1976, *Commentaires de Pappus et de Théon d'Alexandrie sur l'Almageste.* Texte établi et annoté par A. Rome. Rome: Città del Vaticano (Biblioteca Apostolica Vaticana).

THUREAU-DANGIN François (1), 1938, *Textes Mathématiques Babyloniens.* Leiden.

TOOMER G.J. (1), 1974. "The Chord Table of Hipparchus and the Early History of Greek Trigonometry." *Centaurus,* vol. 18, no. 1, pp. 6–28.

TOOMER G.J. transl. (2), 1984. *Ptolemy's* Almagest. London: Duckworth.

TRẦN NGHĨA and GROS François eds. (1), 1993. *Catalogue des livres en hán nôm.* 3 vols. Hanoi: Editions Sciences Sociales (Publications de l'Institut Hán Nôm et de l'Ecole Française d'Extrême-Orient).

TRẦN VĂN GIÁP (1), 1937. Les chapitres bibliographiques de Lê-quí-Đôn et de Phan-huy-Chú. Saigon.

TROPFKE Johannes (1), 1922. *Geschichte der Elementar-Mathematik in Systematischer Darstellung mit Besonderer Berücksichtigung der Fachworter.* vol. 3 (Proportions and equations). Berlin and Leipzig: Walter de Gruyter.

TROPFKE Johannes (2), 1924. *Geschichte der Elementar-Mathematik in systematischer Darstellung mit Besonderer Berücksichtigung der Fachworter.* Vol. 7 (Stereometry). Berlin and Leipzig: Walter de Gruyter.

TROPFKE Johannes (3), 1980. *Geschichte der Elementarmathematik.* 4th ed., vol. 1: *Arithmetik und Algebra,* fully revised by Kurt VOGEL, Karin REICH, Helmuth GERICKE. Berlin/New York: Walter de Gruyter.

TSIEN Tsuen-Hsuin (1), 1954. "Western Impact on China through Translation." *FEQ,* vol. 13, no. 3, pp. 305–327.

TURAN Paul (1), 1954. "A Kínai Matematika történetének egy problémájáról" [On a problem from Chinese mathematics]. *MatLap,* vol. 5, pp. 1–6.

TWITCHETT D.C. (1), 1961. "Chinese Biographical Writing." in W.G. BEASLEY and E.G. PULLEYBLANK. *Historians of China and Japan.* London: School of Oriental and African Studies, pp. 95–114.

UNGURU Sabetai (1), 1979, "History of Ancient Mathematics: Some Reflexions on the State of the Art." *Isis,* vol. 70, no. 254, pp. 555–565.

VANDERMEERSCH Léon (1), 1977–1980. *Wangdao ou La Voie Royale, Recherches sur l'Esprit des Institutions de la Chine Archaïque.* 2 vols. Paris: Ecole Française d'Extrême-Orient.

VAPEREAU G. (1), 1880. *Dictionnaire Universel des Contemporains.* Paris: Hachette (5th edition).

VINOGRADOV I.M. (1), 1954. *Elements of Number Theory.* New York: Dover.

VISSIÈRE Isabelle and Jean-Louis eds. (1), 1979, *Lettres Édifiantes et Curieuses de Chine par des Missionaires Jésuites 1702–1776.* Paris: Garnier-Flammarion.

VOGEL Kurt (1), transl. 1968. *Chiu chang suan shu:* Neun Bücher arithmetischer Technik, Ein Chinesiches Rechenbuch für den praktisschen Gebrauch aus den Frühen Hanzeit. Braunschweig: Friedr. Vieweg Verl. (Ostwalds Klassiker).

VOGEL Kurt (2), 1983. "Ein Vermessungsproblem reist von China nach Paris." *HistMath,* vol. 10, pp. 360–367.

VOLKOV Alexeï (1), 1991. "Dissertation on 'Mathematika v drevnem Kitaie (3–7 vv.)' [Mathematics in Ancient China during the 3rd–7th Centuries AD]." *Histmath.,* vol. 18, no. 2, pp. 185–187.

VOLKOV Alexeï (2), 1994. "Large numbers and counting rods." *EOEO,* no. 16, pp. 71–92.

VOLKOV Alexeï (3), 1994. "Calculations of π in Ancient China: from Liu Hui to Zu Chongzhi." *HS*, vol. 4, no. 2, pp. 139-157.

Van der WAERDEN B.L. (1), 1961. *Science Awakening*. Transl. by Arnold DRESDEN. New York: Oxford University Press.

Van der WAERDEN B.L. (2), 1980, "On Pre-Babylonian Mathematics." *AHES*, vol. 23, pp. 1–25 and 27–46.

Van der WAERDEN B.L. (3), 1983. *Geometry and Algebra in Ancient Civilizations*. Berlin: Springer-Verlag.

WAGNER D.B. (1), 1975. *Proof in Ancient Chinese Mathematics, Liu Hui on the Volumes of Rectilinear Solids*. Unpublished Doctoral Dissertation, Copenhagen University.

WAGNER D.B. (2), 1978. "Doubts Concerning the Attribution of Liu Hui's Commentary on the *Chiu-chang suan-shu*." *ActaOr*, no. 39, pp. 199–212.

WAGNER D.B. (3), 1978. "Liu Hui and Tsu Keng-chih on the Volume of a Sphere." *ChinSci*, no. 3, pp. 59–79.

WAGNER D.B. (4), *1979*. "An Early Derivation of the Volume of a Pyramid: Liu Hui, Third Century AD." *HistMath*, vol. 6, pp. 164–188.

WANG Ling (1), 1956, *The* Chiu chang suan shu *and the History of Chinese Mathematics during the Han Dynasty*. Unpublished PhD Dissertation, Cambridge (Trinity College, PhD Dissertation no. 2917–2918).

WANG Ling (2), 1956. "The Development of Decimal Fractions in China." in *Proc. 8th Int. Congress on the History of Science, Florence, 1956*, pp. 13–17.

WANG Ling (3), 1964. "The Date of the *Sun Tzu Suan Ching* and the Chinese Remainder Theorem." in *Proc. 10th Int. Congress on the History of Science, 1962*. Paris: Hermann, vol. 1, pp. 489–492.

WANG Ling (4), 1977, "A New Suggestion on Tzu Ch'ung-Chih's Method of Finding the value of π and its significance in the History of Mathematics." *PFEH*, vol. 16, pp. 161–165.

WANG Ling and NEEDHAM Joseph (1), 1955. "Horner's Method in Chinese Mathematics; it Origins in the Root-Extraction Procedures of the Han Dynasty." *TP*, vol. 43, pp. 345–401.

WANG Ping (1), 1962. "Alexander Wylie's Influence on Chinese Mathematics." in *Int. Association of Historians of Asia, Second Biennial Conference Proceedings, 6–9 Oct. 1962*, held at Taipei, Taiwan Provincial Museum.

WANG Zhongshu (1), 1982, *Han Civilization*. New Haven and London: Yale University Press.

Wann-Sheng HORNG: See HORNG Wann-Sheng.

WATSON Burton (1), 1961. *Records of the Grand Historian of China, Translated from the* Shih chi [Shiji] *of Ssu-ma Ch'ien*. 2 vols. New York: Columbia University Press.

WILDER R.L. (1), 1952. "The Cultural Basis of Mathematics." in *Proc. Int. Congress of Mathematicians, Aug. 30–Sept. 6, 1950*, Published by the American Mathematical Society, 1952, vol. 1, pp. 258–271.

Wing-tsit CHAN: See CHAN Wing-tsit.

WONG G. (1), 1963. "Some Aspects of Chinese Science before the Arrival of the Jesuits." *CCJ*, vol. 2, pp. 169–180; Rev.: N. SIVIN, *Isis*, 1965, vol. 56, pp. 201–205.

WONG G. (2), 1970. "Wang Jen-tsün: a late Nineteenth Century Obstructor to the Introduction of Western Thought." *CCJ*. vol. 9, no. 2, pp. 210–215.

WONG Ming (1), 1964. "Le Professeur Li Yan." *BEFEO*, vol. 52, no. 1, p. 310.

WORPITZKY J. (1), 1883, "Studien über die Bernoullischen und Eulerschen Zahlen." *JRAM*, vol. 94, pp. 203–232.

WRIGHT Arthur F. (1), *1978. The Sui Dynasty*. New York: Alfred A. Knopf.

WU Wen-tsun [Wu Wenjun] (1), 1986. "Recent Studies of the History of Chinese Mathematics." in *Proc. Int. Congress of Mathematicians, Berkeley, California*, pp. 1657–1667.

WYLIE Alexander (1), 1852/1966. "Jottings on the Science of the Chinese: Arithmetic." First published in the *North China Herald* (Aug.–Nov. 1852, nos. 108–111, 112, 113, 116, 117, 119, 120, 121), reprinted in A. WYLIE, *Chinese Researches*, Taipei: Ch'eng-wen, pp. 159–194.

WYLIE Alexander (2), 1867. *Memorial of Protestant Missionaries to the Chinese*. Shanghai.

WYLIE Alexander (3), 1897/1966. *Chinese Researches*. Taipei: Ch'eng-wen.

YABUUCHI Kiyoshi (1), 1954. "Indian and Arabian Astronomy in China." in *Silver Jubilee Volume of the Zinbun Kagaku Kenkyusyo*. Kyoto: Kyoto University, pp. 585–603.

YABUUCHI Kiyoshi (2), 1963. "The Chiuchih li: an Indian Astronomical Book in the T'ang Dynasty." in K. YABUUCHI (2′), *1963*, pp. 493–538.

YABUUCHI Kiyoshi (3), *1979*. "The Study of Chinese Science in Kyoto." *ActaAsiat*, no. 36, pp. 1–6.

YABUUCHI Kiyoshi (4), 1979, "Researches on the Chiuchih li [*Jiuzhi li*], Indian Astronomy under the Tang Dynasty." *ActaAsiat*, no. 36, pp. 7–48.

YABUUCHI Kiyoshi (5), 1987. "The Influence of Islamic Astronomy in China." in D. A. KING and G. A. SALIBA eds., *From Deferent to Equant: A Volume of Studies in the History of Science in the Ancient and Medieval Near East in Honor of E.S. Kennedy. AnNYAS*, vol. 500, pp. 547–559.

YABUUTI Kiyoshi: See YABUUCHI Kiyoshi.

YAJIMA S. (1), 1953. "Bibliographie du Dr. Yoshio Mikami, Notice Biographique." in *Actes du VIIe Congrès International d'Histoire des Sciences, Jerusalem*, pp. 646–658.

YAMAZAKI Yoemon (1), 1959. "The Origin of the Chinese Abacus." *MRDTB*, no. 18, pp. 91–140.

YAMAZAKI Yoemon (2), 1962. "History of Instrumental Multiplication and Division in China: from the Reckoning Blocks to the Abacus." *MRDTB*, no. 21, pp. 125–148.

YANG Lien-cheng (1), 1961. "Numbers and Units in Chinese Economic History." in YANG Lien-cheng, *Studies in Chinese Institutional History*, Harvard: Harvard University Press, pp. 75–84.

YANO Michio (1), 1992. "Navagraha and Chiu-chih." Unpublished paper presented at the Int. Symposium on the History of Science and Technology in China, Hangzhou, August 1992.

YĀNO Michio and VILADRICH Mercé (1), 1991. "Tasyīr Computation of Kūshyār ibn Labban." *HS*, no. 41, pp. 1–16.

YOUNG D.M. and GREGORY R.T. (1), 1973. *A Survey of Numerical Methods*. Reading (Mass.): Addison-Wesley.

YOUSCHKEVITCH: See YUSHKEVICH.

YU Wang-Luen (1), 1974. "Knowledge of Mathematics and Science in Ching-Hua-Yüan." *OE*, vol. 21, no. 2, pp. 217–236.

YUAN Tong-li (1), 1963. *Bibliography of Chinese Mathematics, 1918–1961.* Washington: published privately.

YUSHKEVICH A.P. (1), 1964. *Geschichte der Mathematik im Mittelalter,* Leipzig: B.G. Teubner.

YUSHKEVICH A.P. (2), 1976, *Les Mathématiques Arabes* (Translated from the German by M. CAZENAVE and K. JAOUICHE, Preface by R. TATON). Paris: J. Vrin.

ZACHER H.J. (1), 1973. *Die Hauptsschriften zur Diadik von G.W. Leibnitz.* Frankfurt: Klostermann.

ZIGGELAAR August (1), 1971. *Le physicien Ignace-Gaston Pardies S.J.* Odense: Odense University Press.

ZURCHER E. (1), 1959. *The Buddhist conquest of China, the Spread and Adaptation of Buddhism in Early and Medieval China.* Leiden: E.J. Brill.

Books and Articles in Chinese or Japanese

Abbreviations (Books and Articles)

ALXB	Anhui lishi xuebao 安徽歷史學報(Hefei)
ASX	Anhui shixue 安徽史學(Hefei)
ASXTX	Anhui shixue tongxun 安徽史學通訊 (Hefei)
BDYK	Beijing daxue yuekan 北京大學月刊(Peking)
BJBH	Beiping Beihai tushuguan yuekan 北平北海圖書館月刊[*Engl. tit.:* Bulletin of the Metropolitan Library] (Peking)
BSF	Beijing shifan daxue xuebao 北京師範大學學報(Peking)
ChugShisKen	Chūgoku shisō shi kenkyū 中國思想史研究 [*Engl. tit.:* Journal of the History of Chinese Thought] (Kyoto)
ChuZhou	Chuban Zhoukan 出版週刊(Shanghai)
DFZZ	Dongfang zazhi 東方雜誌 [*Engl. tit.:* Eastern (The) Miscellany] (Shanghai)
DLZ	Dalu zazhi 大陸雜誌 [*Engl. tit.:* Continent (The) Magazine](Taipei)
Dushu	Dushu 讀書(Peking)
Epistēmē	Epistēmē (Tokyo)
GGBWY	Gugong bowuyuan yuankan故宮博物院院刊 [*Engl. tit.:* Palace Museum Journal](Peking)
GLB	Guoli Beiping tushuguan guankan 國立北平圖書館館刊 [*Engl. tit.:* Bulletin of the National Library of Peping](Peking).
GLZhejiang	Guoli Zhejiang daxue jikan 國立浙江大學季刊(Hangzhou)
GLZY/LY	Guoli zhongyang yanjiuyuan, lishi yuyan yanjiusuo jikan國立中央研究院， 歷史語言研究所季刊 (*Engl. tit.:* Academia Sinica, Bulletin of the National Research Institute of History and Philology] (Peking).
GSKY	*Gu suan kao yuan* 古算考源 (Qian Baocong, 1933)
HangzhouXB	Hangzhou daxue xuebao (ziran kexue ban) 杭州大學學報 (自然科學版) (Hangzhou)
JAS	Journal of Asian Studies (Ann Arbor)
KagakuSK	Kagaku shi kenkyū 科學史研究 (*Engl. tit*: Journal of History of Science, Tokyo)
Kejishi	Kejishi wenji 科技史文集(Shanghai)
Kexue	Kexue 科學[*Engl. tit.:* Science] (Shanghai)
KXSJK	Kexue shi jikan 科學史集刊 (Peking)
KXSYicong	Kexue shi yicong 科學史譯叢 (Huhehot)
KXTB	Kexue tongbao 科學通報 (Peking)
KYue	Kexue yuekan 科學月刊(Taipei)
MeijiGR	Meiji Gakuin Ronsō 明治學院論叢 (Tokyo)
MiyazakiJT	Miyazaki joshi tanki daigaku kiyō 宮崎女子短期大學紀要 (Miyazaki)
NKZK	Nankai zhoukan 南開周刊 (Tianjin)
NMDXXB	Nei Menggu daxue xuebao (ziran kexue ban) 內蒙古大學學報 (自然科學版) (Huhehot)
NMSY	Nei Menggu shiyuan xuebao (ziran kexue ban) 內蒙古師院學報(自然科學版) (Huhehot)
NüshiDa	Nüshi daxue xueshu jikan 女師大學學術季刊 (Peiping)

QBK   *Qian Baocong kexue shi lunwen xuanji* 錢寶琮科學史論文選集
               (Peking, 1963)
QHXB   Qinghua xuebao 清華學報 [*Engl. tit.:* Tsinghua Journal] (Peiping)
RBS   Revue Bibliographique de Sinologie (Paris)
Renwu   Renwu 人物 (Peking)
SGSK   Sūgaku shi kenkyū 數學史研究 [*Engl. tit:* Journal of History of
               Mathematics] (Tokyo)
SK   Sangaku kyōiku 算學教埈 (Tokyo)
SXJZ   Shuxue jinzhan 數學進展 (Peking)
SXTB   Shuxue tongbao 數學通報 (Peking)
SXY   Shuxue yanjiu yu pinglun 數學研究與平論 [*Engl. tit.:* Journal of
               Mathematical Research and Exposition] (Huazhong and Dalian)
SXZZ   Shuxue zazhi 數學雜誌 (Peking)
SY   *Song Yuan shuxue shi lunwen ji* 宋元數學史論文集 (Qian Baocong
               et al., Peking: Kexue Chubanshe, 1966)
TBNüshi   Taibei shili nüzi shifan zhuanke xuexiao xuebao 台北市立女子
               師範專科學校學報 (Taipei)
TBGZ   Tokyo Butsuri Gakkō zasshi 東京物理學校雜誌 (Tokyo)
THG   Tōhō gakuhō 東方學報 [*Engl. tit.:* Journal of Oriental Studies. The
               Institute for Research in Humanities (Jinbun Kagaku Kenkyūsho)]
               (Kyoto)
TNH   Toă no hikari 東亞の光 (Tokyo)
TP   T'oung Pao (Leiden)
TSGX   Tushuguanxue jikan 圖書館學季刊 [*Engl. tit.:* Library Science
               Quarterly] (Peking)
TSJK   Tushu jikan 圖書季刊 (Peking)
TWXB   Tianwen xuebao 天文學報 [*Lat. tit.:* Acta Astronomica Sinica]
               (Peking)
TYG   Tōyō gakuhō 東洋學報 [*Engl. tit.:* The Journal of the Research
               Department of the Toyo Bunko] (Kyoto)
Wenwu   Wenwu 文物 [*Engl. tit.:* Cultural Relics] (Peking)
WHJS   Wenhua jianshe 文化建設 (Shanghai)
WuWenjun   Zhongguo shuxue shi lunwen ji 中國數學史論文集 (Jinan: Shandong
               Jiaoyu Chubanshe, a Journal edited by Wu Wenjun)
WWC   Wenwu cankao ziliao 文物參考資料 [Original title of 文物]
               (Peking)
XBSD   Xibei Shidi jikan 西北史地季刊 (Xi'an)
XJRB   Xijing ribao 西京日報 (Xi'an)
XSJ   Xin shijie 新世界 (Chongqing minsheng shiye gongsi) (重慶民生實
               業公司) (Chongqing)
XueYi   Xue yi zazhi 學藝雜誌 [*German tit.:* Wissen und Wissenschaft]
               (Shanghai)
YJXB   Yanjing xuebao 燕京學報 [*Engl. tit.:* Yenching Journal of Chinese
               Studies] (Peking)
YSB/WSFK   Yishi bao (Wenshi fukan) 益世報 （文史副刊） (Chongqing)
ZDJK   Zhongda jikan 中大季刊 (Peking)
ZHKE   Zhongguo keji shiliao 中國科技史料 [*Engl. tit.:* China Historical
               Materials of Science and technology] (Peking)

ZHWHua   Zhonghua wenhua fuxing yuekan 中華文化復興月刊 [*Engl. tit.:* Chinese Cultural Renaissance Monthly] (Taipei)

ZirBian   Ziran bianzhengfa tongxun 自然辯證法通訊 [*Engl. tit.:* Journal of Dialectics of Nature] (Peking)

ZK   Ziran kexue shi yanjiu 自然科學史研究 [*Engl. tit.:* Studies in the History of Natural Sciences] (Peking)

ZLGBK   Zhongguo lishi bowuguan guankan 中國歷史博物館館刊 (Peking)

ZLZZ   Zhenli zazhi 眞理雜誌 (Chongqing)

ZSSLC-P   *Zhong suan shi luncong* 中算史論叢 (Li Yan, 1954-55, 5 vols., Peking)

ZSSLC-T   *Zhong suan shi luncong* 中算史論叢 [German title: Gesammelte Abhandlüngen über die Geschichte der chinesischen Mathematik] (Li Yan, 1977, 4 vols, Taipei)

ZYY/LY   *Zhongyang yanjiuyuan, lishi yuyan yanjiusuo jikan* 中央研究院歷史語言研究所集刊 [*Engl. tit:* Bulletin of the Institute of History and Philology, Academia Sinica] (Taipei)

ZYY/JS   *Zhongyang yanjiuyuan, jindai shi yanjiusuo jikan* 中央研究院近代史研究所集刊 [*Engl. tit:* Bulletin of the Institute of Modern History Academia Sinica] (Taipei)

Books and Articles

BAI Shangshu 白尚恕 (1'), 1963. "Jieshao wo guo diyibu sanjiaoxue — *Dace*." 介紹我國第一部三角學 ——《大測》 (Introduction to the first Chinese trigonometry — *Dace*). *SXTB*, no. 2, pp. 48–52.

BAI Shangshu 白尚恕 (2'), 1982. "Cong Wang Mang liangqi dao Liu Xin yuanlü." 從王莽量器到劉歆圓率 (From Wang Mang's measure of capacity to Liu Xin's value of π). *BSF*, no. 2, pp. 75–79 (Cf. BAI Shangshu (2), 1988).

BAI Shangshu 白尚恕 (3'), 1983. *Jiuzhang suanshu* zhushi 《九章算術》注釋 (A commentary on the *Jiuzhang suanshu*). Peking: Kexue Chubanshe.

BAI Shangshu 白尚恕 (4'), 1984. "*Celiang quanyi* diben wenti de chutan."《測量全義》底本問題的初探 (Preliminary investigation into the sources of the *Celiang quanyi*). *KXSJK*, no. 11, pp. 133–159.

BAI Shangshu 白尚恕, transl., (5'), 1985. *Ceyuan haijing* jinyi 《測圓海鏡》今譯 (A translation of the *Ceyuan haijing* into modern Chinese). Jinan: Shandong Jiaoyu Chubanshe.

BAI Shangshu 白尚恕 (6'), 1986. "*Jiuzhang suanshu* zhong 'shi' zi tiaoxi."《九章算術》中 " 勢 " 字條析 (A semantical analysis of the character *shi* in the *Jiuzhang suanshu*). *WuWenjun*, no. 2, pp. 39–47.

BAI Shangshu 白尚恕 and LI Di 李迪 (1'), 1980. "Gugong zhencang de yuanshi shouyao jisuanji ." 故宮珍藏的原始手搖計算機 (A hand calculating machine preserved in the collection of the Palace Museum [description of a machine analogous to Pascal's calculating machine]). *GGBWY*, no. 1, pp. 76–82.

CHEN Chun 陳淳 (1'), 1977. "Shuxue shi shang jige wenti de jiantao — guanyu fuhao '0' ji Yindu-Alabo shuzi — ." 數學史上幾個問題的檢討—關於符號 " 0 " 及印度阿拉伯數字 — (An examination of some problems in the history of mathematics — on the symbol '0' and the Hindu-Arab numerals). *TBNüshi*, vol. 9, pp. 1–14.

CHEN Liangzuo 陳良佐 (1'), 1977. "Wo guo chousuan zhong de kongwei — ling — jiqi xiangguan de yixie wenti." 我國籌算中的空位 - 零 - 及其相關的一些問題 (The vacant places in the rod-numeral system — zero — and some related questions). *DLZ*, vol. 54, no. 5, pp. 1–13.

CHEN Liangzuo 陳良佐 (2'), 1978. "Xian Qin shuxue de fazhan jiqi yingxiang." 先秦數學的發展及其影響 (the development and influence of pre-Qin mathematics). *ZYY/LY* vol. 49, part 2, pp. 263–320.

CHEN Liangzuo 陳良佐 (3'), 1982. "Zhao Shuang gougu yuanfang tu zhi yanjiu." 趙爽勾股圓方圖之研究 (Research into the figures of the square, circle and right-angled triangle of Zhao Shuang). *DLZ*, vol. 64, no. 1, pp. 1–19.

CHEN Meidong 陳美東 (1'), 1985. "*Chongxuan, Yitian, Chongtian* san li guichang jisuanfa ji sanci neichafa de yingyong." 崇玄、儀天、崇天三曆晷長計算法及三次差內插法的應用 (The application of interpolation techniques using third-degree polynomials and the computation of the lenght of gnomon shadows in the three calendars *Chongxuan, Yitian* and *Chongtian*). *ZK*, vol. 4, no. 3, pp. 218–228.

CHEN Meidong 陳美東 (2'), 1988. "Zhongguo gudai youguan libiao jiqi suanfa de gongshi hua." 中國古代有關曆表及其算法的公式化 (The [modern] algorithmic formulation of ancient Chinese astronomical tables). *ZK*, vol. 7, no. 3, pp. 232–236.

CHEN Meidong 陳美東 (3'), 1989. "Huangyou, Chongning guichang jisuan fa zhi yanjiu." 皇祐崇寧晷長計算法的研究 (Research into the formulas for the computation of the length of gnomon shadows composed during the Huanyou and Chongning eras). *ZK*, vol. 8, no. 1, pp. 17–27.

CHENG Minde 程民德 ed. (1'), 1994. *Zhongguo xiandai shuxuejia zhuan* 中國現代數學家傳 (Biographies of contemporary Chinese mathematicians). Vol. 1, Nanking: Jiangsu Jiaoyu Chubanshe.

CHENG Te-k'un 鄭德坤 (1'), 1983. "Zhongguo shanggu shuming de yanbian jiqi yingyong." 中國上古數名的演變及其應用 (The evolution and application of numbers in Chinese antiquity). In CHENG Te-k'un, 1983, *Studies in Chinese Art*, Hong Kong: The Chinese University Press, pp. 169–185.

CHEN Yinke 陳寅恪 (1'), 1931. "*Jihe yuanben* manwen yiben ba." 幾何原本滿文譯本跋 (Short account of the Manchu version of Euclid's *Elements*). *GLZY/LY*, vol. 2, no. 3, pp. 281–282.

CHEN Zungui 陳遵嬀 (1'), 1980–1989. *Zhongguo tianwenxue shi* 中國天文學史 (History of Chinese astronomy). 4 vols., Shanghai: Shanghai Renmin Chubanshe. [Only Vol. 3, published in 1984, is cited here].

DAI Nianzu 戴念祖 (1'), 1986. *Zhu Zaiyu — Mingdai de kexue he yishu juxing* 朱載堉－明代的科學和藝術巨星 (Zhu Zaiyu, giant star of sciences and techniques of the Ming dynasty). Peking: Renmin Chubanshe.

DING Fubao 丁富保 and ZHOU Yunqing 周雲青, eds. (1'), 1957. *Sibu zonglu suanfa bian* 四部總錄算法編 (General catalogue of the four departments of litterature, section 'mathematics'). Shanghai: Shangwu Yinshuguan. This compilation is mainly derived from: *(i)* FENG Cheng 馮澂, 1898, *Suanxue kao chubian* 算學考初編 (First compilation of studies on mathematics). *(ii)* LIU Duo 劉鐸, 1898. *Gujin suanxue shulu* 古今算學書錄 (Bibliography of mathematical books, ancient and modern). *(iii)* DING Fubao 丁富保, 1899. *Suanxue shumu tiyao* 算學書目提要 (Critical notices on mathematical books). *(iv)* ZHOU Yunqing 周雲青, 1956. *Buyi*, 補遺 (Complements).

DU Shiran 杜石然 (1'), 1966. "Zhu Shijie yanjiu." 朱世傑研究 (Research into Zhu Shijie). *SY*, pp. 166–209.

DU Shiran 杜石然 (2'), 1988. "Jiangling Zhangjiashan zhujian *Suanshu shu* chutan." 江陵張家山竹簡《算術書》初探 (Preliminary research into the *Suanshu shu*, a mathematical work on bamboo strips found at Zhangjiashan near Jiangling [in Hubei province]). *ZK*, vol. 7, no. 3, pp. 201–204.

DU Shiran 杜石然 (3'), 1989. "Mingdai shuxue jiqi shehui beijing." 明代數學及其社會背景 (Ming mathematics and its social background). *ZK*, vol. 8, no. 1, pp. 9–16.

DU Shiran 杜石然 (4'), 1989. "Suanchou tanyuan." 算籌探源 (Research into the origin of counting-rods). *ZLGBK*, no. 12, pp. 28–36.

DU Shiran 杜石然 ed. (5'), 1992. *Zhongguo gudai kexuejia zhuanji* 中國古代科學家傳記 (Biographies of ancient Chinese scientists). 2 vols., Peking: Kexue Chubanshe.

ENDŌ Toshisada 遠藤利眞 (1'), 1981. *Zōshū Nihon sūgaku shi* 增修日本數學史 (History of Japanese Mathematics, revised and enlarged [by MIKAMI Yoshio and HIRAYAMA Akira]). Tokyo: Kōseisha Kōseikaku.

438   References

FAN Huiguo 範會國 and LI Di 李迪 (1'), 1981. "*Zhongguo shuxue hui de lishi.*" 中國數學會的歷史 (History of the Chinese Mathematical Society). *ZHKE*, no. 3, pp. 72–78.

FENG Ligui 馮禮貴 (1'), 1986. "*Zhen Luan jiqi* Wucao suanjing." 甄鸞及其《五曹算經》 (Zhen Luan and his *Wucao suanjing*). *WuWenjun*, no. 2, pp. 29–38.

FU Pu 傅溥 (1'), 1982. *Zhongguo shuxue fazhan shi* 中國數學發展史 (History of the development of Chinese mathematics). Taipei: Zhongyang Wenwu Gonying She.

FU Tingfang 傅庭芳 (1'), 1985. "*Dui Li Shanlan* Duoji bilei *de yanjiu jian lun 'duoji chafen' de tese.*" 對李善蘭《垛積比類》的研究－兼論＂垛積差分＂的特色 (Research into Li Shanlan's *Duoji bilei* and finite differences). *ZK*, vol. 4, no. 3, pp. 267–283.

GUO Daoyang 郭道揚, ed. (1'), 1988. *Zhongguo kuaiji shi gao* 中國會計史稿 (An outline of the history of Chinese accounting). 2 vols., Beijing: Zhongguo Caizheng Jingji Chubanshe.

GUO Shirong 郭世榮 and LUO Jianjin 羅見今 (1'), 1987. "Dai Xu dui Oula shu de yanjiu." 戴煦對歐拉數的研究 (Dai Xu's research into Eulerian numbers). *ZK*, vol. 6, no. 4, pp. 362–371.

GUO Shuchun 郭書春 (1'), 1983. "Liu Hui *Jiuzhang suanshu zhu* zhong de dingyi ji yanyi luoji shixi." 劉徽《九章算術注》中的定義及演繹邏輯試析 (A tentative analysis of the deductive logic and definitions in Liu Hui's commentary on the *Jiuzhang suanshu*). *ZK*, vol. 2, no. 3, pp. 193–203.

GUO Shuchun 郭書春 (2'), 1984. "*Jiuzhang suanshu* he Liu Hui zhu zhong zhi lü gainian jiqi yingyong shixi." 《九章算術》和劉徽注中之率概念及其應用試析 (A tentative analysis of the notion of *lü* and of its applications in the *Jiuzhang suanshu* and in its commentary by Liu Hui). *KXSJK*, no. 11, pp. 21–36.

GUO Shuchun 郭書春 (3'), 1984. "Liu Hui de jixian lilun." 劉徽的極限理論 (Liu Hui's theory of limits). *KXSJK*, no. 11, pp. 37–46.

GUO Shuchun 郭書春 (4'), 1984. "Liu Hui de tiji lilun."劉徽的體積理論 (Liu Hui's theory of volumes). *KXSJK*, no. 11, pp. 47–62.

GUO Shuchun 郭書春 (5'), 1985. "*Jiuzhang suanshu* fangcheng zhang Liu Hui zhu xin tan." 《九章算術》方程章劉徽注新探 (A new discussion of the *fangcheng* chapter [linear systems] of the *Jiuzhang suanshu* according to Liu Hui's commentary). *ZK*, vol. 4, no. 1, pp. 1–5.

GUO Shuchun 郭書春 (6'), 1985. "*Jiuzhang suanshu* gougu zhang de jiaokan he Liu Hui gougu lilun xitong chutan." 《九章算術》句股章的校勘和劉徽句股理論系統初探 (Textual research into the text of the *gougu* chapter of the *Jiuzhang suanshu* and preliminary investigation of Liu Hui's *gougu* theoretical system). *ZK*, vol. 4, no. 4, pp. 295–304.

GUO Shuchun 郭書春 (7'), 1988. "Liu Hui yu Wang Mang tong hu." 劉徽與王莽銅斛(Liu Hui and the standard bronze measure of Wang Mang). *ZK*, vol. 7, no. 1, pp. 8–15.

GUO Shuchun 郭書春 (8'), 1988. "Jia Xian *Huangdi jiuzhang suanjing xicao* chutan." 賈憲《黃帝九章算經細草》初探 (Preliminary research into Jia Xian's *Huangdi jiuzhang suanjing xicao*). *ZK,* vol. 7, no. 4, pp. 328–334.

GUO Shuchun 郭書春 (9'), 1989. "Li Ji *Jiuzhang suanshu yinyi* chutan." 李籍《九章算術音義》初探 (Preliminary research into Li Ji's *Jiuzhang suanshu yinyi* [Meaning and pronunciation of terms of the *Jiuzhang suanshu*]). *ZK*, vol. 8, no. 3, pp. 197–204.

GUO Shuchun 郭書春 (10'), 1990. *Jiuzhang suanshu* huijiao ben 《九章算術》匯校本 (A critical edition of the *Jiuzhang suanshu*). Shenyang: Liaoning Renmin Chubanshe.

GUO Shuchun 郭書春 (11'), 1992. *Gudai shijie shuxue taidou Liu Hui* 古代世界數學泰斗劉徽 (Liu Hui, an authoritative mathematician from antiquity). Jinan: Shandong Jiaoyu Chubanshe.

GUO Zhengyi 郭正誼 (1'), 1990. "Guanyu qiqiaotu ji qita" 關於七巧圖及其它 (On tangrams and related subjects). *ZHKE*, vol. 11, no. 3, pp. 93–95.

GUO Zhengzhao 郭正昭 ed. (1'), 1974. *Zhongguo kexue shi yuanshi ziliao mulu suoyin, diyi ji* 中國科學史原始資料目錄索引（第一輯）(Index and Catalogue of primary sources for the history of Chinese science, (I)). Taipei: Huanyu Chubanshe.

GUO Zhengzhong 郭正忠 (1'), 1993. *San zhi shisi shiji Zhongguo de quanheng duliang* 三至十四世紀中國的權衡度量 (Chinese weights and measures from the third to the fourteenth century AD). Peking: Zhongguo Shehui Kexue Chubanshe.

HAN Qi 韓琦 (1'), 1991. "Zhong-Yue lishi shang tianwenxue yu shuxue de jiaoliu." 中越歷史上天文學與數學的交流 (Astronomical and mathematical knowledge exchanged between China and Vietnam). *ZHKE*, vol. 12, no. 2, pp. 3–8.

HAN Qi 韓琦 (2'), 1992. "*Shuli jingyun* duishu zaobiaofa yu Dai Xu erxiang zhankaishi yanjiu." 《數理精蘊》對數造表法與戴煦的二項展開式研究 (Research into the binomial expansion studied by Dai Xu and the method of construction of tables of logarithms found in the *Shuli jingyun*). *ZK*, vol. 11, no. 2, pp. 109–119.

HASHIMOTO Keizo 橋本敬造 (1'), 1970. "Bai Buntei no rekisangaku — Kōki nenkan no tenmon rekisangaku —." 梅文鼎の曆算學－康熙年間の天文曆算學 (The mathematical astronomy of Mei Wending — mathematical astronomy under Kangxi). *THG*, vol. 41, pp. 491–518.

HASHIMOTO Keizo 橋本敬造 (2'), 1971. "Da-en hō no tenkai, *Rekishō kōsei kōhen* no naiyō ni tsuite." 橢圓法の展開《曆象考成後編》の內容について (The development of the [Keplerian] method of the ellipse: the content of the *Lixiang kaocheng houbian*). *THG*, vol. 42, pp. 245–272.

HASHIMOTO Keizo 橋本敬造 (3'), 1973. "Bai Buntei no sūgakū kenkyū." 梅文鼎の數學研究 (Research into Mei Wending's mathematics). *THG*, vol. 44, pp. 233–279.

HAYASHI Takao 林隆夫 (1'), 1986. "Hōjinzan: Narayana." 方陳算 ナーラヤナ (A Japanese translation of chapter 14 of Narayana's *Gaṇitakaumudī*). *Epistēmē*, vol. 2, no. 3, pp. 1–34.

HAYASHI Tsuruichi 林鶴一 (1'), 1937. *Wasan kenkyū shūroku* 和算研究集錄 (Collected studies on *wasan*). 2 vols., Tokyo: Kaiseikan.

HE Aisheng 何艾生 and LIANG Chengrui 梁成瑞 (1'), 1984. "*Jihe yuanben* jiqi zai Zhongguo de chuanbo." 《幾何原本》及其在中國的傳播 (The *Jihe yuanben* [Euclid's *Elements*] and its diffusion in China). *ZHKE*, vol. 5, no. 3, pp. 32–42.

HE Shaogeng 何紹庚 (1'), 1982. "Xiang Mingda dui erxiang zhankaishi yanjiu de gongxian." 項名達對二項展開式研究的貢顯 (Xiang Mingda's contribution to the study of the binomial expansion). *ZK*, vol. 1, no. 2, pp. 104–114.

HE Shaogeng 何紹庚 (2'), 1984. "Tuoyuan qiu zhou shu shiyi." 橢圓求周術釋義 (An explanation of [Xiang Mingda's] technique for the computation of the perimeter of an ellipse). *KXSJK*, no. 11, pp. 130–142.

HE Shaogeng 何紹庚 (3'), 1984. "Ming'antu de jishu huiqiu fa." 明安圖的級數回求法 (Ming'antu [Minggatu]'s computational method for the reversion of series). *ZK*, vol. 3, no. 3, pp. 209–216.

HE Shaogeng 何紹庚 (4'), 1989. "Qing dai wuqiong jishu yanjiu zhong de jige guanjian wenti." 清代無窮級數研究中的一個關鍵問題 (A key problem for the infinite series expansions of the Qing dynasty). *ZK*, vol. 8, no. 3, pp. 205–214.

HE Shaogeng 何紹庚 (5'), 1989. "*Jigu suanjing* gougu ti shiwen shibu." 《緝古算經》勾股題佚文試補 (A tentative reconstruction of the lost right-angled triangles problems of the *Jigu suanjing*). *ZLGBK*, no. 12, pp. 37–43.

HIRAYAMA Akira 平山諦 (1'), 1959/1981. Seki Takakazu — sono gyōseki to denki 關孝和その業績と傳記 (Seki Takakazu, his life, his work). Second revised edition, Tokyo: Kōseisha Kōseikaku.

HIRAYAMA Akira 平山諦 (2'), 1980. *Enshuritsu no rekishi* 円周率の歷史 (History of π). Osaka: Kyōiku Tosho.

HIRAYAMA Akira 平山諦 (3'), 1993. *Wasan no tanjō* 和算の誕生 (The birth of *wasan*). Tokyo: Kōseisha Kōseikaku.

HIRAYAMA Akira 平山諦 and MATSUOKA Motohisa 松岡元久 eds., (1'), 1966. *Ajima Naonobu zenshū* 安島直円全集 (Complete works of Ajima Naonobu). Tokyo: Fuji Tanki Daigaku Shuppanbu.

HONG Wansheng: see HORNG Wann-Sheng.

HONG Zhenhuan 洪震寰 (1'), 1986. "*Suanxue bao* yu Huang Qingcheng." 《算學報》與黃慶澄 (Huang Qingcheng and the *Suanxue bao*). *ZHKE*, vol. 7, no. 5, pp. 36–38.

HORNG Wann-sheng 洪萬生 (1'), 1981. "Gudai Zhongguo de jihexue." 古代中國的幾何學 (The geometry of ancient China). *Kyue*, vol. 12, no. 8, pp. 22–30.

HORNG Wann-sheng 洪萬生 (2'), 1982. "Zhongshi zhengming de shidai — Wei Jin Nan Bei chao de keji." 重視證明的時代－魏晉南北朝的科技 (An historical period in which proof is stressed: Sciences and techniques under the Wei, Jin and Northern and Southern dynasties). In HORNG Wann-sheng ed., *Gewu yu chengqi* 格物與成器 (The 'investigation of things' and 'accomplished instruments'). Taipei: Lianjing Chuban Shiye Gongsi.

HORNG Wann-sheng 洪萬生 (3'), 1985. *Cong Li Yuese chufa— shuxue shi, kexue shi wenji* 從李約瑟出發－數學史，科學史文集 (Starting from the works of J. Needham — Collected papers on the history of mathematics and science). Taipei: Jiuzhang Chubanshe.

HORNG Wann-sheng 洪萬生 , ed., (4'), 1993. *Tan tian san you* 談天三友 (The three friends who discuss the heavens [the three mathematicians Jiao Xun 焦循 (1763–1820), Wang Lai 汪萊 (1768–1813) and Li Rui 李銳 (1769–1817)]). Taipei: Mingwen Shuju.

HU Daojing 胡道靜 ed. (1'), 1962/1987. *Mengqi bitan jiaozheng* 夢溪筆談校證 (A critical edition of the *Mengqi bitan*). 2 vols., Shanghai: Guji Chubanshe.

HUA Luogeng 華羅庚 (1'), 1959. "Shi nian lai Zhongguo shuxue yanjiu gongzuo de gaikuang." 十年來中國數學研究工作的概況 (An overview of Chinese mathematical research over the last ten years). *KXTB*, no. 18, pp. 565–567.

HUA Luogeng 華羅庚 (2'), 1978. "Wo guo gudai shuxue chengjiu zhi yipie." 我國古代數學成就之一瞥 (An overview of the achievements of Ancient Chinese Mathematics). *WenWu*, no. 260, pp. 46–49.

HUA Luogeng 華羅庚, GUAN Zhaozhi 關肇直, DUAN Xuefu 段學復 et al. (1'), 1959. "Jiefang qian de huigu." 解放前的回顧 (The pre-1949 period in retrospect). In ZHONGGUO KEXUEYUAN BIANYI CHUBAN WEIYUANHUI 中國科學院編譯出版委員會 ed., 1959. *Shi nian lai de Zhongguo kexue — shuxue — 1949–1959* 十年來的中國科學－數學－1949–1959 (Ten years of Chinese science — mathematics (1949–1959)). Peking: Kexue Chubanshe, pp. 1–3.

HUA Yinchun 華印椿 (1'), 1979. *Jianjie zhusuan fa* 簡捷珠算法 (An introduction to abacus computations). Peking: Zhongguo Caizheng Jingji Chubanshe.

HUA Yinchun 華印椿 (2'), 1987. *Zhongguo zhusuan shi gao* 中國珠算史稿 (Outline of the history of the Chinese abacus). Peking: Zhongguo Caizheng Jingji Chubanshe.

HUA Yinchun 華印椿 and LI Peiye 李培業 eds. (1'), 1990. *Zhonghua zhusuan da cidian* 中華珠算大辭典 (Great dictionary of the Chinese abacus). Hefei: Anhui Jiaoyu Chubanshe.

HUANG Mingxin 黃明信 and CHEN Jiujin 陳久金 (1'), 1989. *Zang li de yuanli yu shijian* 藏曆的原理與實踐 (Theory and practice of Tibetan calendars). Peking: Minzu Chubanshe (bilingual edition composed of Tibetan originals texts translated into Chinese and commented).

I TŌ Shuntarō *et al.* 伊東俊太郎 eds. (1'), 1983. *Kagaku shi gijutsu shi jiten* 科學史技術史事典 (An encyclopaedia of the history of sciences and techniques). Tokyo: Kōbundo.

JI Hongkun 季鴻崑 and WANG Zhihao 王治浩 (1'), 1985. "Wo guo Qing mo aiguo kexuejia Xu Jianyin." 我國清末愛國科學家徐建寅 (Xu Jianyin, a Chinese scientist and patriot from the end of the Qing dynasty). *ZK*, vol. 4, no. 3, pp. 284–294.

JIANG Shuyuan 蔣樹源(1'), 1984. "Xu Shou de liang fen qinbi xin." 徐壽的兩封親筆信 (Two letters by Xu Shou). *ZHKE*, vol. 5, no. 4, pp. 52–54.

JIANG Xiaoyuan 江曉原(1'), 1988. "Shilun Qing dai 'Xi xue Zhong yuan' shuo." 試論清代 " 西學中源 " 說 (A tentative evaluation of the Qing dynasty theory of the Chinese origin of Occidental sciences). *ZK*, vol. 7, no. 2, pp. 101–108.

JIANG Xiaoyuan 江曉原 (2'), 1988. "Cong Taiyang yundong lilun kan Babilun yu Zhongguo tianwenxue zhi guanxi." 從太陽運動理論看巴比倫與中國天文學之關係 (The connection between Chinese and Babylonian mathematics seen from the point of view of the theory of the motion of the Sun).*TWXB*, vol. 29, no. 3, pp. 272–277.

JIANG Zehan 江澤涵 (1'), 1980. "Shuxue mingci zaoqi gongzuo." 數學名詞早期工作 (Early works on Chinese mathematical terminology).*SXTB*, no. 10, pp. 23.

JING Yushu 靖玉樹 ed. (1'), 1994. *Zhongguo lidai suanxue jicheng* 中國歷代算學集成 (Collected Chinese mathematical works), 3 vols., Jinan: Shandong Renmin Chubanshe (contains complete reproductions of ancient editions of Chinese mathematical books).

442   References

JUSCHKEWITSCH: see YUSHKEVICH.

KATANO Zenichirō 片野善一郎 (1'), 1988. *Sotsugyō o tanoshiku suru sūgaku yōgo no yurai* 授業を樂しくする數學用語の由來 (Mathematical studies made pleasant: the origin of [Japanese] mathematical terminology). Tokyo: Meiji Tosho.

KATŌ Heizaemon 加藤平左ェ門 (1'), 1969. "Shina ni okeru han-hai-beki no kyūsū ni tsuite." 支那に於ける半背巾の級數に就て (The Chinese series expansion of the square of the half of an arc). In KATŌ Heizaemon 加藤平左ェ門, 1969. *Wasan no kenkyū* 和算の研究 (Research into *wasan*). Second supplement. Nagoya: Nagoya Daigaku Rikō Gakubu Sūgaku Kyōshitsu, pp. 23–46.

KAWAHARA Hideki 川原秀城 transl. (1'), 1980. "*Kyūshō sanjutsu* kaisetsu." 『九章算術』解說 (The *Jiuzhang suanshu* explained and commented). In YABUUCHI Kiyoshi 藪內清 *et al.*, 1980. *Chūgoku tenmongaku-sūgaku shū* 中國天文學數學集 (A collection of mathematical and astronomical Chinese works). Tokyo: Asahi Shuppansha, pp. 47–262.

KAWAHARA Hideki 川原秀城 (2'), 1984. "Mō hitotsu no Eki zeihō." もう一つの易筮法 (On the method of divination in the *Yijing* newly interpreted by Qin Jiushao). *ChugShisKen*, no. 6, pp. 127–138.

KAWAHARA Hideki 川原秀城 (3'), 1989. "Chūgoku no mugenshō kaiseki — enkansū no kyusū tenkai o megutte." 中國の無限小解析ー圓關數の級數展開をめぐって (The Chinese analysis of infinitesimals from the point of view of infinite expansions of circular functions). In YAMADA Keiji 山田慶兒 ed., 1989. *Chūgoku kodai kagaku shi ron* 中國古代科學史論 (Collected essays on the history of ancient Chinese science). Kyoto: Kyoto Daigaku Jinbun Kagaku Kenkyūsho, pp. 223–316.

KIM Yong-Woon 金容雲 and KIM Yong-Guk 金容局 (1'), 1978. *Kankoku sūgaku shi* 韓國數學史 (History of Korean mathematics). Tokyo: Maki Shoten.

KOBAYASHI Tatsuhiko 小林龍彥 (1'), 1990. "*Rekisan zensho* no sankakuhō to *Sūtei rekisho* no katsuen hassen no hyō no denrai ni tsuite." 『曆算全書』の三角法と『崇禎曆書』の割圓八線之表の傳來について (On the trigonometry in the *Lisuan quanshu* and the introduction of circular functions in the *Chongzhen lishu*). *KagakuSK*, second ser., vol. 29, no. 174, pp. 83–92.

KOBAYASHI Tatsuhiko 小林龍彥 and TANAKA Taoru 田中薫 (1'), 1983. "Wasan ni okeru senkō dai ni tsuite — Seki Takakazu no senkō dai to sono keishō." 和算における穿去題についてー關孝和の穿去題の研究とその繼承 (On problems of pierced objects in *wasan*: the problem of the pierced object [in the works of] Seki Takakazu and after him). *KagakuSK*, 2nd ser., vol. 22, no. 147, pp. 154–159.

KODAMA Akihito 兒玉明人 (1'), 1966. *Jūgo seiki no Chōsen kan dō-katsuji-han sūgaku sho* 十五世紀の朝鮮刊銅活字版數學書 (The Chinese mathematical books printed in Korea, in the fifteenth century, using movable type). Tokyo: privately printed.

KODAMA Akihito 兒玉明人 (2'), 1970. *Jūroku seiki matsu Min kan no shuzan sho* 十六世紀末明刊の珠算書 (The abacus books printed under the Ming dynasty, at the end of the sixteenth century). Tokyo: Fuji Tanki Daigaku Shuppanbu. Review: *Histmath*, 1975, vol. 2, pp. 218–219.

KONG Guoping 孔國平 (1'), 1987. "Dui Li Ye *Yigu yanduan* de yanjiu." 對李冶
《益古演段》的研究(On the *Yigu yanduan* of Li Ye [Li Zhi 李冶]). *WuWenjun*,
no. 3, pp. 58–72.

LI Di 李迪 (1'), 1976. *Guo Shoujing* 郭守敬 (Guo Shoujing). Shanghai. Review:
RBS, 1966, notice no. 1978.

LI Di 李迪 (2'), 1978. *Mengguzu kexuejia Ming'antu* 蒙古族科學家明安圖
(Ming'antu [Minggatu], a Mongol scientist). Huhehot: Nei Menggu Renmin
Chubanshe.

LI Di 李迪 (3'), 1984. *Zhongguo shuxue shi jianbian* 中國數學史簡編 (A concise history of Chinese mathematics). Shenyang: Liaoning Renmin Chubanshe.

LI Di 李迪 (4'), 1986. "Guonei shoucang de Ming kanben yu chaoben *Suanfa
tongzong* yu *Suanfa zuanyao*."國內收藏的明刊本與抄本《算法統宗》與
《算法纂要》(Ming dynasty editions and manuscripts of the *Suanfa tongzong*
and the *Suanfa zuanyao* preserved in China). *WuWenjun*, no. 2, pp. 48–55.

LI Di 李迪 (5'), 1987. "Zhongguo shuxue shi zhong de weijiejue wenti." 中國數
學史中的未解決問題 (Unsolved questions in the history of Chinese
mathematics). *Wuwenjun*, no. 3, pp. 10–27.

LI Di 李迪 ed. (6'), 1990. *Shuxue shi yanjiu wenji, diyi ji* 數學史研究問集,第
一集 (Collected papers on the history of Chinese mathematics, first series).
Huhehot: Nei Menggu Daxue Chubanshe and Jiuzhang Chubanshe.

LI Di 李迪 and BAI Shangshu 白尚恕 (1'), 1984. "Wo guo jindai kexue xianqu
Zhou Boqi."我國近代科學先區鄒佰奇(Zhou Boqi [(1819–1869)] a precur-
sor of modern Chinese science). *ZK*, vol. 3, no. 4, pp. 378–390.

LI Di 李迪 and GUO Shirong 郭世榮 (1'), 1988. *Qingdai zhuming tianwen-
shuxuejia Mei Wending* 清代著名天文數學家梅文鼎 (Mei Wending, famous
astronomer and mathematician of the Qing dynasty). Shanghai: Shanghai Kexue
Jishu Wenxian Chubanshe.

LI Jimin 李繼閔 (1'), 1984. "Cong gougu bilü lun dao chongcha shu." 從勾股比
率論到重差術 (From the theory of the similarity of right-angled triangles to the
chongcha technique). *KXSJK*, no. 11, pp. 96–104.

LI Jimin 李繼閔 (2'), 1990. "*Jiuzhang suanshu* jiqi Liu Hui zhu yanjiu"
《九章算術》及其劉徽注研究 (Research into the *Jiuzhang suanshu* and Liu
Hui's commentary). Xi'an: Shenxi Renmin Jiaoyu Chubanshe.

LI Peiye 李培業 (1'), 1984. *Zhusuan yanjiu* 珠算研究 (Research into the abacus).
Shenxi Zhusuan Xiehui (The Shenxi Abacus Association).

LI Peiye 李培業 (2'), 1986. *Suanfa zuanyao* jiaoshi 算法纂要校釋 ([Cheng
Dawei's] *Suanfa zuanyao* [Essentials of arithmetic] revised and explained). Hefei:
Anhui Jiaoyu Chubanshe.

LI Qun 李群 (1'), 1975. *Mengqi bitan* xuandu 夢溪筆談選讀 (Selected texts
from the *Mengqi bitan*). Peking: Kexue Chubanshe.

LI Wenlin 李文林 and 袁向東 YUAN Xiangdong (1'), 1982. "Zhongguo gudai
buding fenxi ruogan wenti tantao." 中國古代不定分析若干問題探討
(Research into several problems of indeterminate analysis in ancient China).
Kejishi, no. 8, pp. 106–122.

LI Wenlin 李文林 and 袁向東 YUAN Xiangdong (2'), 1986. "Li Shanlan de
jianzhui qiuji shu." 李善蘭的尖錐求積術 (Li Shanlan's method of
computation of the volume of the 'sharp pyramid'). *WuWenjun*, no. 2, pp. 99–106.

444   References

LI Yan 李儼 (1'), 1919–1920. "Zhongguo shuxue yuanliu kaolüe." 中國數學源流考略 (Overview of the origin and evolution of Chinese mathematics). *BDYK*, vol. 1, no. 4, pp. 1–19; no. 5, pp. 59–74; no. 6, pp. 65–94.

LI Yan 李儼 (2'), 1920–1934. "Li Yan suo cang Zhongguo suanxue shu mulu." 李儼所藏中國算學書目祿 (Catalogue of the Chinese mathematical books in the library of Li Yan). *Kexue*, 1920, vol. 5, no. 4, pp. 418–426, no. 5, pp. 525–531; 1925, vol. 10, no. 4, pp. 548–551; 1926, vol. 11, no. 6, pp. 817–820; 1927, vol. 12, no. 12, pp. 1825–1826; 1929, vol. 13, no. 8, pp.1134–1137; 1930, vol. 15, no. 1, pp. 158–160; 1932, vol. 16, no. 5, pp. 856–857; no. 11, pp. 1710–1713; 1933, vol. 17, no. 6, pp.1005–1008; 1934, vol. 18, no. 11, pp. 1547–1556.

LI Yan 李儼 (3'), 1925. "Dayan qiuyi shu de guoqu yu weilai." 大衍求一術的過去與未來 (The past and future of the *dayan* rule [Chinese remainder theorem]). *XueYi*, vol. 7, no. 2, pp. 1–45; *ZSSLC-P*, I, pp. 122–174; *ZSSLC-T*, I, pp. 61–121.

LI Yan 李儼 (4'), 1925. "Zhong-suan shuru Riben zhi jingguo." 中算輸入日本之經過 (The process of introduction of Chinese mathematics into Japan). *DFZZ*, vol. 22, no.18, pp. 82–88; *ZSSLC-P*, V, pp. 168–186; *ZSSLC-T*, I, pp. 349–362.

LI Yan 李儼 (5'), 1925. "Mei Wending nianpu." 梅文鼎年譜 (A yearly biography of Mei Wending). *QHXB*, vol. 2, no. 2, pp. 609–634; *ZSSLC-P*, III, pp. 544–576; *ZSSLC-T*, I, pp. 363–408.

LI Yan 李儼 (6'), 1926. "Chong-cha shu yuanliu jiqi xin zhu." 重差術源流及其新註 (New commentary and research into the origin of the 'Chong-cha' rule [similar triangles]). *XueYi*, vol. 7, no. 8, pp. 1–15; *ZSSLC-T*, I, pp.39–59.

LI Yan 李儼 (7'), 1926. "Dunhuang shishi suanshu." 敦煌石室「算書」(The arithmetical books of the Dunhuang caves). *ZDJK*, vol. 1, no. 2, pp. 1–4; *ZSSLC-T*, I, pp. 123–128.

LI Yan 李儼 (8'), 1926. "Mingdai suanxue shu zhi." 明代算學書志 (A monograph of Ming dynasty mathematical books). *TSGX*, vol. 1, no. 4, pp. 667–682; *ZSSLC-P*, II, pp. 86–102; *ZSSLC-T*, I, pp. 129–147.

LI Yan 李儼 (9'), 1926. "Zhong suanjia zhi Pythagoras dingli yanjiu." 中算家之 Pythagoras 定理研究 (Research into the Pythagorean theorem of Chinese mathematicians). *XueYi*, 1926, vol. 8, no. 2, pp. 1–27; *ZSSLC-P*, I, pp. 44–75; *ZSSLC-T*, I, pp. 1–37.

LI Yan 李儼 (10'), 1927. "Duishu zhi faming jiqi Dong lai. " 對數之發明及其東來 (The discovery of logarithms and their introduction into China). *Kexue*, 1927, vol. 12, no. 2, pp. 109–158; no. 3, pp. 285–325; no. 6, pp. 689–700; *ZSSLC-P*, III, pp. 69–190; *ZSSLC-T*, I, pp. 195–348.

LI Yan 李儼 (11'), 1927. "Sanjiaoshu ji sanjiao hanshubiao zhi dong lai." 三角術及三角函數表之東來 (The introduction of trigonometry and trigonometrical tables in the Orient). *Kexue*, 1927, vol. 12, no. 10, pp. 1393–1445; *ZSSLC-P*, III, pp. 191–253; *ZSSLC-T*, III, pp. 323–400.

LI Yan 李儼 (12'), 1927. "Zhong suanjia zhi zonghengtu yanjiu." 中算家之縱橫圖研究 (Research into the magic squares of Chinese Mathematicians). *XueYi*, vol. 8, no. 9, pp. 1–40; *ZSSLC-P*, I, p. 175–229; *ZSSLC-T*, III, pp. 59–110.

LI Yan 李儼 (13'), 1927. "Ming-Qing zhi ji xi suan shuru Zhongguo nianbiao." 明清之際西算輸入中國年表 (Yearly chronology of the introduction of Occidental mathematics in China during the Ming and Qing dynasties). *TSGX*, vol. 2, no. 1, pp. 21–53; *ZSSLC-P*, III, pp. 10–68; *ZSSLC-T*, I, pp. 149–193.

Lɪ Yan 李儼 (14'), 1927–1928. "Ming-Qing suanjia zhi geyuan shu yanjiu." 明清算家之割圓術研究 (Research into cyclotomy [trigonometry] by Ming-Qing mathematicians). *Kexue*, vol. 12, no. 11, pp. 1487–1520; no. 12, pp. 1721–1766; vol. 13, no. 1, pp. 53–102; no. 2, pp. 201–250; *ZSSLC-P*, III, pp. 254–512; *ZSSLC-T*, II, pp. 129–433.

Lɪ Yan 李儼 (15'), 1928. "*Yongle dadian* suanshu." 永樂大典算書 (The mathematical books of the *Yongle dadian* [Yongle reign-period great encyclopaedia]). *TSGX*, vol. 2, no. 2, pp. 189–195; *ZSSLC-P*, II, pp. 47–53; *ZSSLC-T*, II, pp. 83–91.

Lɪ Yan 李儼 (16'), 1928. "Zhong suan shi zhi gongzuo." 中算史之工作 (Research work into the history of Chinese mathematics). *Kexue*, vol. 13, no. 6, pp. 785–809; *ZSSLC-P*, V, pp. 93–115.

Lɪ Yan 李儼 (17'), 1928. "Li Shanlan nianpu." 李善蘭年譜 (A yearly biography of Li Shanlan). *QHXB*, vol. 5, no. 1, pp. 1625–1651; *ZSSLC-P*, IV, pp. 331–361; *ZSSLC-T*, II, pp. 435–474.

Lɪ Yan 李儼 (18'), 1928. "Zhongguo jinguqi zhi suanxue." 中國近古期之算學 (Chinese mathematics from the pre-modern period [from the Song dynasty onwards]). *XueYi*, vol. 9, no. 4–5, pp. 1–28.

Lɪ Yan 李儼 (19'), 1928–1929. "Jindai Zhong suan zhushu ji." 近代中算著述記 (Bibliographical notes on Chinese mathematical works from the modern period [Qing dynasty]). *TSGX*, 1928, vol. 2, no. 4, pp. 601–640; 1929, vol. 3, no. 1–2, pp. 149–200, no. 3, pp. 367–388, no. 4, pp. 601–617; *ZSSLC-P*, II, pp. 103–308; *ZSSLC-T*, IV (Part 2), pp. 419–638.

Lɪ Yan 李儼 (20'), 1929. "*Jiuzhang suanshu* buzhu." 九章算術補註 Supplementary notes on the *Jiuzhang suanshu*). *BJBH*, vol. 2, no. 2, pp. 127–133; *ZSSLC-T*, III, pp. 1–9.

Lɪ Yan 李儼 (21'), 1929. "Zhong suanjia zhi Pascal sanjiaoxing." 中算家之Pascal三角形 (Chinese mathematicians's Pascal triangle). *XueYi*, vol. 9, no. 9, pp. 1–15; *ZSSLC-P*, I, pp. 230–245; *ZSSLC-T*, III, pp. 111–126.

Lɪ Yan 李儼 (22'), 1929. "Chousuan zhidu kao." 籌算制度考 (Research into the counting-rod system). *YJXB*, no. 6, pp. 1129–1144; *ZSSLC-P*, IV, pp. 1–8; *ZSSLC-T*, III, pp. 29–36.

Lɪ Yan 李儼 (23'), 1930. "Song Yang Hui suanshu kao." 宋楊輝算書考 (Research into the mathematical works of Yang Hui, a mathematician of the Song dynasty). *TSGX*, vol. 4, no. 1, pp. 1–21; *ZSSLC-P*, II, pp. 54–72; *ZSSLC-T*, II, pp. 93–119.

Lɪ Yan 李儼 (24'), 1930. "*Sunzi suanjing* buzhu." 孫子算經補註 (Supplementary notes on the *Sunzi suanjing*). *GLB*, vol. 4, no. 4, pp. 13–29; *ZSSLC-T*, III, pp. 11–27.

Lɪ Yan 李儼 (25'), 1930. "Zhong suanjia zhi fangcheng lun." 中算家之方程論 (The theory of equations of Chinese mathematicians). *Kexue*, vol. 15, no. 1, pp. 7–44; *ZSSLC-P*, I, pp. 246–314; *ZSSLC-T*, III, pp. 127–196.

Lɪ Yan 李儼 (26'), 1931. "Zengxiu Ming dai suanxue shu zhi." 增修明代算學書志 (Critical notes, revised and augmented, on mathematical works of the Ming dynasty). *TSGX*, vol. 5, no. 1, pp. 109–135; *ZSSLC-P*, II, pp. 86–102; See also Lɪ Yan (50'), 1954, pp. 149–178.

Lɪ Yan 李儼 (27'), 1931–1932. "*Ceyuan haijing* yanjiu licheng kao." 測圓海鏡研究歷程考 (Research into the *Ceyuan haijing*). *XueYi*, 1931, vol. 11, no. 2, pp. 1–26; no. 6, pp. 1–15; no. 8, pp. 1–36; no. 9, pp. 1–10; no. 10, pp. 1–14; 1932, vol. 12, no. 1, pp. 117–134; no. 2, pp. 85–101; no. 3, pp. 99–111; no. 4, pp. 83–92; *ZSSLC-P*, IV, p. 32–237; *ZSSLC-T*, IV (Part 1), pp. 27–251.

LI Yan 李儼 (28'), 1931. "Zhusuan zhidu kao." 珠算制度考 (Research into the abacus system). *YJXB*, no. 10, pp. 2123–2138; *ZSSLC-P*, IV, pp. 9–23; *ZSSLC-T*, III, pp. 37–57.

LI Yan 李儼 (29'), 1933. "Ershi nian lai Zhong suan shiliao zhi fajian." 二十年來中算史料之發見 (Discoveries of material concerning the history of Chinese mathematics over the last 20 years). *Kexue*, vol. 17, no. 1, pp. 1–15; *ZSSLC-T*, II, pp. 63–76.

LI Yan 李儼 (30'), 1933. "Dongfang tushuguan shanben suanshu jieti." 東方圖書館善本算書解題 (The dispersion and destruction of the rare books of the Dongfang library). *GLB*, vol. 7, no. 1, pp. 7–11; *ZSSLC-P*, II, pp. 73–76; *ZSSLC-T*, II, pp. 121–127.

LI Yan 李儼 (31'), 1933. "Zhongguo shuxue shi daoyan." 中國數學史導言 (Introduction to the history of Chinese mathematics). *XueYi*, supplementary number (march 1933), pp. 139–160; *ZSSLC-P*, II, pp. 1–31; *ZSSLC-T*, II, pp. 1–39.

LI Yan 李儼 (32'), 1933. "Tang, Song, Yuan, Ming, shuxue jiaoyu zhidu." 唐宋元明數學教育制度 (Mathematical education under the Song, Yuan and Ming dynasties). *Kexue*, vol. 17, no. 10, pp. 1545–1565; *ZSSLC-P*, IV, pp. 238–280; *ZSSLC-T*, IV (Part 1), pp. 253–285.

LI Yan 李儼 (33'), 1933. "Dongfang tushuguan canben *Shuxue juyao* mulu." 東方圖書館殘本數學舉要目錄 (A list of rare books copied from what remains of a rare catalogue of mathematical books entitled *Shuxue juyao*, formerly preserved in the Dongfang library). *TSGX*, vol. 7, no. 4, pp. 721–726; *ZSSLC-P*, II, pp. 77–85; *ZSSLC-T*, IV (Part 2), pp. 359–370.

LI Yan 李儼 (34'), 1934. "*Ceyuan haijing* pijiao." 測圓海鏡批校 (A critical examination of the *Ceyuan haijing*). *GLB*, vol. 8, no. 2, pp. 49–60; *ZSSLC-P*, IV, pp. 24–31; *ZSSLC-T*, IV (Part 1), pp. 13–25.

LI Yan 李儼 (35'), 1934. "Qingdai shuxue jiaoyu zhidu." 清代數學教育制度 (Mathematical education under the Qing dynasty). *XueYi*, vol. 13, no. 4, pp. 37–52; no. 5, pp. 49–59; no. 6, pp. 39–44; *ZSSLC-P*, IV, pp. 281–320; *ZSSLC-T*, IV (Part 1), pp. 287–342.

LI Yan 李儼 (36'), 1934. "Qingji Shenxi shuxuejia jiaoyu shiliao." 清季陝西數學家教育史料 (Historical data on the mathematical education of mathematicians of Shenxi province). *XJRB* (13–19 August issue); *ZSSLC-P*, IV, pp. 321–330; *ZSSLC-T*, IV (Part 2), pp. 343–358.

LI Yan 李儼 (37'), 1934. "Zhongguo de shuli." 中國的數理 (Chinese mathematics). *WHJS*, vol. 1, no. 1, pp. 149–153; *ZSSLC-P*, III, pp. 1–9; *ZSSLC-T*, IV (Part 1), pp. 1–12.

LI Yan 李儼 (38'), 1934. "Yindu lisuan yu Zhongguo lisuan zhi guanxi." 印度歷算與中國曆算之關係 (The relationship between Chinese and Indian astronomy and mathematics). *XueYi*, vol. 13, no. 9, pp. 57–74; no. 10, pp. 51–64; *ZSSLC-T*, IV (Part 2), pp. 371–418.

Li Yan 李儼 (39'), 1935. "Dunhuang shishi suanjing yi juan bing xu." 敦煌石室 算經一卷并序 (A chapter and a preface of an arithmetic from the Dunhuang caves). *GLB*, vol. 9, no. 1, pp. 39–46.

Li Yan 李儼 (40'), 1937. "Zenyang yanjiu Zhongguo suanxue shi." 怎樣研究中 國算學史 (How to study the history of Chinese mathematics). *ChuZhou*, no. 220, pp. 1–7; Reprinted in Li Yan (50'), pp. 17–22.

Li Yan 李儼 (41'), 1937/1977. *Zhong suan shi luncong* 中算史論叢 (Collected papers on the history of Chinese mathematics), 4 vols., Shanghai: Shangwu Yinshuguan. Taipei: Shangwu Yinshuguan.

Li Yan 李儼 (42'), 1937/1978. *Zhongguo suanxue shi* 中國算學史 (History of Chinese mathematics). Taipei: Shangwu Yinshuguan.

Li Yan 李儼 (43'), 1938. "Tang dai suanxue shi." 唐代算學史 (History of Chinese mathematics under the Tang dynasty). *XBSD*, vol. 1, no. 1, pp. 63–95; *ZSSLC-P*, V, pp. 15–56.

Li Yan 李儼 (44'), 1939. "Dunhuang shishi *Licheng suanjing*." 敦煌石室『立 成算經』 (The *Licheng suanjing* of the Dunhuang caves). *TSJK*, new ser., vol. 1, no. 4, pp. 386–396.

Li Yan 李儼 (45'), 1941. "Yiselanjiao yu Zhonggguo lisuan zhi guanxi." 伊斯蘭教 與中國曆算之關係 (The relationship between Chinese mathematical astronomy and Islam). *ZSSLC-P*, V, pp. 57–75.

Li Yan 李儼 (46'), 1943. "Zhong suanjia de fenshu lun." 中算家的分數論 (The theory of fractions of Chinese mathematicians). *Kexue*, vol. 26, no. 2, pp. 183–203; *ZSSLC-P*, I, pp. 15–43.

Li Yan 李儼 (47'), 1944. "Shanggu Zhong suan shi." 上古中算史 (History of Chinese mathematics in early antiquity). *Kexue*, vol. 27, no. 9–12, pp. 16–24; *ZSSLC-P*, V, pp. 1–14.

Li Yan 李儼 (48'), 1947. "Sanshi nian lai de Zhongguo suanxue shi." 三十年來的 中國算學史 (The history of Chinese mathematics over the last thirty years). *Kexue*, vol. 29, no. 4, pp. 101–108; *ZSSLC-P*, V, pp. 146–167.

Li Yan 李儼 (49'), 1947. "Zhong suanjia de yuanzhui quxian shuo." 中算家的圓 錐曲線說 (The theory of conic sections of Chinese mathematicians). *Kexue*, vol. 29, no. 4, pp. 115–120; *ZSSLC-P*, III, pp. 519–537.

Li Yan 李儼 (50'), 1954/1975. *Zhongguo suanxue shi luncong* 中國算學史論叢 (Collected papers on the history of Chinese mathematics). Taipei: Zhengzhong Shuju.

Li Yan 李儼 (51'), 1954–55. *Zhong suan shi luncong.* 中算史論叢 (Collected papers on the history of Chinese mathematics). 5 vols., Peking: Zhongguo Kexueyuan, [note that despite the fact that they have identical titles, the content of this book is not exactly the same as that of Li Yan (41'), 1937].

Li Yan 李儼 (52'), 1957. *Zhong suanjia de neichafa yanjiu* 中算家的內插法研 究 (Research into the interpolation formulae of Chinese mathematicians). Peking: Kexue Chubanshe.

Li Yan (53'), 1957. *Shisan shisi shiji Zhongguo minjian shuxue.* 十三十四世紀民 間數學 (Chinese popular mathematics in the thirteenth and fourteenth centuries). Peking: Kexue Chubanshe.

Li Yan (54'), 1957. "Mei Wending de shengping jiqi zhuzuo mulu." 梅文鼎的生 平及其著作目錄 (The life and works of Mei Wending). *ALXB*, Inaugural issue, pp. 93–94.

LI Yan (55'), 1958. "Alabo shuru de zonghengtu." 阿拉伯輸入的縱橫圖 (Chinese magic squares of islamic origin). *WWC*, no. 7, pp. 17–19.

LI Yan (56'), 1958. *Zhongguo shuxue dagang* 中國數學大綱 (Outlines of Chinese mathematics). 2 vols., Peking: Kexue Chubanshe.

LI Yan (57'), 1958. "*Tong Ling suanfa* de jieshao." 《銅陵算法》的介紹 (Introductory remarks on the *Tong Ling suanfa* [Tong Ling's arithmetic]). *ALXB*, no. 2, pp. 67–70.

LI Yan (58'), 1959. "Cong Zhong suanjia de geyuanshu kan hesuanjia de yuanli he jiaoshu." 從中算家的割圓術看和算家的圓理和角術 (The circular and angular principles of the Japanese seen from the point of view of the cyclotomical techniques of the Chinese). *KXSJK*, vol. 2, pp. 80–125.

LI Yan (59'), 1960. "Hesuanjia 'zengyueshu' yingyong de shuoming." 和算家 "增約術" 應用的說明 (An elucidation of the usage of the *reiyakujutsu* [a computational technique for the reduction of small decimals] of traditional Japanese mathematicians). *KXSJK*, vol. 3, pp. 65–69.

LI Yan (60'), 1960. "*Suanfa zuanyao* de jieshao." 《算法纂要》的介紹 (Introductory remarks on the *Suanfa zuanyao*" [Essentials of arithmetic]). *ASX*, Inaugural issue, pp. 58–61.

LI Yan (61'), 1954/1963. *Zhongguo gudai shuxue shiliao* 中國古代數學史料 (Historical materials for the history of ancient Chinese mathematics). Shanghai: Kexue Jishu Chubanshe, Review: *RBS*, 1963, notice no. 859.

LI Yan (62'), 1982. "Riben shuxuejia (hesuanjia) de pingyuan yanjiu." 日本算學家 — 和算家 — 的平圓研究 (The Japanese traditional mathematicians's researches into the circle). *ZK*, vol. 1, no. 3, pp. 208–214.

LI Yan and DU Shiran (1'), 1963–64. *Zhongguo gudai shuxue jianshi* 中國古代數學簡史 (Chinese mathematics, A concise history). 2 vols., Peking: Zhonghua Shuju (English translation: see CROSSLEY and LUN (1), 1987).

LI Yan and DU Shiran (2'), 1964. *Zhongguo gudai shuxue shi hua* 中國古代數學史話 (Jottings on the history of ancient Chinese mathematics), Peking: Zhonghua Shuju.

LI Yan and LI Di (1'), 1980. *Zhongguo shuxue shi lunwen mulu (1906–1979); Guonei zhi bu* 中國數學史論文目錄 (1906–1979) 國內之部 (A catalogue of papers on the history of Chinese mathematics (Mainland China publications only)). Nei Menggu Shifan Xueyuan.

LI Zhaohua 李兆華 (1'), 1985. "Dai Xu guanyu duishu yanjiu de gongxian." 戴煦關於對數研究的貢獻 (Dai Xu's contribution to research into logarithms). *ZK*, vol. 4, no. 4, pp. 353–362.

LI Zhaohua 李兆華 (2'), 1985. "Dai Xu guanyu erxiang shi he duishu zhankai shi de yanjiu." 戴煦關於二相式和對數展開式的研究 (Dai Xu's research into the binomial expansion and logarithmic series). *WuWenjun*, no. 1, pp. 98–108.

LI Zhaohua 李兆華 (3'), 1987. "Dong Youcheng de duojishu yu geyuanshu shuping." 董祐誠的垛積術與割圓術述評 (A critical appraisal of the finite summation formulae and cyclotomic algorithms of Dong Youcheng). *WuWenjun*, no. 3, pp. 94–112.

LI Zhaohua 李兆華 (4'), 1989. "Guangyu chafenshu de jige wenti." 關於差分術的幾個問題 (Some problems raised by the *chafen* [proportional parts] rules). *ZK*, vol. 8, no. 2, pp. 108–117.

Li Zhaohua 李兆華 (5'), 1990. "*Suanfa tongzong* shitan." 《算法統宗》試探 (A tentative presentation of the *Suanfa tongzong*). *ZK*, vol. 9, no. 4, pp. 308–317.

Li Zhongheng 李仲珩 (1'), 1947. "Sanshi nian lai Zhongguo de suanxue." 三十年來中國的算學 (Chinese mathematics over the last thirty years), *Kexue*, vol. 29, p.68.

Liang Zongju 梁宗巨 (1'), 1981. *Shijie shuxue shi jianbian* 世界數學史簡編 (A handbook for the history of world mathematics). Shenyang: Liaoning Renmin Chubanshe.

Lin Xiashui 林夏水 and Zhang Shangshui 張尚水 (1'), 1983. "Shuli luoji zai Zhongguo." 數理邏輯在中國 (Mathematical logic in China). *ZK*, vol. 2, no. 2, pp. 175–182.

Liu Dun 劉鈍 (1'), 1986. "Tuolemi de he'nalengma yu Mei Wending de 'sanji tongji'." 托勒密的 " 曷捺楞馬 " 與梅文鼎的 " 三極通機 " (Ptolemy's annalemma and Mei Wending' s *sanji tongji*). *ZK*, vol. 5, no. 1, pp. 68–75.

Liu Dun 劉鈍 (2'), 1986. "Qing chu lisuan dashi Mei Wending." 清初曆算大師梅文鼎 (Mei Wending, great astronomer and mathematician of the beginning of the Qing dynasty). *ZirBian*, vol. 8, no. 1, pp. 52–64.

Liu Dun 劉鈍 (3'), 1989. "Li Rui yu Dika'er fuhao faze." 李銳與笛卡兒符號法則 (Li Rui and Descartes's sign rule). *ZK*, vol. 8, no. 2, pp. 127–137.

Liu Dun 劉鈍 (4'), 1989. "Ruogan Ming-Qing biji zhong de shuxue shiliao." 若干明清筆記中的數學史料 (Mathematics in some *biji* [Collections of short literary sketches] from the Ming and Qing dynasties). *ZHKE*, vol. 10, no. 4, pp. 49–56.

Liu Dun 劉鈍 (5'), 1993. *Da zai yan shu* 大哉言數 (Great is the art of number!). Shenyang: Liaoning jiaoyu chubanshe.

Liu Dun 劉鈍 (6') 1995. "Fang Tai suo jian shuxue zhenji." 訪台所見數學珍藉 (Rare books on ancient Chinese mathematics seen during a sojourn in Taiwan), *ZHKE*, vol. 16, no. 1, pp. 8-21.

Lu Tingquan 盧廷權 (1'), 1981. "Shuxuejia Ke Zhao." 數學家可召 (Ke Zhao the mathematician). *Renwu*, no. 8, pp. 183–186 [note that this mathematician is often called 'Ko Chao' in many publications].

Luo Jianjin 羅見今 (1'), 1982. "*Duoji bilei* neirong fenxi." 《垛積比類》內容分析 (An analysis of the content of the *Duoji bilei* [Finite summations formulae]). *NMSY*, no. 1, pp. 89–105.

Luo Jianjin 羅見今 (2'), 1982. "Li Shanlan hengdengshi de daochu — jinian Li Shanlan shishi yibai zhounian — ." 李善蘭恒等式的導出－紀念李善蘭逝世一百周年－ (A derivation of Li Shanlan's formula — in commemoration of the hundredth anniversary of Li Shanlan's death). *NMSY*, no. 2, pp. 42–51.

Luo Jianjin 羅見今 (3'), 1982. "Li Shanlan dui Stirling he Euler shu de yanjiu." 李善蘭對 Stirling 和 Euler 數的研究 (Li Shanlan's research into Stirling and Eulerian numbers). *SXY*, vol. 2, no. 4, pp. 173–182.

Luo Jianjin 羅見今 (4'), 1985. "Qing mo shuxuejia Hua Hengfang." 清末數學家華蘅芳 (Hua Hengfang mathematician of the end of the Qing dynasty). *WuWenjun*, no. 1, pp. 109–120.

Luo Jianjin 羅見今 (5'), 1988. "Ming'antu shi Katalan shu de shouchuangzhe." 明安圖是卡塔蘭數的首創者 (Ming'antu [Minggatu] is the inventor of the Catalan numbers). *NMDXXB*, vol. 19, no. 2, pp. 239–245.

LUO Jianjin 羅見今 (6'), 1990. "(Ming'antu jisuan wuqiong jishu de fangfa fenxi." 明安圖記算無窮級數的方法分析 (Analysis of (Ming'antu [Minggatu]'s method of infinite series). *ZK*, vol. 9, no. 3, pp. 197–207.

MA Guojun 麻國鈞 and MA Shuyun 麻淑云 (1'), 1990. *Zhonghua chuantong youxi daquan* 中華傳統游戲大全 (Encyclopaedia of Chinese Traditional Games), Peking: Nongcun Duwu chubanshe.

MARTZLOFF Jean-Claude (1'), 1983. "Li Shanlan de youxian he gongshi." 李善蘭的有限和公式 (Li Shanlan's finite summation formulae). *KXSYicong*, no. 2, pp. 1–6 (A Chinese translation of a communication to the First International Congress of the History of Chinese Science, Leuven (Belgium)), August 1982.

MEI Rongzhao 梅榮照 (1'), 1960. "Wo guo diyiben weijifenxue de yiben — *Daiweiji shiji* — chuban yibai zhounian." 我國第一本微積分學的譯本《代微積拾級》出版一百周年 (The first Chinese translation of a manual of calculus: the *Dai weiji shiji* — in commemoration of the hundredth anniversary of its publication). *KXSJK*, no. 3, pp. 59–64.

MEI Rongzhao 梅榮照 (2'), 1963. "Xu Guangqi de shuxue gongzuò." 徐光啓的數學工作 (The mathematical works of Xu Guangqi). In ZHONGGUO KEXUEYUAN ZIRAN KEXUE SHI YANJIUSHI 中國科學院中國自然科學史研究室 ed., 1963. *Xu Guangqi jinian lunwenji* 徐光啓紀念論文集 (Collected papers in honour of Xu Guangqi), Peking: Zhonghua Shuju, pp. 143–161.

MEI Rongzhao 梅榮照 (3'), 1984. "Huainian Qian Baocong xiansheng — jinian Qian Baocong xiansheng dansheng jiushi zhounian." 懷念錢寶琮先生－紀念錢寶琮先生誕生九十周年 (In memory of professor Qian Baocong — In commemoration of the ninetieth anniversary of professor Qian Baocong's birth). *KXSJK*, no. 11, pp. 6–11.

MEI Rongzhao 梅榮照 (4'), 1984. "Liu Hui de fangcheng lilun." 劉徽的方程理論 (The *fangcheng* theory according to Liu Hui). *KXSJK*, no. 11, pp. 63–76.

MEI Rongzhao 梅榮照 (5'), 1984. "Liu Hui de gougu lilun." 劉徽的勾股理論 (Liu Hui's theory of right-angled triangles). *KXSJK*, no. 11, pp. 77–95.

MEI Rongzhao 梅榮照 (6'), 1984. "Liu Hui yu Zu Chongzhi fuzi." 劉徽與祖冲之父子 (Liu Hui versus Zu Chongzhi and his son). *KXSJK*, no. 11, pp. 105–129.

MEI Rongzhao 梅榮照 (7'), 1984. "*Jiuzhang suanshu* shaoguang zhang zhong qiu zuixiao gongbeishu de wenti." 《九章算數》少廣章中求最小公倍數的問題 (The problem of the determination of the least common multiple in the *Jiuzhang suanshu*). *ZK*, vol. 3, no. 3, pp. 203–208.

MEI Rongzhao 梅榮照 (8'), 1987. "Qin Jiushao shi ruhe dechu qiu 'dingshu' fangfa de" 秦九韶是如何得出求定數方法的 (What method did Qin Jiushao use to find the *dingshu* [the coprime modulus of the Chinese remainder 'theorem'] ? "). *ZK*, vol. 6, no. 4, pp. 293–298.

MEI Rongzhao 梅榮照 (9'), 1988, "Song-Yuan shuxue de sheng-shuai." 宋元數學的盛衰 (Blossoming and decline of Song-Yuan mathematics). *ZK*, vol. 7, no. 3, pp. 205–213.

MEI Rongzhao 梅榮照 (10'), 1989. "Jia Xian de zeng-cheng kaifang fa." 賈憲的增乘開方法 (Jia Xian's additive-multiplicative root extraction method [Horner's method]). *ZK*, vol. 8, no. 1, pp. 1–8.

MEI Rongzhao 梅榮照 ed. (11'), 1990. *Ming-Qing shuxue shi lunwen ji* 明清數學史論文集 (Collected papers on the history of mathematics under the Ming and Qing Periods). Nanking: Jiangsu Jiaoyu Chubanshe.

MIKAMI Yoshio 三上義夫 (1'), 1917. "Indo no sūgaku to Shina to no kankei." 印度 の 數學 と 支那 と の 關係 (The relationship between Chinese and Indian mathematics). *TNH*, vol. 12, no. 6, pp. 15–25.

MIKAMI Yoshio 三上義夫 (2'), 1921. "Dickson shi no seisū ron rekishijō ni okeru Wa-Kan-jin no kōken." Dickson氏ノ整數論史上ニ於ヶル和漢人ノ貢獻 (The contributions of Chinese and Japanese mathematicians in L. E. Dickson's *History of the Theory of Numbers*). *TBGZ*, no. 354, pp. 192–198; no. 356, pp. 274–282; no. 362, pp. 11–19.

MIKAMI Yoshio 三上義夫 (3'), 1923. "Loria Hakushi no Shina sūgaku ron." Loria 博士 の 支那 數學 論 (The history of Chinese mathematics according to Dr. Loria). *TYG*, vol. 12, no. 4, pp. 500–514.

MIKAMI Yoshio 三上義夫 (4'), 1926. *Shina sūgaku no tokushoku* 支那數學の特色 (The characteristics of Chinese mathematics). *TYG*, vol. 15, no. 4; vol. 16, no. 1; Chinese translation by LIN Ketang 林科棠, *Zhongguo suanxue zhi tese* 中國算學之特色, Shanghai: Shangwu Yinshuguan, 1934.

MIKAMI Yoshio 三上義夫 (5'), 1927. "*Chūjin den ron* — awasete van Hée shi no sho setsu o hyōsu — ." 疇人傳—併せて Van Hée 氏 の 所説 を 評す (Criticism of van Hée's account of (Ruan Yuan's *Chouren zhuan* [Biographies of calendarist-mathematicians] (1799)). *TYG*, vol. 16, no. 2, pp. 185–222; no. 3, pp. 287–333.

MIKAMI Yoshio 三上義夫 (6'), 1928. *Tōzai sūgaku shi* 東西數學史 (History of Oriental and Occidental Mathematics). Tokyo: Kyōritsu Sha.

MIKAMI Yoshio 三上義夫 (7'). 1932–33–34. "Seki Takakazu no gyōseki to Kei-Han no sanka narabini Shina no Sanpō to no kankei oyobi hikaku." 關孝和の業績と京坂の算家並に支那の算法との關係及び比較 (The achievements of Seki Takakazu and his relations with the mathematicians of Osaka and Kyoto and with the Chinese mathematicians) *TYG*, vol. 20, no. 2, pp. 217–249; no. 4, pp. 543–566; vol. 21, no. 1, pp. 54–65; no. 3, pp. 352–373; no. 4, pp. 557–575; vol. 22, no. 1, pp. 54–99.

MIKAMI Yoshio 三上義夫 (8'),1934. *Shina shisō (sūgaku)* 支那思想（數學） (Chinese thought. (Mathematics)). Tokyo.

MIKAMI Yoshio 三上義夫 (9'), 1937. "Sō-Gen sūgaku jo ni okeru endan oyobi shakusa no igi." 宋元數學上における演段及び釋鎖の意義 (The significance of the *yanduan* and *shisuo* methods in the mathematics of the Song and Yuan periods). *SK*, no. 179, pp. 1–15.

MIKAMI Yoshio 三上義夫 (10'), 1937. "Wa-Kan sūgaku shi jō ni okeru senran oyobi gunji no kankei." 和漢に於ける戰亂及び軍事の關係 (Influences of war and military affairs on the history of mathematics in China and Japan). *SK*, no. 182, pp. 1–32.

MINESHIMA Sōichirō 峰島總一郎(1'), 1984. "Kan yaku butten ni okeru sūshi (tai-sū) ni tsuite." 漢譯佛典における數詞（大數）について(Numerals (large numbers) in Chinese translations of Buddhist sutras), *SGSK*, no. 103, pp. 1–31.

MIYAJIMA Kazuhiko 宮島一彦 (1'), 1982. "*Genshi* tenmonshi kisai no isuramu tenmon giki ni tsuite." 『元史』天文志記載のイスラム天文儀器について (On Islamic instruments described in the *Yuanshi*), In *Science and Skills in Asia, A Festschrift for the 77th Anniversary of Professor Yabuuti Kiyoshi [Yabuuchi Kiyoshi]*, Kyoto: Dohosha, pp. 407–427.

MURATA Tamotsu 村田全 (1'), 1981. *Nihon no sūgaku, Seiyō no sūgaku — hikaku sūgaku shi no kokoromi* — 日本の數學，西洋の數學—比較數學史の試み (Japanese mathematics, Occidental mathematics — a tentative comparative history of mathematics —). Tokyo: Chūō Kōron.

NIHON GAKUSHIIN 日本學士院 [Japanese Academy] (1'), 1954–60. *Meijizen Nihon sūgaku shi* 明治前日本數學史 (History of pre-Meiji Japanese mathematics). 5 vols., Tokyo: Iwanami Shoten.

OGURA Kinnosuke 小倉金之助 (1'), 1938. *Sūgaku kyōiku shi* 數學教育史 (History of Mathematical Education). Fourth ed. Tokyo: Iwanami Shoten.

OGURA Kinnosuke 小倉金之助 (2'), 1951. "Mikami Yoshio hakushi (1875–1950) to sono gyōseki." 三上義夫博士とその業績 (The achievements of Dr. Mikami Yoshio). *KagakuSK*, no. 18, pp.1–8.

OGURA Kinnosuke 小倉金之助 (3'), 1978. "Chūgoku, Nihon no sūgaku." 中國・日本の數學 (Chinese and Japanese mathematics). In *Ogura Kinnosuke Chosaku Shū* 小倉金之助著作集 (Collected Works of Ogura Kinnosuke), vol. 3, Tokyo: Keisō Shobō (This volume (350 pp.) is solely devoted to Chinese and Japanese mathematics).

ŌYA Shinichi 大矢眞一 (1'), 1951. "Mikami Yoshio sensei chosaku ronbun mokuroku (sono 1)." 三上義夫先生著作論文目録（その一）(Bibliography of Prof. Mikami Yoshio (Part 1)). *KagakuSK*, no. 18, pp. 9–11 [This bibliography covers only the period 1923–1943. I do not know whether Part 2 has ever been published. But see also the entry 'YAJIMA' in the bibliography of Occidental sources].

ŌYA Shinichi 大矢眞一 (2'), 1979. "Waga kuni ni okeru Chūgoku sūgaku shi kenkyū." わが國における中國數學史の研究 (Japanese researches into the history of Chinese mathematics), *SGSK*, no. 80, pp. 72–75.

ŌYA Shinichi (3') transl., 1975/1980. "*Kyūshō sanjutsu.*" 九章算術 (The *Jiuzhang suanshu* translated into Japanese). In K. YABUUCHI ed., *Chūgoku no kagaku* 中國の科學 (Chinese Science). Tokyo: Chūō Kōron.

PAN Jixing 潘吉星 (1'), 1984. "Kangxi di yu Xiyang kexue." 康熙帝與西洋科學 (Emperor Kangxi and Occidental Science). *ZK*, vol. 3, no. 2, pp. 177–188.

QIAN Baocong 錢寶琮 (1'), 1921. "Jiuzhang wenti fenlei kao." 九章問題分類考 (Research into the classification of the problems in nine chapters'). *XueYi*, vol. 3, no. 1, pp. 1–10; *GSKY*, pp. 11–23; *QBK*, pp. 1–9.

QIAN Baocong 錢寶琮 (2'), 1921. "Jiuzhang suanfa yuanliu kao." 九章算法源流考 (Research into the origin of the computational methods in nine chapters). *XueYi*, vol. 3, no. 2, pp. 1–12.

QIAN Baocong 錢寶琮 (3'), 1921. "Baiji shu yuanliu kao." 百雞術源流考 Research into the origin of the 'hundred fowls' problem). *XueYi*, vol. 3, no. 3, pp. 1–6; *GSKY*, pp. 37–44; *QBK*, pp. 17–21.

QIAN Baocong 錢寶琮 (4'), 1921. "Qiuyi shu yuanliu kao." 求一術源流考 Research into the origin of the rule of determination of the unity). *XueYi*, vol. 3, no.4, pp. 1–16; *GSKY*, pp. 45–66; *QBK*, pp. 22–36.

QIAN Baocong 錢寶琮 (5'), 1921. "Jishufa yuanliu kao." 記數法源流考 (Research into the origin of the notation of numbers). *XueYi*, vol. 3, no. 5, pp. 1–6; *GSKY*, pp. 1–9; *QBK*, pp. 37–42.

QIAN Baocong 錢寶琮 (6'), 1923. "Zhongguo suanshu zhong zhi yuanlü yanjiu." 中國算書中之圓率研究 (Research into the value of π in Chinese mathematics). *Kexue*, vol. 8, no. 2, pp. 114–129, no. 3, pp. 254–265; *QBK*, pp. 50–74.

QIAN Baocong 錢寶琮 (7'), 1925. "Yindu suanxue yu Zhongguo suanxue zhi guanxi." 印度算學與中國算學之關係 (The relationship between Indian and Chinese mathematics). *NKZK*, vol. 1, no. 16, pp. 4–8; *QBK*, pp. 75–82.

QIAN Baocong 錢寶琮 (8'), 1927. "*Jiuzhang suanshu* ying bu zu shu liuzhuan Ouzhou kao." 九章算術盈不足術流傳歐州考 (Research into the transmission of the double-false-position rule (*Jiuzhang suanshu*, ch. 7) from China to Europe). *Kexue*, vol. 12, no. 6, pp. 701–714; *QBK*, pp. 83–96.

QIAN Baocong 錢寶琮 (9'), 1929. "*Zhoubi suanjing* kao." 周髀算經考 (Research into the *Zhoubi suanjing*). *Kexue*, vol. 14, no. 1, pp. 7–29; *QBK*, pp. 119–136.

QIAN Baocong 錢寶琮 (10'), 1929. "*Sunzi suanjing* kao." 孫子算經考 (Research into the *Sunzi suanjing*). *Kexue*, vol. 14, no. 2 , pp. 161–168; *QBK*, 1983, pp. 137–142.

QIAN Baocong 錢寶琮 (11'), 1929. "*Xiahou Yang suanjing* kao." 夏侯陽算經考 (Research into the *Xiahou Yang suanjing*). *Kexue*, vol. 14, no. 3, pp. 311–320; *QBK*, pp. 143–150.

QIAN Baocong 錢寶琮 (12'), 1932. "Mei Wu'an xiansheng nianpu." 梅勿庵先生年譜 (Yearly biography of Master Mei Wu'an [Mei Wending]). *GLZhejiang*, vol. 1, no. 1, pp. 11–44; *QBK*, pp. 608–638.

QIAN Baocong 錢寶琮 (13'), 1932. *Zhongguo suanxue shi* 中國算學史 (History of Chinese mathematics). Peiping: Academia Sinica (The National Research Institute of History and Philology Monographs, serie A, no. 6).

QIAN Baocong 錢寶琮 (14'), 1933. *Gusuan kaoyuan* 古算考源 (Research into the origin of ancient Chinese mathematics). Shanghai: Shangwu Yinshuguan.

QIAN Baocong 錢寶琮 (15'), 1933. "Fangcheng suanfa yuanliu kao." 方程算法源流考 (Research into the origin of the *fangcheng* computational method). *GSKY*, pp. 25–36; *QBK*, pp. 10–16.

QIAN Baocong 錢寶琮 (16'), 1934. "Dai Zhen suanxue tianwen zhuzuo kao." 戴震算學天文著作考 (Research into the astronomical and mathematical works of Dai Zhen). *QBK*, pp. 151–174.

QIAN Baocong 錢寶琮 (17'), 1936. "Wang Lai *Hengzhai suanxue* pingshu." 汪萊《衡齋算學》評述 (A critical examination of Wang Lai's *Hengzhai suanxue*). *QBK*, pp. 235–260.

QIAN Baocong 錢寶琮 (18'), 1936. "Tang dai lijia qiling fenshu jifa zhi yanjin." 唐代曆家奇零分數紀法之演進 (The improvement in the notation for fractions and decimals by the Tang dynasty calendarists). *SXZZ*, vol. 1, no. 1, pp. 65–76; *QBK*, pp. 261–270.

QIAN Baocong 錢寶琮 (19'), 1937. "Zhongguo shuxue zhong zhi zhengshu gouguxing yanjiu." 中國數學中之整數勾股形研究 (Research into right-angled triangles whose sides are integers in Chinese mathematics), *SXZZ*, vol. 1, no. 3, pp. 94–112, *QBK*, pp. 287–303.

QIAN Baocong 錢寶琮 (20'), 1937. "Zhejiang chouren zhushu ji." 浙江疇人著述記 (Note on the works of calendarist-mathematicians of Zhejiang province). *QBK*, pp. 304–316.

454   References

QIAN Baocong 錢寶琮 (21'), 1940, "Jin-Yuan zhi ji shuxue zhi chuanshou." 金元
之際數學之傳授 (The transmission of mathematics from the Jin to the Yuan
dynasty). *QBK*, pp. 317–326.

QIAN Baocong 錢寶琮 (22'), 1956. "*Shoushi li* fa lüelun." 授時曆法略論 (An
overview of the [mathematical] methods of the *Shoushi* calendar). *TWXB*, vol. 4,
no. 2, pp. 193–209; *QBK*, pp. 352–376.

QIAN Baocong 錢寶琮 (23'), 1958. "Gaitian shuo yuanliu kao." 蓋天說源流考
(Research into the origin of the *gaitian* theory). *KXSJK*, no. 1, pp. 29–46; *QBK*,
pp. 377–403.

QIAN Baocong 錢寶琮 (24'), 1959. "Zeng-cheng kaifang fa de lishi fazhan." 增乘
開方法的歷史發展 (The historical development of the additive-multiplicative
root extraction method [Horner's method]). *KXSJK*, no. 2, pp. 126–143; *SY*,
pp. 36–59; *QBK*, pp. 404–430.

QIAN Baocong 錢寶琮 ed. (25'), 1963. *Suanjing shishu* 算經十書 (The Ten
Computational Canons). 2 vols., Peking: Zhonghua Shuju.

QIAN Baocong 錢寶琮 (26'), 1964/1981. *Zhongguo shuxue shi* 中國數學史 (History of Chinese mathematics). Peking: Kexue Chubanshe.

QIAN Baocong 錢寶琮 (27'), 1966. "Qin Jiushao *Shushu jiuzhang* yanjiu." 秦九
韶《數書九章》研究 (Research into Qin Jiushao's *Shushu jiuzhang*). *SY*,
pp. 60–103; *QBK*, pp. 530–578.

QIAN Baocong 錢寶琮 (28'), 1966. "Song-Yuan shiqi shuxue yu daoxue de
guanxi." 宋元時期數學與道學的關係(The relationship between mathematics
and neo-confucianism in the Song and Yuan periods). *SY*, pp. 225–240; *QBK*, pp.
579–596.

QIAN Baocong 錢寶琮 (29'), 1966. "Wang Xiaotong *Jigu suanshu* di'er ti, disan ti,
shuwen shuzheng." 王孝通《緝古算術》第二題第三題術文疏證 (An
emendation of the text of problems no. 2 and 3 of Wang Xiaotong's *Jigu suanshu*
[*Jigu suanjing*]). *KXSJK*, no. 9, pp. 31–52; *QBK*, 1983, pp. 495–529.

QIAN Baocong 錢寶琮 (30'), 1983. "*Jiuzhang suanshu* jiqi Liu Hui zhu yu zhexue
sixiang de guanxi." 《九章算術》及其劉徽注與哲學思想的關係(The
relationship between the *Jiuzhang suanshu*, Liu Hui's Commentary and philosophical thinking). *QBK*, pp. 597–607.

QIAN Baocong 錢寶琮 (31'), 1983. "Jiaozheng yu zengbu." 校正與增補 (Problems of textual criticism). *QBK*, pp. 43–49.

QIAN Baocong 錢寶琮 (32'), 1984. "Wang Lai *Hengzhai suanxue* de yige zhuji."
汪萊《衡齋算學》的一個注記 (A Note on Wang Lai's *Hengzhai suanxue*).
KXSJK, no. 11, pp. 12–13.

QIAN Baocong *et al.* (1'), 1966. *Song Yuan shuxue shi lunwen ji* 宋元數學史論
文集 (Collected papers on the history of mathematics under the Song and Yuan
dynasties). Peking: Kexue Chubanshe.

QIAN Baocong 錢寶琮 and DU Shiran 杜石然 (1'), 1961. "Shilun Zhongguo
gudai shuxue de luoji sixiang." 試論中國古代的邏輯思想 (On the logical
thinking in ancient Chinese mathematics). In *Zhongguo luoji sixiang lunwen xuan
(1949–1979)* 中國邏輯思想論文選 (1949-1979) (Collected papers on the history
of Chinese logical thinking (1949–1979)), 1981, Peking: Xinhua Shudian, pp. 93–
99, reprinted from the 29 May 1961 issue of the *Guangming Ribao*.

QIU Chongman 裘沖曼 (1'), 1926. "Zhongguo suanxue shumu huibian." 中國算
學書目彙編 (A catalogue of Chinese mathematical books). *QHXB*, 1926,
vol. 3, no. 1, pp. 43–96.

REN Jiyu 任繼愈 (1'), 1993. *Zhongguo kexue jishu dianji tonghui — shuxue juan*
中國科學技術典藉通彙— 數學卷 (Comprehensive collection of Chinese
scientific and technical classical works — mathematics), 5 vols., Zhengzhou:
Henan Jiaoyu Chubanshe (contains complete reproductions of many classical Chinese mathematical books).

SAKAI Hiroshi 酒井洋 (1'), 1981. *Kodai Chūgokujin no sū kannen — kōkotsumoji
no kagakuteki kōsatsu o chūshin toshite* — 古代中國人の數觀念—甲骨文
の科學的考察采を中心として—(The ancient Chinese concept of number,
an essay centered on the scientific investigation of inscriptions on bones and
tortoise carapaces). Tokyo: Tsukumo Shuppan.

SHEN Kangshen 沈康身 (1'), 1964. "Wang Xiaotong kaihe zhudi de fenxi." 王孝
通開河築堤題的分析 (An analysis of Wang Xiaotong's problem of the
dike). *HangzhouXB*, vol. 1, no. 4, pp. 43–58.

SHEN Kangshen 沈康身 (2'), 1982. "Geng xiang jian sun shu yuanliu." 更相減損
術源流 (Research into the origin of the rule of mutual subtraction [*antyphairesis*
or 'Euclid's algorithm']). *ZK*, vol. 1, no. 3, pp.193–207.

SHEN Kangshen 沈康身 (3'), 1982. "Hanyu shuxue cihui yuanliu." 漢語數學詞
匯源流 (The origin of Chinese mathematical terminology). Unpublished paper
delivered to the Symposium *Jinian Li Shanlan shishi yibai zhounian xueshu
taolunhui* 紀念李善蘭逝世一百周年學術討論會 (Commemoration of the
centenary of Li Shanlan's death), 30 pp.

SHEN Kangshen 沈康身 (4'), 1986. Zhongsuan daolun 中算導論 (Introduction to
Chinese mathematics). Shanghai: Shanghai Jiaoyu Chubanshe.

SHEN Kangshen 沈康身 (5'), 1986. "Du *Jiuzhang suanshu Liu Hui zhu* xin
yi." 讀《九章算術・劉徽注》新議 (New remarks about Liu Hui's commentary on the *Jiuzhang suanshu*). *ZK*, vol. 5, no. 3, pp. 202–214.

SHIMODAIRA Kazuo 下平和夫 (1'), 1970. *Wasan no rekishi* 和算の歷史
(History of *wasan*). 2 vols., Tokyo: Fuji Tanki Daigaku Shuppanbu.

SONG Jie 宋杰 (1'), 1994. *Jiuzhang suanshu* yu Han dai shehui jingji 《九章算
術》與漢代社會經濟 (The *Jiuzhang suanshu* and the economy of the Han
dynasty). Peking: Shoudu Shifan Daxue Chubanshe.

SU Jing 蘇精 (1'), 1978. *Qingji Tongwen guan* 清季同文館 (The Tongwen guan
of the Qing dynasty), Taipei (edited by the author).

SUGIMOTO Toshio 杉本敏夫 (1'), 1987. "Dui *Shoushi li* de [*sic*] ruogan biaoge de
dingzheng." 對授時曆的若干表格的訂正 (An emendation to several astronomical tables of the *Shoushi* calendar). *MeijiGR*, no. 415, pp. 1–9.

SUGIMOTO Toshio 杉本敏夫 (2'), 1987. "Guanyu yongyu *Shoushi li* de [*sic*] Shen
Gua de nizhengxian gongshi de jingdu." 關於用於授時曆的沈括的逆正弦
公式的精度 (On the precision of Shen Gua's arcsine formula used in the
Shoushi calendar). *MeijiGR*, no. 419, pp. 1–12.

SUN Shuben et al. 孫樹本 (1'), 1983. "Jiang Zehan laoshi." 江澤涵老師
(Professor Jiang Zehan). *SXJZ*, vol. 12, no. 2, pp. 157–160.

SUN Wenqing 孫文青 (1'), 1931. "*Jiuzhang suanshu* yuanliu kao." 九章算術源
流考 (Research into the origin of the *Jiuzhang suanshu*), *NüshiDa*, vol. 2, no. 1,
pp. 1–60.

TAKEDA Kusuo 武田楠雄 (1'), 1953. "Mindai ni okeru sansho keishiki no hensen — Mindai sūgaku no tokushitsu josetsu — ." 明代における算書形式の變遷— 明代數學の特色序説— (Change in the form of Chinese mathematical books of the Ming dynasty — Introductory remarks on the mathematics of the Ming dynasty). *KagakuSK*, no. 26, pp.13–19.

TAKEDA Kusuo 武田楠雄 (2'), 1954. "Mindai sūgaku no tokushitsu: *Sampō tōsō* seiritsu no katei." 明代數學の特質— 算法統宗成立の過成 (The characteristics of Chinese mathematics in the Ming dynasty, the elaboration of the *Suanfa tongzong*). *KagakuSK*, part I: no. 28, pp. 1–12; part II: no. 29, pp. 8–18.

TAKEDA Kusuo 武田楠雄 (3'), 1954. "*Dōbun sanshi* no seiritsu." 同文算指の成立 (The elaboration of the *Tongwen suanzhi*). *KagakuSK*, no. 30, pp. 7–14.

TAKEDA Kusuo 武田楠雄 (4'), 1967. *Chūgoku no sūgaku* 中國の數學 (Chinese mathematics). *SGSK*, vol. 5, no. 2, pp. 1–38.

TAKEDA Tokimasa 武田時昌 (1'), 1984. "*Kyūshō sanjutsu* no kōsei to sūri." 『九章算術』の構成と數理 (The structure and mathematical principles of the *Jiuzhang suanshu*). *ChugShisKen*, no. 6, pp. 69–125.

WANG Daming 王大明 (1'), 1987. "Jingshi Tongwen guan jiqi lishi diwei." 京師同文館及其歷史地位 (The Pekingese Tongwen guan [College of Combined Learning] and its historical position), *ZHKE*, vol. 8, no. 4, pp. 39–47.

WANG Li 王力 (1') 1980. *Hanyu shigao* 漢語史稿 (Materials for the history of the Chinese language). 3 vols., Peking: Zhonghua Shuju.

WANG Ping 王萍 (1'), 1966. *Xifang lisuanxue zhi shuru* 西方曆算學之輸入 (The introduction of Western astronomical and mathematical sciences into China). Nankang/ Taipei: Institute of Modern History, Academia Sinica. Rev.: N. Sivin, *JAS*, 29, pp. 914–916.

WANG Ping 王萍 (2'), 1974. "Ruan Yuan yu *Chouren zhuan*." 阮元與疇人傳 (Ruan Yuan and the *Chouren zhuan*). *ZYY/JS*, vol. 4, Part 2, pp. 601–611.

WANG Ping 王萍 (3'), 1976. "Lidai lisuanjia de chushen yu zhiye zhi tantao." 歷代曆算家的出身與職業之探討 (Research into the social origin and profession of mathematical astronomers), *ZHWHua*, vol. 9, no. 3, pp. 16–21.

WANG Yangzong 王楊宗 (1'), 1988. "Jiangnan zhizaoju fanyiguan shilüe." 江南製造局翻譯館史略 (A brief history of the translation department of the Jiangnan Arsenal). *ZHKE*, vol. 9, no. 3, pp. 65–74.

WANG Yixun 王翼勳 (1'), 1987. "Qin Jiushao, Shi Yuechun, Huang Zongxian de qiu dingshu fangfa." 秦九韶、時曰醇、黃宗憲的求定數方法 (The determination of the *dingshu* [coprime modulus in the Chinese remainder 'theorem'] according to Qin Jiushao, Shi Yuechun and Huang Zongxian), *ZK*, vol. 6, no. 4, pp. 308–313.

WANG Yusheng 王渝生 (1'), 1983. "Li Shanlan de jianzhui shu." 李善蘭的尖錐術 (The 'sharp pyramids' algorithm of Li Shanlan). *ZK*, vol. 2, no. 3, pp. 266–288.

WANG Yusheng 王渝生 (2'), 1987. "Qin Jiushao qiu 'dingshu' fangfa de chengjiu he quexian." 秦九韶求"定數"方法的成就和缺陷 (Weak and strong points of Qin Jiushao's method of determination of the *dingshu* [coprime modulus in the Chinese remainder 'theorem']). *ZK*, vol. 6, no. 4, pp. 299–307.

WANG Yusheng 王渝生 (3'), 1989. "Mianhuai kexue shijia Yan Dunjie xiansheng." 緬懷科學史家嚴敦傑先生 (In memory of Professor Yan Dunjie (1917–1988), a historian of science). *ZHKE*, vol. 10, no. 4, pp. 30–38.

WU Chengluo 吳承洛 (1'), 1937. *Zhongguo dulianggheng shi* 中國度量衡史 (History of Chinese weights and measures). Shanghai: Shangwu Yinshuguan.

WU Wenjun 吳文俊 ed. (1'), 1982. *Jiuzhang suanshu* yu Liu Hui 《九章算術》 與劉徽 (Liu Hui and the *Jiuzhang suanshu*). Peking: Beijing Shifan Daxue Chubanshe; Rev.: J. C. Martzloff, *TP*, 1985, vol. 71, pp. 142–146.

WU Wenjun 吳文俊 ed. (2'), 1985–1986–1987. *Zhongguo shuxue shi lunwenji* 中國數學史論文集 (Collected papers on the history of Chinese mathematics). vols. 1–3. Jinan: Shandong Jiaoyu Chubanshe.

WU Wenjun 吳文俊 ed. (3'), 1987. *Qin Jiushao yu* Shushu jiuzhang 秦九韶與 《數書九章》 (Qin Jiushao and the *Shushu jiuzhang*). Peking: Beijing Shifan Daxue Chubanshe.

WU Xinmou 吳新謀 (1'), 1989. "Jinian Adama lai Hua jiangxue wushi nian." 紀念阿達瑪來華講學五十年 (In commemoration of the fiftieth anniversary of the arrival of Jacques Hadamard in China). *SXJZ*, vol. 18, no. 1, pp. 62–67.

WU Yubin 吳裕賓 (1'), 1986. "Jiao Xun yu *Jiajian chengchu shi*." 焦循與《加減乘除釋》 (Jiao Xun and his *Jiajian chengchu shi* [An explanation of addition, subtraction, multiplication and division]). *ZK*, vol. 5, no. 2, pp. 120–128.

WU Yubin 吳裕賓 (2'), 1992. "*Zhongxi suanxue dacheng* de bianzuan"《中西算學大成》的編纂 (The compilation of the *Zhongxi suanxue dacheng*), *ZHKE*, vol. 13, no. 2, pp. 91-94

XIAO Zuozheng 肖作政 (1'), 1988. "'Wantian' fei qiuguanxing." "宛田" 非球冠形 (The 'tortuous field' is not a spherical segment). *ZK*, vol. 7, no. 2, pp. 109–111.

XU Chunfang 許蒓舫 (1'), 1952. Zhong suanjia de jihexue yanjiu 中算家的幾何學研究 (The study of geometry by Chinese mathematicians). Peking: Kaiming Shudian.

XU Xintong 許鑫銅 (1'), 1986. "*Jiuzhang suanshu* kaifang shu jiqi Liu Hui zhu tantao." 《九章算術》開方術及其劉徽注探討 (Research into square root extraction in the *Jiuzhang suanshu* and Liu Hui's commentary). *ZK*, vol. 5, no. 3, pp. 193–201.

XU Yifu 許義夫 (1'), 1989. "Kong Guangsen guanyu gaoci fangcheng de yingyong." 孔廣森關於高次方程的應用 (Kong Guangsen and the practical applications of high degree equations). *ZK*, vol. 8, no. 2, pp. 118–126.

XU Zongze 徐宗澤 (1'), 1958. *Ming Qing jian Yesuhuishi yizhu tiyao* 明清間耶穌會士譯著提要 (A bibliography of Jesuit translations of Western works into Chinese during the Ming and Qing dynasties), Taipei: Zhonghua Shuju.

YABUUCHI Kiyoshi 藪內清 (1'), 1944. *Shina sūgaku shi* 支那數學史 (History of Chinese Mathematics). Kyoto: Yamaguchi Shoten.

YABUUCHI Kiyoshi 藪內清 (2'), 1963. *Chūgoku chūsei kagaku gijutsu shi no kenkyū* 中國中世科學技術史の研究 (Research into the history of sciences and techniques in mediaeval China). Tokyo: Kadokawa Shoten; Rev.: *RBS*, 1963, notice no. 856.

YABUUCHI Kiyoshi 藪內清 (3'), 1969. *Chūgoku no tenmon rekihō* 中國の天文歷法 (Chinese calendrical astronomy) Tokyo: Heibonsha.

YABUUCHI Kiyoshi 藪內清 (4'), 1967. *Sō Gen jidai no kagaku gijutsu shi* 宋元時代の科學技術史 (History of sciences and techniques in China under the Song and Yuan dynasties). Kyoto: Kyoto Daigaku Jinbun Kagaku Kenkyūsho; Rev.: *RBS*, 1966/67, 12/13, notice no. 1066.

YABUUCHI Kiyoshi 藪内清 (5'), 1970. *Min Shin jidai no kagaku gijutsu shi* 明清時代の科學技術史 (History of Sciences and techniques in China under the Ming and Qing dynasties). Kyoto: Kyoto Daigaku Jinbun Kagaku Kenkyūsho.

YABUUCHI Kiyoshi 藪内清 (6'), 1974. *Chūgoku no sūgaku* 中國の數學 (Chinese mathematics).Tokyo: Iwanami Shoten.

YABUUCHI Kiyoshi 藪内清 (7'), 1977. "Chūgoku no sūgaku." 中國の數學 (Chinese mathematics). *SGSK*, no. 75, pp. 1–8.

YABUUCHI Kiyoshi 藪内清 (8'), 1978. *Chūgoku no kagaku to Nihon* 中國の科學と日本 (Chinese science and Japan), Tokyo: Asahi Shinbunsha.

YAMANE Yukio 山根幸夫 ed. (1'), 1983. *Chūgoku shi kenkyū nyūmon* 中國史研究入門 (Introduction to sinology). 2 vols., Tokyo: Yamagawa Shuppansha.

YAMAUCHI Fumiko 山内芙子 (1'), 1983. "*Sankyo shisho* ni okeru sangaku no karikyuramu no igi." 《算經十書》における算學かりキュラムの意義 (On the meaning of the mathematical curriculum in the *Suanjing shishu*). *MiyazakiJT*, no. 11, 1983, pp. 38–51.

YAMAZAKI Yoemon 山崎與右衛門 (1'), 1962. *Tōzai soroban bunken shū* 東西算盤文獻集 (Collection of articles on the abacus in the Occident and in the Orient). 2 vols., Tokyo: Morikita Shuppansha.

YAN Dunjie 嚴敦傑 (1'), 1936. "Zhongguo suanjia Zu Chongzhi jiqi yuanzhoulü zhi yanjiu." 中國算學家祖冲之及其圓周率之研究 (Research into the Chinese mathematician Zu Chongzhi and his value of π). *XueYi*, vol. 15, no. 5, pp. 37–50.

YAN Dunjie 嚴敦傑 (2'), 1936. "*Suishu* lüli zhi Zu Chongzhi yuanlü jishi shi." 隋書律曆志祖冲之圓率記事釋 (An explanation of quotations concerning Zu Chongzhi's computation of π in the monograph of the *Suishu* on the calendar). *XueYi*, vol. 15, no. 10, pp. 25–57.

YAN Dunjie 嚴敦傑 (3'), 1937. "*Sunzi suanjing* yanjiu." 孫子算經研究 (Research into the *Sunzi suanjing*). *XueYi*, vol. 16, no. 3, pp. 15–31.

YAN Dunjie 嚴敦傑 (4'), 1939. "Shanghai suanxue wenxian shulüe." 上海算學文獻述略 (An overview of mathematical documents preserved in Shanghai). *Kexue*, vol. 23, no. 2, pp. 72–78.

YAN Dunjie 嚴敦傑 (5'), 1939. "Zhupan zakao." 珠盤雜考 (Various remarks on the abacus). *XSJ*, vol. 14, no. 8, pp. 8–10; no. 9, pp. 5–7.

YAN Dunjie 嚴敦傑 (6'), 1940. "Nan Bei chao suanxue shu zhi." 南北朝算學書志 (Note on the mathematical books of the Northern and Southern dynasties). *TSJK*, new ser., vol. 2, no. 2, pp. 196–212.

YAN Dunjie 嚴敦傑 (7'), 1941. "Qingdai Sichuan suanxue zhushu." 清代四川算學著述記 (Notes on the mathematical books of the province of Sichuan in the Qing dynasty). *TSJK*, new ser., vol. 3, no. 3–4, pp. 227–244.

YAN Dunjie 嚴敦傑 (8'), 1943. "Lun *Honglumeng* ji qita xiaoshuo zhong zhi kexue shiliao." 論紅摟夢及其它小說中之科學史料 (The scientific information contained in the *Honglumeng* and other novels). *DFZZ*, vol. 39, no. 9, pp. 59–61.

YAN Dunjie 嚴敦傑 (9'), 1943. "Song Yuan suanshu yu xinyong huobi shiliao." 宋元算書與信用貨幣史料 (The relation between the use of paper money in the Song and Yuan dynasties and the mathematical books of that time). *YSB/WSFK*, no. 38, (no pagination), 2 pp.

YAN Dunjie 嚴敦傑 (10'), 1943. "Oujilide *Jihe yuanben* Yuan dai shuru Zhongguo shuo." 歐幾里得幾何原本元代輸入中國說 (On the coming of Euclid's *Elements* to China in the Yuan dynasty). *DFZZ*, vol. 39, no. 13, pp. 35–36.

YAN Dunjie 嚴敦傑 (11'), 1944. "Suanpan tanyuan." 算盤探源 (On the origin of the abacus). *DFZZ*, vol. 40, no. 2, pp. 33–36.

YAN Dunjie 嚴敦傑 (12'), 1944. "Juyan Han jian suanshu." 局延漢簡算書 (The mathematical writings on the Han bamboo tablets discovered at Juyan [Edsin Gol]). *ZLZZ*, vol. 1, no. 3, pp. 315–319.

YAN Dunjie 嚴敦傑 (13'), 1945. "Chousuan suanpan lun." 籌算算盤論 (On counting-rods and the abacus). *DFZZ*, vol. 41, no. 15, pp. 33–35.

YAN Dunjie 嚴敦傑 (14'), 1945. "*Suanxue qimeng* liuzhuan kao." 算學啓蒙流傳考 (On the transmission and diffusion of the *Suanxue qimeng*). *DFZZ*, vol. 41, no. 9, pp. 26–28.

YAN Dunjie 嚴敦傑 (15'), 1947. "Song-Yuan suanxue congkao." 宋元算學叢考 (Collected research into the mathematics of the Song and Yuan dynasties). *Kexue*, vol. 29, no. 4, pp. 109–114.

YAN Dunjie 嚴敦傑 (16'), 1951. "Jin nian lai Zhong suan zhenji zhi faxian." 近年來中算珍藉之發現 (Recent discoveries of rare Chinese mathematical books). *KXTB*, vol. 2, no. 7, pp. 719–721.

YAN Dunjie 嚴敦傑 (17'), 1954. " Zhong suanjia de sushu lun." 中算家的素數論 (The researches of Chinese mathematicians into prime numbers). *SXTB*, 1954, no. 4, pp. 6–10; no. 5, pp. 12–15.

YAN Dunjie 嚴敦傑 (18'), 1957. "Ming Qing shuxue shi zhong de liangge lunti — Cheng Dawei he Mei Wending —." 明清數史中的兩個論題—程大位和梅文鼎 (Two themes in Ming and Qing mathematics : Cheng Dawei and Mei Wending). *ALXB*, Inaugural issue, 1957, pp. 48–52.

YAN Dunjie 嚴敦傑 (19'), 1959. " 'Jihe' bu shi 'Geo' de yiyin." 幾何不是Geo的譯音 (*Jihe* is not the phonetical tranlitteration of *Geo*). *SXTB*, no. 11, p. 31.

YAN Dunjie 嚴敦傑 (20'), 1959. "Qingdai shuxuejia Mei Juecheng zai shuxue shi shang de gongxian." 清代數學家梅毂成在數學史上的貢獻 (The mathematical contributions of Mei Juecheng, a Qing dynasty mathematician). *ASXTX*, no. 3, pp. 1–5.

YAN Dunjie 嚴敦傑 (21'), 1960. "Fang Zhongtong *Shudu yan* pingshu." 方中通《數度衍》評述 (An appraisal of Fang Zhongtong's *Shudu yan* [Developments on numbers and measures]). *ASX*, Inaugural issue, pp. 52–57.

YAN Dunjie 嚴敦傑 (22'), 1962. "Gugong suo cang Qing dai jisuan yiqi." 古宮所藏清代計算儀器 (The Mathematical instruments of the Qing dynasty preserved in the Imperial Palace (Peking)). *WenWu*, no. 3, pp. 19–22.

YAN Dunjie 嚴敦傑 (23'), 1964. "Jialilüe de gongzuo zaoqi zai Zhongguo de chuanbu." 伽利略的工作早期在中國的傳布 (The early dissemination of Galileo's works in China). *KXSJK*, no. 7, pp. 8–27.

YAN Dunjie 嚴敦傑 (24'), 1965. "Zhongguo shuxue jiaoyu jianshi." 中國數學教育簡史 (Short history of mathematical education in China), *SXTB*, no. 8, pp. 44–48 and no. 9, pp. 46–50.

YAN Dunjie 嚴敦傑 (25'), 1981. "Guanyu Zhongguo shuxue shi er-san shi." 關於中國數學史二三事 (Two or three things about the history of Chinese mathematics). *Dushu*, no. 8, pp. 15–18.

YAN Dunjie 嚴敦傑 (26'), 1982. "Shikiban sōjutsu." 式盤綜述 (A synthetic presentation of the geomantic compass) In *Science and Skills in Asia, A Festschrift for the 77th Anniversary of Professor Yabuuti Kiyoshi* [Yabuuchi Kiyoshi], Kyoto: Dohosha, pp. 62–95.

YAN Dunjie 嚴敦傑 (27'), 1982. "Zhongguo shiyong shumazi de lishi." 中國使用數碼子的歷史 (A history of systems of numeration currently used in China). *Kejishi*, no. 8, pp.31–50.

YAN Dunjie 嚴敦傑 (28'), 1984. "Li Yan yu shuxue shi — jinian Li Yan xiansheng danchen jiushi zhounian." 李儼與數學史－紀念李儼先生誕辰九十周年 (Li Yan and the history of mathematics — in commemoration of the ninetieth anniversary of professor Li Yan's birth). *KXSJK*, no. 11, pp. 1–5.

YAN Dunjie 嚴敦傑 (29'), 1984. "Liu Hui jian zhuan." 劉徽簡傳 (A concise biography of Liu Hui). *KXSJK*, no. 11, pp. 14–20.

YAN Dunjie 嚴敦傑 ed. (30'), 1986. *Zhongguo gudai keji shi lunwen suoyin* 中國古代科技史論文索引 (An index to articles on the history of ancient Chinese sciences and techniques). Jiangsu Kexue Jishu Chubanshe.

YAN Dunjie 嚴敦傑 (31'), 1987. "Ba chongxin faxian zhi *Yongle dadian* suanshu." 跋重新發現之《永樂大典》算書 (Some comments on the newly discovered arithmetical books of the *Yongle dadian* [Yongle reign-period great encyclopaedia]). *ZK*, vol. 6, no. 1, pp. 1–19.

YAN Dunjie 嚴敦傑 (32'), 1988. "*Xi jing lu* ba." 《西鏡錄》跋 (New remarks on the *Xi jing lu* [Records of the Occidental Mirror]). *ZK*, vol. 7, no. 3, pp. 214–217.

YAN Dunjie 嚴敦傑 (33'), 1989. "Mei Wending de shuxue he tianwenxue gongzuo." 梅文鼎的數學和天文學工作 (Mei Wending's mathematical and astronomical works). *ZK*, vol. 8, no. 2, pp. 99–107.

YI Shitong 伊世同 (1'), 1989. "Liangtianchi kao." 量天尺考 (Research into the 'heaven-measuring ruler'). In ZHONGGUO SHEHUI KEXUEYUAN KAOGU YANJIU-SUO 中國社會科學院考古研究所 ed., *Zhongguo gudai tianwen wenwu lunji* 中國古代天文文物論集 (Collection of papers on ancient Chinese astronomical cultural relics). Peking: Wenwu Chubanshe, pp. 358–368.

YOSHIDA Yōichi 吉田洋一 (1'), 1969. *Rei no hakken* 零の發見 (The discovery of zero). Tokyo: Iwanami Shoten.

YU Ningwang 余寧旺 ed. (1'), 1990. *Zhongguo zhusuan daquan* 中國珠算大全 (Encylopaedia of the Chinese abacus). Tianjin: Tianjin Kexue Jishu Chubanshe.

YUAN Xiaoming 袁小明 (1'), 1990. "Lun Zhongguo gudian shuxue de siwei tezheng." 論中國古典數學的思維特徵 (The characteristics of traditional Chinese mathematical thinking). *ZK*, vol. 9, no. 4, pp. 297–307.

YUSHKEVICH A. P. (1'), 1956. "Zhongguo xuezhe zai shuxue lingyu zhong de chengjiu." 中國學者在數學領域中的成就 (The mathematical contributions of Chinese scholars). *SXJZ*, vol. 2, no. 2, pp. 256–278.

ZHANG Yinlin 張蔭鱗 (1'), 1927. "*Jiuzhang suanshu* ji liang Han zhi shu-xue." 《九章算術》及兩漢之數學 (The *Jiuzhang suanshu* and the mathematics of the two Han dynasties). *YJXB*, no. 2, pp. 301–312.

ZHANG Yong 章用 (1'). "Duoji bilei shuzheng." 垛積比類疏證 (Proofs of formulae occuring in the *Duoji Bilei*). *Kexue*, vol. 23, no. 11, pp. 272–277.

ZHAO Zhiyun 趙之云 and XU Wanyun 許宛云 eds. (1'), 1989. *Weiqi cidian* 圍棋詞典 (*Weiqi* dictionary [*weiqi* = the game of go]), Shanghai: Shanghai Cishu Chubanshe.

ZHONGGUO KEXUEYUAN BEIJING TIANWENTAI 中國科學院北京天文臺 ed. (1'), 1989. *Zhongguo Tianwenxue shiliao huibian* 中國天文學史料匯編 (Collected Data on the history of Chinese Astronomy). vol. 1, Peking: Kexue Chubanshe.

ZHONGGUO KEXUEYUAN ZIRAN KEXUE SHI YANJIUSUO 中國科學院自然科學史研究所 ed. (1') 1983. *Qian Baocong kexue shi lunwen xuanji* 錢寶琮科學史論文集 (Collected papers of Qian Baocong on the history of Chinese mathematics). Peking: Kexue Chubanshe.

ZHONGGUO SHEHUI KEXUEYUAN KAOGU YANJIUSUO 中國社會科學院考古研究所 ed. (1'), 1989. *Zhongguo gudai tianwen wenwu lunji* 中國古代天文文物論集 (Collected papers on the history of ancient Chinese astronomy). Peking: Wenwu Chubanshe.

ZHONG-WAI SHUXUE JIANSHI BIANXIE ZU 中外數學簡史編寫組 ed. (1'), 1986. *Zhongguo shuxue jian shi* 中國數學簡史 (A concise history of Chinese mathematics), Jinan: Shandong Jiaoyu Chubanshe.

ZHUSUAN XIAO CIDIAN BIANXIE ZU 珠算小辭典編寫組 ed. (1'), 1988. *Zhusuan xiao cidian* 珠算小辭典 (A concise dictionary of the abacus). Peking: Zhongguo Caizheng Jingji Chubanshe.

Index of Names

Index of Books

Index of Subjects

Druck: STRAUSS OFFSETDRUCK, MÖRLENBACH
Verarbeitung: SCHÄFFER, GRÜNSTADT